INTERACTION OF RADIATION
WITH CONDENSED MATTER
VOL. I

INTERNATIONAL CENTRE FOR THEORETICAL PHYSICS, TRIESTE

INTERACTION OF RADIATION WITH CONDENSED MATTER

LECTURES PRESENTED AT
AN INTERNATIONAL WINTER COLLEGE
AT TRIESTE FROM 14 JANUARY TO 26 MARCH 1976
ORGANIZED BY THE
INTERNATIONAL CENTRE FOR THEORETICAL PHYSICS, TRIESTE

In two volumes

VOL. I

INTERNATIONAL ATOMIC ENERGY AGENCY
VIENNA, 1977

THE INTERNATIONAL CENTRE FOR THEORETICAL PHYSICS (ICTP) in Trieste was established by the International Atomic Energy Agency (IAEA) in 1964 under an agreement with the Italian Government, and with the assistance of the City and University of Trieste.

The IAEA and the United Nations Educational, Scientific and Cultural Organization (UNESCO) subsequently agreed to operate the Centre jointly from 1 January 1970.

Member States of both organizations participate in the work of the Centre, the main purpose of which is to foster, through training and research, the advancement of theoretical physics, with special regard to the needs of developing countries.

INTERACTION OF RADIATION WITH CONDENSED MATTER
IAEA, VIENNA, 1977
STI/PUB/443
ISBN 92–0–130377–7
Printed by the IAEA in Austria
April 1977

INTERACTION OF RADIATION WITH CONDENSED MATTER
VOLUME I

STI/PUB/443

CORRIGENDA

Paper IAEA-SMR-20/1, by T.P. McLean
Page 41, Eq.(10.12)

For $\underline{\overset{\rightrightarrows}{\epsilon}}$ *read* $\underline{\epsilon}$ *(as in Eq.10.9)*

Paper IAEA-SMR-20/51, by R.K. Bullough
Page 383, line 15, first word

For as we shall see *read* As we shall see,

On page 384, the first 5 lines of text should read:

length E and the power P required to sustain the flame. Solitary waves of this character do not have the collision property of solitons: they do not satisfy the local conservation law [10]:

$$\epsilon_t + \mathscr{P}_z = 0$$

where ϵ is the energy density and $\mathscr{P}$ the energy flux (power per square centimetre).

FOREWORD

The International Centre for Theoretical Physics, pursuing its objective of research and training with a comprehensive and synoptic coverage in various disciplines, has already presented four Courses in Solid State Physics: Theory of Condensed Matter (1967), Theory of Imperfect Crystalline Solids (1970), Electrons in Crystalline Solids (1972), and Surface Science (1974). These proceedings have been published by the International Atomic Energy Agency.

The present proceedings constitute a fifth international Course held from 14 January to 26 March 1976 and devoted to the Interaction of Radiation with Condensed Matter, a field always active and still attracting a great deal of interest because of its many applications in such practical and important areas as optoelectronic devices or solar energy. In line with general trends in the policy of the Centre and its Advisory Committee for Condensed Matter Physics, the Course programme, besides dealing with basic theory, also emphasized applications.

The Advisory Committee for Condensed Matter Physics is composed of Professors G. Chiarotti (Italy), F. García-Moliner (Spain), F. Gautier (France), S. Lundqvist (Sweden), N.H. March (United Kingdom), H. Reik (Federal Republic of Germany) and J.M. Ziman (United Kingdom).

Generous grants from the Swedish International Development Authority (SIDA) and the United Nations Development Programme (UNDP) are gratefully acknowledged.

Abdus Salam

EDITORIAL NOTE

The papers and discussions have been edited by the editorial staff of the International Atomic Energy Agency to the extent considered necessary for the reader's assistance. The views expressed and the general style adopted remain, however, the responsibility of the named authors or participants. In addition, the views are not necessarily those of the governments of the nominating Member States or of the nominating organizations.

Where papers have been incorporated into these Proceedings without resetting by the Agency, this has been done with the knowledge of the authors and their government authorities, and their cooperation is gratefully acknowledged. The Proceedings have been printed by composition typing and photo-offset lithography. Within the limitations imposed by this method, every effort has been made to maintain a high editorial standard, in particular to achieve, wherever practicable, consistency of units and symbols and conformity to the standards recommended by competent international bodies.

The use in these Proceedings of particular designations of countries or territories does not imply any judgement by the publisher, the IAEA, as to the legal status of such countries or territories, of their authorities and institutions or of the delimitation of their boundaries.

The mention of specific companies or of their products or brand names does not imply any endorsement or recommendation on the part of the IAEA.

Authors are themselves responsible for obtaining the necessary permission to reproduce copyright material from other sources.

CONTENTS OF VOL. I

GENERAL THEORY

Note

During the Winter College a few courses were given on rather general topics. So much literature is available on these that it was not felt necessary to include in this book lecture notes on, say, general electromagnetic theory, or other fairly standard material. Thus this first section, on General Theory, contains the notes of only two courses, dealing with theoretical areas that are basic and important and yet not so standard: one deals with linear and non-linear optics, and the other is an introduction to quantum optics.

However, the following selected list of basic references is given to help potential readers who might need a first introduction into this field.

Basic electromagnetic theory is covered in:

HEITLER, W., The Quantum Theory of Radiation, 3rd Edn, Oxford Clarendon Press (1954).

LANDAU, L.D., LIFSHITZ, E.M., Electrodynamics of Continuous Media, Pergamon Press (1963).

JACKSON, J.D., Classical Electrodynamics, J. Wiley & Sons (1962).

Besides general textbooks on solid state theory which contain good discussions of basic questions concerning the interaction of radiation with solid matter – for example BORN, M., and HUANG, K., Dynamical Theory of Crystal Lattices, Oxford Clarendon Press (1964), and WEINREICH, G., Solids: Elementary Theory for Advanced Students, J. Wiley & Sons (1965) – there are books devoted to various general problems in this field, such as:

AGRANOVICH, V.M., GINZBURG, V.L., Spatial Dispersion in Crystal Optics and the Theory of Excitons, Interscience, London (1966).

KUBO, R., KAMIMURA, H. (Eds), Dynamical Processes in Solid State Optics, Benjamin (1967).

AZAROFF, L.V. (Ed.), X-ray Spectroscopy of Solids, McGraw-Hill (1974).

Other basic references are quoted in the various lectures making up the two volumes of the present publication.

LINEAR AND NON-LINEAR OPTICS OF CONDENSED MATTER

T.P. McLEAN
Royal Signals and Radar Establishment,
Malvern, Worcestershire,
United Kingdom

Abstract

LINEAR AND NON-LINEAR OPTICS OF CONDENSED MATTER.

Part I – Linear optics: 1. General introduction. 2. Frequency dependence of $\underline{\epsilon}(\omega, \vec{k})$. 3. Wave-vector dependence of $\underline{\epsilon}(\omega, \vec{k})$. 4. Tensor character of $\underline{\epsilon}(\omega, \vec{k})$. Part II – Non-linear optics: 5. Introduction. 6. A classical theory of non-linear response in one dimension. 7. The generalization to three dimensions. 8. General properties of the polarizability tensors. 9. The phase-matching condition. 10. Propagation in a non-linear dielectric. 11. Second harmonic generation. 12. Coupling of three waves. 13. Materials and their non-linearities. 14. Processes involving energy exchange with the medium. 15. Two-photon absorption. 16. Stimulated Raman effect. 17. Electro-optic effects. 18. Limitations of the approach presented here.

INTRODUCTION

This paper is concerned with a discussion of both the linear and non-linear optical properties of condensed matter. Attention is confined to general principles rather than to details of individual materials. (Most of the details of different types of materials, and of the variety of excitations that can be produced by radiation at optical frequencies, are the subject of other contributions to these Proceedings.)

The subject of linear optics deals with the conventional optical effects exhibited by materials, and is generally well understood and documented. I therefore do not intend to discuss it exhaustively. Instead, I shall simply recall the general features of the subject and draw attention to particular aspects of significance in studying and exploiting non-linear phenomena.

The subject of non-linear optics really began with the advent of the laser, whose power and coherence properties allow, as will be seen, non-linear phenomena to be revealed and exploited. I shall deal with this subject in much more detail and shall try to develop a logical treatment of it which exhibits the large variety of effects that can be produced.

Part I

LINEAR OPTICS

1. GENERAL INTRODUCTION

The behaviour of electromagnetic radiation in any medium is governed by Maxwell's equations, which in conventional notation can be written as

$$\left.\begin{aligned} \text{curl } \vec{E} &= -\frac{1}{c}\frac{\partial}{\partial t}\vec{B} \\ \text{curl } \vec{H} &= \frac{1}{c}\frac{\partial}{\partial t}\vec{D} + \frac{4\pi}{c}\vec{j} \\ \text{div } \vec{D} &= 4\pi\rho \\ \text{div } \vec{B} &= 0 \end{aligned}\right\} \tag{1.1}$$

Here and elsewhere in this paper, mixed Gaussian units are used, in which magnetic quantities are measured in e.m.u. and electrostatic and current density in e.s.u. Although against current trends, this allows a direct correlation with most of the literature on non-linear optics, and also much of that on conventional linear optics.

The properties of the medium in which the radiation is propagating are characterized by the relationships between the electric displacement $\vec{D}$, electric current $\vec{j}$, magnetic induction $\vec{B}$ and the electric and magnetic fields $\vec{E}$ and $\vec{H}$ — the so-called constitutive relations. To describe conventional optical phenomena, it is sufficient to assume linear relationships. We are consequently by definition dealing with linear optics.

In this regime we need only consider the field and associated quantities oscillating in time at a single angular frequency ω. We can then write the electric field, for example, as

$$\vec{E} = \vec{E}(\omega)\, e^{-i\omega t} + E^*(\omega)\, e^{i\omega t} \tag{1.2}$$

which is just a convenient way of writing the more conventional expression $\vec{E} = \vec{\mathscr{E}}\cos(\omega t - \phi)$ for such a field with amplitude $\mathscr{E}$ and phase constant ϕ in terms of the complex quantity $\vec{E}(\omega) = \frac{1}{2}\vec{\mathscr{E}}\exp(-i\phi)$. The linear constitutive relations are then normally expressed in terms of tensors describing the frequency-dependent, second-order, dielectric permeability $\underline{\kappa}$, magnetic permeability $\underline{\mu}$ and conductivity $\underline{\sigma}$ as follows:

$$\vec{D}(\omega) = \underline{\kappa}(\omega) \cdot \vec{E}(\omega) \tag{1.3}$$

$$\vec{B}(\omega) = \underline{\mu}(\omega) \cdot \vec{H}(\omega) \tag{1.4}$$

$$\vec{j}(\omega) = \underline{\sigma}(\omega) \cdot \vec{E}(\omega) \tag{1.5}$$

Here we have introduced a notation which will be particularly convenient when we consider non-linear effects, and where, by $\underline{\sigma}(\omega) \cdot \vec{E}(\omega)$, for example, we mean the vector whose component in the direction α (α = x,y, or z) is

$$\sum_{\beta = x,y,z} \sigma_{\alpha\beta}(\omega)\, E_\beta(\omega)$$

Well-known manipulations on Maxwell's and these constitutive equations yield the wave equation

$$\mathrm{curl\,curl}\,\vec{E}(\omega) - \frac{\omega^2}{c^2}\underline{\mu}(\omega)\cdot[\underline{\epsilon}(\omega)\cdot\vec{E}(\omega)] = 0 \quad (1.6)$$

where we have defined a general complex "dielectric" tensor

$$\underline{\epsilon}(\omega) = \underline{\kappa}(\omega) + \frac{4\pi i}{\omega}\underline{\sigma}(\omega) \quad (1.7)$$

and assumed the charge density ρ in the medium to be zero.

The complex tensor $\underline{\epsilon}(\omega)$ formally describes both the dielectric and conduction properties of the medium. The dominant property depends on the relative magnitudes of the oscillation period of the field $T = 1/\omega$ and the dielectric relaxation time $\tau_d = \kappa/4\pi\sigma$, i.e. the time it takes for a charge imbalance to correct itself. For $T \gg \tau_d$, charges follow the field and the conductivity contribution to $\vec{\epsilon}(\omega)$ dominates; this is the case in metals. For $T \ll \tau_d$, dielectric polarization effects characterized by $\vec{\kappa}(\omega)$ dominate; this is the case in insulators.

General solutions of this wave equation (1.6) can be constructed from the plane wave form

$$\vec{E}(\omega) = E_\omega \vec{a}_\omega e^{i\vec{k}\cdot\vec{r}} \quad (1.8)$$

where E_ω is a constant scalar factor which is a measure of the amplitude of the field in the simple case when the wave-vector $\vec{k}$ is real and $\vec{a}_\omega$ is a unit vector in the direction of polarization of the wave. For any direction of the wave-vector $\vec{k}$, its magnitude and the associated direction of polarization are given by solutions of the Fresnel equations

$$k^2\vec{s}\times(\vec{s}\times\vec{a}_\omega) + \frac{\omega^2}{c^2}\underline{\mu}(\omega)\cdot[\underline{\epsilon}(\omega)\cdot\vec{a}_\omega] = 0 \quad (1.9)$$

obtained by inserting (1.8) into (1.6). We have used $\vec{s}$ to denote a unit vector in direction of propagation, so that $\vec{k} = k\vec{s}$.

We shall have more to say about these Fresnel equations later. For the present it is perhaps helpful to derive a familiar result for a non-magnetic ($\underline{\mu}(\omega) = \underline{1}$), isotropic ($\underline{\epsilon}(\omega) = \epsilon(\omega)\,\underline{1}$) medium. Then we have

$$k^2\vec{s}\times(\vec{s}\times\vec{a}_\omega) + \frac{\omega^2}{c^2}\epsilon(\omega)\,\vec{a}_\omega = 0$$

i.e.

$$\vec{s}(\vec{s}\cdot\vec{a}_\omega)k^2 + \left[\frac{\omega^2}{c^2}\epsilon(\omega) - k^2\right]\vec{a}_\omega = 0 \quad (1.10)$$

By taking scalar products in turn with $\vec{s}$ and $\vec{a}_\omega$, we find that $\vec{s}\cdot\vec{a}_\omega = 0$ if $\epsilon(\omega) \neq 0$, i.e. the electric field vibrates transversely to the direction of propagation, and $k^2 = \epsilon(\omega)\,\omega^2/c^2$ — a well-known result; for $\epsilon(\omega) = 0$, $\vec{s}\cdot\vec{a}_\omega \neq 0$, and the electric field has a component parallel to the direction of propagation — a longitudinal component. Normally $\epsilon(\omega) \neq 0$.

Since $\epsilon(\omega)$ is in general complex, k will be likewise. Writing $k = N\omega/c = (n + i\kappa)\omega/c$, the spatial dependence of the plane wave becomes

$$\exp\left(i\frac{\omega}{c}n - \frac{\omega}{c}\kappa\right)\vec{s}\cdot\vec{r}$$

where n and κ are, respectively, the real refractive index and the extinction coefficient, and N is the complex refractive index. They are related to the real $\epsilon'(\omega)$ and imaginary $\epsilon''(\omega)$ parts of $\epsilon(\omega)$ by the relations

$$\begin{aligned} \mathrm{Re}\, N^2 &= n^2 - \kappa^2 = \epsilon'(\omega) \\ \mathrm{Im}\, N^2 &= 2n\kappa = \epsilon''(\omega) \end{aligned} \tag{1.11}$$

The imaginary part $\epsilon''(\omega)$ of $\epsilon(\omega)$ is a measure of the amount of energy absorbed from the electromagnetic wave by the medium. This is readily seen by considering the mean Joule heating of the medium $\overline{\vec{J}\cdot\vec{E}}$, where the total current $\vec{J} = \vec{j} + (1/4\pi)(\partial/\partial t)\vec{D}$ is the sum of the conduction and displacement currents. With (1.2), (1.7) and (1.11), this turns out to be

$$\frac{\omega}{2\pi}\epsilon''(\omega)\left|\vec{E}(\omega)\right|^2 = \frac{\omega}{\pi}n\kappa\left|\vec{E}(\omega)\right|^2$$

At frequencies where no absorption takes place, we must therefore have $\epsilon''(\omega) = 0$. It then follows from (1.11) that either κ or n is zero, depending on whether $\epsilon'(\omega)$ is positive or negative. As we shall see later, $\epsilon'(\omega)$ is usually positive and then, in the absence of absorption, the wave propagates with constant amplitude E_ω and real wave-vector $(\omega/c)\,n$.

In and around the anomalous regions, where $\epsilon'(\omega)$ can be negative, $n = 0$ and the wave is evanescent, decaying with a decay length $c/\omega\kappa$. This decay is not associated with the absorption of energy from the wave by the medium but with the medium's inability to support a propagating wave at all.

The linear constitutive relations (1.3) - (1.5) are adequate when only electric and magnetic dipole interactions between the field and the medium are significant. This is the case most of the time since at optical frequencies, the wavelength of the radiation (~ 5000 Å), and so the distance over which the field changes appreciably, is very much larger than electronic orbits in atoms and lattice spacings in crystals, which are the order of a few ångströms. Over these atomic and lattice dimensions, the field is essentially constant, and polarization effects depend only on that constant value. In certain special circumstances, however, optical effects manifest themselves which are dependent on the spatial rate of change of the field for their existence. These will be discussed later. For the present it is sufficient to note that when this is the case, our simple constitutive relations must be replaced by more general ones, in which the various tensors of (1.3) - (1.5) are replaced by more general ones depending not only on the frequency ω but also on the wave-vector $\vec{k}$. When this is done, the general tensors $\underline{\kappa}(\omega,\vec{k})$, $\underline{\mu}(\omega,\vec{k})$ and $\underline{\sigma}(\omega,\vec{k})$ describe completely the linear optical properties of the medium concerned. No other information is required; they contain all the physics of the medium which is needed to define or explain these properties.

The physics of the medium may be based on a classical or a quantum-mechanical model. But the final products from the model, be it classical or quantum-mechanical, are appropriate expressions for the tensors. It is then usually sufficient to use these with the classical treatment of Maxwell's equations (just described) to explain linear optical phenomena. When the classical treatment of Maxwell's equations is allied to a quantum-mechanical description of the medium, the treatment is said to be semiclassical.

Later in these Proceedings, much will be contributed about the optical properties of a variety of media, e.g. metals, insulators, etc., and also about many types of excitations which can exist in these media, such as phonons, excitons, polarons, etc. In all of this, it is well to remember that no matter what the excitation is, in whatever the medium, its effect on the linear optical properties of the medium is described by its contribution to the complex dielectric or magnetic tensors $\underline{\epsilon}(\omega,\vec{k})$ and $\underline{\mu}(\omega,\vec{k})$.

The vast majority of linear optical phenomena can be explained in terms of the electrical properties of a medium, characterized by $\underline{\epsilon}(\omega,\vec{k})$. Magnetic effects are rarely of much importance, since magnetic resonant frequencies are typically low compared to optical frequencies. Henceforth these effects will therefore be neglected and $\underline{\mu}(\omega,\vec{k})$ treated as a unit tensor. There are then three aspects of the dielectric permeability tensor $\underline{\epsilon}(\omega,\vec{k})$ which I should now like to consider in turn — its dependence on frequency, its dependence on wave-vector and its tensor character.

2. FREQUENCY DEPENDENCE OF $\underline{\epsilon}(\omega,\vec{k})$

The dependence of $\underline{\epsilon}(\omega,\vec{k})$ on frequency is called its frequency dispersion law. Since we shall not be concerned in this section with its $\vec{k}$-dependence and tensor character, the discussion will be carried out, for simplicity, in terms of a dielectric constant $\epsilon(\omega)$.

The details of the frequency dependence of $\epsilon(\omega)$, as will be seen, reflect the detailed physical properties of the medium it describes. In saying this, I refer mainly to dielectric rather than conduction effects. As we have already seen, the conductivity of a medium contributes an imaginary part $4\pi i\sigma(\omega)/\omega$ to $\epsilon(\omega)$. The electrical polarizability of the medium makes a more complicated contribution to $\epsilon(\omega)$ and it is with this that the rest of this section will be concerned.

However, before discussing this in detail there are two general relations governing the frequency dependence which are universal and are a direct consequence of the principle of causality. They are the Kramers-Kronig relations and arise as follows. The time-dependent electric induction D(t) is quite generally related to the time-varying electric field E(t) by

$$D(t) = E(t) + \int_{-\infty}^{t} f(t-\tau)\, E(\tau)\, d\tau \qquad (2.1)$$

where f(t) is a function, characterized by the response times of the various physical processes which can occur in the medium. Using (1.2) and (1.3), it is readily shown that the dielectric constant is related to f(t) by

$$\epsilon(\omega) = 1 + \int_{0}^{\infty} f(\tau)\, e^{i\omega\tau}\, d\tau \qquad (2.2)$$

The significant point to note is that the limits of integration in (2.1) extend not to infinity but only to t. This is simply a mathematical expression of causality, which says that the electric induction at any time can be a result only of the previous history of the electric field and cannot depend on future values of the field. This apparently simple and obvious statement leads to the Kramers-Kronig relations between the real and imaginary parts of $\epsilon(\omega)$ [1]. For a material with zero conductivity, these are

$$\epsilon'(\omega) = 1 + \frac{2}{\pi} P \int_0^\infty \frac{x\epsilon''(x)}{x^2 - \omega^2} dx$$
$$\epsilon''(\omega) = -\frac{2\omega}{\pi} P \int_0^\infty \frac{\epsilon'(x) - 1}{x^2 - \omega^2} dx \tag{2.3}$$

where we have $\epsilon(\omega) = \epsilon'(\omega) + i\epsilon''(\omega)$. Thus a knowledge of the complete frequency dependence of either the real or the imaginary part of $\epsilon(\omega)$ is sufficient to define the whole function completely.

The detailed dispersion law of $\epsilon(\omega)$ arising from the electrical polarizability of a medium depends on the nature of the medium it describes. An instructive classical model of such a medium was due originally to Lorentz. This pictures each electron in a dielectric medium as being held in its equilibrium position by a harmonic restoring force. When an electric field E is applied to the system, each electron is forced to move according to the equation of motion:

$$\frac{d^2r}{dt^2} + 2\gamma \frac{dr}{dt} + \omega_0^2 r = -\frac{e}{m} E \tag{2.4}$$

Here r is the displacement of the electron from its equilibrium position, e is the magnitude of its charge, ω_0 is the natural frequency of its motion and γ is a damping parameter; for simplicity we have also assumed a one-dimensional model. With the expression (1.2) for E, this equation is readily solved to give

$$r = -\frac{e}{m} E(\omega) \frac{e^{-i\omega t}}{\omega_0^2 - 2i\gamma\omega - \omega^2} + \text{complex conjugate} \tag{2.5}$$

Associated with this deplacement r is an electric polarization,

$$P = -Ner \tag{2.6}$$

where N is the electron density, and the electric induction $D = E + 4\pi P$. It follows, using (2.5), that for this system

$$\epsilon(\omega) = 1 + \frac{4\pi Ne^2}{m} \frac{1}{\omega_0^2 - 2i\gamma\omega - \omega^2} \tag{2.7}$$

Thus

$$\epsilon'(\omega) = 1 + \frac{4\pi Ne^2}{m} \frac{\omega_0^2 - \omega^2}{(\omega_0^2 - \omega^2)^2 + 4\gamma^2\omega^2} \tag{2.8}$$

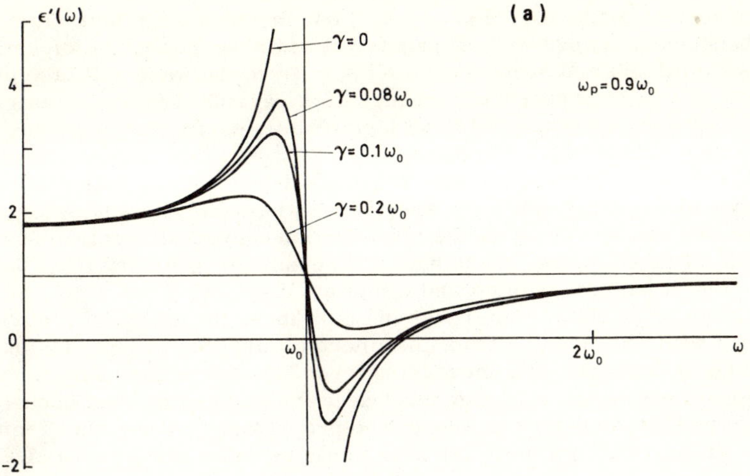

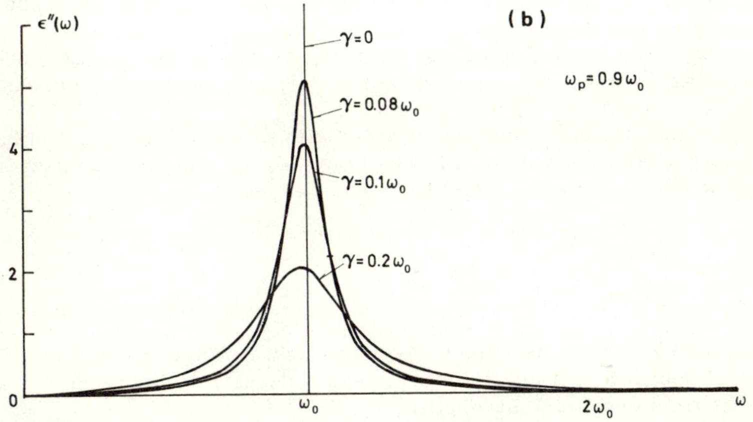

FIG. 1. Frequency dependence of $\epsilon'(\omega)$ (part (a)) and $\epsilon''(\omega)$ (part (b)) for a classical Lorentzian oscillator model through the resonant frequency ω_0 for various values of the damping parameter γ.

and

$$\begin{aligned} \epsilon''(\omega) &= \frac{8\pi Ne^2}{m} \frac{\gamma\omega}{(\omega_0^2 - \omega^2)^2 + 4\gamma^2\omega^2} \\ &\approx \frac{2\pi Ne^2}{m\omega_0} \frac{\gamma}{(\omega_0 - \omega)^2 + \gamma^2}, \text{ for } \omega \approx \omega_0 \end{aligned} \tag{2.9}$$

We need only consider positive frequencies and then we notice that $\epsilon''(\omega)$ is a function with a peak at $\omega \approx \omega_0$ and around which its shape is Lorentzian with a half-width γ at half-height. For frequencies ω removed from ω_0 by more than a few line-widths γ, $\epsilon''(\omega) \ll \epsilon'(\omega)$ and we can neglect it. Thus outside the immediate neighbourhood of ω_0, $\epsilon(\omega)$ can be considered entirely real, and consequently a wave will propagate without absorption; but close to ω_0, $\epsilon''(\omega)$ is significant and absorption of energy from the wave will take place. Physically, ω_0 is the resonant frequency of the system and, in the neighbourhood of this frequency, the field can significantly excite the electronic oscillators and lose energy to them.

More detailed sketches of the frequency dependence of both $\epsilon'(\omega)$ and $\epsilon''(\omega)$ are shown in Fig.1 (a) and (b), respectively, where the quantity $4\pi Ne^2/m$ has been denoted by ω_p^2; ω_p is the plasma frequency of an electron gas of density N. For all values of the damping parameter γ, we see that $\epsilon'(\omega) = 1$ at $\omega = \omega_0$ and it has a maximum and minimum displaced from ω_0 by $\sim\gamma$. At all frequencies outside the range between these turning points, $\epsilon'(\omega)$ increases with frequency — it exhibits normal dispersion; only inside the range does it decrease with increasing frequency and is then said to exhibit anomalous dispersion. $\epsilon'(\omega)$ can adopt negative values for weak enough damping, and the medium can support no propagating waves; the frequency range over which $\epsilon'(\omega)$ is negative is sometimes called the stop band. For $\gamma = 0$, it extends from ω_0 to $\omega_L = (\omega_0^2 + \omega_p^2)^{\frac{1}{2}}$, at which frequency $\epsilon'(\omega) = 0$ and longitudinal waves can propagate. As γ increases, the stop band decreases in size; for $\gamma \geqq (\omega_L - \omega_0)/2$, no stop band exists and the medium can support propagating waves at all frequencies.

For $\gamma = 0$, $\epsilon''(\omega)$ is a delta function which broadens out in a Lorentzian-type curve as γ increases. As it broadens out, its peak height decreases to maintain a constant area under the curve.

In general, one can think of a system with different groups of electrons being held with different harmonic restoring forces and having different effective masses. Equation (2.7) then generalizes to

$$\epsilon(\omega) = 1 + \sum_p \frac{4\pi N_p e^2}{m} \frac{f_p}{\omega_p^2 - 2i\gamma_p\omega - \omega^2} \tag{2.10}$$

where ω_p and γ_p are the resonant frequency and damping parameter of the pth type of oscillator, which is occupied by a density of electrons N_p, and f_p is its so-called oscillator strength.

A more rigorous quantum-mechanical treatment of an atomic system yields an expression for $\epsilon(\omega)$ very similar in form to (2.10) — the Kramers-Heisenberg formula:

$$\epsilon(\omega) = 1 + \sum_{p,q} \frac{4\pi N_p e^2}{m} \frac{f_{pq}}{\omega_{pq}^2 - 2i\gamma_{pq}\omega - \omega^2} \tag{2.11}$$

Here p and q label the electronic quantum states and N_p is the density of electrons in state p; $\hbar\omega_{pq}$ is the energy difference between states p and q, and the damping parameter γ_{pq} is related to the reciprocals of their lifetimes. Finally, f_{pq} is the oscillator strength of the transition from state p to state q and can be expressed in terms of the dipole moment matrix element $\mathscr{P}_{pq}$ between these states as

$$f_{pq} = \frac{2m\,\omega_{qp}}{3e^2\hbar} |\mathscr{P}_{pq}|^2 \tag{2.12}$$

It follows from this discussion that $\epsilon(\omega)$ has in general a series of resonant frequencies in the neighbourhood of which it has a significant imaginary part, but that for all other frequencies it is essentially real. Only around the resonant frequencies therefore does the medium absorb energy from the radiation field. This absorption takes place through the excitation of the appropriate resonant transition, and the absorbed energy is dissipated to the rest of the system through the lifetime-limiting mechanisms of the various states. Far from resonance, no absorption of energy from the field takes place, although the associated excitations or oscillators are still driven by the field; this leads to a modification of the velocity of propagation of the radiation in the medium characterized by its refractive index $n = \sqrt{\epsilon'(\omega)}$.

The frequency dispersion law of a medium thus arises from the existence of resonant transitions in the medium, around whose frequency the medium can absorb energy from the field. Such transitions exist in any medium and can occur over a wide range of frequencies, from the infra-red where the transitions are associated with the excitation of molecular rotational or vibrational motion or with lattice vibrations, through the visible and into the ultra-violet and X-ray regions, where the transitions are electronic in nature. It follows that there is always a certain amount of frequency dispersion in any medium other than vacuum; the amount of dispersion, i.e. the strength of the frequency dependence of $\epsilon(\omega)$, is a function of the proximity to a resonant frequency of the medium. Far away from resonance, the dispersion is weak; close to it, the dispersion can be very strong.

The relationship between the frequency dispersion of $\epsilon'(\omega)$ and the existence of a resonant absorption frequency can also be seen from the Kramers-Kronig relations (2.3). In the ideal case of a resonant frequency ω_0 with zero associated damping, (2.9) becomes in the limit $\gamma \to 0$,

$$\epsilon''(\omega) = \frac{8\pi Ne^2}{m} \frac{\pi}{4\omega_0} [\delta(\omega - \omega_0) - \delta(\omega + \omega_0)]$$

Using this in (2.3) gives

$$\epsilon'(\omega) = 1 + \frac{2}{\pi} P \int_0^\infty \frac{x}{x^2 - \omega^2} \frac{2\pi^2 Ne^2}{m\omega_0} \delta(x - \omega_0)\, dx$$

$$= 1 + \frac{4\pi Ne^2}{m} \frac{1}{\omega_0^2 - \omega^2}$$

which agrees with (2.8) in this same limit.

A sketch of the typical frequency dependence of the real part $\epsilon'(\omega)$ of the dielectric constant of a non-conducting medium is shown in Fig.2. Between resonant frequencies, $\epsilon'(\omega)$ always increases with frequency — this is called normal dispersion. But through each resonant frequency over a region about two line-widths wide, $\epsilon'(\omega)$ decreases very rapidly as ω increases — this is called anomalous dispersion. The general level of each region of slowly increasing $\epsilon'(\omega)$ between resonances decreases from one region to the next at higher frequency, till eventually beyond the highest resonance $\epsilon'(\omega) \to 1$. We shall find that this ever-present frequency dispersion is an

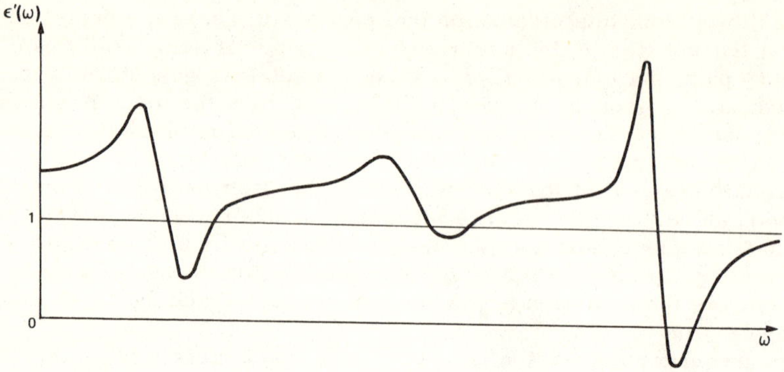

FIG.2. Sketch of the general frequency dependence of $\epsilon'(\omega)$, covering several resonant frequencies with different associated oscillator strengths and damping parameters.

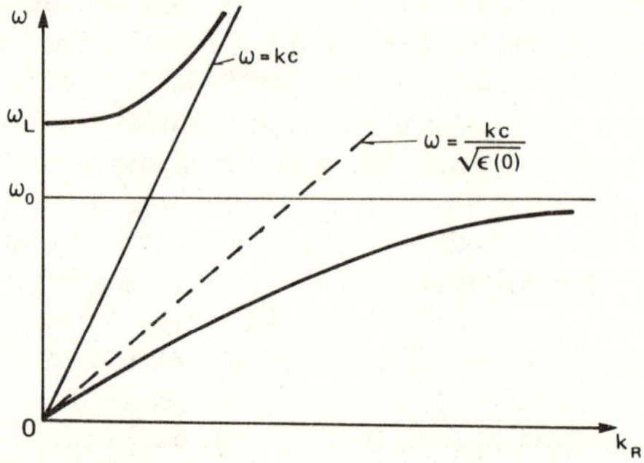

FIG.3. Frequency ω of a propagating wave as a function of the real part of its propagation constant k_R in a medium with a classical Lorentzian resonance frequency ω_0. Also shown, by straight lines, are the corresponding relations for the free-space electromagnetic wave, $\omega = kc$, and the oscillators, $\omega = \omega_0$; and by the dashed line the low-frequency behaviour of the dispersion relation.

effect which can limit severely the study and effective exploitation of many non-linear optical effects. The possibility of $\epsilon'(\omega)$ becoming negative at the high frequency end of a region of anomalous dispersion can also be seen. The smaller the damping parameter γ associated with a transition, the more violent is the anomalous dispersion and the more likely is $\epsilon'(\omega)$ to go negative. As discussed earlier, in these circumstances the medium cannot support the propagation of a wave and the region of negative $\epsilon'(\omega)$ is in consequence often called a stop band.

Before leaving the subject of the frequency dependence of $\epsilon(\omega)$, it is worth returning to the simple expression (2.7) and looking at it in another way. For zero damping

$$\epsilon(\omega) = 1 + \frac{\omega_p^2}{\omega_0^2 - \omega^2}, \quad \omega_p^2 = \frac{4\pi N e^2}{m} \qquad (2.13)$$

The form of this frequency dependence is shown in Fig.1(a) by the curve labelled $\gamma = 0$. Note that we denote by $\omega_L = (\omega_0^2 + \omega_p^2)^{\frac{1}{2}}$ the frequency at which

$\epsilon'(\omega) = 0$ when $\gamma = 0$ and consequently waves with a longitudinal component can propagate (as discussed following Eq.(1.10)).

In this simple isotropic case, we showed from (1.10) that the magnitude of the wave vector $\vec{k}$ is given by

$$k^2 = \epsilon(\omega)\,\frac{\omega^2}{c^2} \tag{2.14}$$

Combining this with (2.13), it is a simple matter to plot the real part of k, i.e. k_R, against ω or more usefully ω against k_R. This is shown in Fig.3 by the curved lines. On this diagram are also shown, by the straight lines, the relations for electromagnetic radiation propagating in the absence of any interaction, i.e. $\omega = kc$, and for the electronic oscillators, again in the absence of any interaction, i.e. $\omega = \omega_0$. When we allow interactions to take place between these two modes of the independent systems, the usual mode-coupling effects take place which produce a repulsion of modes at the cross-over point and give us our curves, defined by (2.13) and (2.14). This way of looking at transitions of a medium which can be excited by the electromagnetic field and their effects on the propagation of the field is common in discussions of excitons and optical phonons, the coupled modes of polarization and field being called polaritons. However, it is only another way of looking at the origin of the frequency dispersion of the dielectric constant.

3. WAVE-VECTOR DEPENDENCE OF $\underline{\epsilon}(\omega,\vec{k})$

I have already pointed out that wave-vector dependence of $\underline{\epsilon}(\omega,\vec{k})$, or its spatial dispersion, can be ignored most of the time, since at optical frequencies the wavelength of the radiation is very much larger than the typical dimensions of electronic orbits and lattice spacings. This is exactly analogous to our ability to neglect frequency dispersion at frequencies well below the lowest resonant frequency where $\epsilon(\omega)$ tends to its constant d.c. value; in this region the vibrational period of the field is much longer than any typical resonant frequency of the medium. At optical frequencies, we cannot usually achieve this situation in the time domain but it is usually achieved in the spatial domain. Thus at these frequencies, frequency dispersion is always important whilst spatial dispersion is usually unimportant.

Just as frequency dispersion arises from the dependence of the electric induction at a given time on the electric field strengths at other earlier times, so spatial dispersion arises from a dependence of the induction at a given point in a medium on field strengths at neighbouring points. This reveals itself as a dependence of the induction on not only the field strength, but also on its spatial derivatives, at a point. This, we have already seen, leads to a wave-vector dependence of the dielectric tensor.

Spatial dispersion becomes important in two cases: (a) when it gives rise to new effects which, however small, are solely dependent for their existence on its presence or, (b) when it is anomalously large and can produce effects which compete directly with more conventional optical phenomena.

Natural optical activity is the simplest phenomenon dependent for its existence on spatial dispersion [1, 2]. It manifests itself as a rotation of the plane of polarization of a monochromatic beam of radiation as it traverses certain optically isotropic materials; the amount of rotation is proportional to the thickness traversed. It can be explained on the basis of the following relationship between the electric induction and field:

$$\vec{D}(\omega) = \epsilon^{(0)}(\omega)\vec{E}(\omega) - \frac{c}{\omega} f(\omega)\, \mathrm{curl}\, \vec{E}(\omega) \tag{3.1}$$

which with (1.8) becomes

$$\vec{D}(\omega) = \epsilon^{(0)}(\omega)\vec{E}(\omega) - \frac{ic}{\omega} f(\omega)[\vec{k} \times \vec{E}(\omega)] \tag{3.2}$$

Here $\epsilon^{(0)}(\omega)$ is used to denote the dielectric constant required in the absence of optical activity and $f(\omega)$ is a function of ω chosen to account for the frequency dependence of the activity, as will be seen later. Equation (3.2) clearly leads to a wave-vector dependent dielectric tensor:

$$\epsilon_{\alpha\beta}(\omega,\vec{k}) = \epsilon^{(0)}(\omega)\,\delta_{\alpha\beta} + \frac{ic}{\omega} f(\omega) \sum_{\gamma=x,y,z} e_{\alpha\beta\gamma} k_\gamma \tag{3.3}$$

where $e_{\alpha\beta\gamma}$ is the antisymmetric unit tensor of rank three. Using this expression (3.3) in the Fresnel equations, and assuming that

$$f(\omega) \ll n_0(\omega) = \sqrt{\epsilon^{(0)}(\omega)}$$

the refractive indices $n_\pm(\omega)$ for left and right circularly polarized light are readily shown to be given by

$$n_\pm^2(\omega) = n_0^2(\omega) \pm f(\omega)\, n_0(\omega) \tag{3.4}$$

These different refractive indices for different senses of circularly polarized radiation give rise to the effect described.

In the way we have discussed optical activity, it characterizes the class of effects (a) described above; no rotation of the plane of polarization of radiation takes place in an isotropic material in the absence of spatial dispersion. Optical activity can also be found in the same way in some anisotropic materials when radiation propagates along the optic axis — α-quartz is the classic example. But when propagation takes place along a direction other than the optic axis, the effects of optical activity are lost in the similar, but much larger, effects of birefringence.

An absorption effect that requires spatial dispersion for its explanation is found in the exciton absorption spectrum of cuprous oxide (Cu_2O). This is a cubic (and thus normally optically isotropic) crystal whose excitonic properties have been particularly well studied. It turns out that in this material, optical excitation of the ground exciton state is forbidden by electric dipole transitions. But a weak absorption line is found at the appropriate frequency which has a magnitude varying with the propagation direction and polarization of the exciting radiation. These variations are consistent with electric quadrupole excitation [3, 4]. They therefore depend on spatial derivatives of the electric field and are a further manifestation of spatial dispersion.

Finally, we have seen how the dielectric constant and consequently the refractive index can become very large close to regions of anomalous dispersion. A large refractive index implies a short propagation wavelength in the medium. Thus for large enough refractive index the approximation, on which the neglect of spatial dispersion depends, breaks down. We can therefore expect spatial dispersion effects to become particularly significant in these high refractive index regions close to regions of anomalous frequency dispersion. Much theoretical work was done on this by Pekar [5], and Agranovich and Ginzburg [6], who showed that in these regions, additional modes of propagation become allowed, which can interfere with the normal modes to produce new effects. Hopfield and Thomas [7] have also studied these effects both theoretically and experimentally in some detail, and more recent relevant work has been carried out by Frova [8] and his colleagues. It would be too large a digression to discuss them in detail here. Suffice it to say that the Fresnel equations (1.9) are quadratic in k^2 and, as we shall see in the next section, normally give rise to two modes of propagation at any frequency. But when $\underline{\epsilon}(\omega)$ becomes dependent on $\vec{k}$, additional roots of the equations can be generated, giving rise to additional modes of propagation.

4. TENSOR CHARACTER OF $\underline{\epsilon}(\omega,\vec{k})$

The tensor nature of $\underline{\epsilon}(\omega,\vec{k})$ arises from the internal spatial symmetry of the medium it describes. The way in which this tensor character affects wave propagation in the medium is a mathematical expression of the influence the medium's spatial symmetry has on the propagation modes, which can be supported.

The basic points I wish to make in this section can be seen by limiting our attention as before to a non-magnetic medium ($\underline{\mu} = \underline{1}$) with zero charge density ($\rho = 0$). Negligible spatial dispersion is assumed. Maxwell's equations (1.1) for a wave of frequency ω and wave-vector $\vec{k}$ then become

$$\begin{aligned} \vec{k} \times \vec{E}(\omega) &= \frac{\omega}{c}\vec{H}(\omega) \\ \vec{k} \times \vec{H}(\omega) &= -\frac{\omega}{c}\vec{D}(\omega) \qquad (4.1) \\ \vec{k}\cdot\vec{D}(\omega) &= \vec{k}\cdot\vec{H}(\omega) = 0 \end{aligned}$$

It follows that quite generally, $\vec{k}$, $\vec{D}$ and $\vec{H}$ are mutually perpendicular vectors and, since $\vec{H}$ is also perpendicular to $\vec{E}$, that $\vec{k}$, $\vec{D}$ and $\vec{E}$ must be coplanar in a plane normal to $\vec{H}$. Furthermore, the Poynting or energy flux vector $\vec{S} \sim \vec{E} \times \vec{H}$ must be coplanar with $\vec{k}$, $\vec{D}$ and $\vec{E}$ and perpendicular to $\vec{E}$. The geometry is illustrated in Fig.4 where $\vec{H}$ points out towards the reader. It should be noted that only in a so-called isotropic medium, where $\epsilon(\omega)$ is proportional to the unit tensor so that $\vec{D}$ and $\vec{E}$ are parallel, do the directions of propagation and energy flow coincide.

Maxwell's equations, along with the appropriate constitutive equation connecting $\vec{D}$ and $\vec{E}$, lead to the already derived Fresnel equations:

$$k^2\,\vec{s} \times (\vec{s} \times \vec{a}_\omega) + \frac{\omega^2}{c^2}\,\underline{\vec{\epsilon}}(\omega)\cdot\vec{a}_\omega = 0 \qquad (4.2)$$

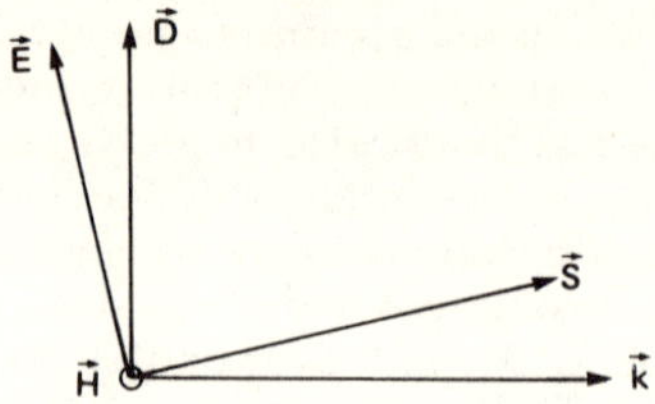

FIG. 4. Illustration of the relative directions, in the most general case, of the electric field $\vec{E}$, magnetic field $\vec{H}$, electric induction $\vec{D}$, propagation vector $\vec{k}$ and Poynting vector $\vec{S}$ of a wave. $\vec{H}$ points out of the diagram.

which, in terms of complex refractive index $N = kc/\omega$, become

$$N^2[\vec{s}\,(\vec{s}\cdot\vec{a}_\omega) - \vec{a}_\omega\,] + \underline{\epsilon}(\omega)\cdot\vec{a}_\omega = 0 \tag{4.3}$$

These are three homogeneous linear equations for the components of $\vec{a}_\omega$. Non-zero solutions require the determinant of coefficients to be zero, and this defines values of N for given directions of propagation $\vec{s}$.

In deriving this requirement on the determinant, it is convenient to proceed in not the most obvious fashion and consequently to derive another result en route. It follows from the constitutive relation between $\vec{D}$ and $\vec{E}$ that $\vec{d}_\omega = \underline{\epsilon}(\omega)\cdot\vec{a}_\omega$ is a vector parallel to $\vec{D}$. Thus we can write (4.3) as

$$\vec{d}_\omega = N^2\,[\vec{a}_\omega - \vec{s}(\vec{s}\cdot\vec{a}_\omega\,)] \tag{4.4}$$

We now use the fact that $\underline{\epsilon}(\omega)$ is a symmetric tensor. This symmetry is not a trivial thing to show and I shall not go into details here. In the absence of any damping, the basic microscopic equations of motion in a medium are invariant under time reversal; in this case, the symmetry follows from this invariance [9]. But in general, Onsager's principle in irreversible thermodynamics must be used to establish the symmetry [10]. There therefore exists a set of co-ordinate axes Oxyz with respect to which $\underline{\epsilon}(\omega)$ is diagonal with components ϵ_x, ϵ_y, ϵ_z — the frequency dependence will be dropped for the rest of this section since it is unnecessary. In this co-ordinate reference frame $d_\alpha = \epsilon_\alpha a_\alpha$, where α = x,y or z, and (4.4) can be written

$$d_\alpha = N^2\left[\frac{d_\alpha}{\epsilon_\alpha} - s_\alpha(\vec{s}\cdot\vec{a})\right]$$

i.e.

$$d_\alpha = \frac{(\vec{s}\cdot\vec{a})s_\alpha}{\dfrac{1}{\epsilon_\alpha} - \dfrac{1}{N^2}} \tag{4.5}$$

But we have shown that $\vec{D}\cdot\vec{k} = 0$ so that $\vec{d}\cdot\vec{s} = 0$. It follows that

$$\frac{s_x^2}{\dfrac{1}{\epsilon_x} - \dfrac{1}{N^2}} + \frac{s_y^2}{\dfrac{1}{\epsilon_y} - \dfrac{1}{N^2}} + \frac{s_z^2}{\dfrac{1}{\epsilon_z} - \dfrac{1}{N^2}} = 0 \tag{4.6}$$

This is essentially the relation we seek defining the values of N for given directions of propagation, characterized by (s_x, s_y, s_z).

Equation (4.6) will be discussed more fully in a moment. But first of all note that this equation is a quadratic in N^2 defining two values N_1 and N_2 which through (4.5) define two associated values of $\vec{d}$, viz. $\vec{d}_1$ and $\vec{d}_2$. It follows that

$$\vec{d}_1 \cdot \vec{d}_2 = (\vec{s} \cdot \vec{a})^2 \sum_{\alpha=x,y,z} \frac{s_\alpha^2}{\left(\frac{1}{\epsilon_\alpha} - \frac{1}{N_1^2}\right)\left(\frac{1}{\epsilon_\alpha} - \frac{1}{N_2^2}\right)}$$

$$= (\vec{s} \cdot \vec{a})^2 \frac{N_1^2 N_2^2}{N_1^2 - N_2^2} \sum_\alpha \left(\frac{s_\alpha^2}{\frac{1}{\epsilon_\alpha} - \frac{1}{N_1^2}} - \frac{s_\alpha^2}{\frac{1}{\epsilon_\alpha} - \frac{1}{N_2^2}} \right) \tag{4.7}$$

and consequently using (4.6) that

$$\vec{d}_1 \cdot \vec{d}_2 = 0 \tag{4.8}$$

This shows that in the regime of linear optics, a medium (with negligible spatial dispersion) can sustain only two electromagnetic waves at a given frequency and this arises purely from the symmetric nature of the dielectric tensor, which follows from the Onsager principle. These waves propagate with different refractive indices and so different wavelengths and velocities, and are plane polarized orthogonally to one another in the sense that their associated electric inductions are mutually orthogonal. It also follows from our earlier results in this section that their magnetic fields are also mutually orthogonal. These two waves therefore propagate independently of one another and do not interfere.

To determine the refractive indices associated with the two waves we must return to (4.6) which can be cast into the now more convenient form:

$$(N^2 - \epsilon_y)(N^2 - \epsilon_z)\epsilon_x s_x^2 + (N^2 - \epsilon_z)(N^2 - \epsilon_x)\epsilon_y s_y^2$$
$$+ (N^2 - \epsilon_x)(N^2 - \epsilon_y)\epsilon_z s_z^2 = 0 \tag{4.9}$$

Although any material has in general three independent components in its dielectric tensor, viz. ϵ_x, ϵ_y and ϵ_z, the internal spatial symmetry of the medium determines relations between these. In fact, materials fall into three groups. The first contains isotropic media like liquids and crystals with cubic symmetry; in this group the high spatial symmetry forces the components to be equal so that $\epsilon_x = \epsilon_y = \epsilon_z = \epsilon$. Equation (4.9) then reduces to $(N^2 - \epsilon)^2 = 0$ for all values of ω and this has two coincident roots $N^2 = \epsilon$. It follows from (4.5) that $\vec{D}$ is finite only if $\vec{s} \cdot \vec{a} = 0$; consequently $\vec{E}$ is normal to $\vec{k}$ but is otherwise unrestricted. This group of materials are therefore optically isotropic, since in any direction of propagation they support two orthogonally polarized plane waves whose propagation is characterized by the same refractive index.

The second group of materials contains crystalline solids belonging to the tetragonal, trigonal and hexagonal crystal classes. In this group, two

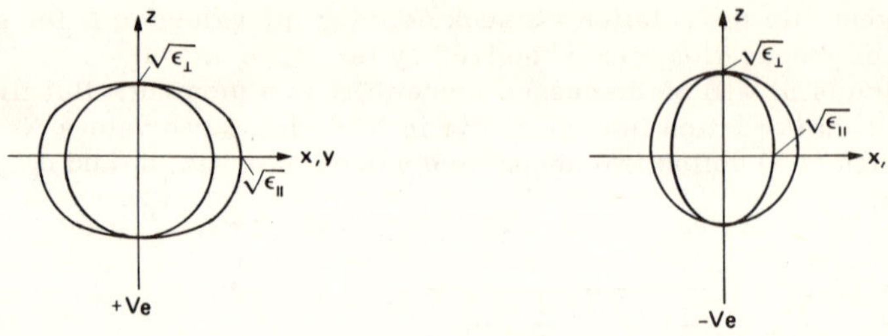

FIG. 5. Plane sections through the optic axes of the refractive index surfaces in positive and negative uniaxial materials.

diagonal elements of $\underline{\epsilon}$ must be equal but can be different from the third. Materials in this group are called uniaxial. It is conventional to choose $\epsilon_x = \epsilon_y = \epsilon_\perp$ and $\epsilon_z = \epsilon_\parallel$. Then (4.9) becomes (since $\vec{s}^2 = 1$)

$$(N^2 - \epsilon_\perp)[N^2(\epsilon_\perp(s_x^2 + s_y^2) + \epsilon_\parallel s_z^2) - \epsilon_\parallel \epsilon_\perp] = 0 \tag{4.10}$$

whose roots are given by

$$N^2 = \epsilon_\perp \tag{4.11}$$

and

$$\frac{N^2(s_x^2 + s_y^2)}{\epsilon_\parallel} + \frac{N^2 s_z^2}{\epsilon_\perp} = 1 \tag{4.12}$$

In the space of the vector $\vec{N} = N\vec{s}$, which is proportional to the wave-vector $\vec{k} = \vec{N}\omega/c$, these equations define two refractive index surfaces whose intersections with a line from the origin in any direction define the refractive indices of the two waves which can propagate in that direction. One surface is a sphere and describes a wave, the ordinary wave, which, as in an isotropic material, propagates with the same refractive index $N = \epsilon_\perp^{1/2}$ in any direction. The other surface is an ellipsoid of revolution about the z-axis; it describes a wave, the extraordinary wave, whose refractive index is a function of the angle its propagation direction makes with the z-axis. Sections through the two possible pairs of surfaces are shown in Fig.5. For propagation along the z-axis, both waves are described by the same refractive index $\epsilon_\perp^{1/2}$; this direction is called the optic axis. For propagation in the plane normal to this axis, the ordinary wave is unchanged but the extraordinary wave is described by a refraction index $\epsilon_\parallel^{1/2}$. For $\epsilon_\parallel > \epsilon_\perp$, the crystal is said to be positive uniaxial and for $\epsilon_\parallel < \epsilon_\perp$, it is negative uniaxial.

It should be noted that waves at different frequencies will be characterized by different pairs of surfaces. With normal dispersion, the surfaces will enlarge themselves with increasing frequency. It is therefore possible for two waves at different frequencies to propagate in certain special directions with the same refractive index. For example, it can be seen from Fig.5 that, in a positive uniaxial crystal, the ellipsoid of the extraordinary wave at one frequency could be cut in two circular sections by the sphere of the ordinary wave at a slighly higher frequency. In the directions defined by these two circular sections, the two waves would propagate with the same refractive index. This is an important concept, which we shall find is made considerable use of in studying non-linear optical effects.

It is conventional to discuss these linear propagation effects in terms of refractive indices and this convention is followed here. A similar discussion could of course be carried out in wave-vector space, which differs only by a scaling factor ω/c. We shall find in discussing non-linear optics that the wave-vector is a more useful concept to use than the refractive index. The above concept obviously carries over, showing that in certain directions waves of different frequency can propagate with the same wave-vector. The frequency-dependent scaling factor ω/c means that the directions with common refractive index and common wave-vector are different.

The third and last group of materials contains crystalline solids belonging to the triclinic, monoclinic and orthorhombic crystal classes. For these, all elements of $\underline{\epsilon}$ can be different and the materials are called biaxial. The refractive indices which describe the propagation are fairly complex functions of direction in the crystal. As always, there are two orthogonally polarized waves which can propagate in any direction.

Finally, it should be pointed out that all the discussions in this section have assumed that there is no spatial dispersion. The existence of certain types of such dispersion can upset some of our conclusions; for example, a third mode of propagation sometimes can be supported and quadrupole effects can lead to anisotropic effects in a cubic crystal, as mentioned in cuprous oxide.

Part II

NON-LINEAR OPTICS

5. INTRODUCTION

The high intensity of radiation available from a laser can profoundly alter the normal optical properties of any system illuminated by it. These normal optical properties belong to the realm of linear optics insofar as they can be understood in terms of the linear response of a medium, as characterized by relations of the type (1.3), (1.4), (1.5). However, a linear relationship between two physical quantities, describing a driving force and a response, is almost invariably an approximation of limited validity, of which the breakdown of Hooke's law for large enough stresses and the temperature dependence of the thermal expansion coefficient of a solid, are familiar examples. The linear relationship between polarization and field, used in linear optics, is no exception to this and when strong enough fields are applied, non-linearities in the relationship inevitably become significant. That lasers can produce fields of sufficient strength for this to happen is perhaps not surprising when it is realised that the electric field strength in the beam from a laser can be at least some two or three orders of magnitude greater than the field strength available from a conventional source of radiation.

These non-linearities would be almost insignificant if they resulted only in modifications to effects already arising from the linear response, which is generally several orders of magnitude larger than its non-linear counterpart. Their importance lies in the fact that they give rise, as non-linearities always do, to effects which do not take place in a completely linear system and which can therefore be readily discriminated. They open up the possibilities, for example, of amplifying, modulating and mixing beams of radiation at infra-red and optical frequencies. Such effects are well known in the microwave and radiofrequency ranges of the spectrum but, until the advent of the laser, they had never been observed at higher frequencies. These processes are important for several reasons. By the generation of the harmonics of a single laser beam or by the mixing of two different beams, they provide a means of extending the frequency range of intense monochromatic sources available and generally allow us much more versatility in the ways in which we can manipulate radiation at these high frequencies. Last, and most obvious, they extend the range of information which can be obtained about a material from a study of its response to applied electromagnetic radiation.

In this second part, we develop the theoretical background to these non-linear effects, and review and discuss some experimental observations in terms of this background. Just as, in describing conventional linear optical effects, we were able to neglect magnetic effects, so too in the non-linear regime we can concentrate our attention on electric field effects.

6. A CLASSICAL THEORY OF NON-LINEAR RESPONSE IN ONE DIMENSION

We have seen that within the properties of the dielectric tensor $\underline{\epsilon}(\omega,\vec{k})$ resides all the information about how the optical radiation field will propagate in and interact with a medium, within the confines of linear optics. To understand the details of the propagation and interaction, we had to understand how the physics of the interaction controlled the mathematical properties of $\vec{\underline{\epsilon}}(\omega,\vec{k})$. Similarly, in the non-linear regime, we must understand the properties, imposed by the physics of the interaction on the parameters, analogous to $\vec{\underline{\epsilon}}(\omega,\vec{k})$, which describe the non-linear response. We do so in this and the following two sections before going on to study the problems associated with the propagation of the field.

The classical theory of linear dispersion is based on a model due to Lorentz which pictures each electron in a dielectric medium as being held in its equilibrium position by a harmonic restoring force. When an electric field E is applied to the system, each electron is forced to move according to the equation of motion

$$\frac{d^2 r}{dt^2} + 2\gamma \frac{dr}{dt} + \omega_0^2 r = -\frac{e}{m} E \tag{6.1}$$

where r is the displacement of the electron from its equilibrium position, m its mass, e the magnitude of its charge, ω_0 the natural frequency of its motion and γ a damping parameter. This equation is the same as the one considered earlier. For simplicity, we limit ourselves for the present to motion in one dimension only; this will be generalized later. Equation (6.1) is a linear equation and as such has solutions which depend only linearly on the electric field E. The model which gives rise to this result is therefore of no value to us for a description of non-linear effects. It can, however, be readily extended in such a way that non-linear effects are produced. We do this by considering small anharmonic contributions to be present in the restoring force of the oscillator. This is most simply done by replacing the harmonic force $-\omega_0^2 r$ by $-\omega_0^2 r + \lambda r^2 + \eta r^3$ where λ and η are parameters which characterize the anharmonic effects. It is desirable to include both the terms λr^2 and ηr^3 although either one is sufficient to make the total force anharmonic. Our basic problem is now that of solving the extended version of Eq.(6.1):

$$\frac{d^2 r}{dt^2} + 2\gamma \frac{dr}{dt} + \omega_0^2 r - \lambda r^2 - \eta r^3 = -\frac{e}{m} E \qquad (6.2)$$

from which the polarization density P = -Ner induced by the field E can be calculated, N being the electron density in the medium.

To keep the discussion as simple as possible, we consider an electric field E which oscillates at a single angular frequency ω. We can then write

$$E = E(\omega)\, e^{-i\omega t} + E^*(\omega)\, e^{i\omega t} \qquad (6.3)$$

Even with this simple time dependence for the electric field, Eq.(6.2) is by no means easy to solve in general as a result of its non-linear character. However, a solution in the form of a power series in E is readily found. This is quite adequate for our purposes for, as mentioned already and as will be seen in more detail later, the non-linear effects normally produced, although significant, are always much smaller than any similar linear effects. We therefore try to find a solution to Eq.(6.2) of the form

$$r = \sum_{n=1}^{\infty} r_n \qquad (6.4)$$

where r_n is proportional to the nth power of the electric field. The r_n then satisfy a series of equations each of which involves terms which are all of the same order in the field. The first three equations of this set, which are all we shall consider, are

$$\ddot{r}_1 + 2\gamma \dot{r}_1 + \omega_0^2 r_1 = -\frac{e}{m}[E(\omega)\, e^{-i\omega t} + E^*(\omega)\, e^{i\omega t}] \qquad (6.5)$$

$$\ddot{r}_2 + 2\gamma \dot{r}_2 + \omega_0^2 r_2 = \lambda r_1^2 \qquad (6.6)$$

$$\ddot{r}_3 + 2\gamma \dot{r}_3 + \omega_0^2 r_3 = 2\lambda r_1 r_2 + \eta r_1^3 \qquad (6.7)$$

The method of solution is now obvious. Equation (6.5) is identical to (6.1) whose solution we know — see Eq.(2.5). This solution can now be used in the right side of (6.6) which then takes on the same form as (2.5) except that the terms on the right side are either constant or oscillate at frequency 2ω. The solutions for r_1 and r_2 can be substituted in the right side of (6.7) to produce again an equation of similar form to (6.5) but with terms on the right side oscillating at frequencies ω and 3ω. It is not difficult therefore to obtain the following results:

$$r_1 = -\frac{e}{m} E(\omega) \frac{e^{-i\omega t}}{\omega_0^2 - 2i\gamma\omega - \omega^2} + \text{complex conjugate} \tag{6.8}$$

$$r_2 = \frac{e^2}{m^2} \lambda E^2(\omega) \frac{e^{-2i\omega t}}{(\omega_0^2 - 4i\gamma\omega - 4\omega^2)(\omega_0^2 - 2i\gamma\omega - \omega^2)^2}$$

$$+ \frac{e^2}{m^2} \lambda E(\omega) E^*(\omega) \frac{1}{\omega_0^2(\omega_0^2 - 2i\gamma\omega - \omega^2)(\omega_0^2 + 2i\gamma\omega - \omega^2)}$$

$$+ \text{complex conjugate} \tag{6.9}$$

$$r_3 = -\frac{e^3}{m^3} E^3(\omega) \left[\frac{4\lambda^2}{\omega_0^2 - 4i\gamma\omega - \omega^2} + \eta\right] \frac{e^{-3i\omega t}}{(\omega_0^2 - 6i\gamma\omega - 9\omega^2)(\omega_0^2 - 2i\gamma\omega - \omega^2)^3}$$

$$- \frac{e^3}{m^3} E^2(\omega) E^*(\omega) \left[\frac{4\lambda^2}{\omega_0^2 - 4i\gamma\omega - 4\omega^2} + \frac{8\lambda^2}{\omega_0^2} + 3\eta\right]$$

$$\times \frac{e^{-i\omega t}}{(\omega_0^2 - 2i\gamma\omega - 4\omega^2)(\omega_0^2 - 2i\gamma\omega - \omega^2)^2(\omega_0^2 + 2i\gamma\omega - \omega^2)} + \text{complex conjugate} \tag{6.10}$$

We see that, as expected, only the term linear in the field, i.e. r_1, is independent of the degree of anharmonicity. All the higher order terms depend on λ and η and tend to zero as the anharmonicity disappears.

If we now express the polarization density P in the same form as r in (6.4) i.e.

$$P = \sum_{n=1}^{\infty} P_n, \quad P_n = -Ner_n \tag{6.11}$$

it is easy to obtain from (6.8) - (6.10) expressions for the contributions to the induced polarization density proportional to the first, second and third powers of the electric field. We find that

$$P_1 = \chi^{(1)}(\omega) E(\omega) e^{-i\omega t} + \text{complex conjugate} \tag{6.12}$$

where

$$\chi^{(1)}(\omega) = \frac{Ne^2}{m} \frac{1}{\omega_0^2 - 2i\gamma\omega - \omega^2} \tag{6.13}$$

$\chi^{(1)}(\omega)$ is just the linear or first-order polarizability of the medium and this expression for $\chi^{(1)}(\omega)$ is identical to the expression (2.7) previously obtained on the basis of a simple harmonic restoring force. It is convenient to introduce at this point an abbreviated notation by defining

$$F(\omega) = \frac{1}{\omega_0^2 - 2i\gamma\omega - \omega^2} \tag{6.14}$$

Expression (6.13) then becomes

$$\chi^{(1)}(\omega) = \frac{Ne^2}{m} F(\omega) \tag{6.15}$$

The second-order polarization density can now be written as

$$P_2 = \chi^{(2)}(\omega,\omega) E^2(\omega) e^{-2i\omega t} + \chi^{(2)}(\omega,-\omega) E(\omega) E^*(\omega) + \text{complex conjugate} \tag{6.16}$$

where $\chi^{(2)}(\omega,\omega)$ and $\chi^{(2)}(\omega,-\omega)$ can be obtained from the general function

$$\chi^{(2)}(\omega_1, \omega_2) = -N\lambda \frac{e^3}{m^2} F(\omega_1) F(\omega_2) F(\omega_1+\omega_2) \tag{6.17}$$

by setting $\omega_1 = \omega$ and $\omega_2 = \pm\omega$. Finally, the third-order polarization density can be written as

$$P_3 = \chi^{(3)}(\omega,\omega,\omega) E^3(\omega) e^{-3i\omega t} + 3\chi^{(3)}(\omega,\omega,-\omega) E^2(\omega) E^*(\omega) e^{-i\omega t} + \text{complex conjugate} \tag{6.18}$$

where $\chi^{(3)}(\omega,\omega,\omega)$ and $\chi^{(3)}(\omega,\omega,-\omega)$ can be obtained from the general function

$$\chi^{(3)}(\omega_1, \omega_2, \omega_3) = N \frac{e^4}{m^3}\left[\eta + \frac{4}{3}\lambda^2\{F(\omega_1+\omega_2) + F(\omega_2+\omega_3) + F(\omega_3+\omega_1)\}\right] \times F(\omega_1) F(\omega_2) F(\omega_3) F(\omega_1+\omega_2+\omega_3) \tag{6.19}$$

by setting $\omega_1 = \omega_2 = \omega$ and $\omega_3 = \pm\omega$.

The basic feature of the non-linear response is seen from expressions (6.12), (6.16) and (6.18) for P_1, P_2 and P_3. A field oscillating at the single frequency ω can induce in the system a polarization having components oscillating not only at frequency ω but at frequencies 2ω and 3ω as well as one which is constant in time. These polarizations, of course, radiate at their frequency of oscillation in a way described by Maxwell's equations and so give rise to radiation fields containing the second and third harmonics of the applied frequency. These remarks are based on our calculation which was valid to third order in the field. However, it is clear that higher order terms would give rise to the possibility of generating higher order harmonics. We shall not be concerned in general with effects higher than third order in the field but the fact that they can exist should be borne in mind.

In general we want to consider radiation fields consisting of several components each of different frequency. Our calculation can again be carried through but becomes much more complicated to write out in detail. As well as inducing polarizations oscillating at the second and third harmonic frequencies, we find that components oscillating at all the possible sum and difference frequencies are also induced. The magnitudes of these additional components are controlled by the quantities $\chi^{(2)}(\omega_1,\omega_2)$ and $\chi^{(3)}(\omega_1,\omega_2,\omega_3)$ evaluated at appropriate values of the frequencies ω_1, ω_2 and ω_3. For example, for an applied field containing two components with frequencies ω and ω', i.e.

$$E = E(\omega)\, e^{-i\omega t} + E(\omega')\, e^{-i\omega' t} + \text{complex conjugate} \tag{6.20}$$

we find that the induced second-order polarization density is

$$\begin{aligned} P_2 = {} & \chi^{(2)}(\omega,\omega)\, E^2(\omega)\, e^{-2i\omega t} + \chi^{(2)}(\omega,-\omega)\, E(\omega)\, E^*(\omega) \\ & + \chi^{(2)}(\omega',\omega')\, E^2(\omega')\, e^{-2i\omega' t} + \chi^{(2)}(\omega'-,\omega')\, E(\omega')\, E^*(\omega') \\ & + 2\chi^{(2)}(\omega,\omega')\, E(\omega)\, E(\omega')\, e^{-i(\omega+\omega')t} \\ & + 2\chi^{(2)}(\omega,-\omega')\, E(\omega)\, E^*(\omega')\, e^{-i(\omega-\omega')t} + \text{complex conjugate} \end{aligned} \tag{6.21}$$

The general expressions for the components of the induced polarization can be written in a simple compact form when it is noticed from (6.15), (6.17) and (6.19) that (i) since, from (6.14), $F^*(\omega) = F(-\omega)$,

$$\begin{aligned} \chi^{(1)*}(\omega) &= \chi^{(1)}(-\omega) \\ \chi^{(2)*}(\omega_1,\omega_2) &= \chi^{(2)}(-\omega_1,-\omega_2) \\ \chi^{(3)*}(\omega_1,\omega_2,\omega_3) &= \chi^{(3)}(-\omega_1,-\omega_2,-\omega_3) \end{aligned} \tag{6.22}$$

and (ii) the values of $\chi^{(2)}(\omega_1,\omega_2)$ and $\chi^{(3)}(\omega_1,\omega_2,\omega_3)$ are independent of the order in which the frequencies are written inside these functions so that, e.g.,

$$\chi^{(2)}(\omega_1,\omega_2) = \chi^{(2)}(\omega_2,\omega_1)$$

and

$$\chi^{(3)}(\omega_1,\omega_2,\omega_3) = \chi^{(3)}(\omega_3,\omega_1,\omega_2) \tag{6.23}$$

Let us label the frequencies of the components of the applied field by $\omega_1, \omega_2, \omega_3$ etc. and in general ω_n. If we allow n to take negative as well as positive values, defining ω_{-n} to be equal to $-\omega_n$, and furthermore define $E(\omega_{-n}) = E(-\omega_n)$ to be equal to $E^*(\omega_n)$, we see first of all that a field containing components of several frequencies can then be expressed as

$$E = \sum_n E(\omega_n)\, e^{-i\omega_n t} \tag{6.24}$$

the sum being over all the required pairs of positive and negative values of n. Thus, for example, by allowing n to take on only the values ±1, ±2, and identifying ω_1 with ω and ω_2 with ω', we obtain (6.20) from (6.24). As well as this, however, we can now write P_1, P_2 and P_3 in the following simple way:

$$P_1 = \sum_n \chi^{(1)}(\omega_n)\, E(\omega_n)\, e^{-i\omega_n t}$$

$$P_2 = \sum_{n,m} \chi^{(2)}(\omega_n, \omega_m)\, E(\omega_n)\, E(\omega_m)\, e^{-i(\omega_n+\omega_m)t} \qquad (6.25)$$

$$P_3 = \sum_{n,m,l} \chi^{(3)}(\omega_n, \omega_m, \omega_l)\, E(\omega_n)\, E(\omega_m)\, E(\omega_l)\, e^{-i(\omega_n+\omega_m+\omega_l)t}$$

the sums over n, m and l all running over the same required pairs of positive and negative values. This result, (6.25), is easily checked for those special cases discussed in detail. Its validity in the general case would seem to be eminently reasonable and it can in fact be shown to be correct by a calculation of a more general nature than we have performed [11, 12].

The significance of the three quantities $\chi^{(1)}(\omega)$, $\chi^{(2)}(\omega_1,\omega_2)$ and $\chi^{(3)}(\omega_1,\omega_2,\omega_3)$ introduced in (6.15), (6.17) and (6.19) is now clear. They characterize, respectively, according to (6.25), the first, second and third order polarization densities induced by the field and so are known as the first, second and third order polarizabilities or susceptibilities of the system. They are the basic quantities with which we shall be concerned and in terms of which we shall discuss the various non-linear effects which are observable experimentally.

7. THE GENERALIZATION TO THREE DIMENSIONS

To generalize our results to three dimensions, we must take into account the fact that the electric field and polarization density take on a vector character. The quantity $E(\omega_n)$ is then also a vector, which we write as $\vec{E}(\omega_n)$ and the expressions (6.25) relate the vectors $\vec{P}_1$, $\vec{P}_2$ and $\vec{P}_3$ to the products of several of these vectors $\vec{E}(\omega_n)$ so that the polarizabilities $\chi^{(1)}$, $\chi^{(2)}$ and $\chi^{(3)}$ become tensors. $\chi^{(1)}(\omega)$ becomes a second-rank tensor $\underline{\chi}^{(1)}(\omega)$ with components $\chi^{(1)}_{\mu\alpha}(\omega)$ for μ and α = x,y or z, $\chi^{(1)}_{xz}(\omega)\, E_z(\omega)\, e^{-i\omega t}$ being, e.g., the component P_{1x} of the first-order polarization density induced by the z-component of a field of frequency ω. In general we have

$$\vec{P}_1 = \sum_n \underline{\chi}^{(1)}(\omega_n) \cdot \vec{E}(\omega_n)\, e^{-i\omega_n t} \qquad (7.1)$$

where by $\underline{\chi}^{(1)}(\omega_n) \cdot \vec{E}(\omega_n)$ we mean a vector whose component in the direction μ (μ = x, y or z) is

$$\sum_{\alpha=x,y,z} \chi^{(1)}_{\mu\alpha}(\omega_n)\, E_\alpha(\omega_n)$$

Similarly $\chi^{(2)}$ and $\chi^{(3)}$ become, respectively, tensors of the third and fourth rank, so that in general we have

$$\vec{P}_2 = \sum_{n,m} \underline{\chi}^{(2)}(\omega_n, \omega_m) : \quad \vec{E}(\omega_n)\vec{E}(\omega_m)\, e^{-i(\omega_n+\omega_m)t}$$

$$\vec{P}_3 = \sum_{n,m,l} \underline{\chi}^{(3)}(\omega_n, \omega_m, \omega_l) \vdots\, \vec{E}(\omega_n)\vec{E}(\omega_m)\vec{E}(\omega_l)\, e^{-i(\omega_n+\omega_m+\omega_l)t} \tag{7.2}$$

where the quantity to be summed in each case is a vector and the dot-products mean that the components of these vectors in the direction μ are

$$\sum_{\alpha,\beta=x,y,z} \chi^{(2)}_{\mu\alpha\beta}(\omega_n, \omega_m)\, E_\alpha(\omega_n)\, E_\beta(\omega_m)\, e^{-i(\omega_n+\omega_m)t}$$

$$\sum_{\substack{\alpha,\beta,\gamma\\=x,y,z}} \chi^{(3)}_{\mu\alpha\beta\gamma}(\omega_n, \omega_m, \omega_l)\; E_\alpha(\omega_n)\, E_\beta(\omega_m)\, E_\gamma(\omega_l)\, e^{-i(\omega_n+\omega_m+\omega_l)t} \tag{7.3}$$

8. GENERAL PROPERTIES OF THE POLARIZABILITY TENSORS

In a real dielectric medium, polarization arises from both electronic and ionic motion. An exact quantum-mechanical calculation of the linear polarization can be carried out, giving rise to an expression for the first order or linear polarizability of a medium. This type of calculation is easily extended, by working to higher orders of perturbation theory, to yield expressions for the higher order polarizabilities of a medium in terms of parameters characteristic of both the electronic and ionic motion in the system [11, 12]. However, that will not be done here since the results of interest which emerge from such a calculation can be obtained by fairly obvious generalizations of the results found from our classical model. We shall discuss these results one by one. Full quantum-mechanical expressions for the tensors will be given in Subsection (vi), and it is readily verified that they satisfy our generalized conditions.

(i) Reality condition

The property (6.22) generalizes directly to the general case so that it holds for each element of the polarizability tensors. Thus we have

$$\left.\begin{aligned} \underline{\chi}^{(1)*}(\omega) &= \underline{\chi}^{(1)}(-\omega) \\ \underline{\chi}^{(2)*}(\omega_1, \omega_2) &= \underline{\chi}^{(2)}(-\omega_1, -\omega_2) \\ \underline{\chi}^{(3)*}(\omega_1, \omega_2, \omega_3) &= \underline{\chi}^{(3)}(-\omega_1, -\omega_2, -\omega_3) \end{aligned}\right\} \tag{8.1}$$

This is called the reality condition since it ensures that the polarizations given by (7.1) and (7.2) are real quantities, as they should be.

(ii) Intrinsic symmetry

The property which emerged from the one-dimensional model and which is exemplified by (6.23) is also easily generalized. In three dimensions, a component of the electric field has not only a frequency but also a polarization direction associated with it. The general property therefore relates to the independence of, e.g., the element $\chi^{(2)}_{\mu\alpha\beta}(\omega_1,\omega_2)$ to the order in which $\alpha\omega_1$ and $\beta\omega_2$ are written rather than simply the frequencies ω_1 and ω_2; i.e. we have $\chi^{(2)}_{\mu\alpha\beta}(\omega_1\omega_2) = \chi^{(2)}_{\mu\beta\alpha}(\omega_2,\omega_1)$. This property arises from the fact that the polarizations $\vec{P}_2$ and $\vec{P}_3$ appearing in the relationships (7.2) which define $\underline{\chi}^{(2)}$ and $\underline{\chi}^{(3)}$ must clearly be independent of the order in which the various complex electric field amplitudes appear in these expressions. Thus any pair of elements $\chi^{(2)}_{\mu\alpha\beta}(\omega_1,\omega_2)$ and $\chi^{(2)}_{\mu\beta\alpha}(\omega_2,\omega_1)$ of $\underline{\chi}^{(2)}$ can be chosen to be equal, as can all the elements of the set generated from $\chi^{(3)}_{\mu\alpha\beta\gamma}(\omega_1,\omega_2,\omega_3)$ by any rearrangement of the pairs $\alpha\omega_1$, $\beta\omega_2$ and $\gamma\omega_3$, e.g. $\chi^{(3)}_{\mu\alpha\beta\gamma}(\omega_1,\omega_2,\omega_3) = \chi^{(3)}_{\mu\gamma\alpha\beta}(\omega_3,\omega_1,\omega_2)$. Since this property is inherent in the defining relations of the polarizabilities, it is called the intrinsic symmetry. This symmetry property is the source of the factor 3 in expression (6.18) and the factors 2 in (6.21).

(iii) General form and frequency dependence

The polarizabilities $\chi^{(1)}$, $\chi^{(2)}$ and $\chi^{(3)}$ arising from our classical model are expressed in (6.15), (6.17) and (6.19) in terms of the function $F(\omega)$ which is defined in (6.14). This function is complex and can be written in terms of its real and imaginary parts as

$$F(\omega) = \frac{\omega_0^2 - \omega^2}{(\omega_0^2 - \omega^2)^2 + 4\gamma^2\omega^2} + i\,\frac{2\gamma\omega}{(\omega_0^2 - \omega^2)^2 + 4\gamma^2\omega^2} \qquad (8.2)$$

We notice that, as in Section 2, the imaginary part of $F(\omega)$ is a function which has peaks at $\omega = \pm\omega_0$. Around each peak its shape is Lorentzian with a half-width γ at half-height. For frequencies ω removed from ω_0 by more than a few line-widths γ, the imaginary part becomes much smaller than the real part and can be neglected. Thus, outside the immediate neighbourhood of the frequencies $\pm\omega_0$, $F(\omega)$ can be considered real. Physically, ω_0 is the resonant frequency of the system and, in the neighbourhood of this frequency, energy can be exchanged between the radiation field and the system. We have previously seen in Eq.(1.11) that the amount of energy absorbed by a medium is proportional to the imaginary part of $\chi^{(1)}$ and so, by (6.15), to the imaginary part of $F(\omega)$.

From (6.17) and (6.19) we see that the second and third order polarizabilities $\chi^{(2)}(\omega_1,\omega_2)$ and $\chi^{(3)}(\omega_1,\omega_2,\omega_3)$ can be considered real quantities except for relatively small frequency ranges in which any single one or the sum of any of the frequencies appearing in $\chi^{(2)}$ and $\chi^{(3)}$ is within a few line-widths γ of ω_0. Within these ranges the polarizabilities have significant imaginary parts.

The way in which these results can be generalized to a real system has been discussed for the case of the linear response. The generalization for the non-linear response follows in the same way. Instead of having one resonant frequency ω_0, there are many such frequencies in a real system, one being associated with every pair of energy levels in the system. The resonant frequency associated with the pair of levels with energies E_1 and E_2 is $\omega_{21} = E_2 - E_1/\hbar$. Every resonant frequency has its own line-width and its own weighting factor which depends on the wave functions of the levels concerned and which varies from one order of the polarizability to another. This weighting factor measures the strength of the contribution made to the polarizability by a pair of levels and for many pairs of levels it will be zero. The resonant frequencies can range from the far infra-red region of the spectrum where they are associated with vibrational motions of ions and molecules right through to the optical region where they are associated with electronic motions. Providing that none of the frequencies nor the sum of any set of the frequencies appearing in a polarizability lies within a few line-widths of any resonant frequency, the polarizability can be taken to be real. We shall see later that there can be no exchange of energy between the radiation field and the system in these circumstances so that these conditions are equivalent to the condition that the system be completely lossless.

(iv) Overall symmetry [9, 11, 13]

It is noticeable that the intrinsic symmetry property allows one to make interchanges involving all but the first subscript μ of a general tensor element. An overall symmetry exists which allows one to interchange any of the subscripts including the first. The simplest form of this overall symmetry and the only one we shall discuss occurs when all the polarizabilities are real so that for the frequencies of interest the medium is completely lossless. We then note from (6.20) that, since the imaginary part of $F(\omega)$ is negligible, $F(\omega) = F(-\omega)$. It follows from this fact that $\chi^{(1)}(\omega)$ in (6.15) is unaltered by replacing ω by $-\omega$; $\chi^{(2)}(\omega_1,\omega_2)$ in (6.17) is unaltered by replacing ω_1 or ω_2 by $-(\omega_1 + \omega_2)$ and $\chi^{(3)}(\omega_1,\omega_2,\omega_3)$ in (6.19) is unaltered by replacing ω_1, ω_2 or ω_3 by $-(\omega_1+\omega_2+\omega_3)$. In the general case, this additional frequency, i.e. minus the sum of the frequencies already present, is associated with the first subscript in a tensor element. Then one finds that all the elements of the set generated from $\chi^{(2)}_{\mu\alpha\beta}(\omega_1,\omega_2)$ by interchanging $\alpha\omega_1$ or $\beta\omega_2$ with $\mu(-\omega_1-\omega_2)$ are equal; thus, e.g. $\chi^{(2)}_{\mu\alpha\beta}(\omega_1,\omega_2) = \chi^{(2)}_{\alpha\mu\beta}(-\omega_1-\omega_2,\omega_2)$. Coupled with the intrinsic symmetry, this leads to the equality of all elements of $\underline{\chi}^{(2)}$ generated from $\chi^{(2)}_{\mu\alpha\beta}(\omega_1,\omega_2)$ by any permutations of the pairs $\mu(-\omega_1-\omega_2)$, $\alpha\omega_1$, $\beta\omega_2$. Similarly all the elements of $\underline{\chi}^{(3)}$ generated from $\chi^{(3)}_{\mu\alpha\beta\gamma}(\omega_1,\omega_2,\omega_3)$ by a permutation of the pairs $\mu(-\omega_1-\omega_2-\omega_3)$, $\alpha\omega_1$, $\beta\omega_2$, $\gamma\omega_3$ are equal.

This overall symmetry property has several important consequences. We shall discuss three of them, two immediately and another later. One consequence which we discuss now is that it can provide relationships between several apparently unconnected physical processes. This is best seen by an example. It is seen from (6.16) that the presence of a second-order polarizability leads to a constant polarization being set up from an oscillating electric field — so-called optical rectification. This d.c. polarization was observed by Bass et al. [14] in potassium dihydrogen phosphate, KDP, illuminated by a ruby laser, and they obtained a value of $\underline{\chi}^{(2)}(\omega,-\omega)$. Now the overall symmetry of $\vec{\chi}^{(2)}$ in a lossless region says that $\chi^{(2)}_{\mu\alpha\beta}(\omega,-\omega) = \chi^{(2)}_{\beta\alpha\mu}(\omega,0)$. $\underline{\chi}^{(2)}(\omega,0)$ gives the polarization $\vec{P}_2$ which would be set up at frequency ω in the presence of a d.c. electric field by another field oscillating at this same frequency ω. This second-order polarization appears in addition to the normal first-order polarization $\vec{P}_1$ at the same frequency. It is therefore detectable as a change in refractive index with d.c. electric field — the electro-optic effect. Overall symmetry says that there should be a close connection between these two apparently unrelated effects as indeed Bass and his associates found when they compared their measured value of $\underline{\chi}^{(2)}(\omega,-\omega)$ and the measured electro-optic coefficient of KDP. In general, the overall symmetry provides relationships between experiments involving the generation of sum and difference frequencies, e.g. ω_1 and $\omega_2 \rightarrow (\omega_1+\omega_2)$, and ω_1 and $(\omega_1+\omega_2) \rightarrow \omega_2$.

A second consequence of the overall symmetry leads to a symmetry property originally discussed by Kleinman [15] in 1962, which is now known by his name. At frequencies well below any resonance, the susceptibility tensors become frequency independent. Overall symmetry then implies a general invariance of each of the tensors under any permutations of its subscripts. Kleinman originally proposed the presence or absence of this symmetry as a way to reveal whether or not a given non-linearity arose primarily from electronic effects characterized by resonant frequencies higher than optical, or ionic effects with much lower resonant frequencies. The symmetry is frequently shown in the second-order tensor $\vec{\vec{\chi}}^{(2)}$ of crystals [16] and to a good approximation it can often be assumed to apply.

(v) Spatial symmetry

The spatial symmetry of a medium places restrictions on the form that the tensors $\underline{\chi}^{(1)}$, $\underline{\chi}^{(2)}$ and $\underline{\chi}^{(3)}$ can have. This arises from the fact that the relations (7.1) and (7.2) which define these tensors must have exactly the same form when expressed in all co-ordinate systems relative to which the medium has the same appearance. The effect of this condition is to force some elements of these tensors to be zero and to set up relationships among others. The higher the symmetry of the medium, the fewer are the non-zero elements in each tensor and the more relationships there are between them. For example, in the simplest case of $\underline{\chi}^{(1)}$, it is found that for an orthorhombic crystal the element $\chi^{(1)}_{\mu\alpha}(\omega)$ is zero unless $\mu = \alpha$; the three non-zero elements $\chi^{(1)}_{xx}(\omega)$, $\chi^{(1)}_{yy}(\omega)$ and $\chi^{(1)}_{zz}(\omega)$ are all different. However, for a cubic crystal, the higher symmetry forces all these non-zero elements to be equal. Similarly for an isotropic medium like a gas, which has the same appearance in all co-ordinate systems, the only non-zero elements of $\underline{\chi}^{(1)}$ are the diagonal ones and they are all equal.

The effects of spatial symmetry are more complicated in detail as far as $\underline{\chi}^{(2)}$ and $\underline{\chi}^{(3)}$ are concerned. A particularly important case is that in which the medium has a centre of inversion, i.e. is centro-symmetric. The relations (7.1) and (7.2) must then be exactly the same in the two co-ordinate systems which are obtained from one another by inversion, i.e. by putting $x \rightarrow -x$, $y \rightarrow -y$ and $z \rightarrow -z$. Relative to these two co-ordinate systems, the vectors $\vec{P}$ and $\vec{E}$ point in opposite directions, so that, in transforming from one system to the other, $\vec{P} \rightarrow -\vec{P}$ and $\vec{E} \rightarrow -\vec{E}$. The expressions (7.1) and (7.2) for $\vec{P}_1$ and $\vec{P}_3$ therefore do not change under this transformation, irrespective of the forms of $\underline{\chi}^{(1)}$ and $\underline{\chi}^{(3)}$. Inversion symmetry therefore places no restrictions on $\underline{\chi}^{(1)}$ or $\underline{\chi}^{(3)}$. However, under the transformation the expression (7.2) for $\vec{P}_2$ becomes

$$\vec{P}_2 = -\sum_{n,m} \underline{\chi}^{(2)}(\omega_n, \omega_m) : \vec{E}(\omega_n)\vec{E}(\omega_m)\, e^{-i(\omega_n+\omega_m)t}$$

TABLE I. RESTRICTIONS PLACED ON $\underline{\chi}^{(2)}$ BY SPATIAL SYMMETRY AND BY A COMBINATION OF SPATIAL AND KLEINMAN SYMMETRY, FOR A VARIETY OF CRYSTAL CLASSES, CONTAINING MATERIALS OF INTEREST

Elements of $\underline{\chi}^{(2)}$ are simply denoted by their subscripts so that $\chi^{(2)}_{\mu\alpha\beta}$ is denoted by $\mu\alpha\beta$. The relation between reference and symmetry axes is used as recommended in Ref. [17]. The z-direction here is therefore not the same as the propagation direction of our theory, which we choose quite arbitrarily to call the z-direction

Crystal class and conventional classification with some representative materials	Non-zero elements under spatial symmetry	Non-zero elements under spatial and Kleinman symmetry
$\bar{4}3m$ (cubic) - isotropic GaAs and all III-V semiconductors	6 equal, non-zero elements xyz = zxy = yzx = yxz = zyx = xzy	6 equal, non-zero elements xyz = zxy = yzx = yxz = zyx = xzy
$\bar{4}2m$ (tetragonal) - uniaxial KH_2PO_4 (KDP) $(NH_4)H_2PO_4$ (ADP)	6 non-zero elements 3 independent xyz = yxz zxy = zyx xzy = yzx	6 equal, non-zero elements xyz = yxz = zxy = zyx = xzy = yzx
3m (trigonal) - uniaxial $LiNbO_3$ Ag_3AsS_3 (proustite)	11 non-zero elements 5 independent zzz yyy = -yxx = -xyx = -xxy zxx = zyy xzx = yzy xxz = yyz	11 non-zero elements 3 independent zzz yyy = -yxx = -xyx = -xxy zxx = zyy = xzx = yzy = xxz = yyz
mm2 (orthorhombic) - biaxial $LiIO_3$	7 non-zero, independent elements zzz, zxx, xzx, xxz, zyy, yzy, yyz	7 non-zero elements 3 independent zzz zxx = xzx = xxz zyy = yzy = yyz
6 (hexagonal) - uniaxial $Ba_2NaNb_5O_{15}$(BSN)	13 non-zero elements 7 independent zzz zxx = zyy xyz = -yxz xzx = yzy zxy = -zyx xxz = yyz yzx = -xyz	7 non-zero elements 2 independent zzz zxx = zyy = xzx = yzy = xxz = yyz

which can be the same as the original expression only if $\underline{\chi}^{(2)} = 0$. This shows the important result that any system which has a centre of inversion possesses no second-order polarizability. A useful point to remember is that crystals which exhibit a piezo-electric effect cannot have a centre of inversion and so they possess non-zero second-order polarizabilities. One obvious but significant consequence of a system having an inversion centre is that second harmonic generation can never take place in it; this follows from (6.12), (6.16) and (6.15).

The effects of inversion symmetry can be seen clearly by considering our one-dimensional classical model. The force $-\omega_0^2 r + \lambda r^2 + \mu r^3$ arises from a potential

$$V(r) = \tfrac{1}{2}\omega_0^2 r^2 - \tfrac{1}{3}\lambda r^3 - \tfrac{1}{4}\eta r^4$$

which has inversion symmetry, i.e. $V(-r) = V(r)$ only if $\lambda = 0$. We see from (6.17) and (6.19) that $\lambda = 0$ implies that $\chi^{(2)} = 0$ but $\chi^{(3)}$ remains non-zero.

In general when $\underline{\chi}^{(2)}$ exists, it can have 27 non-zero, independent coefficients. These are greatly reduced in number in many cases. Some examples of this for a few crystal classes containing materials of interest are given in Table I. Further restrictions on the elements are also imposed if Kleinman symmetry is also satisfied; these are also shown in Table I.

$\underline{\chi}^{(3)}$ can in general have 81 non-zero, independent elements whose number and independence is also reduced by spatial symmetry. But even in the simplest case of crystals belonging to the highest symmetry classes of the cubic system, there are still 21 non-zero elements belonging to four independent groups of related elements. Details can be found in Ref.[18].

(vi) Quantum-mechanical expressions

General quantum-mechanical expressions for the susceptibilities can be derived and written in a variety of forms [11, 12]. For simplicity, the expressions are given which are appropriate to an assembly of similar atoms with bound electrons, forming some state of condensed matter. Then, if we label the quantum-mechanical states of the atoms by p, q, r, etc., call N_p the average density of atoms in the state p, use $\hbar\omega_{pq}$ to denote the energy difference between state p and state q, and denote by $\mathscr{P}^{\mu}_{pq}$ the μth component of the matrix element of the dipole moment operator

$$\vec{\mathscr{P}} = -e\sum_j \vec{r}_j$$

(j being summed over all electrons in an atom) between states p and q, we have the following expressions:

$$\chi^{(1)}_{\mu\alpha}(\omega_1) = -\sum_P P(\mu\omega_0 = -\omega_1, \alpha\omega)\sum_{p,q}\frac{N_p}{\hbar}\,\frac{\mathscr{P}^{\mu}_{pq}\mathscr{P}^{\alpha}_{qp}}{\omega_{pq}+\omega_1} \tag{8.3}$$

$$\chi^{(2)}_{\mu\alpha\beta}(\omega_1,\omega_2) = \sum_P P(\mu\omega_0 = -\omega_1-\omega_2, \alpha\omega_1, \beta\omega_2)$$

$$\times\sum_{p,q,r}\frac{N_p}{2\hbar^2}\,\frac{\mathscr{P}^{\mu}_{pq}\mathscr{P}^{\alpha}_{qr}\mathscr{P}^{\beta}_{rp}}{(\omega_{pq}+\omega_1+\omega_2)(\omega_{pr}+\omega_2)} \tag{8.4}$$

$$\chi^{(3)}_{\mu\alpha\beta\gamma}(\omega_1, \omega_2, \omega_3) = -\sum_P P(\mu\omega_0 = -\omega_1 - \omega_2 - \omega_3, \alpha\omega_1, \beta\omega_2, \gamma\omega_3)$$

$$\times \sum_{p,q,r,s} \frac{N_p}{6\hbar^3} \frac{\mathscr{P}^{\mu}_{pq} \mathscr{P}^{\alpha}_{qr} \mathscr{P}^{\beta}_{rs} \mathscr{P}^{\gamma}_{sp}}{(\omega_{pq}+\omega_1+\omega_2+\omega_3)(\omega_{pr}+\omega_1+\omega_2)(\omega_{ps}+\omega_3)} \qquad (8.5)$$

where

$$\sum_P$$

is a sum over all permutations of appropriate label pairs.

This expression for $\underline{\chi}^{(1)}(\omega)$ is equivalent to that given for $\epsilon(\omega)$ for an isotropic medium in (2.11) — the Kramers-Heisenberg formula. This can be seen as follows. For an isotropic medium, $\underline{\chi}^{(1)}$ must be diagonal and $\mathscr{P}^{\mu}_{pq} \mathscr{P}^{\mu}_{qp}$ must have the same value

$$|\mathscr{P}^{\mu}_{pq}|^2 = \tfrac{1}{3} |\vec{\mathscr{P}}_{pq}|^2 \quad \text{for } \mu = x, y \text{ or } z$$

The Hermitian character of $\mathscr{P}$ ensures that $\vec{\mathscr{P}}_{pq} = \vec{\mathscr{P}}^{*}_{qp}$. Thus,

$$\chi^{(1)}_{\mu\alpha}(\omega) = -\delta_{\mu\alpha} \sum_{p,q} \frac{N_p}{3\hbar} \left|\vec{\mathscr{P}}_{pq}\right|^2 \left[\frac{1}{\omega_{pq}+\omega} + \frac{1}{\omega_{pq}-\omega}\right]$$

i.e.

$$\chi^{(1)}_{\mu\alpha}(\omega) = -\delta_{\mu\alpha} \sum_{p,q} \frac{2N_p}{3\hbar} \omega_{pq} \left|\vec{\mathscr{P}}_{pq}\right|^2 \frac{1}{\omega^2_{pq} - \omega^2}$$

$$= \delta_{\mu\alpha} \sum_{p,q} \frac{N_p e^2}{m} \frac{f_{pq}}{\omega^2_{pq} - \omega^2}$$

where

$$f_{pq} = \frac{2m\omega_{qp}}{3e^2\hbar} |\vec{\mathscr{P}}_{pq}|^2$$

This is identical to (2.12) in the limit of zero damping.

9. THE PHASE-MATCHING CONDITION

Before considering the mixing and harmonic generation processes which can take place in a non-linear dielectric, we shall discuss an important condition which governs the efficiency of these processes and which arises from the optical dispersion of the medium. The nature of this condition is most easily seen by considering the simplest process: second harmonic generation [19]. Let us therefore calculate the intensity of radiation at frequency 2ω which emerges from one side of a slab of non-linear dielectric of thickness ℓ when radiation at frequency ω is entering from the other side. The radiation at the fundamental frequency ω will propagate through the medium with a wavelength $\lambda_1 = 2\pi c/n_1\omega$, i.e. with propagation constant $k_1 = 2\pi/\lambda_1 = n_1\omega/c$, where n_1 is the refractive index at frequency ω. We can therefore write the electric field at the fundamental frequency as $\vec{E}_1 \cos(k_1 z - \omega t)$ so that in the expression

$$\vec{E} = \vec{E}(\omega)\, e^{-i\omega t} + \vec{E}^*(\omega)\, e^{i\omega t}$$

we see that $\vec{E}(\omega) = \frac{1}{2}\vec{E}_1 \exp(ik_1 z)$. It follows from (7.2) that the polarization induced at frequency 2ω is

$$\tfrac{1}{4}\underline{\chi}^{(2)}(\omega,\omega)\colon\ \vec{E}_1\vec{E}_1\, e^{i(2k_1 z - 2\omega t)} + \tfrac{1}{4}\underline{\chi}^{(2)*}(\omega,\omega)\colon\ \vec{E}_1\vec{E}_1\, e^{-i(2k_1 z - 2\omega t)}$$

$$= \tfrac{1}{2}\underline{\chi}^{(2)}(\omega,\omega)\colon\ \vec{E}_1\vec{E}_1 \cos(2k_1 z - 2\omega t)$$

where we have used the reality of $\underline{\chi}^{(2)}$ in a lossless region. The point to note here is that the spatial variation of this polarization is controlled by twice the propagation constant of the fundamental wave and not by the propagation constant $k_2 = 2n_2\omega/c$ of radiation at the second harmonic frequency 2ω, for which the medium has refractive index n_2. Radiation at frequency 2ω will be emitted by this oscillating polarization, the electric field at the exit surface of the slab arising from a slice of the medium of thickness dz at z being proportional to

$$\cos[2k_1 z - 2\omega(t-t')]\,dz$$

Here t' is the time taken for the radiation at frequency 2ω to propagate the distance $\ell - z$ from its point of origin to the exit surface. Therefore

$$t' = (\ell - z)/v_2 = (\ell - z)\frac{k_2}{2\omega}$$

since the phase velocity of the wave is $v_2 = 2\omega/k_2$. The total harmonic electric field at the exit surface is therefore proportional to

$$\int_0^\ell \cos[2k_1 z - 2\omega(t-t')]\,dz$$

$$= \int_0^\ell \cos[(2k_1 - k_2)z + k_2\ell - 2\omega t]\,dz$$

$$= \frac{\sin(2k_1\ell - 2\omega t) - \sin(k_2\ell - 2\omega t)}{2k_1 - k_2}$$

$$= 2\cos[\tfrac{1}{2}(2k_1 + k_2)\ell - 2\omega t]\,\frac{\sin\frac{1}{2}(2k_1 - k_2)\ell}{2k_1 - k_2}$$

and consequently the intensity I of harmonic radiation at the exit surface is such that

$$I \propto \frac{\sin^2 \frac{1}{2}(2k_1-k_2)\ell}{(2k_1-k_2)^2} \propto \frac{\sin^2 \frac{\omega}{c}(n_1-n_2)\ell}{(n_1-n_2)^2} \tag{9.1}$$

The role played by the optical dispersion of the medium in the process of second harmonic generation is clearly seen from this expression. In general, $n_1 \neq n_2$ and the maximum possible intensity is generated in slabs of material of thickness

$$\ell_c = \frac{\pi}{|2k_1-k_2|} = \frac{\pi c}{2\omega|n_1-n_2|} = \frac{\lambda_0}{4|n_1-n_2|} \tag{9.2}$$

or any odd integral multiples of ℓ_c. In (9.2), λ_0 is the free-space wavelength at the fundamental frequency ω. With slabs whose thickness is an even integral multiple of ℓ_c, no harmonic radiation is obtained. Only in the special case of $n_1 = n_2$, i.e. $2k_1 = k_2$, does the intensity increase continually with increasing thickness ℓ, the rate of increase from (9.1) being proportional to ℓ^2.

Perhaps a more instructive and physical way of seeing these results is as follows. Due to the non-linearities, a polarization is set up which oscillates and so radiates at frequency 2ω. However, this polarization has superimposed on it a periodic spatial variation with wavelength $\lambda_p = 2\pi/2k_1$. It is clear that the radiation emitted from any pair of small regions of this polarization field separated by a distance d will add together constructively only if the phase change $k_2 d$ in the radiation field as it propagates from one region to the other is equal to the phase difference $2k_1 d$ in the polarization field at these two regions, i.e. only if $2k_1 = k_2$. In general, there will be a phase difference $(2k_1-k_2)\,d$ at any point in space between the radiation emitted from the two regions. When $|2k_1 - k_2|\,d \ll \pi$, the two waves combine together constructively but when $|2k_1-k_2|\,d = \pi$, they are exactly out of phase and cancel one another. We can therefore think of our non-linear medium as being divided up into slices normal to the direction of propagation and each of thickness $\ell_c = \pi/|2k_1-k_2|$. The harmonic radiation emitted from any point adds more or less constructively to the radiation from any other point within the same slice. However, the phase relations between the radiation emitted from points in neighbouring slices is such that the total contribution from two such slices is exactly zero. Maximum intensity is obtained, therefore, from a slab of material which contains exactly an odd number of slices. On the basis of this discussion, ℓ_c is referred to as the coherence length for the process and the condition $2k_1 = k_2$, which allows full use to be made of the available thickness of material, as the phase-matching condition. There is some ambiguity in the literature over the definition of the coherence length ℓ_c. It is sometimes taken to be twice the quantity defined here [20, 21], although the one used here would seem to have the more physical appeal.

The periodic variation with specimen thickness of second harmonic intensity has been shown by Maker et al. [21]. They measured the intensity of the second harmonic of a ruby laser produced in a plate of quartz. The effective thickness of the plate was varied by rotating it about an axis in the

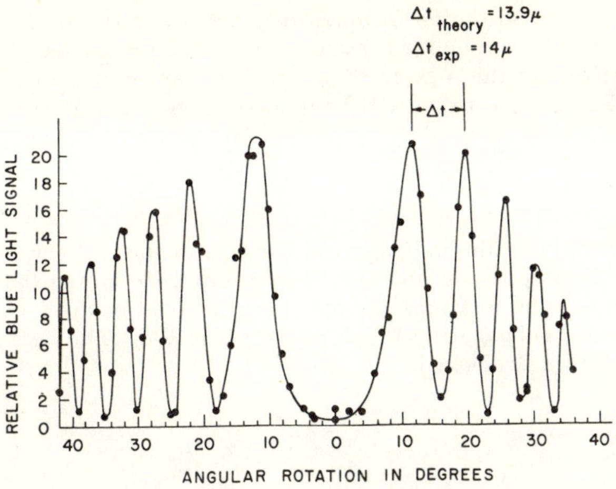

FIG. 6. Oscillations observed by Maker et al. [21] in the intensity of the blue second harmonic light generated from ruby laser radiation in a thin plate of quartz whose effective thickness was varied by rotation about an axis normal to the laser beam. Successive intensity maxima occur for effective thicknesses differing by $\Delta t = 14$ μm, indicating a coherence length of 7 μm.

plane normal to the propagation direction of the light. Oscillations in the intensity as shown in Fig.6 were observed whose period corresponded to a coherence length of 7 μm. This value agreed well with the value of 6.95 μm calculated from (9.2) using the known values of the refractive indices n_1 and n_2 in quartz.

The desirability of achieving the phase-matching condition can be readily seen from another point of view. If we describe the radiation quantum mechanically, then the fundamental and harmonic beams consist of streams of photons which have momenta $\hbar k_1$ and $\hbar k_2$ respectively along the direction of propagation. Second harmonic generation on this picture is simply the process of two photons from the fundamental beam combining as a result of the non-linearity of the medium to form a single photon. Energy conservation

$$\hbar\omega + \hbar\omega = \hbar(2\omega) \tag{9.3}$$

shows that the photon produced is at the second harmonic frequency and strict momentum conservation

$$\hbar k_1 + \hbar k_1 = \hbar k_2 \tag{9.4}$$

shows that the process should take place only when phase matching, i.e. $2k_1 = k_2$, is achieved. The fact that the process takes place in a region of space which is of finite size, i.e. within the thickness ℓ of the medium, relaxes the strict momentum conservation relation (9.4) by an Uncertainty Principle argument. The degree to which (9.4) can be violated is described by the function in (9.1) which concentrates itself more and more around the point $2k_1 = k_2$ as $\ell \to \infty$.

This discussion shows us immediately how to generalize the phase-matching condition to processes involving several frequencies. For example, any process involving the mixing of any pair of the frequencies ω_1, ω_2 and $\omega_3 = \omega_1 + \omega_2$ to produce the third will have associated with it the phase-matching condition

$$\vec{k}_1 + \vec{k}_2 = \vec{k}_3 \tag{9.5}$$

where $\vec{k}_1$, $\vec{k}_2$ and $\vec{k}_3$ are the propagation vectors of the three waves at frequencies ω_1, ω_2 and ω_3 in the non-linear medium. We notice that the condition is in general a vector relationship and takes account of the fact that all the waves need not be propagating in the same direction. The coherence length associated with this process is

$$\ell_c = \frac{\pi}{|\vec{k}_1 + \vec{k}_2 - \vec{k}_3|} \tag{9.6}$$

It should perhaps be pointed out that nothing in our discussion of the phase-matching condition has depended on the fact that we are interested in radiation of optical and infra-red frequencies. The same condition arises and is well known in the theory of travelling-wave devices operating in the microwave region of the spectrum [22]. In this context, our non-linear dielectric is simply a travelling-wave structure for optical radiation.

It should be realized that phase-matching conditions can be satisfied, if at all, only under certain special circumstances. If one is interested simply in the optical properties of a dielectric when it exhibits non-linearities, then one finds that these properties apparently become abnormal under special conditions which correspond to phase matching being achieved for some particular process. If, however, one wishes to exploit the non-linear properties of a system to generate harmonics or mix light beams, then it is precisely under these conditions, which give abnormally large responses, that one would like to work. We must therefore discuss the problem of how phase-matching conditions can be satisfied.

For processes involving waves at two or three different frequencies, phase matching cannot be achieved in an optically isotropic medium under conditions of normal dispersion. For under these conditions the refractive index n is an increasing function of frequency. Clearly then the phase-matching condition for second harmonic generation, $n_1 = n_2$, cannot be satisfied. Neither can the condition (9.5) for processes involving three waves since the maximum value of the left side of this expression, obtained when all the wave vectors are parallel, is $(n_1\omega_1 + n_2\omega_2)/c$, which is always less than magnitude $n_3(\omega_1 + \omega_2)/c$ of the right side.

The most frequently employed technique for overcoming this difficulty is to work with an optically anisotropic crystal, i.e. a uniaxial or biaxial crystal. Such crystals are birefringent and support in general two waves with different propagation characteristics at any frequency — the ordinary and extraordinary waves. These waves have different refractive indices and it is often possible, by using a selection of the ordinary and extraordinary waves at the various frequencies involved, to achieve phase matching. The birefringence of the crystal is used to counterbalance the effects of normal dispersion. Giordmaine [23] and Maker et al. [21] demonstrated an increase

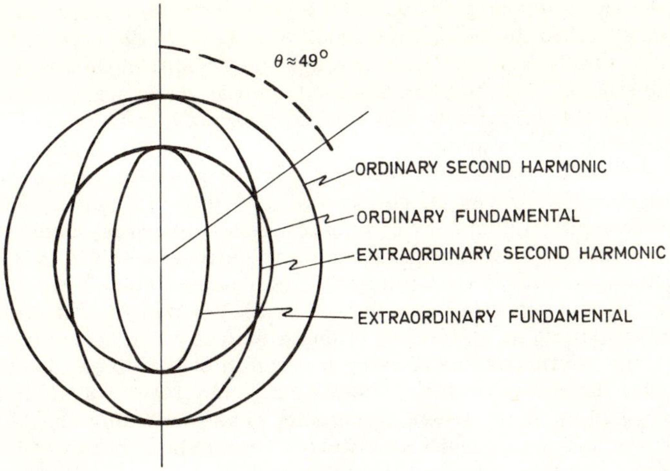

FIG. 7. Section through optic z-axis of refractive index surface in KDP for radiation at ruby laser frequency and its second harmonic.

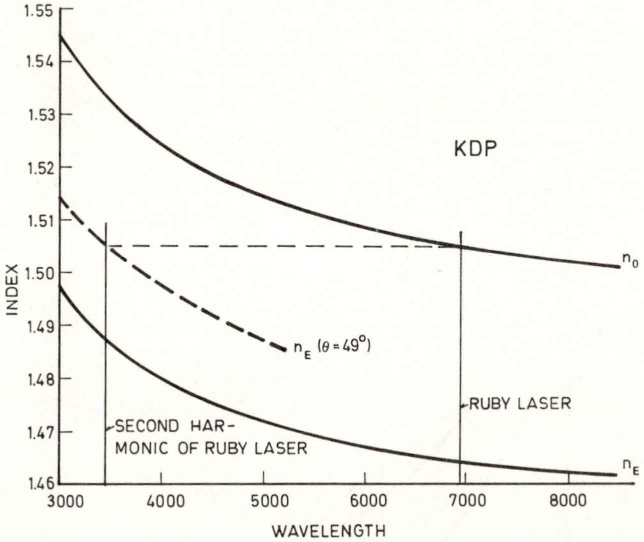

FIG. 8. The ordinary, n_0, and extraordinary, n_E, refractive indices of KDP as a function of wavelength. The effective refractive index for an extraordinary wave, propagating at an angle of 49° to the optic axis, shows how phase matching can be achieved in this direction for second harmonic generation of a ruby laser.

of several orders of magnitude in the power of the second harmonic generated from a ruby laser beam in potassium dihydrogen phosphate, KDP, when the phase-matching condition has been satisfied by this technique (see Section 11). KDP is a uniaxial crystal and its refractive index surfaces are surfaces of revolution about its optic axis — the z-axis. A section of these surfaces is shown in Fig.7 for the ruby laser frequency and its second harmonic. At each frequency there are two surfaces, one for the ordinary refractive index and one for the extraordinary index. The length of the line from the centre to a surface is equal to the refractive index of a wave of the type associated with the surface which is propagating through the crystal in the direction of the line. The surface for the ordinary fundamental wave, i.e. the laser beam, is seen to cut the surface for the extraordinary second harmonic beam for a direction which makes an angle $\theta \approx 49°$ with the optic axis. For propagation along this direction and this direction only the phase-matching condition $n_1 = n_2$ can therefore be satisfied. It was indeed for propagation along this direction that greatly enhanced harmonic generation was observed.

An alternative way of understanding the effect is shown in Fig.8, where the ordinary, n_0, and extraordinary, n_E, refractive indices of KDP are plotted as a function of wavelength and show normal dispersion. An extraordinary wave can propagate at a given wavelength with any refractive index between n_0 and n_E, the particular value being a function of the direction of propagation (see Fig.7). Since the ordinary index at the ruby laser wavelength lies between n_0 and n_E at its second harmonic, phase matching can be achieved.

This technique is not successful in all materials, however. In quartz, for example, the two index surfaces for radiation at the ruby laser frequency lie completely inside the two surfaces for the second harmonic, so that phase matching cannot be achieved in quartz for second harmonic generation at this frequency. The lack of sufficient birefringence to compensate for the frequency dispersion in this case is illustrated in Fig.9. This must not be taken to imply, however, that the phase-matching condition cannot be satisfied

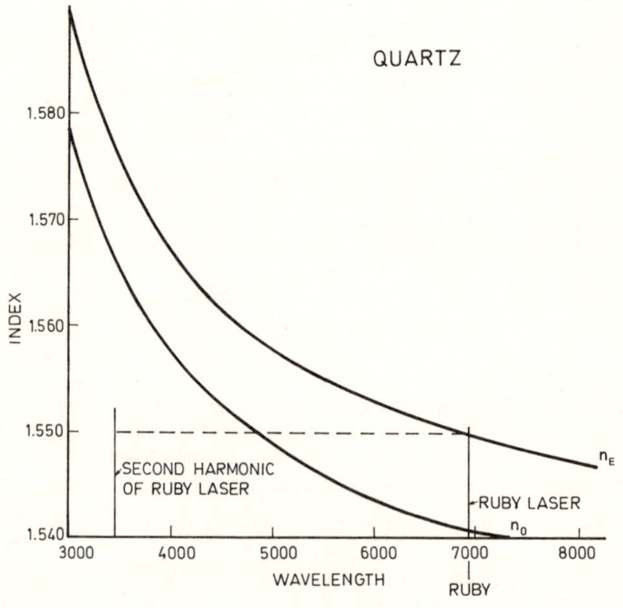

FIG.9. The ordinary, n_0, and extraordinary, n_E, refractive indices of quartz as a function of wavelength. The fact that phase matching cannot be achieved for the second harmonic generation of a ruby laser, due to lack of sufficient birefringence, is illustrated.

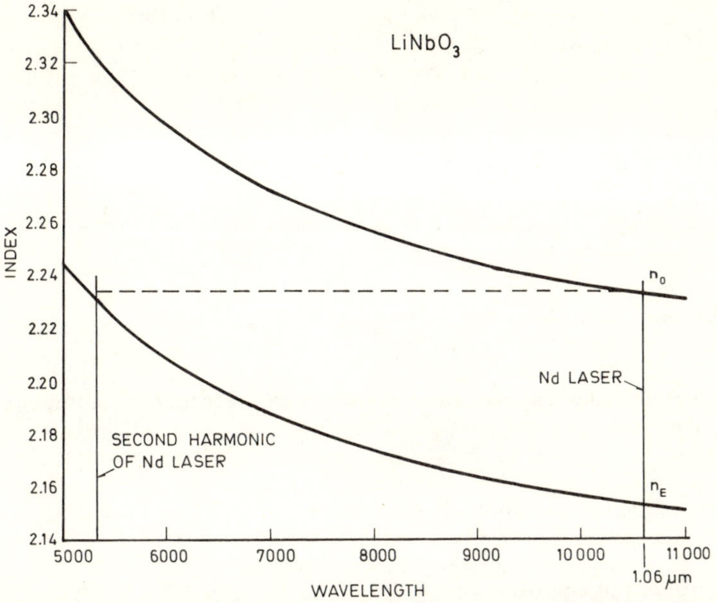

FIG.10. The ordinary, n_0, and extraordinary, n_E, refractive indices of $LiNbO_3$ as a function of wavelength. At this temperature, the birefringence is just sufficient to provide phase matching for second harmonic generation of a Nd laser.

in quartz for second harmonic generation at other frequencies or for processes involving waves at three frequencies.

Another possible way of achieving phase matching is illustrated for lithium niobate in Fig.10. In this case, by using the temperature dependence of the refractive indices, a situation can be achieved where the phase-matching condition is only just satisfied for second harmonic generation from a Nd laser. This calls for propagation in a plane normal to the optic axis (see Fig.7) and the two refractive index surfaces touch rather than cross; less sensitivity to the propagation direction and beam divergence can clearly be achieved in this way [24]. This situation is referred to as non-critical phase matching whereas the more usual situation, resulting from the crossing of the two refractive index surfaces, is called critical phase matching. Another advantage associated with non-critical phase matching will be discussed in Section 17.

Yet another possible way of achieving phase matching is to make use of the presence in any medium of regions of anomalous dispersion. If these happen to be suitably placed relative to the frequencies of interest, phase matching could be obtained.

10. PROPAGATION IN A NON-LINEAR DIELECTRIC

The propagation of electromagnetic radiation in a dielectric material, whose magnetic polarizability is negligible, is described by solutions of Maxwell's equations

$$\text{curl}\,\vec{E} = -\frac{1}{c}\frac{\partial}{\partial t}\vec{H}, \qquad \text{curl}\,\vec{H} = \frac{1}{c}\frac{\partial}{\partial t}\vec{D} \tag{10.1}$$

where

$$\vec{D} = \vec{E} + 4\pi\vec{P}$$

Eliminating the magnetic field $\vec{H}$ form these equations in the usual way produces the equation

$$\text{curl curl}\,\vec{E} + \frac{1}{c^2}\frac{\partial^2}{\partial t^2}\vec{E} = -\frac{4\pi}{c^2}\frac{\partial^2}{\partial t^2}\vec{P} \tag{10.2}$$

If $\vec{E}$ and $\vec{P}$ are now expressed in terms of their frequency components as in (6.24), i.e.

$$\vec{E} = \sum_n \vec{E}(\omega_n)\, e^{-i\omega_n t} \quad \text{and} \quad \vec{P} = \sum_n \vec{P}(\omega_n)\, e^{-i\omega_n t} \tag{10.3}$$

(10.2) yields the equation

$$\text{curl curl}\,\vec{E}(\omega_n) - \frac{\omega_n^2}{c^2}\vec{E}(\omega_n) = \frac{4\pi\omega_n^2}{c^2}\vec{P}(\omega_n) \tag{10.4}$$

for each component.

For a medium whose response is linear

$$\vec{P}(\omega_n) = \vec{P}_1(\omega_n) = \underline{\chi}^{(1)}(\omega_n)\cdot\vec{E}(\omega_n) \tag{10.5}$$

from (7.1). Equation (10.4) then becomes the familiar wave equation

$$\text{curl curl}\,\vec{E}(\omega_n) - \frac{\omega_n^2}{c^2}\underline{\epsilon}(\omega_n)\cdot\vec{E}(\omega_n) = 0 \tag{10.6}$$

where the dielectric tensor is defined as

$$\underline{\epsilon}(\omega_n) = \underline{1} + 4\pi\underline{\chi}^{(1)}(\omega_n) \tag{10.7}$$

General solutions of (10.6) can be built up from the plane wave solutions

$$\vec{E}(\omega_n) = E_{\omega_n}\vec{a}_{\omega_n}\, e^{i\vec{k}_n\cdot\vec{r}} \tag{10.8}$$

already used (see (1.8)). We recall that E_{ω_n} is a constant scalar factor which measures the amplitude of the field as long as $\vec{k}_n$ is real, which it is in the absence of loss, and $\vec{a}_{\omega_n}$ is a unit vector in the direction of polarization of the wave. For any direction of the propagation vector $\vec{k}_n$, its magnitude and the associated direction of polarization are given by solutions of the Fresnel equations

$$k_n^2 \vec{z} \times (\vec{z} \times \vec{a}_{\omega_n}) + \frac{\omega_n^2}{c^2} \underline{\epsilon}(\omega_n) \cdot \vec{a}_{\omega_n} = 0 \tag{10.9}$$

Here we have called the direction of propagation the z-direction along which points a unit vector $\vec{z}$ rather than the vector $\vec{s}$ used in Part I. Equation (10.9) is simply obtained by substituting (10.8) into (10.6).

For a medium whose response is non-linear, we can express the polarization field $\vec{P}(\omega_n)$ as

$$\vec{P}(\omega_n) = \vec{P}_1(\omega_n) + \vec{P}_{NL}(\omega_n) \tag{10.10}$$

where

$$\vec{P}_{NL}(\omega_n) = \vec{P}_2(\omega_n) + \vec{P}_3(\omega_n) + \ldots \tag{10.11}$$

$\vec{P}_{NL}(\omega_n)$ is that part of the polarization which depends on the electric field in a non-linear fashion. Equation (10.4) now becomes

$$\text{curl curl}\,\vec{E}(\omega_n) - \frac{\omega_n^2}{c^2} \vec{\underline{\epsilon}}(\omega_n) \cdot \vec{E}(\omega_n) = \frac{4\pi\omega_n^2}{c^2} \vec{P}_{NL}(\omega_n) \tag{10.12}$$

which is the linear equation (10.6) with an additional source or driving term on the right side. This is basically the equation which we must solve. The presence of the additional source term does not in itself make (10.12) much more difficult to solve than (10.6). It is the fact that this additional term is a non-linear function of the electric field that adds considerably to the difficulties of solving this equation in anything like a general fashion.

We have previously pointed out that, with most available laser sources, modifications which are made to existing linear effects as a result of the non-linear response are very small. It is in the additional effects which a non-linear response produces that its importance lies. One is therefore tempted to try and solve Eq.(10.12) by treating the non-linear source term on the right side as a perturbation. With this idea in mind, some progress can be made towards a solution of the equation. We express a solution of (10.12) in the form

$$\vec{E}(\omega_n) = E_{\omega_n}(z)\,[\vec{a}_{\omega_n} + \vec{b}_{\omega_n}(z)]\, e^{ik_n z} \tag{10.13}$$

which is similar to the solution (10.8) of the linear equation (10.6) for a wave propagating in the z-direction. In the linear case the amplitude factor and polarization vector are constants independent of the distance z travelled by the wave through the medium. In the non-linear problem, both these quantities must be considered to vary with z. However, since we are treating the non-linear source term in (10.12) as a perturbation to the linear problem, we can consider the factor $E_{\omega_n}(z)$ to be a slowly varying function of z and the modification $\vec{b}_{\omega_n}(z)$ to the polarization vector to be both small and slowly varying. By slowly varying in this context, we mean that little change occurs in the quantity over a distance $\Delta z = 2\pi/k_n$, the wavelength of the radiation. When (10.13) is substituted into (10.12), first-order effects in the non-linearity are therefore obtained by neglecting all terms in $d^2E_{\omega_n}/dz^2$, $d\vec{b}_{\omega_n}/dz$ and $d^2\vec{b}_{\omega_n}/dz^2$ on the left-hand side and all terms in $\vec{b}_{\omega_n}$ on the right. We then obtain the equation

$$\left(2ik_n \frac{dE_{\omega_n}}{dz} - k_n^2 E_{\omega_n}\right)\vec{z} \times [\vec{z} \times (\vec{a}_{\omega_n} + \vec{b}_{\omega_n})] - \frac{\omega_n^2}{c^2} \underline{\epsilon}(\omega_n) \cdot (\vec{a}_{\omega_n} + \vec{b}_{\omega_n}) E_{\omega_n}$$

$$= \frac{4\pi\omega_n^2}{c^2} \vec{P}'_{NL}(\omega_n)\, e^{-ik_n z} \qquad (10.14)$$

where the prime on $\vec{P}_{NL}$ indicates that all incremental polarization vectors $\vec{b}_{\omega_n}$ are to be neglected in this expression.

If the scalar product of this equation (10.14) with $\vec{a}_{\omega_n}$ is now added to the scalar product of (10.9) with $(\vec{a}_{\omega_n} + \vec{b}_{\omega_n})E_{\omega_n}$, considerable simplification results. Using the vector identity $\vec{a} \cdot (\vec{z} \times \vec{c}) = (\vec{a} \times \vec{z}) \cdot \vec{c}$. we see that the terms in $k_n^2 E_{\omega_n}$ cancel one another as do those involving $\underline{\epsilon}(\omega_n)$ due to the symmetrical nature of this tensor. We therefore obtain the equation

$$2ik_n \frac{dE_{\omega_n}}{dz} (\vec{a}_{\omega_n} \times \vec{z}) \cdot [\vec{z} \times (\vec{a}_{\omega_n} + \vec{b}_{\omega_n})] = \frac{4\pi\omega_n^2}{c^2} \vec{a}_{\omega_n} \cdot \vec{P}'_{NL}(\omega_n)\, e^{-ik_n z} \qquad (10.15)$$

Since both dE_{ω_n}/dz and $\vec{b}_{\omega_n}$ are being treated as small quantities, the term in this equation involving their product can be neglected, giving us the differential equation

$$\frac{dE_{\omega_n}}{dz} = \frac{2\pi i\omega_n^2}{k_n c^2 (\vec{z} \times \vec{a}_{\omega_n})^2} \vec{a}_{\omega_n} \cdot \vec{P}'_{NL}(\omega_n)\, e^{-ik_n z} \qquad (10.16)$$

for the scalar amplitude factor E_ω. This equation is considerably simpler to work with than (10.12) and we shall use it as the basis for all discussions to follow. It can in fact be simplified slightly further by noting that in most materials $|\vec{z} \times \vec{a}_{\omega_n}|$ is close to unity, i.e. the electric vector of a plane electromagnetic wave does not usually deviate far from being normal to the direction of propagation or, equivalently, the induced linear polarization usually lies in a direction close to that of the electric field. We make this simplification to (10.16) and obtain

$$\frac{dE_{\omega_n}}{dz} = \frac{2\pi i\omega_n^2}{k_n c^2} \vec{a}_{\omega_n} \cdot \vec{P}'_{NL}(\omega_n)\, e^{-ik_n z} \qquad (10.17)$$

as the equation on which we shall base all following discussions.

11. SECOND HARMONIC GENERATION

Second harmonic generation is the simplest non-linear process of interest. Since many of its features are common both to it and to more complicated mixing processes, we consider it in some detail. The problem is to find how radiation of frequency $\omega_2 = 2\omega$ is generated in a non-linear medium from an initially present beam of radiation at the fundamental frequency $\omega_1 = \omega$.

In general, radiation at the single frequency ω will set up in such a medium a polarization which has components vibrating at all the many harmonic frequencies 2ω, 3ω .... One would therefore expect the radiation generated to contain components at all those frequencies. This is in general true. However, if we are interested in generating radiation at the second harmonic frequency, we should attempt to concentrate if possible all the power in the generated radiation into the component at this frequency, at the expense of the other components. This can indeed be done experimentally by working with an arrangement which ensures that the phase-matching condition $2k_1 = k_2$ is satisfied as nearly as possible. The chance that the phase-matching condition $nk_1 = k_n (n > 2)$ is simultaneously satisfied for a higher harmonic is extremely small. With such an arrangement, second harmonic generation is the only process we need consider.

The phase-matching condition can be used quite generally in this way to isolate from all possible processes the one which is of particular interest. This is very convenient both experimentally and theoretically. On the experimental side, it allows us to exploit to the full any particular process and, on the theoretical side, it allows us to give special consideration to simple forms of solution to our non-linear equations.

With this discussion in mind, we must consider our radiation field to contain components of appreciable magnitude only at the frequencies $\omega_1 = \omega$ and $\omega_2 = 2\omega$ when the phase-matching condition $2k_1 = k_2$ is close to being satisfied. Thus, from (7.2), (10.3) and (10.11),

$$\vec{P}_{NL}(2\omega) = \underline{\chi}^{(2)}(\omega,\omega) : \vec{E}(\omega)\vec{E}(\omega) \tag{11.1}$$

and

$$\vec{P}_{NL}(\omega) = \underline{\chi}^{(2)}(2\omega,-\omega) : \vec{E}(2\omega)\vec{E}^*(\omega) + \underline{\chi}'^{(2)}(-\omega,2\omega) : \vec{E}^*(\omega)\vec{E}(2\omega)$$

$$= 2\underline{\chi}^{(2)}(2\omega,-\omega) : \vec{E}(2\omega)\vec{E}^*(\omega) \tag{11.2}$$

using the intrinsic symmetry of $\underline{\chi}^{(2)}$. With (10.13), these expressions give for $\vec{P}'_{NL}(2\omega)$ and $\vec{P}'_{NL}(\omega)$

$$\vec{P}'_{NL}(2\omega) = \underline{\chi}^{(2)}(\omega,\omega) : \vec{a}_\omega \vec{a}_\omega E_\omega^2 \, e^{2ik_\omega z} \tag{11.3}$$

$$\vec{P}'_{NL}(\omega) = 2\underline{\chi}^{(2)}(2\omega,-\omega) : \vec{a}_{2\omega}\vec{a}_\omega E_{2\omega} E_\omega^* \, e^{i(k_{2\omega}-k_\omega)z} \tag{11.4}$$

denoting by k_ω and $k_{2\omega}$ the propagation constants of the waves at frequencies ω and 2ω. Thus from (10.17), we obtain the two coupled equations:

$$\frac{dE_{2\omega}}{dz} = \frac{8\pi i\omega^2}{k_{2\omega}c^2}\left[\underline{\chi}^{(2)}(\omega,\omega) : \vec{a}_{2\omega}\vec{a}_\omega\vec{a}_\omega\right]E_\omega^2\, e^{i\Delta kz} \tag{11.5}$$

$$\frac{dE_\omega}{dz} = \frac{4\pi i\omega^2}{k_\omega c^2}\left[\underline{\chi}^{(2)}(2\omega,-\omega) : \vec{a}_\omega\vec{a}_{2\omega}\vec{a}_\omega\right]E_{2\omega}E_\omega^*\, e^{-i\Delta kz} \tag{11.6}$$

where

$$\Delta k = 2k_\omega - k_{2\omega} \tag{11.7}$$

and by the dot products we mean that

$$\underline{\chi}^{(2)} : \vec{a}\vec{b}\vec{c} \equiv \sum_{\substack{\alpha,\beta,\gamma\\ = x,y,z}} \chi^{(2)}_{\alpha\beta\gamma}\, a_\alpha b_\beta c_\gamma \tag{11.8}$$

One would normally use for second harmonic generation a medium lossless at both the frequencies ω and 2ω. We maintained earlier that in such a medium the tensor $\underline{\chi}^{(2)}(\omega,\omega)$ would be real. We can now show this to be true. For if $\underline{\chi}^{(2)}(\omega,\omega)$ is real, it also possesses the overall symmetry discussed in Section 8(iv), so that

$$\chi^{(2)}_{\mu\alpha\beta}(\omega,\omega) = \chi^{(2)}_{\alpha\mu\beta}(-2\omega,\omega) = \chi^{(2)*}_{\alpha\mu\beta}(2\omega,-\omega) = \chi^{(2)}_{\alpha\mu\beta}(2\omega,-\omega) \tag{11.9}$$

Hence Eqs (11.1) and (11.6) can both be expressed in terms of the single real quantity

$$\chi_{2\omega} \equiv \underline{\chi}^{(2)}(\omega,\omega) : \vec{a}_{2\omega}\vec{a}_\omega\vec{a}_\omega = \underline{\chi}^{(2)}(2\omega,-\omega) : \vec{a}_\omega\vec{a}_{2\omega}\vec{a}_\omega \tag{11.10}$$

so that we now have

$$\frac{dE_{2\omega}}{dz} = \frac{8\pi i\omega^2}{k_{2\omega}c^2}\chi_{2\omega}E_\omega^2\, e^{i\Delta kz} \tag{11.11}$$

$$\frac{dE_\omega}{dz} = \frac{4\pi i\omega^2}{k_\omega c^2}\cdot\chi_{2\omega}E_{2\omega}E_\omega^*\, e^{-i\Delta kz} \tag{11.12}$$

As a result of this, we can eliminate $\chi_{2\omega}$ from these two equations and obtain the result that

$$\tfrac{1}{2}k_{2\omega}\left|E_{2\omega}\right|^2 + k_\omega\left|E_\omega\right|^2 = \text{constant} \tag{11.13}$$

The left side of this expression is proportional to the sum of the magnitudes of the Poynting vectors of the two waves. The fact that this is a constant at all points in the medium shows that the medium is lossless.

The two coupled non-linear differential equations (11.11) and (11.12) can in fact be solved generally in terms of Jacobian elliptic functions for arbitrary initial values of the amplitude and phase of the two waves [11]. We shall, however, simply discuss the case which is both the most interesting and one of the simplest to work with. This is the case in which no radiation is initially present at the harmonic frequency, i.e. $E_{2\omega} = 0$ at $z = 0$. Any increase which may occur in $E_{2\omega}$ through power being converted from the fundamental to the harmonic frequency as the radiation travels through the medium in the direction of increasing z will cause a negligibly small change in E_ω for small enough values of z. For these small z values, we can treat E_ω as a constant in (11.11). Then, when phase matching is perfect, i.e. $\Delta k = 0$, we find that

$$E_{2\omega} = \frac{8\pi i\omega^2}{k_{2\omega}c^2}\chi_{2\omega}E_\omega^2 z \tag{11.14}$$

Writing

$$E_\omega = \tfrac{1}{2}\mathscr{E}_\omega e^{-i\phi_\omega} \tag{11.15}$$

so that the electric field at frequency ω is $\mathscr{E}_\omega \cos(\omega t + \phi_\omega)$, and a similar expression for $E_{2\omega}$, (11.14) becomes

$$\mathscr{E}_{2\omega} = \frac{4\pi\omega^2}{k_{2\omega}c^2}\left|\chi_{2\omega}\right|\mathscr{E}_\omega^2 \exp\left[i\left(\phi_{2\omega} - 2\phi_\omega \pm \frac{\pi}{2}\right)\right] z \tag{11.16}$$

where the sign of $\pi/2$ in the exponent is the same as that of $\chi_{2\omega}$. Since everything in (11.16) is real with the possible exception of the exponential, its imaginary part must be zero. Hence we must have

$$\phi_{2\omega} - 2\phi_\omega \pm \frac{\pi}{2} = 0 \tag{11.17}$$

in which case (11.16) can be written in the form

$$\mathscr{E}_{2\omega} = \mathscr{E}_\omega \frac{z}{\ell_{SH}}, \quad \ell_{SH}^{-1} = \frac{4\pi\omega^2}{k_{2\omega}c^2}\left|\chi_{2\omega}\right|\mathscr{E}_\omega \tag{11.18}$$

This shows that a wave is generated at the second harmonic frequency whose amplitude increases linearly with distance z for small z at a rate which is characterized by the length ℓ_{SH}. The phase constant of this generated wave is related to that of the fundamental by (11.17) where the sign is controlled by the sign of $\chi_{2\omega}$.

The field strength in the second harmonic wave cannot continue to increase in a linear fashion indefinitely. Clearly it can never have more

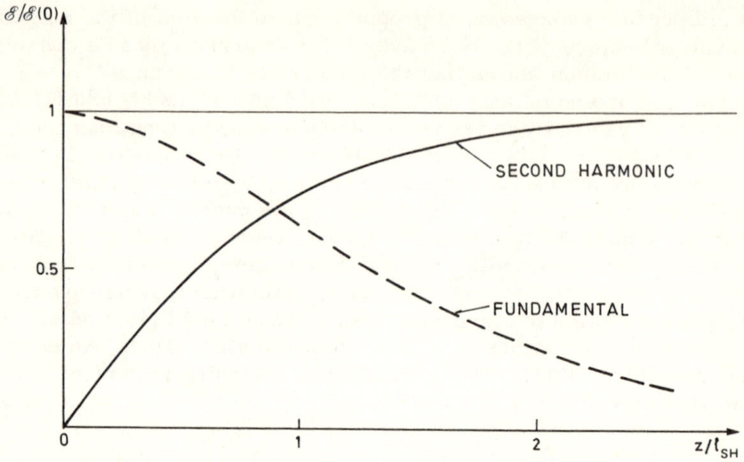

FIG. 11. Build-up of the second harmonic field strength at the expense of the fundamental wave field strength in a non-linear medium, under conditions of perfect phase matching.

power transferred to it than was originally present in the fundamental wave. A saturation effect takes place which shows up in the solution of our non-linear equations when account is taken of the changes in the fundamental field strength which arise from the build-up of the harmonic wave. For the case we are considering, the general solution of (11.11) and (11.12) takes a particularly simple form. It is found [11] that

$$\begin{aligned} \mathscr{E}_{2\omega} &= \mathscr{E}_{\omega}(0) \tanh \frac{z}{\ell_{SH}} \\ \mathscr{E}_{\omega} &= \mathscr{E}_{\omega}(0) \operatorname{sech} \frac{z}{\ell_{SH}} \end{aligned} \qquad (11.19)$$

where $\mathscr{E}_{\omega}(0)$ is the initial value of $\mathscr{E}_{\omega}$ at $z = 0$ and ℓ_{SH} is evaluated from (11.18) using this value of $\mathscr{E}_{\omega}$. This solution is shown in Fig.11. We see then that under conditions of perfect phase matching, a wave at the second harmonic frequency builds up from zero amplitude and eventually all the power originally in the fundamental wave is transferred to it. The characteristic length for this second harmonic generation, ℓ_{SH}, is such that in this distance about half of the power in the fundamental wave has been converted to the second harmonic. We see from (11.18) that this length decreases both as the field strength in the fundamental wave and the degree of non-linearity of the medium, measured by $\chi_{2\omega}$, increase.

The solution (11.19) is not a typical type of solution of our equations (11.11) and (11.12). Generally, solutions exhibit an oscillatory character [11] with energy being periodically transferred back and forth between the fundamental and second harmonic frequencies. The direction of energy flow at any point is determined by the relative phases of the waves at the two frequencies; in particular, whether an existing second harmonic wave will originally grow or decay is determined by its phase constant relative to that of the simultaneously imposed fundamental wave. In the special case just considered, no second harmonic wave is originally present; the only possibility available is for it to grow from this zero value, and it can only do so with a phase constant determined by (11.17). It turns out, as will be seen in Section 12(i), that the normal oscillatory character of the solution has an infinite period in this case and is given by (11.19).

Our treatment of non-linear processes is based on a classical description of the electromagnetic field. Most of the time this is quite valid since the field modes in which we are interested contain many photons, and spontaneous processes are quite negligible compared to those induced by the field. However, in certain cases more care is required and the special solution (11.19) is one of them. For large z this predicts a stable situation in which all the energy in the field is at the second harmonic frequency and the fundamental frequency mode is completely depleted. This is not physically sensible, for the spontaneous decay of a second harmonic photon into two photons at the fundamental frequency should be able to trigger off the reverse process. Walls and Tindle [25] have studied this problem and predict aperiodic oscillations of the energy between the two frequencies.

When perfect phase matching is not achieved, i.e. $\Delta k \neq 0$, Eq.(11.11) can still be solved in the "small signal approximation", i.e. when any change in the amplitude of the fundamental wave can be neglected. It is not difficult to show that in that case (11.11) has the solution

$$\mathscr{E}_{2\omega} = \mathscr{E}_{\omega}(0)\,\frac{\sin\dfrac{\Delta k z}{2}}{\dfrac{\Delta k}{2}\,\ell_{SH}} \tag{11.20}$$

This result verifies the form of the expression (9.1) for the second harmonic intensity which we calculated on the basis of a simple argument. We see from (11.20) that the amplitude of the second harmonic wave oscillates between zero and a maximum value of $2\mathscr{E}_{\omega}(0)/\Delta k \ell_{SH}$ as the fundamental and harmonic waves propagate through the medium, the period of this oscillation being twice the coherence length $\pi/|\Delta k|$. Thus, as the degree of phase mismatch increases, i.e. as Δk increases, the maximum amplitude of the harmonic wave decreases. The justification for omission of higher harmonics in our original equations is in fact based on this result. In general, they will be more severely mismatched to the fundamental wave than the second harmonic, and since their build-up will be described by an expression of similar form to (11.20), their amplitude can never increase nearly as much as that of the second harmonic wave. We also notice that as the phase matching becomes more perfect, i.e. as $\Delta k \to 0$, (11.20) reduces, as it should, to (11.18).

Second harmonic generation is the most extensively studied of all the non-linear effects which can arise in bulk materials. It was the first non-linear mixing effect to be observed; Franken et al. [26] detected in 1961 weak radiation at 3470 Å emerging from the back surface of a plate of crystalline quartz on whose front surface was focused the output from a ruby laser (6940 Å). The electric field strength in the focused laser radiation was estimated to be of the order of 10^5 V/cm and the estimated 10^{19} incident laser photons produced $\sim 10^{11}$ photons of second harmonic radiation. Soon after this, both Giordmaine [23] and Maker et al. [21] successfully carried out

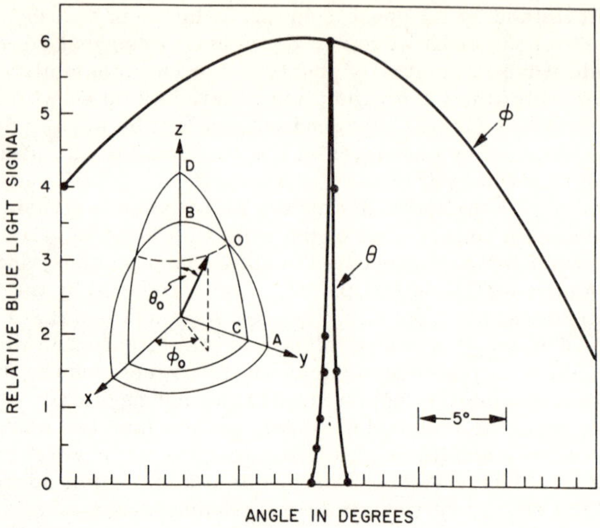

FIG. 12. Intensity of ruby laser second harmonic radiation as a function of propagation direction in KDP [21]. The rapid variation with θ shows the marked effect of phase matching; the slow variation with φ is due to the change of the effective non-linear susceptibility.

a similar experiment using the lower field strength available from an unfocused laser beam. They also demonstrated the greater efficiency of conversion which can be obtained when the phase-matching condition is approached. This cannot be done in quartz since, as already pointed out, the phase-matching condition can never be satisfied in this material for second harmonic generation from ruby laser light. However, the condition can be satisfied e.g. in KDP as discussed in Section 9, and this material was used to demonstrate the effect, as shown in Fig.12, where the second harmonic output is plotted as a function of propagation direction in the crystal; the critical dependence of the output on θ is due to phase matching, the much less violent dependence on ϕ is due to the variation of the value of $\chi_{2\omega}$ with propagation direction. Complete phase matching was not and seldom can be achieved as a result of the presence of several modes in the output of the ruby laser. However, the condition was approached closely enough to obtain an increased conversion efficiency of $\sim 10^{-6}$ in spite of the lower field strength in the unfocused beam. These experiments all employed laser beams in which the power level was of the order of 10 kW. Terhune et al. [27] achieved an efficiency as high as 20% for second harmonic generation in ammonium dihydrogen phosphate, ADP, from a Q-spoiled ruby laser beam having a peak power of the order of 1 MW. This was obtained under conditions in which the maximum possible degree of phase matching was used, and is fairly commonly achieved nowadays.

Extreme exploitation of the phase-matching condition was achieved by Ashkin, Boyd and Dziedzic [28]. They studied the second harmonic generation in KDP of the 1.1526 μm line of a He-Ne laser. In contrast to the ruby laser, this gas laser worked continuously, giving a much narrower spectral line-width, a beam whose divergence was diffraction limited and was easily capable of single-mode operation. All these properties combined to allow a much higher proportion of the beam to propagate in a medium along a phase-matching direction. The degree of phase matching attainable with such a system was therefore much greater than with the ruby laser. Ashkin and his associates in fact calculated an effective coherence length for second harmonic generation from their laser beam in KDP of 20 cm which was much larger than could be achieved with the ruby laser beam. It was also much larger than the thickness of any of the crystals of KDP used in the experiments. With a cyrstal of thickness 1.23 cm, they obtained with phase matching a second harmonic power of 8.1×10^{-14} W from a fundamental power of 1.48×10^{-3} W. Furthermore, it was found that the second harmonic power varied closely as the square of the crystal thickness for three different thicknesses up to 1.23 cm. This fact, along with the small fraction of power converted, indicated that the approximate expression (11.18) for the second harmonic output was valid. This is confirmed by a calculation of ℓ_{SH} on the basis of this expression, which gives the large value of

$$\ell_{SH} = \frac{\mathscr{E}_\omega}{\mathscr{E}_{2\omega}} z = \left(\frac{1.48 \times 10^{-3}}{8.1 \times 10^{-14}}\right)^{\frac{1}{2}} \times 1.23 = 1.66 \times 10^5 \text{ cm}$$

A knowledge of the effective beam width is sufficient for the value $\chi_{2\omega} = (3 \pm 1) \times 10^{-9}$ e.s.u. to be obtained from (11.18), using this value of ℓ_{SH}. For the polarizations of the waves used in the experiment, it was found that $\chi_{2\omega} = \chi^{(2)}_{zxy}(\omega,\omega)$ for KDP. They were also able to reproduce the dependence of the second harmonic output on Δk as given in (11.20) by departing by varying amounts from the phase-matching condition $\Delta k = 0$.

12. COUPLING OF THREE WAVES

Second harmonic generation can be considered as a special case of the mixing together of components of a radiation field at two different frequencies ω_1 and ω_2 to produce a component at the sum frequency $\omega_3 = \omega_1 + \omega_2$. We are therefore led naturally on from a study of second harmonic generation to a study of the more general process of the coupling produced in a non-linear medium amongst the components of a radiation field with frequencies ω_1, ω_2 and $\omega_3 = \omega_1 + \omega_2$.

As in the case of second harmonic generation, we again assume that the phase-matching condition for the processes involved is approximately satisfied, i.e. $\vec{k}_1 + \vec{k}_2 - \vec{k}_3 \sim 0$. We can then neglect the coupling of the waves at frequencies ω_1, ω_2 and ω_3 to waves at all other frequencies and therefore restrict our attention to these three components of the radiation field. These waves are coupled together in a non-linear medium by the fact that any pair together induce a polarization which oscillates at the frequency of the third. This is readily seen using (7.2), (10.3) and (10.11), for it is found that

$$\begin{aligned}
\vec{P}_{NL}(\omega_1) &= \underline{\chi}^{(2)}(\omega_3,-\omega_2):\vec{E}(\omega_3)\vec{E}^*(\omega_2) + \underline{\chi}^{(2)}(-\omega_2,\omega_3):\vec{E}^*(\omega_2)\vec{E}(\omega_3) \\
&= 2\underline{\chi}^{(2)}(\omega_3,-\omega_2):\vec{E}(\omega_3)\vec{E}^*(\omega_2) \\
\vec{P}_{NL}(\omega_2) &= 2\underline{\chi}^{(2)}(\omega_3,-\omega_1):\vec{E}(\omega_3)\vec{E}^*(\omega_1) \\
P_{NL}(\omega_3) &= 2\underline{\chi}^{(2)}(\omega_1,\omega_2):\vec{E}(\omega_1)\vec{E}(\omega_2)
\end{aligned} \tag{12.1}$$

where the intrinsic symmetry of $\underline{\chi}^{(2)}$ has been used. We notice in passing that no coupling involving all three waves can arise from the third-order polarizability $\underline{\chi}^{(3)}$. Our problem is now to solve the three coupled equations for the amplitude factors E_{ω_1}, E_{ω_2} and E_{ω_3} which are obtained when the expressions (12.1) are inserted into our basic equation (10.17). Before making this substitution, however, we use (10.13) to express the non-linear polarizations in (12.1) in terms of the amplitude factors E_{ω_1}, E_{ω_2} and E_{ω_3} of interest, obtaining

$$\vec{P}'_{NL}(\omega_1) = 2\underline{\chi}^{(2)}(\omega_3,-\omega_2):\vec{a}_{\omega_3}\vec{a}_{\omega_2}E_{\omega_3}E^*_{\omega_2}\,e^{i(k_3-k_2)z}$$

$$\vec{P}'_{NL}(\omega_2) = 2\underline{\chi}^{(2)}(\omega_3,-\omega_1):\vec{a}_{\omega_3}\vec{a}_{\omega_1}E_{\omega_3}E^*_{\omega_1}\,e^{i(k_3-k_1)z} \qquad (12.2)$$

$$\vec{P}'_{NL}(\omega_3)\,2\underline{\chi}^{(2)}(\omega_1,\omega_2):\vec{a}_{\omega_1}\vec{a}_{\omega_2}E_{\omega_1}E_{\omega_2}\,e^{i(k_1+k_2)z}$$

Then using (10.17), we obtain the three coupled equations:

$$\frac{dE_{\omega_1}}{dz} = \frac{4\pi i\omega_1^2}{k_1c^2}[\underline{\chi}^{(2)}(\omega_3,-\omega_2)\vdots\vec{a}_{\omega_1}\vec{a}_{\omega_3}\vec{a}_{\omega_2}]E_{\omega_3}E^*_{\omega_2}\,e^{-i\Delta kz}$$

$$\frac{dE_{\omega_2}}{dz} = \frac{4\pi i\omega_2^2}{k_2c^2}[\underline{\chi}^{(2)}(\omega_3,-\omega_1)\vdots\vec{a}_{\omega_2}\vec{a}_{\omega_3}\vec{a}_{\omega_1}]E_{\omega_3}E^*_{\omega_1}\,e^{-i\Delta kz} \qquad (12.3)$$

$$\frac{dE_{\omega_3}}{dz} = \frac{4\pi i\omega_3^2}{k_3c^2}[\underline{\chi}^{(2)}(\omega_1,\omega_2)\vdots\vec{a}_{\omega_3}\vec{a}_{\omega_1}\vec{a}_{\omega_2}]E_{\omega_1}E_{\omega_2}\,e^{i\Delta kz}$$

where

$$\Delta k = k_1 + k_2 - k_3 \qquad (12.4)$$

and the dot products are defined as in (11.8).

In circumstances in which the three polarizabilities appearing in (12.4) are real, these equations can be simplified. For the overall symmetry discussed in Section 8(iv) then applies, and so we have

$$\chi^{(2)}_{\mu\alpha\beta}(\omega_1,\omega_2) = \chi^{(2)}_{\alpha\mu\beta}(-\omega_3,\omega_2) = \chi^{(2)*}_{\alpha\mu\beta}(\omega_3,-\omega_2) = \chi^{(2)}_{\alpha\mu\beta}(\omega_3,-\omega_2)$$

$$= \chi^{(2)}_{\beta\mu\alpha}(-\omega_3,\omega_1) = \chi^{(2)*}_{\beta\mu\alpha}(\omega_3,-\omega_1) = \chi^{(2)}_{\beta\mu\alpha}(\omega_3,-\omega_1) \qquad (12.5)$$

Hence, the equations (12.3) can all be written in terms of the single real quantity

$$\chi_c = \underline{\chi}^{(2)}(\omega_1, \omega_2) \vdots \vec{a}_{\omega_3} \vec{a}_{\omega_1} \vec{a}_{\omega_2} = \underline{\chi}^{(2)}(\omega_3, -\omega_2) \vdots \vec{a}_{\omega_1} \vec{a}_{\omega_3} \vec{a}_{\omega_2}$$

$$= \underline{\chi}^{(2)}(\omega_3, -\omega_1) \vdots \vec{a}_{\omega_2} \vec{a}_{\omega_3} \vec{a}_{\omega_1} \tag{12.6}$$

which gives a measure of the strength of the coupling amongst the three waves. We now have

$$\frac{dE_{\omega_1}}{dz} = \frac{4\pi i \omega_1^2}{k_1 c^2} \chi_c E_{\omega_3} E^*_{\omega_2} e^{-i\Delta kz} \tag{12.7}$$

$$\frac{dE_{\omega_2}}{dz} = \frac{4\pi i \omega_2^2}{k_2 c^2} \chi_c E_{\omega_3} E^*_{\omega_1} e^{-i\Delta kz} \tag{12.8}$$

$$\frac{dE_{\omega_3}}{dz} = \frac{4\pi i \omega_3^2}{k_3 c^2} \chi_c E_{\omega_1} E_{\omega_2} e^{i\Delta kz} \tag{12.9}$$

It follows from these equations and the fact that $\omega_3 = \omega_1 + \omega_2$ that

$$\frac{k_1}{\omega_1} E^*_{\omega_1} \frac{dE_{\omega_1}}{dz} + \frac{k_2}{\omega_2} E^*_{\omega_2} \frac{dE_{\omega_2}}{dz} + \frac{k_3}{\omega_3} E_{\omega_3} \frac{dE^*_{\omega_3}}{dz} = 0$$

Adding this equation to its complex conjugate, we can integrate and obtain the relationship

$$\frac{k_1}{\omega_1} |E_{\omega_1}|^2 + \frac{k_2}{\omega_2} |E_{\omega_2}|^2 + \frac{k_3}{\omega_3} |E_{\omega_3}|^2 = \text{constant} \tag{12.10}$$

This relationship simply expresses the fact that there is no loss of energy from the electromagnetic field to the non-linear medium. For $(k/\omega)|E_\omega|^2$ is easily shown to be proportional to the energy flux S_ω (i.e. the mean value of the Poynting vector) carried through a plane normal to the direction of propagation by that component of the radiation field with frequency ω and propagation constant k. Equation (12.10) therefore says that the total energy flux carried by the three coupled waves is the same at all points in the medium, so that there can be no exchange of energy between the radiation field and the medium. This result arises from the assumed reality and consequent overall symmetry of the polarizabilities which couple the three waves together. It therefore justifies again our earlier statement that real polarizabilities are associated with regions in which the medium can be considered lossless.

Our equations (12.7) - (12.9) give rise to other relations governing the way in which energy is transported through the medium by the radiation field. These relations are obtained by eliminating χ_c from any pair of the equations. Taking the first two equations, for example, we find that

$$\frac{k_1}{\omega_1^2} E^*_{\omega_1} \frac{dE_{\omega_1}}{dz} - \frac{k_2}{\omega_2^2} E^*_{\omega_2} \frac{dE_{\omega_2}}{dz} = 0$$

which, when added to its complex conjugate, can be integrated to yield

$$\frac{1}{\omega_1} \frac{k_1}{\omega_1} |E_{\omega_1}|^2 - \frac{1}{\omega_2} \frac{k_2}{\omega_2} |E_{\omega_2}|^2 = \text{constant} \qquad (12.11)$$

We can rewrite this in terms of the energy fluxes S_{ω_1} and S_{ω_2} which are respectively proportional to

$$\frac{k_1}{\omega_1} |E_{\omega_1}|^2 \quad \text{and} \quad \frac{k_2}{\omega_2} |E_{\omega_2}|^2$$

obtaining

$$\frac{1}{\omega_1} S_{\omega_1} - \frac{1}{\omega_2} S_{\omega_2} = \text{constant} \qquad (12.12)$$

Similar relations are obtained from the other pairs of equations. These are

$$\frac{1}{\omega_1} S_{\omega_1} + \frac{1}{\omega_3} S_{\omega_3} = \text{constant} \qquad (12.13)$$

$$\frac{1}{\omega_2} S_{\omega_2} + \frac{1}{\omega_3} S_{\omega_3} = \text{constant} \qquad (12.14)$$

Any pair of these relations implies the third and the three together imply the energy conservation condition (12.10). The relations (12.12) - (12.14) are known as Manley-Rowe relations after J.M. Manley and H.E.Rowe, who first obtained relationships of this type governing the power flow in lossless non-linear lumped circuit elements [29]. Equations (12.12) - (12.14) are the analogous relations for extended lossless non-linear media.

Although our treatment of the radiation field is completely classical, the Manley-Rowe relations are most easily thought of in terms of photons. For $S_\omega/\hbar\omega$ is simply the mean flux N_ω of photons of frequency ω in the field so that (12.12) - (12.14) can be written as

$$N_{\omega_1} - N_{\omega_2} = \text{constant} \qquad (12.15)$$

$$N_{\omega_1} + N_{\omega_3} = \text{constant} \qquad (12.16)$$

$$N_{\omega_2} + N_{\omega_3} = \text{constant} \qquad (12.17)$$

and in this same notation, the energy conservation equation (12.10) becomes

$$\omega_1 N_{\omega_1} + \omega_2 N_{\omega_2} + \omega_3 N_{\omega_3} = \text{constant} \tag{12.18}$$

The physical interpretation of these relations is now almost absurdly simple. Equation (12.15) shows that if N_{ω_1} changes as the radiation propagates through the medium, then N_{ω_2} must change in exactly the same way. That is, a photon at frequency ω_1 can never be gained or lost as the result of the non-linear interaction without another photon at frequency ω_2 being simultaneously gained or lost. This is fairly obvious, for, in the absence of any exchange of energy between the radiation field and the medium, photons at ω_1 and ω_2 can only be produced or destroyed together as a result of one photon at frequency ω_3 being simultaneously destroyed or generated. The relations (12.16 and (12.17) simply say this same thing in another way.

The consequences of the Manley-Rowe relations are of some significance. They show, for instance, that when the frequency of some signal at ω_1 is converted up to ω_3 by mixing with radiation at ω_2, there is an apparent gain in power by a factor of ω_3/ω_1. For each photon at ω_1 gives rise to one photon at ω_3 whose energy is greater than the original photon by just this factor. This extra power is of course sucked from the radiation at frequency ω_2. In a similar way, if we attempt to generate radiation at a low frequency ω_1 from a high frequency beam at ω_3 by mixing it with ω_2 to produce the difference frequency $\omega_3 - \omega_2 = \omega_1$, there is an apparent loss in power by a factor of ω_1/ω_3. The power, which is apparently lost, goes into the field at ω_2 which, according to the relations, must be amplified at the same time as ω_1 is being generated. It is worth noting that this factor can be quite small, $\sim 10^{-3}$, when the generation of far infra-red radiation from radiation at optical frequencies is envisaged. The Manley-Rowe relations provide upper limits to the power conversion efficiency which can be achieved in any of the lossless non-linear media we consider.

The energy conservation and Manley-Rowe relations arise because of the reality and consequent overall symmetry of $\underline{\chi}^{(2)}(\omega_1, \omega_2)$. They provide the third significant feature which arises out of the overall symmetry of the non-linear polarizabilities. The other significant features, discussed earlier in Section 8(iv), were (a) the connection it provided amongst several apparently independent physical processes and (b) Kleinman symmetry.

Experimentally one would normally wish to exploit the coupling between the three components of the radiation field with frequencies ω_1, ω_2 and $\omega_3 = \omega_1 + \omega_2$ in a non-linear medium which is lossless at all these frequencies. We shall therefore take Eqs (12.7) - (12.9) as the basic equations describing the coupling between the three components, for, as we have just seen, the lossless character of the medium is built into them. Like those governing second harmonic generation, these equations can be solved quite generally in terms of Jacobian elliptic functions for arbitrary initial values of the amplitudes and phases of the three waves [11, 30]. However, we shall not derive the general solution here but restrict our attention to various cases which are both physically interesting and relatively simple to treat.

Basically the only processes which the non-linear coupling amongst the three waves can produce are the generation of the sum or difference frequency of two given frequencies. The actual process that takes place is governed by which of the two associated phase-matching conditions is the more nearly satisfied. However, it is useful, both from a physical point of view and as an approach to solving the basic equations in different approximations, to subdivide these two processes by considering what happens to the incident radiation at the two given frequencies. We consider first of all the possibility of generating, from two beams of monochromatic radiation incident on a non-linear medium, radiation at either the sum or the difference frequency. Our main concern will be with the generation process; what happens to the incident radiation will be of no concern. In fact, we shall regard the incident beams as being of high-intensity laser radiation of comparable magnitudes and as not being affected appreciably by the generation process. We shall, however, also indicate the form of the general solutions of our equations in these two cases from which our approximate solutions emerge in the appropriate limits.

We shall then consider the problem from another point of view and discuss what happens when a monochromatic beam of relatively weak radiation is passed through a non-linear medium which is being simultaneously irradiated by a beam of intense laser radiation. The change in intensity of the weak incident beam that accompanies the generation of the sum or difference frequency cannot be neglected and significant attenuation or amplification of the beam is obtained. Changes in the intensity of the intense laser beam can still be neglected, of course, to a first approximation. Whether amplification or attenuation of the weak incident beam is obtained depends on whether the sum or difference frequency is being generated and on the relative magnitudes of the three frequencies involved. The Manley-Rowe relations can be used to decide which situation obtains. For example, let us denote by ω_s the frequency of the weak incident or signal beam and by ω_L that of the laser radiation and consider the generation of the difference frequency $|\omega_L - \omega_s|$. If the frequencies are such that $\omega_L - \omega_s < \omega_s < \omega_L$, we can identify $\omega_L - \omega_s$, ω_s and ω_L respectively with our frequencies ω_1, ω_2 and ω_3. The Manley-Rowe relations (12.15) - (12.17) then show that the difference frequency ω_1 can be generated only with the accompanied amplification of the signal beam at ω_2. However, if $\omega_s - \omega_L < \omega_L < \omega_s$, we can now identify $\omega_s - \omega_L$, ω_L and ω_s respectively with ω_1, ω_2 and ω_3 and see that generation of the difference frequency ω_1 is accompanied by the attenuation of the signal beam at ω_3. The complete list of possibilities is shown in Table II with appropriate identifications of ω_1, ω_2

TABLE II. PARAMETRIC PROCESSES

Identification			Signal	Frequency generated	Nomenclature
ω_1	ω_2	ω_3			
$\omega_L - \omega_s$	ω_s	ω_L	Amplified	Difference	Parametric amplifier
ω_s	$\omega_L - \omega_s$	ω_L			
ω_L	$\omega_s - \omega_L$	ω_s	Attenuated	Difference	Parametric down-converter
$\omega_s - \omega_L$	ω_L	ω_s			
ω_s	ω_L	$\omega_L + \omega_s$	Attenuated	Sum	Parametric up-converter
ω_L	ω_s	$\omega_L + \omega_s$			

and ω_3 with ω_s, ω_L and $|\omega_L - \omega_s|$ for the Manley-Rowe relations (12.15) - (12.17) to be applied. The generation of the sum frequency is seen to be accompanied always by an attenuation of the signal beam but difference frequency generation can be accompanied by either amplification or attenuation of the signal.

The names of parametric amplifier, down-converter and up-converter, which we have given to the various processes that can take place, arise in the following way. A parametric device is obtained when some property or parameter of a system can be varied, usually periodically, in time so that the response of the system to some disturbance is modified. The most familiar example of this effect is a mechanical one; it is a child on a swing who varies the effective length of the swing, by raising himself up and down, at a frequency — the pump frequency — which is twice that of the swing. The initial disturbance or signal that set the swing in motion is amplified as a result of this variation of the length of the swing. However, this simple parametric amplifier can be considered in another way. Idealizing the swing as a simple pendulum, its normal equation of motion is

$$\ddot{\theta} = -\frac{g}{\ell}\theta$$

where θ is its angular displacement and ℓ its length. This shows the swing normally to be a device whose angular acceleration is a linear function of its displacement. When the child periodically alters the length of the swing, he is effectively making the length ℓ a function of θ so that the acceleration becomes a non-linear function of the displacement. From this point of view the parametric amplification arises from the non-linear response of the swing. This is quite a general result and, in a converse way, any system whose response is non-linear can be shown to provide the basis of a parametric device [22]. In our case the laser beam acts as a pump in the non-linear medium which results in the effective dielectric constant of the medium varying periodically in time at the pump frequency. It is this periodic variation of the effective dielectric constant of the medium which leads to the various processes described above. In the parametric amplifier, as the name implies, radiation at the signal frequency is amplified. In the parametric converters, radiation at the signal frequency is attenuated and reappears at a higher or lower frequency in the form of the generated sum or difference frequency.

(i) Sum frequency generation

We consider first of all the generation of the initially absent component of the radiation field at ω_3 by the mixing of the components at ω_1 and ω_2. This is the direct generalization of second harmonic generation which is obtained in the special case of $\omega_1 = \omega_2$. As in that special case, our equations are easily solved in the "small signal approximation" which is valid for small enough values of z when any change produced in the intensities of the fields at ω_1 and ω_2 can be neglected. We can then treat E_{ω_1} and E_{ω_2} in (12.9) as constants so that, when perfect phase matching is achieved, i.e. $\Delta k = 0$, this equation yields the solution

$$E_{\omega_3} = \frac{4\pi i \omega_3^2}{k_3 c^2} \chi_c E_{\omega_1} E_{\omega_2} z \qquad (12.19)$$

When the real amplitudes and phase constants of the three components are introduced by a substitution of the type (11.15), (12.19) becomes

$$\mathscr{E}_{\omega_3} = \frac{2\pi\omega_3^2}{k_3c^2}\,|\chi_c|\,\mathscr{E}_{\omega_1}\mathscr{E}_{\omega_2}\exp\left[i\left(\phi_3-\phi_1-\phi_2\pm\frac{\pi}{2}\right)\right]z \tag{12.20}$$

where the sign of $\pi/2$ in the exponent is the same as that of χ_c. This result is very similar to that obtained for second harmonic generation. It shows that the component with the sum frequency ω_3 is initially generated with an amplitude which increases linearly with z and a phase constant ϕ_3 which is related to ϕ_1 and ϕ_2 by the fact that the imaginary part of the exponential in (12.20) must be zero, i.e.

$$\phi_3 - \phi_1 - \phi_2 \pm \frac{\pi}{2} = 0 \tag{12.21}$$

The field strength at the sum frequency ω_3 does not, of course, continue to increase indefinitely with increasing z. It will saturate at a value which is reached when all the power, initially carried by the weaker of the two components at ω_1 and ω_2, has been exhausted by the generation process and has been converted to power at ω_3. Physically this is the result one would expect and formally it arises directly from the Manley-Rowe relations. The general solution of (12.7) - (12.9) for the generation process exhibits this feature; for it is found that [1, 18]

$$\mathscr{E}_{\omega_3} = \left(\frac{\omega_3^2}{\omega_2^2}\frac{k_2}{k_3}\right)^{\frac{1}{2}}\mathscr{E}_{\omega_2}(0)$$

$$\times\,\mathrm{sn}\left[\frac{2\pi}{c^2}\left(\frac{\omega_2^2\omega_3^2}{k_2k_3}\right)^{\frac{1}{2}}|\chi_c|\,\mathscr{E}_{\omega_1}(0)\,z;\quad\left(\frac{\omega_1^2}{\omega_2^2}\frac{k_2}{k_1}\right)^{\frac{1}{2}}\frac{\mathscr{E}_{\omega_2}(0)}{\mathscr{E}_{\omega_1}(0)}\right] \tag{12.22}$$

where we have labelled with ω_2 the frequency of the weaker of the two components initially present. The Jacobian elliptic function [31] sn in (12.22) is periodic with a maximum value of unity so that the maximum value of $\mathscr{E}_{\omega_3}$ is seen to be controlled by this weaker component. Furthermore, $\mathrm{sn}(x;\gamma) \simeq x$ for small values of x, i.e. $x \ll 1$, so that the general solution (12.22) reduces to our special solution, given by (12.20) and (12.21), in the limit of small values of z.

The form taken by our solutions is more easily understood when they are expressed in terms of the photon fluxes carried by the various components rather than in terms of field strengths. The mean photon flux N_ω in a component with frequency ω is, as we have seen, proportional to $(k/\omega^2)\mathscr{E}_\omega^2$ so that (12.22) can be written in the form

$$N_{\omega_3} = N_{\omega_2}(0)\,\mathrm{sn}^2\left[\frac{z}{\ell_M};\ \frac{N_{\omega_2}^{\frac{1}{2}}(0)}{N_{\omega_1}^{\frac{1}{2}}(0)}\right] \tag{12.23}$$

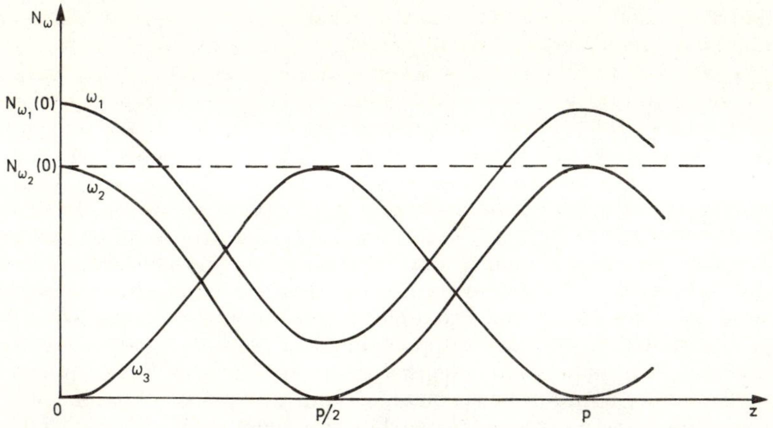

FIG.13. Dependence on the distance z travelled through the non-linear medium of the mean photon fluxes in the waves initially present, with frequencies ω_1 and ω_2, and in the wave generated at the sum frequency ω_3, under conditions of perfect phase matching. The period of oscillation p is proportional to ℓ_M, the constant of proportionality depending on the value of the ratio $N_{\omega_2}(O)/N_{\omega_1}(O)$. In the limit of this ratio becoming very small, $p \to \pi\ell_M$ and the mixing process is regarded as parametric up-conversion from ω_2 to ω_3. In the limit of the ratio becoming unity, $p \to \infty$, and the mixing process becomes almost identical to the second harmonic generation process, as discussed in the text.

where

$$\ell_M = \left[\frac{2\pi}{c^2}\left(\frac{\omega_2^2}{k_2}\frac{\omega_3^2}{k_3}\right)^{\frac{1}{2}} |\chi_c| \, \mathscr{E}_{\omega_1}(0)\right]^{-1} \tag{12.24}$$

is a length which characterizes the rate at which the mixing process takes place. This general solution for N_{ω_3}, along with the associated solutions for N_{ω_1} and N_{ω_2}, is shown as a function of z in Fig.13. In the limit of small z, (12.23) reduces to

$$N_{\omega_3} = N_{\omega_2}(0)\left(\frac{z}{\ell_M}\right)^2 \tag{12.25}$$

which is equivalent to the solution given by (12.20) and (12.21). We see immediately from (12.23) that the maximum flux of photons generated at the sum frequency is limited to $N_{\omega_2}(0)$ which is the photon flux initially present in the weaker of the two components.

The period, p, of the periodic variation with x of the function $sn(x;\gamma)$ depends on the value of γ [31]. As $\gamma \to 1$, the period becomes larger and larger. In the limit when $\gamma = 1$, the period is infinitely large and the function effectively loses its periodic character. It in fact becomes tanh x. This implies that when $N_{\omega_1}(0) = N_{\omega_2}(0)$, (12.22) becomes

$$\mathscr{E}_{\omega_3} = \left(\frac{\omega_3^2}{\omega_2^2}\frac{k_2}{k_3}\right)^{\frac{1}{2}} \mathscr{E}_{\omega_2}(0) \tanh\frac{z}{\ell_M} \tag{12.26}$$

This has the same form as the result (11.19) for the growth of a second harmonic wave under conditions of perfect phase matching. In fact when $\omega_1 = \omega_2 = \omega$ and $\omega_3 = 2\omega$, the two results are almost identical. They only differ because we have omitted from our present treatment the process of second harmonic generation from ω_1 and ω_2 which is, of course, equivalent to and should be considered in addition to the production of the sum frequency of ω_1 and ω_2 when these frequencies are equal.

Experimental work on the mixing of light beams at two different frequencies is much less extensive than the corresponding work on the special case of second harmonic generation. The first observation of the effect was made by Bass et al. [32] who examined the light emerging from a crystal of triglycine sulphate which was illuminated with the beams from two ruby lasers, one operating at room temperature and the other at liquid nitrogen temperature. The output wavelengths of two such lasers differ by about 10 Å. Around the wavelength of the second harmonics of the two lasers (3470 Å), three spectral lines were observed. The two lines at the highest and lowest frequencies corresponded to the second harmonics of the two lasers, the middle line to their sum frequency. No attempt was made in this work to achieve phase matching. Miller and Savage [33] observed the sum frequency at 4200 Å obtained by mixing the beams from ruby and $CaWO_3$: Nd^{3+} lasers in a variety of crystals, e.g. KDP, ADP, $BaTiO_3$ and tourmaline, of which KDP proved to be the most efficient. As much phase matching as possible was used in this work. A measure of the non-linear susceptibility χ_c which controls the mixing process in GaAs has been obtained by Garfinkel and Engeler [34]. They observed and measured the intensity of a large group of spectral lines which was found in emission from a GaAs laser. These lines fell into two groups. One group around 8500 Å corresponded to the various fundamental modes of the laser. The other group of lines had a wavelength in the region of 4200 Å and could be interpreted as the second harmonics and sum frequencies of the fundamental lines. They were able to deduce from their measurements a value of χ_c which gave a value for the only non-zero component of $\underline{\chi}^{(2)}$ in GaAs. This gave $\chi^{(2)}_{xyz}(\omega_1, \omega_2) = 2.6 \times 10^{-6}$ e.s.u., ω_1 and ω_2 being frequencies of the order of the GaAs laser frequency. This was the largest value of a component of $\underline{\chi}^{(2)}$ so far measured at that time (1963), being 10^3 times larger than the value of $\chi^{(2)}_{zxy}(\omega, \omega)$ in KDP at the ruby laser frequency.

Non-linear coefficients as large as this are not common. It is unfortunate that GaAs, and other III-V semiconductors with similar large non-linearities, are optically isotropic in the linear regime. Phase matching is therefore not easily achieved, and it is difficult to exploit the large non-linearity.

(ii) Difference frequency generation

The solution of Eqs (12.7) - (12.9) for this case goes through in the "small signal approximation" in almost exactly the same way as for sum frequency generation. Initially we now have radiation at frequencies ω_3 and ω_1 and investigate the possible generation of a wave at frequency $\omega_2 = \omega_3 - \omega_1$. For small values of z we can treat E_{ω_3} and E_{ω_1} as constants and integrate (12.8) immediately. For perfect phase matching we find that

$$\mathscr{E}_{\omega_2} = \frac{2\pi\omega_2^2}{k_2 c^2} |\chi_c| \mathscr{E}_{\omega_3} \mathscr{E}_{\omega_1} z \qquad (12.27)$$

and

$$\phi_1 + \phi_2 - \phi_3 \pm \frac{\pi}{2} = 0 \tag{12.28}$$

where the sign associated with $\pi/2$ is that of χ_c. Thus a wave at the difference frequency grows with an amplitude which increases linearly with z. The Manley-Rowe relations show that the maximum flux of photons which can be generated in this wave is $N_{\omega_3}(0)$ which is attained when all the power initially in the wave at ω_3 has been sucked from it. In analogy with our discussion of sum frequency generation (see Eq.(12.25)), (12.27) can be cast into the form

$$N_{\omega_2} = N_{\omega_3}(0)\left(\frac{z}{\ell_M}\right)^2 \tag{12.29}$$

where ℓ_M is defined in (12.24) and is again the length which characterizes the rate at which the mixing process takes place. The general solution of our equations in this case [1, 18] gives a generalization of (12.29) to

$$N_{\omega_2} = \frac{N_{\omega_1}(0)N_{\omega_3}(0)}{N_{\omega_1}(0)+N_{\omega_3}(0)}\, f^2\left[\left(\frac{N_{\omega_1}(0)+N_{\omega_3}(0)}{N_{\omega_1}(0)}\right)^{\frac{1}{2}} \frac{z}{\ell_M};\ \left(\frac{N_{\omega_3}(0)}{N_{\omega_1}(0)+N_{\omega_3}(0)}\right)^{\frac{1}{2}}\right] \tag{12.30}$$

where the function $f(x;\gamma)$ is defined as the ratio of the Jacobian elliptic functions $\mathrm{sn}\,(x;\gamma)$ and $\mathrm{dn}\,(x;\gamma)$ i.e.

$$f(x;\gamma) = \frac{\mathrm{sn}\,(x;\gamma)}{\mathrm{dn}\,(x;\gamma)} \tag{12.31}$$

This general solution for N_{ω_2} is illustrated as a function of z in Fig.14 along with the corresponding solutions for N_{ω_1} and N_{ω_3} which can be obtained using the Manley-Rowe relations (12.15) - (12.17). In terms of the mixing length ℓ_M, (12.27) becomes

$$\mathscr{E}_{\omega_2} = \left(\frac{\omega_2^2}{\omega_3^2}\frac{k_3}{k_2}\right)^{\frac{1}{2}} \mathscr{E}_{\omega_3}(0)\frac{z}{\ell_M} \tag{12.32}$$

There is interest in using this process to generate long wavelength radiation in the infra-red. Faries et al. [35] have succeeded in producing radiation in the 1-8 cm^{-1} wave number range by mixing in $LiNbO_3$ the output from two ruby lasers at slightly different temperatures and consequently output frequencies. Tunable radiation has been generated by Hanna and his colleagues [36] by mixing the outputs of a ruby laser and a tunable dye laser in proustite (Ag_3AsS_3); frequencies around 5 μm and 10 μm have been generated in this way.

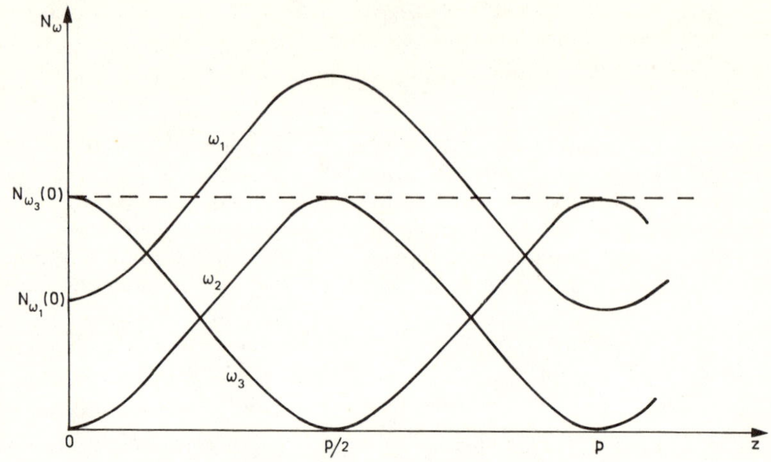

FIG. 14. Dependence on the distance z travelled through the non-linear medium of the mean photon fluxes in the waves initially present, with frequencies ω_1 and ω_3, and in the wave generated at the difference frequency ω_2, under conditions of perfect phase matching. The period of oscillation p is proportional to ℓ_M, the constant of proportionality depending on the value of the ratio $N_{\omega_3}(O)/N_{\omega_1}(O)$. In the limit of this ratio becoming very small, $p \to \pi \ell_M$, and the mixing process is regarded as parametric down-conversion from ω_3 to ω_2. In the limit of the ratio becoming very large, $p \to \infty$, and the mixing process is regarded as parametric amplification of the wave at frequency ω_1, the pump frequency being ω_3 and the idler frequency ω_2.

(iii) Parametric conversion

We can consider parametric up-conversion and down-conversion simultaneously and distinguish finally between the two processes by using the different boundary conditions which apply. We see from our earlier discussion that, for these two processes, the frequency of the intense laser radiation — the pump frequency — is always identified with one of the two lower frequencies ω_1 or ω_2. For definiteness, we shall identify it with ω_1 for the present. Results with the other identification are easily obtained from our final expressions by interchanging the labels 1 and 2.

We have already seen that, in the conversion process, radiation at the signal frequency ω_s is attenuated in favour of the generation of radiation at the sum or difference frequencies $\omega_s \pm \omega_L$. The maximum power generated at these frequencies is limited by the power initially carried at the signal frequency, a fact which follows from the Manley-Rowe relations. Since the initial radiation at the signal frequency is always assumed to be much less intense than the laser or pumping radiation, the amount of power gained or lost by the pump frequency radiation in the conversion process is therefore only a small fraction of the total power carried by this radiation. It is consequently a good approximation to neglect this change in intensity of the pumping radiation and so consider E_{ω_1} as a constant for this process in our equations not only for small values but for all values of z.

With this approximation, Eqs (12.8) and (12.9) can be solved quite generally. Elimination of E_{ω_3} from the equations leads to the equation

$$\frac{d^2}{dz^2} E_{\omega_2} + i\Delta k \frac{d}{dz} E_{\omega_2} + \frac{1}{\ell_M^2} E_{\omega_2} = 0 \qquad (12.33)$$

for E_{ω_2} where ℓ_M is again the length defined in (12.24). The solution of this equation is a linear combination of the two functions

$$\exp\left(-\frac{i}{2}\Delta kz\right)\sin\left[1+\left(\frac{\Delta k}{2}\ell_M\right)^2\right]^{\frac{1}{2}}\frac{z}{\ell_M} \tag{12.34}$$

and

$$\exp\left(-\frac{i}{2}\Delta kz\right)\cos\left[1+\left(\frac{\Delta k}{2}\ell_M\right)^2\right]^{\frac{1}{2}}\frac{z}{\ell_M} \tag{12.35}$$

The solution of the corresponding equation for E_{ω_3} is an associated linear combination of the complex conjugates of these two functions. Now the wave at the generated frequency must depend only on the sine function (12.34) since it starts from zero at $z = 0$. The coefficient of the sine function is determined from the rate of change of amplitude at $z = 0$ which is given by either (12.8) or (12.9) in terms of the initial amplitudes. Thus we find for the down-conversion process in which the frequency ω_2 is generated from ω_3 that

$$\mathscr{E}_{\omega_2} = \left(\frac{\omega_2^2\, k_3}{\omega_3^2\, k_2}\right)^{\frac{1}{2}} \frac{\mathscr{E}_{\omega_3}(0)}{\left[1+\left(\frac{\Delta k}{2}\ell_M\right)^2\right]^{\frac{1}{2}}}\sin\left[1+\left(\frac{\Delta k}{2}\ell_M\right)^2\right]^{\frac{1}{2}}\frac{z}{\ell_M} \tag{12.36}$$

As usual, this result is more simply expressed in terms of the mean photon flux N_{ω_2}. It becomes

$$N_{\omega_2} = \frac{N_{\omega_3}(0)}{1+\left(\frac{\Delta k}{2}\ell_M\right)^2}\sin^2\left[1+\left(\frac{\Delta k}{2}\ell_M\right)^2\right]^{\frac{1}{2}}\frac{z}{\ell_M} \tag{12.37}$$

and from (12.17) it follows that

$$N_{\omega_3} = \frac{N_{\omega_3}(0)}{1+\left(\frac{\Delta k}{2}\ell_M\right)^2}\left\{\left(\frac{\Delta k}{2}\ell_M\right)^2 + \cos^2\left[1+\left(\frac{\Delta k}{2}\ell_M\right)^2\right]^{\frac{1}{2}}\frac{z}{\ell_M}\right\} \tag{12.38}$$

The corresponding expressions which govern the up-conversion process are obtained from (12.37) and (12.38) simply by interchanging the frequency labels ω_2 and ω_3 on N.

We see from our solutions that after an initial increase as z^2 of the photon flux in the generated wave, this flux varies, as z increases, in an oscillatory fashion with period $\pi\ell_M/[1+((\Delta k/2)\ell_M)^2]^{\frac{1}{2}}$. The maximum value it attains is $[1+((\Delta k/2)\ell_M)^2]^{-1}$ of the original flux in the signal beam. Thus, as the degree of phase matching increases, i.e. as $\Delta k \to 0$, the period of the variation increases towards the value $\pi\ell_M$ and the maximum flux obtainable at the generated frequency increases towards its greatest possible value, i.e. that which corresponds to a complete conversion of all photons at the signal frequency. We see here explicitly the increase in efficiency of the conversion process as the phase-matching condition is approached. The dependence of the up and down conversion processes on distance travelled through the medium when phase matching is perfect is illustrated by Figs 13 and 14 respectively when $N_{\omega_1}(0)$ becomes very large so that $p \to \pi \ell_M$.

Parametric up and down conversion were first observed by Smith and Braslau [37]. The pumping frequency was that of a ruby laser with an average power per pulse of 1 kW. A crystal of KDP was illuminated with the beam of this laser and with this arrangement they were able to convert both up and down the frequency of a small part of some spectral lines from a mercury lamp, the power in each line being of the order of 10 mW. A factor of 10^5 therefore existed between the pump and signal powers. They took steps to achieve as much phase matching as possible at all times. Powers of the order of 10^{-9} W were obtained in the generated waves so that only a very small conversion efficiency was achieved. The linear relationship between input signal strength and the converted output intensity predicted from (12.37) was verified by altering the power radiated by the mercury lamp.

There has been considerable interest in up-converting weak signals and images in the infra-red region of the spectrum up to visible frequencies, where faster and much more sensitive detectors and image pick-up devices are available; a comprehensive review was published by Warner [38] in 1970. In certain circumstances, the inefficiency of the up-conversion process is more than compensated by the improved detection capability at the shorter visible wavelengths and an overall improvement in detection sensitivity is achieved. For example, Smith and Mahr [39] obtained an improved sensitivity for radiation in the 2.5 - 4.5 μm region by up-converting in $LiNbO_3$ pumped at 5150 Å by radiation from an argon ion laser.

The possibility of image up-conversion is a consequence jointly of the phase-matching condition and the linear relationship between the signal and up-converted intensities. To an observer, an image, for simplicity at infinity, is characterized by the intensity of radiation arriving from a scene as a function of the angular position of its point of origin. In the parametric up-conversion process, the linear relationship between signal and up-converted intensities preserves relative intensities and the phase-matching condition preserves the angular information. This latter point is readily understood from Fig.15, where the phase-matching triangle implied by the phase-matching condition

$$\vec{k}_{ir} + \vec{k}_L = \vec{k}_V$$

is shown, $\vec{k}_{ir}$, $\vec{k}_L$ and $\vec{k}_V$ being respectively the wave-vectors of the infra-red signal from a particular direction, of the pumping laser and of the up-converted visible radiation. The condition clearly defines a unique relationship between the angles θ_{ir} and θ_v made by the propagation directions of the infra-red and visible beams with the fixed propagation direction of the laser radiation. Since $k_V \sim k_L \gg k_{ir}$, $\theta_v \approxeq \theta_{ir}\, k_{ir}/k_v$, which shows that the visible up-converted image is simply demagnified relative to the original infra-red image by the ratio λ_v/λ_{ir} of the wavelengths at the two frequencies.

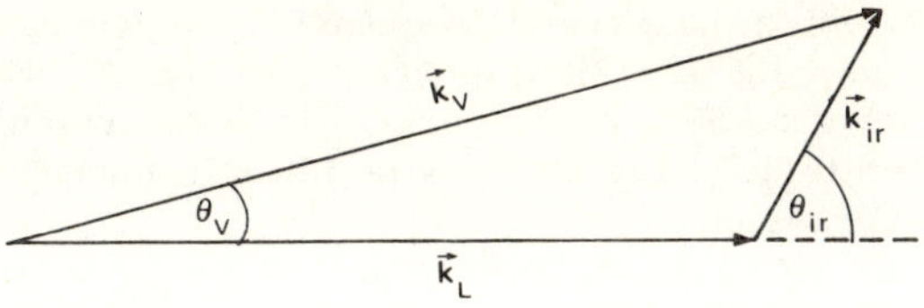

FIG.15. Phase-matching triangle for the three wave vectors in a parametric up-conversion process from the infra-red to visible.

(iv) Parametric amplification

It is seen from the previous discussion that for parametric amplification of a small signal to take place we must identify the pump or laser frequency with the highest frequency ω_3 of our notation. The frequency of the small signal can be identified with ω_1 or ω_2; for definiteness we identify it with ω_1. The difference frequency $\omega_2 = \omega_3 - \omega_1$ which, as we have seen, is generated simultaneously with the amplification of the signal is known in the theory of parametric amplifiers as the idler frequency. In contrast to the parametric converter, where the initial intensity of the signal wave limits the amount of power that can be transferred from one wave to another, it is the intensity at the pump or laser frequency that limits the power transfer which can take place in a parametric amplifier. A much larger transfer of power is therefore possible in the amplifier. It follows from these arguments that we cannot consider any of the field amplitudes E_{ω_1}, E_{ω_2} or E_{ω_3} in our equations to be constant in this case for all values of z, for in principle it is possible to reduce E_{ω_3} to zero and to greatly increase both E_{ω_1} and E_{ω_2}. However, if we restrict ourselves to small enough values of z, no appreciable decrease in the pump intensity can take place, although a significant increase can take place at the signal and idler frequencies. For such small values of z, we can neglect the depletion of power at the pump frequency and treat E_{ω_3} as a constant.

With this approximation, the solution of Eqs (12.7) and (12.8) can be carried through in close analogy with the solution of (12.8) and (12.9) for the parametric conversion process. For elimination of E_{ω_1} from (12.7) and (12.8) leads to the equation

$$\frac{d^2}{dz^2} E_{\omega_2} + i\Delta k \frac{d}{dz} E_{\omega_2} - \frac{1}{\ell_{PA}^2} E_{\omega_2} = 0 \qquad (12.39)$$

for E_{ω_2} where

$$\ell_{PA} = \left[\frac{4\pi}{c^2}\left(\frac{\omega_1^2\omega_2^2}{k_1k_2}\right)^{\frac{1}{2}} |\chi_c E_{\omega_3}|\right]^{-1} = \left[\frac{2\pi}{c^2}\left(\frac{\omega_1^2\omega_2^2}{k_1k_2}\right)^{\frac{1}{2}} |\chi_c|\, \mathscr{E}_{\omega_3}\right]^{-1} \qquad (12.40)$$

is a length characteristic of the parametric amplification process. We notice immediately that (12.39) can be obtained from (12.33) by replacing ℓ_M^2 by $-\ell_{PA}^2$, i.e. ℓ_M by $i\ell_{PA}$. A solution of (12.39) can therefore be obtained from a solution of (12.33) by a similar replacement. The solution for the problem in hand in which the wave at frequency ω_2 is generated can therefore be obtained from the solution (12.36) or (12.37) of (12.33), which gives the corresponding generation of ω_2 in the parametric conversion process. Thus we find that the amplitude of the wave at the idler frequency increases according to the expression

$$\mathscr{E}_{\omega_2} = \left(\frac{\omega_2^2}{\omega_1^2}\frac{k_1}{k_2}\right)^{\frac{1}{2}} \frac{\mathscr{E}_{\omega_1}(0)}{\left[1 - \left(\frac{\Delta k}{2}\ell_{PA}\right)^2\right]^{\frac{1}{2}}} \sinh\left[1 - \left(\frac{\Delta k}{2}\ell_{PA}\right)^2\right]^{\frac{1}{2}} \frac{z}{\ell_{PA}} \qquad (12.41)$$

The hyperbolic sine function comes from the corresponding trigonometric function in (12.36) and its coefficient is fixed by necessity of satisfying (12.8) at $z = 0$. As usual, this result simplifies when expressed in terms of mean photon fluxes; it becomes

$$N_{\omega_2} = \frac{N_{\omega_1}(0)}{1 - \left(\frac{\Delta k}{2}\ell_{PA}\right)^2} \sinh^2\left[1 - \left(\frac{\Delta k}{2}\ell_{PA}\right)^2\right]^{\frac{1}{2}} \frac{z}{\ell_{PA}} \qquad (12.42)$$

From (12.15) it follows that the mean photon flux in the signal wave is amplified according to

$$N_{\omega_1} = \frac{N_{\omega_1}(0)}{1 - \left(\frac{\Delta k}{2}\ell_{PA}\right)^2} \left\{\cosh^2\left[1 - \left(\frac{\Delta k}{2}\ell_{PA}\right)^2\right]^{\frac{1}{2}} \frac{z}{\ell_{PA}} - \left(\frac{\Delta k}{2}\ell_{PA}\right)^2\right\} \qquad (12.43)$$

For large enough values of z but still small enough for no appreciable depletion in the pump power, N_{ω_1} increases exponentially as $\exp \alpha z$ where

$$\alpha = \left[\frac{4}{\ell_{PA}^2} - \Delta k^2\right]^{\frac{1}{2}} \qquad (12.44)$$

α is the amplification constant for the process; it achieves a maximum value of $2/\ell_{PA}$ when phase matching is perfect.

Equations (12.42) and (12.43) predict a continuous increase with z of the power at the signal and idler frequencies as long as $\Delta k < 2/\ell_{PA}$. This is a result of our neglect of any depletion of the power at the pump frequency. The more general solution of Eqs (12.7) - (12.9) which takes this depletion into account shows the expected saturation of the power at the signal and idler frequencies which corresponds to a complete depletion of the power at the pump frequency [11, 30]. This is obtained when phase matching is perfect from the general solution illustrated in Fig.14 when $N_{\omega_1}(0)$ becomes very large and with it the period p of the variation. In the limit of $N_{\omega_3}(0) \to \infty$, the general solution (12.30) becomes identical to our approximate solution

(12.42) with $\Delta k = 0$. In practice, as we shall see, the length ℓ_{PA} is likely to be very large and our approximate solutions (12.42) and (12.43) are likely to be quite adequate.

The parametric amplification process is widely exploited in the parametric oscillator which can be used to provide sources of intense laser-like radiation; this can be tuned over a range of frequencies. In the oscillator, it is usually arranged that the pump-signal and idler beams propagate collinearly, and oscillation is achieved by siting the amplifying material between two mirrors, as in a laser cavity, to provide the feedback which converts the amplifier into an oscillator. The mirrors are normal to the common direction of propagation and the pumping laser radiation enters the cavity from outside, through one of the mirrors. Tuning is achieved by altering the phase-matching condition by geometry or, more conveniently, by temperature, in a material like lithium niobate with temperature-sensitive refractive indices. The first observation of parametric oscillation was by Giordmaine and Miller [40] in 1965. By pumping a lithium niobate crystal with 0.53 μm radiation, produced as the second harmonic of a Nd-laser in another piece of lithium niobate, they produced an oscillator with an output, tunable by temperature, over the wavelength range 0.92 - 1.15 μm. Much work has been done since then, much of it aimed at extending the wavelengths available well into the infra-red. A good recent review of the situation is given in an article on tunable sources by Colles and Pidgeon [41].

(v) Characteristic mixing lengths

The rate at which each of the mixing processes we have been considering takes place is characterized by an appropriate length parameter ℓ_M defined in (12.24) or ℓ_{PA} defined in (12.40) or indeed ℓ_{SH} defined in (11.18). The expressions for these lengths are all similar and are generally of the form

$$\ell^{-1} = \frac{2\pi}{c^2}\left(\frac{\omega_a^2\,\omega_b^2}{k_a\,k_b}\right)^{\frac{1}{2}} \left|\chi\right| \mathscr{E}_{\omega_c}(0) \tag{12.45}$$

where we have labelled the three frequencies involved as a, b and c, with ω_c being the frequency either of one of the intense incident laser beams or of the only one. It is more realistic and useful to work not in terms of the field strength $\mathscr{E}_{\omega_c}(0)$ of this beam but in terms of its power density

$$S_{\omega_c}(0) = \frac{c^2 k_c}{8\pi\omega_c}\,\mathscr{E}_{\omega_c}^2(0)$$

which follows from evaluating the Poynting vector. Then, converting propagation constants to refractive indices, we find that

$$\ell^{-2} = \frac{32\pi^3}{c^3}\,\omega_a\,\omega_b\,\frac{|\chi|^2}{n_a n_b n_c}\,S_{\omega_c}(0) \tag{12.46}$$

In this expression $|\chi|^2/n_a n_b n_c$ contains all the parameters characteristic of the medium. It is consequently often referred to as the figure of merit of a material, as it provides a measure of how effectively a given mixing process will take place in that material. It should be noted that a good figure of merit calls not only for large non-linearities but also low refractive indices. We shall see later that these two requirements tend to oppose one another.

It is useful at this stage to obtain some estimate of the magnitude of this mixing length to ensure that we are discussing processes that can take place in reasonably sized pieces of material. This is easily done by assuming that all refractive indices are about unity and the frequencies of interest correspond to a visible wavelength of 7000 Å. It was pointed out in Section 11 that for KDP, $\chi \simeq 3 \times 10^{-9}$ e.s.u. We shall see later that this is a lower than average value and it is reasonable to take $\chi \simeq 10^{-8}$ e.s.u. Then (12.46) gives $\ell \sim 2/\sqrt{S}$ cm when S is measured in MW/cm². Power densities of ~ 1 MW/cm² are readily obtained from lasers, giving rise to a characteristic mixing length of centimetre dimensions which is quite acceptable.

13. MATERIALS AND THEIR NON-LINEARITIES

The desire to exploit fully the frequency mixing and conversion processes discussed in the last two sections has led to the search for materials with as large a non-linear response as possible. A large number of crystalline materials have therefore been studied and measurements made of the non-zero elements of their second-order susceptibility tensors $\underline{\chi}^{(2)}$, principally by studying second harmonic generation in a variety of ways. The techniques used are usefully reviewed in an article by Hulme [42] and values are comprehensively tabulated by Bechmann and Kurtz [43]. Measured values of components of $\underline{\chi}^{(2)}$ now range over three orders of magnitude. At the lower end of this range, we have a material like quartz (SiO_2) with $\chi^{(2)}_{xxx} = 2.4 \times 10^{-9}$ e.s.u.; an intermediate value is found in gallium arsenide with $\chi^{(2)}_{xyz} = 320 \times 10^{-9}$ e.s.u., and a maximum value of $\chi^{(2)}_{xxx} = 4400 \times 10^{-9}$ e.s.u. in tellurium.

A word of warning is appropriate here to point out that a variety of notations and definitions exist in the literature on second-order susceptibilities [43]. d is often used in place of χ and a contracted notation is frequently used for the last two subscripts, which automatically allows intrinsic symmetry (see Section 8 (ii)) to be taken into account. There is also some ambiguity about whether the susceptibilities are defined through relations between the coefficients of complex exponentials like $e^{-i\omega t}$, as we have used, or of real functions like $\cos(\omega t - \phi)$. All these differences lead to ambiguities in the values of components of the susceptibility quoted by different workers by factors of two. The definitions used in this paper lead to a difference of a factor of two between the values quoted here and those quoted in Refs [42] and [43].

Some rationalization of the wide range of values found for components of $\underline{\chi}^{(2)}$ was achieved by Miller [44]. He noticed that, in spite of the wide range covered, the quantity

$$\delta_{\mu\alpha\beta}(\omega_1, \omega_2) \equiv \frac{\chi^{(2)}_{\mu\alpha\beta}(\omega_1, \omega_2)}{\chi^{(1)}_{\mu\mu}(\omega_1 + \omega_2)\, \chi^{(1)}_{\alpha\alpha}(\omega_1)\, \chi^{(1)}_{\beta\beta}(\omega_2)}$$

was virtually constant over all materials. Indeed, over a range of 72 materials whose $\chi^{(2)}$ components, measured by second harmonic generation, are listed by Bechmann and Kurtz, they quote an average value for δ of 2.9×10^{-6} e.s.u., with a standard deviation from that average of only 1.8×10^{-6} e.s.u. Miller's rule thus provides a considerable empirical simplification of our understanding of $\vec{\chi}^{(2)}$ values. It says quite simply that a material will have a strong non-linear response if its linear response is likewise strong; since, in transparent region, $\chi^{(1)} \approx (n^2 - 1)/4\pi$ materials with strong non-linearities must have large refractive indices.

In spite of its empirical foundation, Miller's rule has been important in the search for good non-linear materials, as it gives a strong indication of the type of material in which a good non-linear response can be expected. Qualitatively it can be understood, insofar as a large refractive index arises from charge distributions in a material which are easily distorted by an applied field; one would expect such easily distorted charge distributions to be readily driven beyond the region of simple linear response. No general theoretical proof of the rule exists, although much work has been done which indicates some basis for it in theory; this is reviewed by Hulme [42] and Bloembergen [45], where more detailed references can be found. It is found that δ can be related to the dipole moment of the chemical bonds in a material, this moment being a measure of the asymmetry of the charge distribution in the bonds. This dipole moment does not vary greatly from one material to another and hence the near constancy of δ.

A strong non-linear response is clearly necessary for a material to be an efficient medium for optical frequency mixing and conversion. However, it is far from sufficient, and a good material must have a variety of other properties for us to be able to exploit its non-linear response.

First of all, it must allow phase matching to take place and usually this means that it must be birefringent enough to compensate for normal dispersion. This condition immediately forces us to reject the cubic III-V semiconductors like gallium arsenide, all of which have attractively large non-linear coefficients of the order of 10^{-7} e.s.u. Their cubic symmetry precludes any birefringence and it is difficult to induce sufficient birefringence by applied electric fields or uniaxial strains to achieve phase matching. Another technique is available to achieve phase matching in non-birefringent materials if the frequencies of interest, although lying in regions of transparency and consequently normal dispersion, can be arranged to straddle a region of anomalous dispersion. This has been made use of in metal vapours but has so far not been exploited in any solids or liquids.

Secondly, as I have just hinted, the material must be transparent at the frequencies of interest, otherwise any mixing process is overwhelmed by simultaneous linear absorption. The transparency of materials in the visible and infra-red region of the spectrum is limited at short wavelengths by the onset of electronic transitions and at long wavelengths by lattice or molecular vibrations. Wide transparent windows thus demand large energy-band gaps in materials with heavy atoms and consequently low vibrational frequencies. Good insulating materials like potassium or ammonium dihydrogen phosphate (KH_2PO_4 and $(NH_4)H_2PO_4$) are transparent between 0.2 μm and about 1.0 μm. Their transparent window can be extended to about 1.5 μm if their hydrogen is replaced by deuterium (KH_2PO_4 is referred to as KDP and KD_2PO_4 as KD*P). Oxide materials have long wavelength cut-offs at about 4.5 μm and to achieve transparency further into the infra-red, sulphur- or selenium-bearing materials are required.

TABLE III. MAIN SIGNIFICANT PROPERTIES OF KDP, $LiNbO_3$, BSN, $LiIO_3$ AND Ag_3AsS_3

Material	$\underline{\chi}^{(2)}$ (10^{-9} esu)	Phase-matching technique	Transparency range (μm)	Damage resistance	Crystal growth problems
KDP	$\chi^{(2)}_{xyz} = 3.0 \pm 20\%$	Angle	0.2 - 1.0 (KD*P : 0.2 - 1.5) [−]	Excellent [+]	Easy [+]
$LiNbO_3$	$\chi^{(2)}_{zzz} = 260 \pm 60\%$ $\chi^{(2)}_{yyy} = 20 \pm 30\%$ $\chi^{(2)}_{zxx} = 36 \pm 30\%$	Angle [+] Temperature	0.4 - 4.5	Problems [−]	Difficult, but now solved.
BSN	$\chi^{(2)}_{zzz} = 100 \pm 30\%$ $\chi^{(2)}_{zxx} = 90 \pm 30\%$	Angle	0.4 - 4.0	Good [+]	Very difficult [−]
$LiIO_3$	$\chi^{(2)}_{zzz} = 28 \pm 65\%$ $\chi^{(2)}_{zxx} = 34 \pm 35\%$ $\chi^{(2)}_{zyy} = ?$	Angle	0.3 - 5.5	Good [+]	Not too difficult
Ag_3AsS_3	$\chi^{(2)}_{zzz} = ?$ $\chi^{(2)}_{yyy} = 120 \pm ?$ $\chi^{(2)}_{zxx} = 80 \pm ?$	Angle	0.6 - 13 [+]	Not very good [−]	Difficult, but now solved

Plus and minus signs indicate definite advantages and disadvantages, respectively.

Thirdly, since the efficiency of mixing increases with pumping laser power, high powers tend to be used and so the material must be resistant to damage from this. Damage can result from a variety of factors and can originate either on the surface or in the bulk of the material.

Finally, it must be possible to grow the crystalline material in large enough volumes to obtain samples of typically centimetre dimensions, of high optical quality and homogeneity and oriented in particular crystalline directions. Naturally occurring samples are rarely of high enough optical quality and one must depend on laboratory produced material. The main problems are associated with the relatively large uniform volumes of material usually required. Uniformity is, of course, required to maintain control over propagation directions and phase matching.

Hulme [42] gives a good review of materials of interest in visible and infra-red frequency mixing. From the several hundred materials which have now been studied, he singles out as the most important: potassium dihydrogen phosphate (KH_2PO_4), known as KDP, and its isomorphs; lithium niobate ($LiNbO_3$); barium sodium niobate ($Ba_2NaN_5O_{15}$), known as BSN; lithium iodate ($LiIO_3$); and silver thioarsenate (Ag_3AsS_3), known from its naturally occurring form as proustite. The main significant properties of these materials are given in Table III; they are also associated with their appropriate crystal class in Table I. Kleinman symmetry is fairly well satisfied in each material which explains the reduced number of components of $\underline{\chi}^{(2)}$ quoted. A more detailed discussion of each material is given Ref.[24], and I have tried to summarize this in Table III in qualitative fashion and, in addition, indicate definite advantages and disadvantages of each material by plus and minus signs.

14. PROCESSES INVOLVING ENERGY EXCHANGE WITH THE MEDIUM

So far, we have been concerned with non-linear effects which involve no exchange of energy between the radiation field and the medium. Furthermore, we have discussed only effects of this type which arise from the second-order polarizability $\underline{\chi}^{(2)}$. Such effects are present, of course, only in a medium which lacks inversion symmetry, for otherwise $\underline{\chi}^{(2)}$ is zero. However, similar effects can also arise out of the third-order polarizability $\underline{\chi}^{(3)}$ which always possesses some non-zero components no matter what symmetry the medium may have. These similar effects consist of third harmonic generation and generally any process that depends on the coupling of four components of the radiation field with frequencies ω_1, ω_2, ω_3 and $\omega_4 = \omega_1 + \omega_2 + \omega_3$. Although slightly more complicated, these processes can be described in a way very similar to that which we used for processes involving the coupling of three waves [11, 46]. We therefore do not discuss them in detail here. What we shall do is to discuss some effects which depend on the exchange of energy between the radiation field and the medium.

The simplest of these effects is well known and arises from the linear response of the medium. It is the process through which radiant energy is normally considered to be absorbed by a system. Absorption takes place whenever any frequency ω of the radiation field is close enough to one of the transition frequencies ω_0 of the medium considered. Formally, the effect arises from the existence of an imaginary part of $\underline{\chi}^{(1)}(\omega)$, the strength of the absorption being proportional to the magnitude of this imaginary part. The absorption is also proportional to the difference in population of the lower and upper energy states of the medium between which a transition is being induced by the radiation. In normal circumstances this population difference is positive and absorption takes place; however, if by some means this difference can be made negative, by an inversion of populations, amplification instead of absorption of the radiation takes place.

When non-linearities in the response of a medium are significant, energy can be exchanged between the radiation field and the medium by more complicated processes. We shall consider the simplest of these processes, which arises when the frequencies of two components of the radiation field are such that their sum or difference is close to a transition frequency of the medium. As in the linear case, we shall find that under normal conditions, i.e. when no population inversion exists, energy is always absorbed by the medium from the radiation field. However, the significant point which will emerge from our discussions is that this overall absorption of energy from the radiation field need not imply an attenuation of both components of the field which contribute to the process. It is quite possible to have the low-frequency component amplified simultaneously with an attenuation of the high-frequency component. When this occurs, the process is known as the stimulated Raman effect, for reasons which will be seen later, and gives rise to what has been called a Raman laser [47]. This Raman laser operates under normal equilibrium conditions and does not require any population inversion. It is therefore fundamentally different from the conventional laser.

Raman laser action takes place when the difference of the two frequencies involved is close to a transition frequency of the medium. The process which takes place when the sum of the two frequencies is close to a transition frequency is perhaps simpler to understand. It is referred to as two-photon absorption and we consider it first.

15. TWO-PHOTON ABSORPTION

We focus attention on two components of the radiation field with frequencies ω_1 and ω_2. In a medium with inversion symmetry, the lowest order coupling which can exist between these components takes place through the third-order susceptibility $\underline{\chi}^{(3)}$. In a medium which lacks this symmetry, a lower order coupling through the second-order susceptibility $\underline{\chi}^{(2)}$ also exists. However, this lower order coupling, as we have seen, always involves a component at a third frequency $\omega_1 + \omega_2$ or $|\omega_1 - \omega_2|$ and is of significance only when the appropriate phase-matching condition is satisfied. We can discriminate against this lower order coupling, when it occurs, by assuming that the more usual state of affairs exists in which neither of these phase-matching conditions is satisfied. Similarly we can assume that none of the phase-matching conditions required for efficient second or third harmonic generation is satisfied. We need not take into account therefore any components of the radiation field other than those at frequencies ω_1 and ω_2. Furthermore we need only consider the components $\vec{P}_{NL}(\omega_1)$ and $\vec{P}_{NL}(\omega_2)$ of the non-linear polarization field at frequencies ω_1 and ω_2 which arise from the radiation field components at these same frequencies. It follows from (7.2), (10.3) and (10.11) that these are

$$\vec{P}_{NL}(\omega_1) = 6\underline{\chi}^{(3)}(\omega_2, -\omega_2, \omega_1) \vdots \vec{E}(\omega_2)\vec{E}^*(\omega_2)\vec{E}(\omega_1) \qquad (15.1)$$

and

$$\vec{P}_{NL}(\omega_2) = 6\underline{\chi}^{(3)}(\omega_1,-\omega_1,\omega_2) \vdots \vec{E}(\omega_1)\vec{E}^*(\omega_1)\vec{E}(\omega_2) \quad (15.2)$$

The factor 6 in these expressions is a result of the intrinsic symmetry of $\underline{\chi}^{(3)}$ and represents the 3! ways in which the three different frequencies occurring in $\underline{\chi}^{(3)}$ can be arranged. Using (10.13) and neglecting the small polarization correction terms in $\vec{b}_{\omega_1}(z)$ and $\vec{b}_{\omega_2}(z)$, (15.1) and (15.2) give

$$\vec{P}'_{NL}(\omega_1) = 6\underline{\chi}^{(3)}(\omega_2,-\omega_2,\omega_1) \vdots \vec{a}_{\omega_2}\vec{a}_{\omega_2}\vec{a}_{\omega_1} |E_{\omega_2}|^2 E_{\omega_1} e^{ik_1 z} \quad (15.3)$$

and

$$\vec{P}'_{NL}(\omega_2) = 6\underline{\chi}^{(3)}(\omega_1,-\omega_1,\omega_2) \vdots \vec{a}_{\omega_1}\vec{a}_{\omega_1}\vec{a}_{\omega_2} |E_{\omega_1}|^2 E_{\omega_2} e^{ik_2 z} \quad (15.4)$$

Using these expressions in (10.17), we obtain the two coupled equations:

$$\frac{dE_{\omega_1}}{dz} = \frac{12\pi i\omega_1^2}{k_1c^2}\underline{\chi}^{(3)}(\omega_2,-\omega_2,\omega_1) \vdots \vec{a}_{\omega_1}\vec{a}_{\omega_2}\vec{a}_{\omega_2}\vec{a}_{\omega_1} |E_{\omega_2}|^2 E_{\omega_1} \quad (15.5)$$

$$\frac{dE_{\omega_2}}{dz} = \frac{12\pi i\omega_2^2}{k_2c^2}\underline{\chi}^{(3)}(\omega_1,-\omega_1,\omega_2) \vdots \vec{a}_{\omega_2}\vec{a}_{\omega_1}\vec{a}_{\omega_1}\vec{a}_{\omega_2} |E_{\omega_1}|^2 E_{\omega_2} \quad (15.6)$$

where the dot products are defined by an obvious generalization of (11.8).

The first thing to notice about these equations, compared to our previous equations (11.5), (11.6) and (12.3) of this type, is that no exponential factor appears in (15.5) or (15.6) which depends on the propagation constants k_1 and k_2 of the two waves. There is therefore no question in this case of considering various degrees of phase mismatching. This is basically because the polarizations given by (15.3) and (15.4) have the same exponential dependence on kz as the components of the field which oscillate at the same frequency. The two parts of $\vec{P}'_{NL}(\omega_1)$ and $\vec{P}'_{NL}(\omega_2)$ which depend on the real and the imaginary parts of $\vec{\chi}^{(3)}$ are therefore always respectively in-phase and 90° out-of-phase with the corresponding electric field components $\vec{E}(\omega_1)$ and $\vec{E}(\omega_2)$.

Since we are interested in the situation where $\omega_1 + \omega_2$ is close to a transition or resonance frequency of the medium, the polarizabilities $\underline{\chi}^{(3)}(\omega_2,-\omega_2,\omega_1)$ and $\underline{\chi}^{(3)}(\omega_1,-\omega_1,\omega_2)$ appearing in (15.5) and (15.6) have not only real but imaginary parts which are finite and must be taken into account in our equations. The real parts of these polarizabilities possess the overall symmetry discussed earlier, so that

$$\mathrm{Re}[\underline{\chi}^{(3)}(\omega_2,-\omega_2,\omega_1) \vdots \vec{a}_{\omega_1}\vec{a}_{\omega_2}\vec{a}_{\omega_2}\vec{a}_{\omega_1}]$$

$$= \mathrm{Re}[\underline{\chi}^{(3)}(\omega_1,-\omega_1,\omega_2) \vdots \vec{a}_{\omega_2}\vec{a}_{\omega_1}\vec{a}_{\omega_1}\vec{a}_{\omega_2}] \equiv \chi \quad (15.7)$$

To find out something about their imaginary parts, we consider the case of the simple classical model used as a basis for evaluating the polarizabilities. It follows from (6.19) that, when $\omega_1 + \omega_2 \sim \omega_0$, i.e. $|\omega_1 + \omega_2 - \omega_0| \lesssim \gamma$,

$$\operatorname{Im} \chi^{(3)}(\omega_2, -\omega_2, \omega_1) = \frac{4\pi N e^4 \lambda^2}{3m^3} F^2(\omega_1) F^2(\omega_2) \operatorname{Im} F(\omega_1 + \omega_2)$$

$$= \operatorname{Im} \chi^{(3)}(\omega_1, -\omega_1, \omega_2) \tag{15.8}$$

Not only are these two imaginary parts equal but they are seen to have the same sign as N, for $F(\omega_1)$ and $F(\omega_2)$ are both real and $\operatorname{Im} F(\omega_1 + \omega_2) > 0$, a fact which follows from (8.2). This result is true in general when it is found that $\operatorname{Im} \underline{\chi}^{(3)}(\omega_2, -\omega_2, \omega_1)$ and $\operatorname{Im} \underline{\chi}^{(3)}(\omega_1, -\omega_1, \omega_2)$ are equal and have the sign of the difference in populations between the lower and upper energy levels whose transition frequency is close to $\omega_1 + \omega_2$. This population difference is the parameter in the real case to which the quantity N of our model is equivalent. We can now define

$$\chi_{TA} = \operatorname{Im} [\underline{\chi}^{(3)}(\omega_2, -\omega_2, \omega_1) \,\vdots\, \vec{a}_{\omega_1} \vec{a}_{\omega_2} \vec{a}_{\omega_2} \vec{a}_{\omega_1}]$$

$$= \operatorname{Im} [\underline{\chi}^{(3)}(\omega_1, -\omega_1, \omega_2) \,\vdots\, \vec{a}_{\omega_2} \vec{a}_{\omega_1} \vec{a}_{\omega_1} \vec{a}_{\omega_2}] \tag{15.9}$$

using a notation which arises from the fact that we shall find that χ_{TA} controls the magnitude of the two-photon absorption process. We note that χ_{TA} will be positive under normal equilibrium conditions but will become negative under conditions of population inversion.

With (15.7) and (15.9), our two equations (15.5) and (15.6) can now be written simply as

$$\frac{dE_{\omega_1}}{dz} = \frac{12\pi\omega_1^2}{k_1 c^2} (i\chi - \chi_{TA}) |E_{\omega_2}|^2 E_{\omega_1} \tag{15.10}$$

$$\frac{dE_{\omega_2}}{dz} = \frac{12\pi\omega_2^2}{k_2 c^2} (i\chi - \chi_{TA}) |E_{\omega_1}|^2 E_{\omega_2} \tag{15.11}$$

It follows then that

$$\frac{k_1}{\omega_1^2} E^*_{\omega_1} \frac{dE_{\omega_1}}{dz} - \frac{k_2}{\omega_2^2} E^*_{\omega_2} \frac{dE_{\omega_2}}{dz} = 0 \tag{15.12}$$

If this expression is added to its complex conjugate, the result can be integrated to yield the result that

$$\frac{k_1}{\omega_1^2} |E_{\omega_1}|^2 - \frac{k_2}{\omega_2^2} |E_{\omega_2}|^2 = \text{constant} \tag{15.13}$$

When written in terms of the mean photon fluxes N_{ω_1} and N_{ω_2}, which, as we have seen, are, respectively, proportional to

$$\frac{k_1}{\omega_1^2}|E_{\omega_1}|^2 \quad \text{and} \quad \frac{k_2}{\omega_1^2}|E_{\omega_2}|^2$$

this result takes the form

$$N_{\omega_1} - N_{\omega_2} = \text{constant} = N_{\omega_1}(0) - N_{\omega_2}(0) \tag{15.14}$$

where $N_{\omega_1}(0)$ and $N_{\omega_2}(0)$ are the initial values of N_{ω_1} and N_{ω_2} at $z = 0$. This is seen to be a relationship of the Manley-Rowe type (see Eqs (12.15) - (12.17)) and shows that any gain or loss of photons by the radiation field at frequency ω_1 must be accompanied by an exactly equal gain or loss of photons at frequency ω_2. Thus amplification or attenuation of either component of the field, through the coupling we are considering here, can take place only with the simultaneous amplification or attenuation of the other component.

The general solution of Eqs (15.10) and (15.11) is not difficult to obtain. However, some of its characteristics are readily seen from the equations and it is instructive to discuss it first of all from this point of view. The equations show that both E_{ω_1} and E_{ω_2} vary at rates which are, among other things, proportional to their own magnitudes. An exponential type of solution is therefore to be expected. In fact, if for the moment we neglect the effect of the variation of each field on the other, we can treat $|E_{\omega_2}|^2$ in (15.10) and $|E_{\omega_1}|^2$ in (15.11) as constants. Then we see that

$$E_{\omega_1} \sim \exp\left[\frac{12\pi\omega_1^2}{k_1 c^2}|E_{\omega_2}|^2 (i\chi - \chi_{TA})z\right] \tag{15.15}$$

and

$$E_{\omega_2} \sim \exp\left[\frac{12\pi\omega_2^2}{k_2 c^2}|E_{\omega_1}|^2 (i\chi - \chi_{TA})z\right] \tag{15.16}$$

These expressions show that under normal conditions when χ_{TA} is positive both E_{ω_1} and E_{ω_2} will decrease with increasing z at a rate proportional to the strength of the other field and to χ_{TA}. The imaginary part of the exponent shows that the non-linear coupling between the waves also results in a modification to the propagation constant of each wave proportional to χ, i.e. to the real part of $\underline{\chi}^{(3)}$ and to the intensity in the other wave. Far from resonance, only χ will remain non-zero and the waves will no longer be attenuated but will simply propagate through the medium with slightly modified propagation constants. This modification to the propagation constants of each wave resulting from the presence of the other is detectable as a change in the effective refractive index of the medium. The change is proportional to the square of the amplitude of the wave causing the change. It is therefore the generalization to waves of finite frequency of the quadratic or Kerr electro-optic effect (see Section 16) which is the zero frequency limit of the effect we have here.

The attenuation of the two waves which results from the presence of the term in χ_{TA} near a resonance cannot continue indefinitely for both waves. It should cease when the weaker of the two waves is exhausted and this is precisely what the Manley-Rowe relation (15.14) tells us will happen. The general solution of our equations, which is best expressed in terms of the fluxes N_{ω_1} and N_{ω_2}, also bears this out.

Equations (15.10) and (15.11) are respectively multiplied by $E^*_{\omega_1}$ and $E^*_{\omega_2}$, added to their complex conjugates, and the general relation

$$N_\omega = \frac{c^2}{2\pi\hbar}\,\frac{k}{\omega^2}\,|E_\omega|^2 \tag{15.17}$$

used, to yield

$$\frac{dN_{\omega_1}}{dz} = \frac{dN_{\omega_2}}{dz} = -\alpha_{TA} N_{\omega_1} N_{\omega_2} \tag{15.18}$$

where

$$\alpha_{TA} = \frac{48\pi^2\hbar}{c^4}\,\frac{\omega_1^2\omega_2^2}{k_1 k_2}\,\chi_{TA} \tag{15.19}$$

(15.18) are readily integrated using (15.14) to give

$$N_{\omega_1} = N_{\omega_1}(0)\,\frac{N_{\omega_1}(0)-N_{\omega_2}(0)}{N_{\omega_1}(0)-N_{\omega_2}(0)\,e^{-z/\ell_{TA}}} \tag{15.20}$$

$$N_{\omega_2} = N_{\omega_1} - N_{\omega_1}(0) + N_{\omega_2}(0) \tag{15.21}$$

where ℓ_{TA} is a length which is characteristic of the two-photon absorption process and is defined by

$$\ell^{-1}_{TA} = \alpha_{TA}[N_{\omega_1}(0)-N_{\omega_2}(0)] = \frac{6\pi\omega_1^2\omega_2^2}{k_1 k_2 c^2}\left[\frac{k_1}{\omega_1^2}\,\mathscr{E}^2_{\omega_1}(0) - \frac{k_2}{\omega_2^2}\,\mathscr{E}^2_{\omega_2}(0)\right]\chi_{TA} \tag{15.22}$$

If we adopt the convention of labelling the frequency of the weaker of the two waves by ω_2, then we see that ℓ_{TA} is normally positive so that, for large values of z, $N_{\omega_1} \to N_{\omega_1}(0) - N_{\omega_2}(0)$ and $N_{\omega_2} \to 0$. The spatial dependence of N_{ω_1} and N_{ω_2} as given by (15.20) and (15.21) is illustrated in Fig.16.

The origin of the name — two-photon absorption — given to this process is fairly obvious from our results. The waves at frequencies ω_1 and ω_2 are attenuated by the non-linear medium when their sum approximates to a transition frequency of the medium. Neither wave is attenuated on its own, but when both are present the attenuation arises from a simultaneous absorption of two photons, one from each wave, whose total energy is just sufficient to cause an excitation of the medium. The energy lost by the two waves thus appears as excitation energy in the medium.

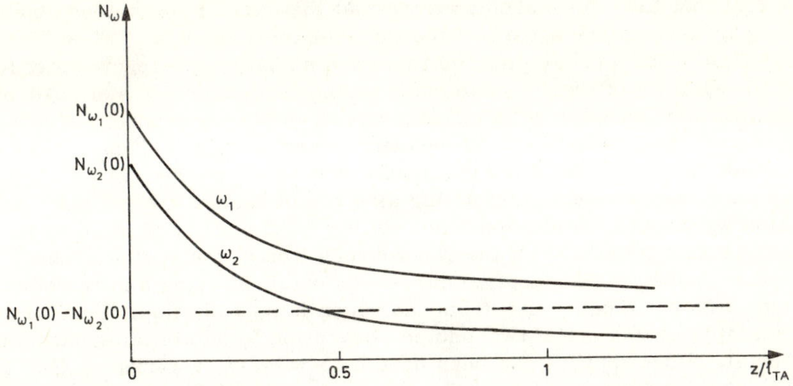

FIG.16. Attenuation due to the two-photon absorption process of two waves with frequencies ω_1 and ω_2 as they propagate through a non-linear medium with a transition frequency $\omega_0 \sim \omega_1 + \omega_2$.

Two-photon absorption was first observed by Kaiser and Garrett [48], who found evidence, from the detection of fluorescence, for the excitation by ruby laser radiation of a level of Eu^{++} in CaF_2 whose transition frequency from the ground state was of the order of twice the ruby laser frequency. The process of successive absorption of two ruby laser photons cannot explain this effect, for such a process is not possible in this system since Eu^{++} has no levels close enough to the ground state for a single ruby laser photon to produce any excitation. Furthermore, the inversion symmetry of the CaF_2 crystal prohibits the generation of the second harmonic of the laser frequency in the crystal so that the excitation cannot arise from the absorption of a previously generated second harmonic photon. The only explanation of the effect is that the process of two-photon absorption is taking place.

This experiment is a special case of our general treatment, obtained from it by putting $\omega_1 = \omega_2$. Under these conditions $N_{\omega_1}(0) = N_{\omega_2}(0)$, and our general solution apparently behaves in a peculiar way, for ℓ_{TA} becomes zero. However, it is not difficult to show by taking the limit of $\omega_2 \to \omega_1$ that (15.20) becomes

$$N_{\omega_1} = \frac{N_{\omega_1}(0)}{1 + N_{\omega_1}(0)\alpha_{TA} z} \tag{15.23}$$

where α_{TA}, defined by (15.19), is seen to be independent of $N_{\omega_1}(0)$. For small values of $\alpha_{TA}z$, (15.23) becomes

$$N_{\omega_1} = N_{\omega_1}(0)[1 - N_{\omega_1}(0)\alpha_{TA} z] \tag{15.24}$$

and the number of photons absorbed in a depth z of the medium is seen to be proportional to $N^2_{\omega_1}(0)$, i.e. to the square of the incident intensity. The magnitude of the effect observed by Kaiser and Garrett, viz. the order of one in 10^7 laser photons absorbed by this process, indicates that the small α_{TA} z condition is satisfied in their experimental arrangement. This is confirmed by the fact that they did indeed observe that the number of photons absorbed was proportional to the square of the laser beam intensity.

Hopfield et al. [49] studied the two-photon absorption spectrum of KI and CsI in the vicinity of their fundamental absorption edges. These absorption edges in KI and CsI were not well understood and the two-photon spectrum was measured to obtain additional information to that available from the normal one-photon spectrum about the electronic processes producing the absorption. A ruby laser and a xenon lamp acting as a broad band source of ultraviolet radiation were used. Radiation from neither is directly absorbed by KI or CsI but the sum frequency of the two sources lies in the region of the absorption edges. An absorption curve was obtained using a spectrometer to pick out different frequencies from the xenon lamp radiation. This was the first example of the use of two-photon absorption to obtain information about a material. Since then, it has been used by a variety of workers. The technique is well reviewed by Gold [50].

16. STIMULATED RAMAN EFFECT

We consider the same set of circumstances in this section as in the last, except that we now assume that the difference, instead of the sum, of the frequencies of two components of the radiation field is close to a resonance frequency of the medium. As before, all phase-matching conditions for sum, difference and harmonic frequency generation are taken to be badly satisfied. The previous discussion applies just as well to this situation up to the point where Eqs (15.5) and (15.6) were derived. These equations are still valid but it is convenient to label our two frequencies ω_1 and $\omega_3(\omega_3 > \omega_1)$ now for a reason which will emerge later, so that these equations become

$$\frac{dE_{\omega_1}}{dz} = \frac{12\pi i\omega_1^2}{k_1c^2}\underline{\chi}^{(3)}(\omega_3,-\omega_3,\omega_1)\vdots\vec{a}_{\omega_1}\vec{a}_{\omega_3}\vec{a}_{\omega_3}\vec{a}_{\omega_1}|E_{\omega_3}|^2 E_{\omega_1} \tag{16.1}$$

$$\frac{dE_{\omega_3}}{dz} = \frac{12\pi i\omega_3^2}{k_3c^2}\underline{\chi}^{(3)}(\omega_1,-\omega_1,\omega_3)\vdots\vec{a}_{\omega_3}\vec{a}_{\omega_1}\vec{a}_{\omega_1}\vec{a}_{\omega_3}|E_{\omega_1}|^2 E_{\omega_3} \tag{16.2}$$

Again, as before, we notice that no phase-matching condition controls the solution of these equations.

The polarizabilities $\underline{\chi}^{(3)}(\omega_3,-\omega_3,\omega_1)$ and $\underline{\chi}^{(3)}(\omega_1,-\omega_1,\omega_3)$ are again complex, this time for the reason that $\omega_3-\omega_1$ is close to a resonance frequency. Their real parts have overall symmetry and in fact are identical to the real parts of the polarizabilities appearing in (15.5) and (15.6) when ω_2 is replaced by ω_3. We again investigate the properties of the imaginary parts of the polarizabilities by appealing to the result (6.19) we obtained from our simple classical model. We see that, when $|\omega_3-\omega_1-\omega_0| \lesssim \gamma$,

$$\begin{aligned}\operatorname{Im}\chi^{(3)}(\omega_1,-\omega_1,\omega_3) &= \frac{4Ne^4\lambda^2}{3m^3}F^2(\omega_1)F^2(\omega_3)\operatorname{Im}F^2(\omega_3-\omega_1)\\ &= -\frac{4Ne^4\lambda^2}{3m^3}F^2(\omega_1)F^2(\omega_3)\operatorname{Im}F(\omega_1-\omega_3)\\ &= -\operatorname{Im}\chi^{(3)}(\omega_3,-\omega_3,\omega_1)\end{aligned} \tag{16.3}$$

which follows using (8.2). We also see from (8.2) that, since $\omega_3-\omega_1>0$, $\mathrm{Im}\,F(\omega_3-\omega_1)>0$. It follows that $\mathrm{Im}\,\chi^{(3)}(\omega_1,-\omega_1,\omega_3)$ has the same sign as N and the opposite sign to $\mathrm{Im}\,\chi^{(3)}(\omega_3,-\omega_3,\omega_1)$. The generalization of this result to a real system preserves the difference in sign of the imaginary parts of these two polarizabilities and the sign of $\mathrm{Im}\,\underline{\chi}^{(3)}(\omega_1,-\omega_1,\omega_3)$ is given by the sign of the difference in populations between the lower and upper energy levels whose transition frequency is close to $\omega_3-\omega_1$. We therefore define

$$\chi_R = \mathrm{Im}\,[\underline{\chi}^{(3)}(\omega_1,-\omega_1,\omega_3)\,\vdots\,\vec{a}_{\omega_3}\,\vec{a}_{\omega_1}\,\vec{a}_{\omega_1}\,\vec{a}_{\omega_3}]$$

$$= -\mathrm{Im}\,[\underline{\chi}^{(3)}(\omega_3,-\omega_3,\omega_1)\,\vdots\,\vec{a}_{\omega_1}\,\vec{a}_{\omega_3}\,\vec{a}_{\omega_3}\,\vec{a}_{\omega_1}] \qquad (16.4)$$

since we shall find that it is this quantity χ_R which controls the Raman effect. It will be positive under normal equilibrium conditions but will become negative under conditions of population inversion.

Our Eqs (16.1) and (16.2) now take the simplified form:

$$\frac{dE_{\omega_1}}{dz} = \frac{12\pi\omega_1^2}{k_1c^2}(i\chi+\chi_R)|E_{\omega_3}|^2\,E_{\omega_1} \qquad (16.5)$$

$$\frac{dE_{\omega_3}}{dz} = \frac{12\pi\omega_3^2}{k_3c^2}(i\chi-\chi_R)|E_{\omega_1}|^2\,E_{\omega_3} \qquad (16.6)$$

They are therefore very similar to Eqs (15.10) and (15.11) of the two-photon absorption case but contain an important and significant change of sign in the first equation. This alternative sign has the effect of changing the sign which appears in the expression analogous to (15.12), viz.,

$$\frac{k_1}{\omega_1^2}E^*_{\omega_1}\frac{dE_{\omega_1}}{dz} + \frac{k_3}{\omega_3^2}E_{\omega_3}\frac{dE^*_{\omega_3}}{dz} = 0 \qquad (16.7)$$

By a similar argument to that used before, this leads to the result that

$$N_{\omega_1} + N_{\omega_3} = \text{constant} = N_{\omega_1}(0) + N_{\omega_3}(0) \qquad (16.8)$$

This is again a relationship of the Manley-Rowe type but in this case it shows that any gain or loss of photons by the radiation field at frequency ω_1 must be accompanied by an exactly equal loss or gain of photons at frequency ω_3. We see immediately therefore that, of the two components at frequencies ω_1 and ω_3, one must be amplified and one attenuated as a result of the non-linear coupling. This contrasts with the previous case when (15.4) showed that both components must either be amplified or attenuated together.

The general nature of the solution of (16.5) and (16.6) can again be seen from the equations. The imaginary terms in χ again lead to a modification of the propagation constants of each wave due to the presence of the other.

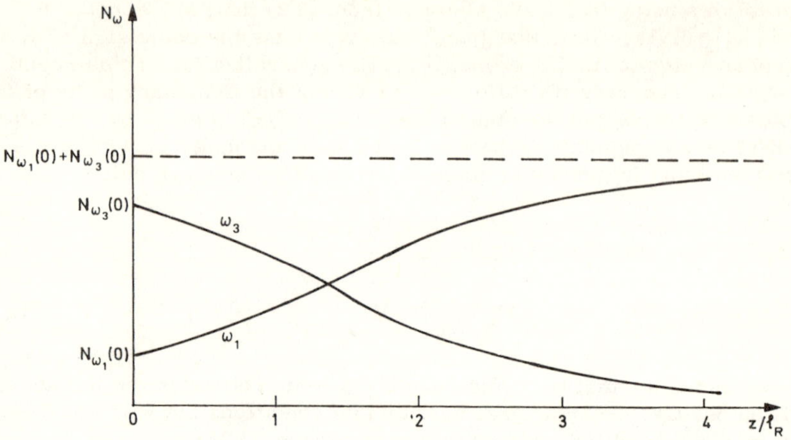

FIG. 17. Respective attenuation and amplification due to the stimulated Raman effect of two waves with frequencies ω_1 and ω_3 as they propagate through a non-linear medium with a transition frequency $\omega_0 \sim \omega_3 - \omega_1$.

The real terms in χ_R lead, under normal conditions when χ_R is positive, to an exponential type of growth of the component with the lower frequency ω_1 and a similar type of decay of the high-frequency component at ω_3.

The general solution is again most readily expressed in terms of the mean photon fluxes carried by the two waves. Using the techniques of the previous section, it is found to be

$$N_{\omega_1} = N_{\omega_1}(0)\,\frac{N_{\omega_1}(0)+N_{\omega_3}(0)}{N_{\omega_1}(0)+N_{\omega_3}(0)\,e^{-z/\ell_R}} \tag{16.9}$$

$$N_{\omega_3} = N_{\omega_1}(0)+N_{\omega_3}(0) - N_{\omega_1} \tag{16.10}$$

where ℓ_R is a length which characterizes the Raman process and is defined by

$$\ell_R^{-1} = \frac{6\pi\omega_1^2\,\omega_3^2}{k_1 k_3 c^2}\left[\frac{k_1}{\omega_1^2}\,\mathscr{E}^2_{\omega_1}(0) + \frac{k_3}{\omega_3^2}\,\mathscr{E}^2_{\omega_3}(0)\right]\chi_R \tag{16.11}$$

This general solution which is illustrated in Fig.17, confirms the exponential dependence of N_{ω_1} and N_{ω_3} on z. It also shows that, for small values of z, N_{ω_1} increases linearly and N_{ω_3} decreases linearly with z under normal conditions when χ_R is positive according to the relations:

$$N_{\omega_1} = N_{\omega_1}(0)\left[1 + \frac{N_{\omega_3}(0)}{N_{\omega_1}(0)+N_{\omega_3}(0)}\,\frac{z}{\ell_R}\right] \tag{16.12}$$

$$N_{\omega_3} = N_{\omega_3}(0)\left[1 - \frac{N_{\omega_1}(0)}{N_{\omega_1}(0)+N_{\omega_3}(0)}\,\frac{z}{\ell_R}\right] \tag{16.13}$$

As z increases, N_{ω_1} eventually saturates at a maximum value of $N_{\omega_1}(0)+N_{\omega_3}(0)$ and at the same time N_{ω_3} decreases to zero. Thus we see that it is possible by this process to extract all the power from the higher frequency component of the field. Part of this power reappears at the lower frequency ω_1, i.e. a fraction ω_1/ω_3 of it, and serves to amplify this component. The remainder of the power is lost to the radiation field and serves to excite the non-linear medium which provides the coupling. Although there is a net loss of energy by the radiation field, the lower frequency component of the field is seen to be amplified. If a condition of population inversion existed, there would be a net gain of energy by the field; in this case, the high frequency component would be amplified and the low frequency component attenuated.

The name of stimulated Raman effect which is given to this process arises in the following way. The Raman effect or Raman scattering is well known. It is the effect whereby radiation is scattered with a change of frequency by a medium. The change in frequency of the radiation is equal to a transition frequency of the medium; when the frequency change is downwards, the energy lost by the radiation is used to excite the medium upwards across the appropriate transition, and, when the change is upwards, the energy gained by the radiation comes from the appropriate de-excitation of the medium. The Raman effect as usually observed is associated with the excitation or de-excitation of a vibrational mode of the medium but in general any possible excitation of the medium with an appropriate transition frequency can give rise to the effect. Quantum mechanically, the process proceeds in two steps. A photon of incident radiation is absorbed by the medium which simultaneously makes a transition by which energy is not generally conserved — a virtual transition. This is immediately followed by a second transition of the medium with the simultaneous emission of a photon of scattered radiation. This second transition of the medium is to a final state which permits an overall conservation of energy in the process between the field and the medium. Alternatively to this, a photon of scattered radiation can first be emitted and be followed by the absorption of a photon of incident radiation. The point to notice here is that an emission process takes place and such a process can either take place spontaneously or be induced or stimulated to take place by the presence of radiation of exactly the same type as is emitted. In the Raman effect it is the spontaneous process which completely dominates, for no effort is made to irradiate the medium with radiation at the scattered frequency. We could therefore appropriately rename what is usually known as the Raman effect as the spontaneous Raman effect.

The process we have been discussing in this section is clearly very similar to the (spontaneous) Raman effect. Radiation at frequency ω_3 is incident on a medium. Some of it disappears and reappears as scattered radiation at frequency ω_1 which is lower than ω_3 by a transition frequency of the medium. However, we see from Eq.(16.5) that the rate at which radiation at frequency ω_1 builds up is proportional to the amount of radiation present at that same frequency. We are therefore discussing the process in which the medium is induced or stimulated to emit radiation at ω_1. For obvious reasons, this is called the stimulated Raman effect. The fact that only the stimulated and not the spontaneous effect emerges from our equations is a result of our classical treatment of the radiation field.

It might seem from this discussion that we have been wasting our time calculating the characteristics of a process which turns out to be essentially the stimulated Raman effect. For we pointed out that this effect can usually be regarded as quite negligible compared with the spontaneous effect. However, this is not always true for, by using an intense laser beam to provide the incident radiation at ω_3, enough spontaneous Raman scattering can take place to provide intense enough scattered radiation to stimulate the proces still further. In other words, the induced effect can build up from noise. The effect can be still further enhanced by confining the radiation inside a high-Q cavity. The radiation can then be thought of as continually passing and repassing through the medium in a variety of directions. Since no phase-matching condition controls the stimulated Raman effect, as pointed out earlier, the low-frequency radiation is amplified on each pass so that an appreciable intensity can eventually be built up. This is sometimes called Raman laser action.

It will be realized from the discussion that there is a certain similarity between the stimulated Raman effect and the parametric amplification effect, discussed in Section 12. In both cases, a low-frequency component of the radiation field at ω_1 is amplified at the expense of a higher frequency component at ω_3. The exchange of energy between the components in both cases is photon by photon so that there is a net loss of energy of $\hbar(\omega_3-\omega_1)$ per photon exchange from the two components of the field at ω_1 and ω_3. In the Raman effect, this energy corresponds to an energy level separation in the non-linear medium; the energy lost by the radiation is used to excite a transition in the medium between these two levels. In the parametric effect, this energy reappears in radiant form at the difference frequency $\omega_2 = \omega_3 - \omega_1$. We have discriminated against the parametric process in the present discussion by considering either a medium with inversion symmetry in which it cannot take place or by assuming that the process is badly phase-matched if it is not prohibited by spatial symmetry. Regarding the problem from the other point of view, a complicated state of affairs exists if we consider parametric amplification to be taking place close to its phase-matching condition when the difference or idler frequency ω_2 is near to a resonance of the non-linear medium. A contribution to the amplification of the signal wave at ω_1 then arises not only from the parametric effect which depends on $\underline{\chi}^{(2)}$ but also on the stimulated Raman effect which depends on $\underline{\chi}^{(3)}$. The way in which these two contributions combine was first analysed by Butcher, Loudon and McLean [51] and later more correctly by Henry and Garrett [52]. The results are by no means straightforward and they will not be discussed further here. An analogous situation arises when the sum of two frequencies is close to a resonant absorption frequency in a material. The two processes of sum frequency generation and two photon absorption, normally considered as distinct, become closely related. This problem has been treated by Boggett and Loudon [53].

The first observation of the stimulated Raman effect was made by accident by Woodbury and Ng [54] in the output beam of a Q-spoiled ruby laser in which the Q-spoiling was done by means of a Kerr cell filled with nitrobenzene. As well as radiation at the ruby laser frequency being present in the beam, radiation only about 20% less intense was found at a lower frequency. This frequency was lower than that of the ruby laser radiation by an amount which agreed well with a transition frequency of nitrobenzene which was known to give rise to a strong spontaneous Raman effect. The explanation of the result was made in terms of the laser radiation being Raman-scattered by the nitrobenzene; some of the scattered radiation is trapped by the laser cavity and so continues to pass and repass through the Kerr cell along with the laser radiation. On each pass, more Raman scattering is stimulated and a large build-up in power of the scattered radiation travelling along the axis of the laser cavity is achieved. Subsequent to this work, many other organic liquids e.g. benzene, cyclohexane, acetone, etc., and liquid gases like H_2, N_2, etc., all of which are known to be strong Raman scatterers, have been used as the non-linear medium in which Raman laser action has been observed [46, 55-58] as well as crystalline solids such as diamond, calcite and α-sulphur [59]. The effect has been observed both with the medium inside the cavity of a Q-spoiled ruby laser as in the original experiment and with the high-intensity laser beam focused into the medium outside the cavity. Not only has a strong line been seen carrying about 1% to 10% of the power of the laser line beam observed at frequencies lower than the laser frequency by known Raman frequencies of the medium in use but additional, progressively weaker, lower frequency lines have also been seen at frequencies lower than the laser frequency by integral multiples of the Raman frequencies. Furthermore, radiation has also been observed with frequencies higher than the laser frequency by the Raman frequencies of the medium. A detailed discussion of all these effects can be found in the original papers; we merely wish to point out their existence here.

One of the most actively exploited forms of Raman laser action is in the so-called spin-flip Raman laser, which has been worked on fairly energetically since its first demonstration in 1970 by three groups of workers [60]. In this type of laser, stimulated Raman scattering is used involving the excitation of electrons in a semiconductor, between their lower and upper spin states in a d.c. magnetic field. Tuning is achieved by varying the magnetic field. By working with a semiconductor with a small energy gap like indium antimonide (~ 0.24 eV), the conduction electrons have a small effective mass (~ $0.014m_0$) and large effective g-factors (~ -50). The small mass enhances the Raman cross-section and the large g-factor keeps the magnetic field required for a given frequency shift down to manageable values of a few tens of kilogauss. Further enhancement of the effect is achieved by pumping with radiation from a carbon monoxide or a frequency-doubled carbon dioxide laser (~ 5.3 μm); this has a photon energy close to but less than the energy gap and a resonant enhancement is obtained. There is much interest in this effect for the production of tunable infra-red radiation (a good recent review of the subject is given by Colles and Pidgeon [41]).

Before leaving the subject of two-photon absorption and the stimulated Raman effect, we should point out that, although no phase-matching condition of the type encountered in the harmonic generation and mixing effects controls the efficiency of these processes, a phase-matching condition of different type can be important. We saw earlier that our usual phase-matching condition was equivalent to momentum conservation in the process of frequency conversion. The same type of argument can be applied to the two-photon absorption process and the stimulated Raman effect. In these cases, energy is normally lost by the radiation field to the non-linear medium and it is the energy conservation conditions $\hbar\omega_1 + \hbar\omega_2 = \hbar\omega_0$ and $\hbar\omega_3 - \hbar\omega_1 = \hbar\omega_0$ where ω_0 is a transition frequency of the medium which controls the frequencies in the radiation field which are effective in these processes. At the same time, momentum conservation requires that $\vec{k}_1 + \vec{k}_2 = \vec{k}_0$ and $\vec{k}_3 - \vec{k}_1 = \vec{k}_0$ where $\vec{k}_0$ is the wave vector associated with the excited state of the medium. These relations are the generalization to higher order processes of the well-known result that in the usual one-photon absorption processes only excitations with a wave vector $\vec{k}_0(\sim 0)$ can be excited. The two generalized phase-matching conditions $\vec{k}_1 + \vec{k}_2 = \vec{k}_0$ and $\vec{k}_3 - \vec{k}_1 = \vec{k}_0$ control respectively the two-photon absorption and Raman processes. However, they are of significance and place limitations on the processes only when the energy $\hbar\omega_0$ of the excitation in the medium is a strongly dependent function of its wave vector $\vec{k}_0$. This is certainly not the case for excitations involving bound electronic states, for example. For the wave functions of such states contain components with all wave vectors and the associated energy of the state considered as a function of $\vec{k}_0$ is constant. Even with excitations like optical lattice vibrations with which one normally associates a wave vector, the dependence of the vibrational energy on this wave vector is insignificant in general for the small wave vectors which arise in optically excited states. Cases in which the phase-matching condition imposes a limitation can be found in the Brillouin scattering effect and the Raman effect in a polar crystal for scattering in a predominantly forward direction. Brillouin scattering is the name given to Raman scattering when the excitation in the medium is an acoustical vibrational mode; such modes have energies that are directly proportional to their wave vector, the constant of proportionality being essentially the velocity of sound. In polar crystals, the optical modes of lattice vibrations have energies which are rapidly varying functions of wave vector at the very small wave vector values $|\vec{k}_3 - \vec{k}_1|$ which are of significance for forward scattering when $\vec{k}_1$ and $\vec{k}_3$ are virtually parallel [61].

The usual phase-matching conditions involve only wave vectors of components of the radiation field and arise directly out of a discussion of Maxwell's equations. In the cases considered here, where the wave vector of an excitation of the medium is involved, the phase-matching condition arises out of the detailed properties of the polarizability tensors.

17. ELECTRO-OPTIC EFFECTS

Two electro-optic effects have been known for many years. They reveal themselves by a change in the refractive index of a material, when it is subjected to a d.c. or low-frequency electric field; in this context, low frequency means up to microwave frequencies but still well below optical frequencies. The linear electro-optic, known as the Pockels effect, after its discoverer, gives rise to a refractive index change proportional to the applied field strength; the quadratic effect, known for similar reasons as the Kerr effect, gives rise to a change proportional to the square of the applied field. The Pockels and Kerr effects can be related respectively to the second- and third-order susceptibility tensors $\underline{\chi}^{(2)}$ and $\underline{\chi}^{(3)}$. It is the purpose of this section to discuss the relationship in some detail.

(i) The Pockels effect

When a d.c. electric field is applied to an acentric material, simultaneously with an optical frequency field a non-linear polarization is established at the optical frequency in addition to the normal linear polarization. It is given by

$$\vec{P}_{NL}(\omega) = 2\underline{\chi}^{(2)}(\omega,0) : \vec{E}(\omega)\vec{E}(0) \tag{17.1}$$

the factor of two arising from intrinsic symmetry. The total polarization to this order is then

$$\vec{P}(\omega) = [\underline{\chi}^{(1)}(\omega) \cdot + 2\underline{\chi}^{(2)}(0,\omega) : \vec{E}(0)]\vec{E}(\omega) \tag{17.2}$$

The first-order response of the material to the optical radiation is thus modified by a term linear in the applied d.c. field. This is the linear electro-optic or Pockels effect. It clearly exists only in acentric materials for which $\underline{\chi}^{(2)} \neq 0$. It need not be induced by a strictly d.c. electric field; up to microwave frequencies $\underline{\chi}^{(2)}$ will be insensitive to any frequency change from zero, and the change in frequency of the induced polarization from that of the applied optical field will be negligible.

Although simple in concept, the magnitude of the effect is usually quoted in what at first appears to be a complicated fashion; this is in terms of the change produced in the refractive index ellipsoid or indicatrix of the material. This ellipsoid is defined by the equation

$$\underline{\epsilon}_0^{-1} : \vec{r}\,\vec{r} = 1 \tag{17.3}$$

using the type of notation defined in (11.8). Here $\underline{\epsilon}_0^{-1}$ is the inverse of the dielectric tensor in the absence of the d.c. field, and in the "natural" reference axes of the material, in which $\underline{\epsilon}_0$ is diagonal with components ϵ_{0x}, ϵ_{0y} and ϵ_{0z}, (17.3) is simply

$$\frac{x^2}{\epsilon_{0x}} + \frac{y^2}{\epsilon_{0y}} + \frac{z^2}{\epsilon_{0z}} = 1 \tag{17.4}$$

The indicatrix has the property that the refractive indices associated with the two plane wave modes of propagation in any direction are given by the semi-axes of its central elliptical section by a plane normal to the propagation direction; furthermore, the $\vec{D}$-vectors of the two waves vibrate along the directions of the associated semi-axes.

When a d.c. electric field is applied to the material, the indicatrix will generally change size and orientation. To first order in the field, this change is characterized by the linear electro-optic coefficient $r_{\mu\alpha\beta}$, which is used to quantify the Pockels effect. It is defined by the expression

$$\epsilon_{\mu\alpha}^{-1} = \epsilon_{0\mu\alpha}^{-1} + \sum_{\beta=x,y,z} r_{\mu\alpha\beta}\, E_\beta(0) \tag{17.5}$$

Since $\vec{D} = \underline{\epsilon} \cdot \vec{E} = \vec{E} + 4\pi\vec{P}$, it follows from (11.1) and (17.5) that

$$r_{\mu\alpha\beta} = -8\pi \sum_{\gamma\zeta} \epsilon_{0\mu\gamma}^{-1}\, \chi_{\gamma\zeta\beta}^{(2)}(\omega,0)\, \epsilon_{0\zeta\alpha}^{-1} \tag{17.6}$$

which in the "natural" axes of the material becomes

$$r_{\mu\alpha\beta} = -\frac{8\pi}{\epsilon_{0\mu}(\omega)\,\epsilon_{0\alpha}(\omega)}\, \chi_{\mu\alpha\beta}^{(2)}(\omega,0) \tag{17.7}$$

(ii) The Kerr effect

When a field of any frequency ω' is applied to a material, we saw in Section 14 how the response at another applied optical frequency ω can be modified. In addition to the normal linear polarization

$$\vec{P}_L(\omega) = \underline{\chi}^{(1)}(\omega) \cdot \vec{E}(\omega)$$

a polarization $\vec{P}_{NL}(\omega)$ is set up depending on the square of the field strength at the other frequency

$$\vec{P}_{NL}(\omega) = 6\underline{\chi}^{(3)}(\omega,\omega',-\omega') \vdots \vec{E}(\omega)\vec{E}(\omega')\vec{E}^*(\omega') \qquad (17.8)$$

This is the origin of the Kerr effect. In addition to its quadratic rather than linear dependence on the field strength, it differs from the Pockels effect in two other respects. First of all, it can occur for any frequency ω' of the applied field, for irrespective of the value of this frequency a polarization $\vec{P}_{NL}(\omega)$ is established at precisely the optical frequency. Secondly, it can occur in any material, for there are no materials for which symmetry considerations force $\underline{\chi}^{(3)}$ to be zero.

The effect was first discovered in 1875 by Kerr in glass, which is an optically isotropic medium. He found an optical anisotropy, characterized by a difference in refractive indices $n_{\parallel}$ and $n_{\perp}$ for light polarized parallel and perpendicular to an applied d.c. electric field, whose magnitude was proportional to the square of the applied field strength. The effect has been traditionally studied and exploited in organic liquids like benzene, carbon disulphide, etc., which show a strong effect and are of course also optically isotropic. Thus although the effect can reveal itself in any material, it is most readily observed, studied and exploited in optically isotropic materials since a birefringence, not ordinarily present, is induced. In other materials, only a birefringence additional to that already present is induced. The Kerr constant which measures the magnitude of the effect is therefore defined relative to an isotropic medium as

$$K_{\omega'}(\omega) = \frac{\Delta n_{\parallel}(\omega) - \Delta n_{\perp}(\omega)}{\lambda \mathscr{E}^2_{\omega'}} \qquad (17.9)$$

where $\Delta n_{\parallel}(\omega)$ and $\Delta n_{\perp}(\omega)$ are the changes to the refractive index $n(\omega)$ of the material at frequency ω and wavelength λ, respectively, for radiation polarized parallel and perpendicular to the polarization of the applied field of strength $\mathscr{E}_{\omega'}$, at frequency ω'. Defined in this way, $K_{\omega'}(\omega)$ is independent of direction in an isotropic material; for an anisotropic material, it depends on the direction of the applied field.

Our attention will be restricted to the relationship between $K_{\omega'}(\omega)$ and $\underline{\chi}^{(3)}$ in an isotropic material. For propagation of the optical wave at frequency ω in the z-direction, we have, for example, from (15.3)

$$\vec{P}_{NL}(\omega) = 6\underline{\chi}^{(3)}(\omega,\omega',-\omega') \vdots \vec{a}_{\omega}\vec{a}_{\omega'}\vec{a}_{\omega'}|E_{\omega'}|^2 \; E_{\omega} e^{ikz} \qquad (17.10)$$

If we take the applied field at ω' to be polarized in the y-direction, $\vec{a}_{\omega'} = (0,1,0)$ and we have, with $|E_{\omega'}|^2 = (1/4)\mathscr{E}^2_{\omega'}$,

$$\vec{P}'_{NL\mu}(\omega) = \frac{3}{2}\mathscr{E}^2_{\omega'} E_\omega e^{ikz} \sum_{\alpha=x,y,z} \chi^{(3)}_{\mu\alpha yy}(\omega,\omega',-\omega')\, a_{\omega\alpha} \tag{17.11}$$

Now for an isotropic medium, of the 81 possible components of $\underline{\chi}^{(3)}$, only 21 can be non-zero and,of these, only 3 are independent [18]. Denoting $\chi_{\mu\alpha\beta\gamma}$ simply by $\mu\alpha\beta\gamma$, the non-zero components are

$$\begin{aligned} &xxxx = yyyy = zzzz \\ &yyzz = zzyy = zzxx = xxzz = xxyy = yyxx \\ &yzyz = zyzy = zxzx = xzxz = xyxy = yxyx \\ &yzzy = zyyz = zxxz = zxxz = xyyx = yxxy \\ &\text{with} \quad xxxx = xxyy + xyxy + xyyx \end{aligned} \tag{17.12}$$

Combining this with the fact that $a_{\omega z} = 0$, we find from (17.11) that the non-zero components of $\vec{P}_{NL}(\omega)$ are

$$\begin{aligned} P'_{NLx}(\omega) &= \frac{3}{2}\mathscr{E}^2_{\omega'}\chi^{(3)}_{xxyy}(\omega,\omega',-\omega')\, a_{\omega x} E_\omega e^{ikz} \\ P'_{NLy}(\omega) &= \frac{3}{2}\mathscr{E}^2_{\omega'}\chi^{(3)}_{yyyy}(\omega,\omega',-\omega')\, a_{\omega y} E_\omega e^{ikz} \end{aligned} \tag{17.13}$$

With our basic propagation equation (10.17), we therefore find that for radiation polarized parallel to $\mathscr{E}_{\omega'}$, i.e. $\vec{a}_\omega = (0,1,0)$,

$$\frac{dE_\omega}{dz} = \frac{2\pi i\omega^2}{kc^2} P'_{NLy}(\omega) = \frac{3\pi i\omega^2}{kc^2}\mathscr{E}^2_{\omega'}\chi^{(3)}_{yyyy}(\omega,\omega',-\omega') E_\omega \tag{17.14}$$

Thus

$$E_\omega \sim \exp\left[i\frac{\omega}{c}\left(\frac{3\pi\omega}{kc}\mathscr{E}^2_{\omega'}\chi^{(3)}_{yyyy}(\omega,\omega',-\omega')\right)z\right] \tag{17.15}$$

Since the normal propagation of the wave at frequency ω is taken account of by the exponential in

$$\vec{E}(\omega) = \vec{a}_\omega E_\omega e^{ikz} = \vec{a}_\omega E_\omega \exp\left(i\frac{\omega}{c}n(\omega)z\right) \tag{17.16}$$

the exponential in (17.15) describes a modified propagation characterized by a modification to the refractive index n(ω) by

$$\Delta n_{\parallel}(\omega) = \frac{3\pi\omega}{kc}\mathscr{E}^2_{\omega'}\chi^{(3)}_{yyyy}(\omega,\omega',-\omega') \tag{17.17}$$

Similarly for radiation polarized perpendicular to $\mathscr{E}_{\omega'}$, a polarization $P'_{NLx}(\omega)$ is induced, and using (17.13), this gives

$$\Delta n_{\perp}(\omega) = \frac{3\pi\omega}{kc}\,\mathscr{E}^2_{\omega'}\,\chi^{(3)}_{xxyy}(\omega,\omega',-\omega') \tag{17.18}$$

As one would expect, the same expression for $\Delta n_{\perp}(\omega)$ is obtained by considering the geometry with the propagation direction at frequency ω to be parallel to $\mathscr{E}_{\omega'}$. It follows from (17.9), (17.17), (17.18) and the relation $\lambda = 2\pi/k$, that

$$K_{\omega'}(\omega) = \frac{3\omega}{2c}\left[\chi^{(3)}_{yyyy}(\omega,\omega',-\omega') - \chi^{(3)}_{xxyy}(\omega,\omega',-\omega')\right] \tag{17.19}$$

The usual d.c. Kerr constant $K_0(\omega)$ is obtained by putting $\omega' = 0$.

We have so far treated the optical radiation field and the field producing the effect as different. However, an interesting situation arises when a strong optical field produces the effect itself. It will then propagate with a refractive index, which follows from (17.17) with $\omega' = \omega$,

$$n(\omega) + \frac{3\pi}{n(\omega)}\chi^{(3)}_{yyyy}(\omega,\omega,-\omega)\,\mathscr{E}^2_{\omega} \tag{17.20}$$

For a wavefront of uniform intensity, this modified refractive index gives rise simply to an intensity-dependent phase velocity. More interesting and significant effects are produced on a beam with an intensity variation across its wavefront; for rays bend towards regions of higher refractive index and away from regions of lower. Thus a beam with a maximum of intensity at its centre, which is frequently the case, will induce a lens-like effect due to a refractive index variation across the beam with a maximum or minimum at the centre according as $\chi^{(3)}_{yyyy}(\omega,\omega,-\omega)$ is positive or negative. When positive, a positive lens effect is produced and the beam will focus itself down on to the high-intensity region more and more as the intensity maximum builds up; this is called self-focusing. When $\chi^{(3)}_{yyyy}(\omega,\omega,-\omega)$ is negative, a negative lens effect is produced and the beam will defocus away from the region of high intensity and will tend to produce a more uniform intensity distribution; this is called blooming.

Self-focusing and blooming are both important effects. Away from absorption bands, $\chi^{(3)}_{yyyy}(\omega,\omega,-\omega)$ is normally positive and self-focusing can occur. This limits the intensity of radiation that can be propagated through a material. For it is very difficult to maintain uniform intensity wavefronts, and the more intense the radiation, the smaller is the deviation from uniformity that will produce focusing. Once begun, it is a catastrophic effect usually leading to physical damage of the material. It can be initiated either by inability to produce uniform intensity wavefronts or, if they have been successfully produced, by built-in optical inhomogeneities in the material or even by dust on its surface. It is a major factor limiting the design of high-power solid-state lasers. A number of reviews on self-focusing are available [62].

Blooming usually occurs in the wings of an absorption band. With intense enough radiation, even extremely weak absorption will give rise to a significant absorption of energy. This will produce heating and, particularly in a gas, associated expansion and lowering of the refractive index; we effectively have a negative $\chi^{(3)}_{yyyy}(\omega,\omega,-\omega)$. Any attempt to focus a beam of radiation down to smaller spot sizes and higher intensities will be self-defeating in these circumstances, as the self-induced lowering of the refractive index will compete with the focusing action. This can, for example, be a major factor in the propagation and control of powerful CO_2 laser beams through the atmosphere since it has a weak but not insignificant absorption at 10.6 μm due to the presence of CO_2 itself in the atmosphere.

18. LIMITATIONS OF THE APPROACH PRESENTED HERE

In this presentation of the theory of non-linear optical effects, I have concentrated on getting over the general principles. The theory has therefore been simplified to rid it of unnecessary detail and complication, so as to ensure that the general principles involved are more clearly revealed. In this last section, I shall point out some general limitations of the treatment given and, as far as possible, indicate briefly how they can be removed.

(i) Plane waves of infinite extent

We have assumed that the radiation at each frequency can be described by a plane wave of infinite extent, and that all those plane waves propagate in the same direction. The formal generalization of this to plane waves propagating in different directions presents no problem and introduces no additional concepts. However, apart from the practical problem of producing radiation propagating in such plane wave modes, there is a basic attraction to focusing beams down to narrow widths. This is readily seen, for example, from the equations governing second harmonic generation, but the conclusion is true in general. From (11.18), the second harmonic energy flux $S_2(z)$ grows initially from a flux $S_1(0)$ at the fundamental frequency as $S_2(z) = S_1(0)\, z^2/\ell_{SH}^2$ where $\ell_{SH}^{-2} \sim S_1(0)$; thus $S_2(z) \sim S_1^2(0)$. However, it is the total energy flow P rather than the flux S we are normally interested in, and $P \approx SA$ where A is the beam-width. It follows that $P_2(z) \sim P_1^2(0)/A$, which indicates that more second harmonic power can be generated by narrowing the beam-widths. However, there are limitations on what can be gained which are not seen from this simple argument. These limitations arise from the effects of crossing beams, diffraction, and walk-off. These will be discussed in turn.

The beam crossing limitation is quite simple. Beams of finite width, travelling in different directions, will only couple through non-linear effects in the region of overlap. The narrower the beams, the smaller will this region become.

Diffraction limits our ability to maintain a narrow beam. A beam of width D at a given position will have an angular divergence due to diffraction of $\sim \lambda/D$. Over a propagation distance ℓ, this will be negligible only if $\lambda\ell/D \ll D$, i.e. $D^2 \gg \lambda\ell$. For propagation over 1 cm, this condition limits us to widths greater than 10^{-2} cm, for radiation of wavelength 1 μm.

Walk-off is the effect originating from the fact that in general, the directions of propagation $\vec{k}$ and energy flow $\vec{S}$ in a plane wave are not necessarily coincident (see Fig.4). Thus, for example, with collinear phase matching of fundamental and second harmonic waves, the directions of energy flow will in general be different. This can give rise to problems with beams of finite width. For an ordinary wave, $\vec{k}$ and $\vec{S}$ are always parallel, but for an extraordinary wave the angle α between them is given in a uniaxial crystal by

$$\tan\alpha = \frac{(n_E^2 - n_0^2)\sin 2\theta}{2(n_E^2 \sin^2\theta + n_0^2\cos^2\theta)}$$

Here θ is the angle of propagation relative to the optic axis and

$$n_0 = \epsilon_\perp^{\frac{1}{2}} \text{ and } n_E = \epsilon_\parallel^{\frac{1}{2}}$$

are the ordinary and extraordinary refractive indices. Note that $\alpha = 0$ for $\theta = 0$ and $\theta = 90°$. For small birefringence,

$$\tan\alpha \approxeq \left(\frac{n_E}{n_0} - 1\right)\sin 2\theta$$

and α is typically a few degrees. Walk-off is negligible only as long as $D \gg \alpha\ell$, so that for $\ell = 1$ cm we must have beam-widths greater than ~ 0.1 cm to avoid this problem.

It will be recalled that non-critical phase matching is achieved with propagations in the plane normal to the optic axis, i.e. with $\theta = 90°$ (see Section 9). We see now an additional advantage of achieving this type of phase matching over the advantages previously discussed. For with $\theta = 90°$, $\alpha = 0$ and the walk-off problem is removed.

General treatments of second harmonic generation and three-frequency mixing processes for collinear, focused, Gaussian beams in uniaxial crystals have been carried through by Boyd and Kleinman [16]. They incorporate in a general but, by necessity, complicated fashion the basic effects of diffraction and walk-off just discussed. They base their treatment on the analysis of the beams into plane-wave components. The non-linear interaction effects between all these components can then be described by our general plane-wave theory and the overall result obtained by the appropriate combination of these, based on the structure of the input beams. This technique could clearly be used in more general cases than those treated by Boyd and Kleinman, as long as the input beams were of a known, well-defined structure.

(ii) Coherent radiation

A classical plane-wave mode of the electromagnetic field, and indeed any mathematically well-defined combination of such modes, forms fully coherent states of the field, in the most general sense [63]. This follows from the fact that, in such circumstances, the field is uniquely defined at all points in space at all times, without any statistical uncertainty. The development of our theory using plane-wave modes, and its possible extension to more complex situations discussed in the previous subsection, implicitly assume that this high degree of coherence exists in fields of interest. This is rarely the case in practice, the nature of the source of the field imposing fluctuations on it, which can be described only statistically. Thus a so-called chaotic source, like a gas discharge or incandescent lamp, from which radiation is emitted independently from a large number of microscopic sources, produces a field strength at any point with a Gaussian probability distribution; its mean value is zero and it has a variance proportional to the square root of the mean field intensity. Even radiation from a well-controlled laser can contain fluctuations; in amplitude, these are less significant than from a chaotic source and can be negligibly small, but in phase they can be just as large. These effects are discussed in the paper by F.T Arecchi in these Proceedings; they are also well described by Loudon [64].

For non-linear optical effects, these statistical fluctuations in the input field strengths give rise to effects not contained in our treatment. This is most easily seen by considering the process of two-photon absorption from a single beam, which allows us to restrict attention to one frequency component of the field. Section 15 shows that the rate of change of photon flux due to this absorption process is proportional to the square of the flux, so that enhanced absorption will occur in a beam whose intensity fluctuates to give values both larger and smaller than the average. Thus an input beam, whose average photon flux $N_\omega(0)$ is achieved by having twice that flux for half the time and zero for the other half, will be attenuated on average at twice the rate of a beam having a constant input photon flux $N_\omega(0)$ at all times. Moreover, because of the preferential absorption from high-intensity fluctuations in the beam, these fluctuations will be smoothed out, whilst little change will be effected in low-intensity fluctuations [65]. An overall change will therefore be made to the statistical nature of the fluctuations in the beam, which in turn decrease the rate of attenuation towards that for a non-fluctuating beam of the same mean intensity. More generally, in non-linear processes involving input beams at different frequencies, fluctuations in these beams will be important through the way in which they occur relative to one another; this is expressed by appropriate field correlation functions [64].

Non-linear processes are mostly studied and exploited using well-controlled laser beams to achieve high intensity and good phase matching. Intensity fluctuations in these beams are usually very small. Insofar as this is the case, the theory developed here can be applied. However, when experiments of the type carried out by Hopfield et al. [49] (see Section 15) employing a conventional source are considered, some care must be exercised in interpreting absolute levels of absorption.

(iii) Quantum effects

It is rare for non-linear mixing processes to be carried experimentally to the stage where depletions occur in the input beam powers to the extent that discrete photon effects become significant. If, however, this does occur, we clearly have a situation in which our classical description of the radiation field becomes inadequate. This was pointed out in Section 11 in connection with the classical result, which says that, in second harmonic generation, all the input power can be transferred permanently to the second harmonic frequency. This is true only when the spontaneous break-up of a second harmonic photon into two fundamental photons is neglected. Walls and

Tindle [25] have shown how a complete quantum-mechanical description of the problem avoids this difficulty and gives a significantly different result from the classical one, when appreciable power is transferred from the fundamental to the second harmonic frequency. Walls [66] has also considered quantum effects in other non-linear mixing processes.

We have just seen how intensity fluctuation phenomena can affect non-linear processes. For weak enough beams, fluctuations arising from the individual photons comprising the beam become important and influence the processes in a way which can be described only quantum mechanically. This has been discussed in connection with two-photon absorption by Simaan and Loudon [67]. More recently, Loudon [68] has described how some uniquely quantum-mechanical effects can be obtained associated with the anti-bunching of photons.

(iv) Transient, coherent non-linear effects

Finally, there is a class of effects that lack of space prevents me from dealing with. These are concerned with the passage through matter of pulses of high intensity radiation with a frequency very close to a resonant absorption frequency. Self-induced transparency, photon echoes and π-pulses are examples of some of the effects which can arise. They are extensively discussed by Allen and Eberly [69]. A recent review by Shen [70] discusses the up-to-date situation on these and in fact on all non-linear optical effects.

REFERENCES

[1] LANDAU, L.D., LIFSHITZ, E.M., Electrodynamics of Continuous Media, Pergamon (1960).
[2] BALDWIN, G.C., Introduction to Non-Linear Optics, Pergamon (1969).
[3] GROSS, A.F., KAPLYANSKII, A.A., Sov. Phys. – Solid State 2 (1960) 253.
[4] ELLIOTT, R.J., Phys. Rev. 124 (1961) 340.
[5] PEKAR, S.Y., Sov. Phys.– JETP 6 (1958) 785; 7 (1958) 813.
[6] AGRANOVICH, V.M., GINZBURG, V.L., Sov. Phys. - Usp. 5 (1962) 323; 5 (1963) 675.
[7] HOPFIELD, J.J., THOMAS, D.G., Phys. Rev. 132 (1963) 563.
[8] EVANGELISTI, F., FISCHBACH, J.U., FROVA, A., Phys. Rev. B9 (1974) 1516.
[9] BUTCHER, P.N., McLEAN, T.P., Proc. Phys. Soc. 83 (1964) 579.
[10] LANDAU, L.D., LIFSHITZ, E.M., Statistical Physics, Pergamon (1958).
[11] AMSTRONG, J.A., BLOEMBERGEN, N., DUCUING, J., PERSHAN, P.S., Phys. Rev. 127 (1962) 1918.
[12] BUTCHER, P.N., McLEAN, T.P., Proc. Phys. Soc. 81 (1963) 219.
[13] PERSHAN, P.S., Phys. Rev. 130 (1963) 919.
[14] BASS, M., FRANKEN, P.A., WARD, J.F., WEINREICH, G., Phys. Rev. Lett. 9 (1962) 446.
[15] KLEINMAN, D.A., Phys. Rev. 126 (1962) 1977.
[16] BOYD, G.D., KLEINMAN, D.A., J. Appl. Phys. 39 (1968) 3597.
[17] Standards in piezoelectric crystals, Proc. IRE 37 (1949) 1378.
[18] BUTCHER, P.N., Non-Linear Optical Phenomena, Ohio State Univ. Press, Columbia (1965).
[19] FRANKEN, P.A., WARD, J.F., Rev. Mod. Phys. 35 (1963) 23.
[20] KLEINMAN, D.A., Phys. Rev. 128 (1962) 1761.
[21] MAKER, P.D., TERHUNE R.W., NISENOFF, M., SAVAGE, C.M., Phys. Rev. Lett. 8 (1962) 21.
[22] LOUISELL, W.H., Coupled Mode and Parametric Electronics, Wiley, New York (1960).
[23] GIORDMAINE J.A., Phys. Rev. Lett. 8 (1962) 19.
[24] ZERNIKE, F., MIDWINTER, J., Applied Non-linear Optics, Wiley, New York (1973) 68.
[25] WALLS, D.F., TINDLE C.T., J. Phys. A 5 (1972) 534.
[26] FRANKEN, P.A., HILL, A.E., PETERS, C.W., WEINREICH, G., Phys. Rev. Lett. 7 (1961) 118.
[27] TERHUNE, R.W., MAKER, P.D., SAVAGE, C.M., Appl. Phys. Lett. 2 (1963) 54.

[28] ASHKIN, A., BOYD, G.D., DZIEDZIC, J.M., Phys. Rev. Lett. 11 (1963) 14.
[29] MANLEY, J.M., ROWE, H.E., Proc. IRE 44 (1956) 904.
[30] JURKUS, A., ROBSON, P.N., Proc. IEE 107B (1960) 119.
[31] MORSE, P.M., FESHBACH, H., Methods of Theoretical Physics, McGraw-Hill, New York (1953).
[32] BASS, M., FRANKEN, P.A., HILL, A.E., PETERS, C.W., WEINREICH, G., Phys. Rev. Lett. 8 (1962) 18.
[33] MILLER, R.C., SAVAGE, C.M., Bull. Am. Phys. Soc. 7 (1962) 397.
[34] GARFINKEL, M., ENGELER, W.E., Appl. Phys. Lett. 3 (1963) 178.
[35] FARIES, D.W., GEHRING, K.A., RICHARDS, P.L., SHEN, Y.R., Phys. Rev. 180 (1969) 363.
[36] HANNA, D.C., SMITH, R.C., STANLEY, C.R., Optics Commun. 4 (1971) 300.
[37] SMITH, A.W., BRASLAU, N., IBM J. Res. Dev. 6 (1962) 361; J. Appl. Phys. 34 (1963) 2105.
[38] WARNER, J., Quantum Electronics 1 (1975) 703.
[39] SMITH, H.A., MAHR, H., in Optroelectronics (Proc. 6th Int. Conf. Quantum Electronics Kyoto, 1970) (1971), paper 5.10.
[40] GIORDMAINE, J.A., MILLER, R.C., Phys. Rev. Lett. 14 (1965) 973.
[41] COLLES, M.J., PIDGEON, C.R., Rep. Prog. Phys. 38 (1975) 329.
[42] HULME, K.F., Rep. Prog. Phys. 36 (1973) 497.
[43] BECHMANN, R., KURTZ, S.K., Landolt-Bornstein, Numerical Data and Functional Relationships in Science and Technology, New Series Group III, Vol. 2, Springer-Verlag, Berlin (1969) 167.
[44] MILLER, R.C., Appl. Phys. Lett. 5 (1964) 17.
[45] BLOEMBERGEN, N., Comments Solid State Phys. 2 5 (1969) 161.
[46] MAKER, P.D., TERHUNE, R.W., Phys. Rev. 137 (1965) A801.
[47] JAVAN, A., Am. Phys. Soc. 3 (1958) 213; WEBER, J., Rev. Mod. Phys. 31 (1959) 681.
[48] KAISER, W., GARRETT, C.G.B., Phys. Rev. Lett. 7 (1961) 229.
[49] HOPFIELD, J.J., WORLOCK, J.M., PARK, K., Phys. Rev. Lett. 11 (1963) 414; HOPFIELD, J.J., WORLOCK, J.M., Phys. Rev. 137 (1965) A1455.
[50] GOLD, A., Quantum Optics, p.397 (GLANBER, R.J., Ed.) Academic Press, New York, 1969).
[51] BUTCHER, P.N., LOUDON, R., McLEAN, T.P., Proc. Phys. Soc. 85 (1965) 565.
[52] HENRY, C.H., GARRETT, C.G.B., Phys. Rev. 171 (1968) 1058.
[53] BOGGETT, D., LOUDON, R., J. Phys. C. 6 (1973) 1763.
[54] WOODBURY, E.J., NG, W.K., Proc. IRE 50 (1962) 2367; Quantum Electronics III, 1575 (Dunod, Paris, 1964).
[55] ECKHARDT, G., HELLWARTH, R.W., McCLUNG, F.J., SCHWARZ, S.E., WEINER, D., WOODBURY, E.J., Phys. Rev. Lett. 9 (1962) 455.
[56] GELLER, M., BORTFIELD, D.P., SOOY, W.R., Appl. Phys. Lett. 3 (1963) 36.
[57] STOICHEFF, B., Physics Letters 7 (1963) 186.
[58] ZEIGER, H.J., TANNEWALD, P.E., KERN, S., HERENDEEN, R., Phys. Rev. Lett. 11 (1963) 419.
[59] ECKHARDT, G., BORTFIELD, D.P., GELLER, M., Appl. Phys. Lett. 3 (1963) 137.
[60] PATEL, C.K.W., SHAW, E.D., Phys. Rev. Lett. 24 (1970) 451; ALLWOOD, R.L., DENNIS, R.B., SMITH, S.D., WHERRETT, B.S., WOOD, R.A., J. Phys. C 3 (1970) L186; MOORADIAN, A., BRUECH, S.R.J., BLUM, F.A., Appl. Phys. Lett. 17 (1970) 481.
[61] LOUDON, R., Proc. Phys. Soc. 82 (1963) 527.
[62] AKHMANOV, S.A., KHOKHLOV, R.V., SUKHORUHOV, A.P., Laser Handbook (ARECCHI, F.T., SCHULZ-DUBOIS, E.O., Eds), North-Holland, Amsterdam (1972) 1151; SVELTO, O., Progress in Optics 12 (WOLF, E., Ed), North-Holland, Amsterdam (1970).
[63] GLAUBER, R.J., Quantum Optics and Electronics (Proc. Grenoble University Summer School of Theoretical Physics, Les Houches, 1964) (DeWITT, C.M., BLANDIN, A., COHEN-TANNOUDJI, C., Eds), Gordon and Breach, New York (1965) 63.
[64] LOUDON, R., The Quantum Theory of Light, Oxford Univ. Press (1973).
[65] WEBER, H.P., IEEE, J. Quant. Electron. QE-7 (1971) 189.
[66] WALLS, D.F., J. Phys. A 4 (1971) 813.
[67] SIMAAN, H.D., LOUDON, R., J. Phys. A 8 (1975) 539.
[68] LOUDON, R., Phys. Bull. 27 (1976) 21.
[69] ALLEN, L., EBERLY, J.H., Optical Resonance and Two-level Atoms, Wiley, New York (1975).
[70] SHEN, Y.R., Rev. Mod. Phys. 48 (1976) 1.

IAEA-SMR-20/16

AN INTRODUCTION TO QUANTUM OPTICS

F.T. ARECCHI*
Università di Pavia and CISE,
Milano,
Italy

Abstract

AN INTRODUCTION TO QUANTUM OPTICS.

Part 1. Quantum optics: A heuristic approach. (Terminology and numerology); 1.1. Definition of quantum optics; 1.2. Physics of the stimulated emission processes; 1.3. Stimulated emission and non-linear optics; 1.4. Coherence and cooperative phenomena. *Part 2.* Photon statistics; 2.1. Relevance of photon statistics; 2.2. Limits of classical optics; 2.3. Characterization of random processes; 2.4. Gaussian processes and the Hanbury-Brown and Twiss effect; 2.4.1. Gaussian distribution with zero average; 2.4.2. 'Coherent' field without fluctuations; 2.5. Measurement of photon statistics; 2.6. Laser fluctuations; 2.6.1. Review of the theory; 2.6.2. Stationary experiments (ensemble distributions and time correlation); 2.6.3. Transient experiments; 2.7. Distortion of photon statistics owing to attenuation; 2.8. The photomultiplier as a statistical device. *Part 3.* Quantum optics: Coherent resonant spectroscopy; 3.1. Introduction; 3.2. The interaction model; 3.3. The two-level atom; 3.4. The Bloch equations; 3.5. Irreversible processes in the presence of dampings; 3.6. Saturation and non-linear spectroscopy; 3.7. Two-photon spectroscopy. Comparison with saturation; 3.8. Perturbed fluorescence spectroscopy; 3.9. Dynamic Stark shift. *Part 4.* Field and atomic coherent states; 4.1. Introduction; 4.2. Description of the free field; 4.2.1. The harmonic oscillator states; 4.2.2. Coherent states of the field; 4.2.3. The coherent states as a basis; 4.2.4. Statistical operator for the field; 4.3. Description of the free atoms; 4.3.1. The angular momentum states; 4.3.2. Coherent atomic states; 4.3.3. The Bloch states as a basis; 4.3.4. Statistical operators for the atoms.

PART 1. QUANTUM OPTICS: A HEURISTIC APPROACH
Terminology and numerology

1.1. DEFINITION OF QUANTUM OPTICS

The term 'quantum electronics' was first used in 1959 at a conference dealing with the physical and engineering uses of the MASER (Microwave Amplifier by Stimulated Emission of Radiation) and MASER oscillators in high-resolution spectroscopy and in the handling of electromagnetic signals at $\lambda \sim 1$ cm.

Half a year later, the first LASER was operated. (L stands for light; the L has replaced the M because this time the generated radiation is at $\lambda \lesssim 1$ μm.) Since 1960 the laser has become a useful device in many areas of physics and technology.

In the spectral range of interest, it is more convenient to speak of quantum optics rather than quantum electronics. Quantum optics can be approached from three points of view:

(a) Physics of the stimulated emission processes
(b) Coherence and cooperative phenomena in radiation/matter interaction
(c) Application of the laser related to its spectral purity

* Now director of Istituto Nazionale d'Ottica, Firenze, Italy.

We shall discuss the three aspects, defining the terms and giving the orders of magnitude. Aspect (a) is covered in Sections 1.2 and 1.3 and aspect (b) in Section 1.4, where we mention also that the application (c) makes only partial use of quantum optics and should rather be considered as optical engineering.

1.2. PHYSICS OF THE STIMULATED EMISSION PROCESSES

If the e.m. cavity, where we are considering the radiation/atom interaction, is a rectangular cavity of sides X, Y, Z (volume V = X Y Z), then the solution of the wave equation, with periodic boundary conditions, yields the plane wave expansion for the field

$$E(x, y, z, t) = \sum E_{\vec{k}}(t) e^{i(k_1 x + k_2 y + k_3 z)} \tag{1.1}$$

where $k_1 = n_1 \cdot 2\pi/X$

$k_2 = n_2 \cdot 2\pi/Y \quad (n_i = 1, 2, \ldots)$

$k_3 = n_3 \cdot 2\pi/Z$

For each set of $k_{1,2,3}$ we have a different field configuration or *mode.*

The dispersion relation imposes a constraint between frequency ω and amplitude $k = \sqrt{k_1^2 + k_2^2 + k_3^2}$ of the $\vec{k}$ vector

$$\omega = ck \tag{1.2}$$

In $\vec{k}$ space (Fig. 1.1) each mode occupies an elementary volume

$$\delta^3 k = \delta k_1 \delta k_2 \delta k_3 = \frac{(2\pi)^3}{V} \tag{1.3}$$

In a spherical shell of radius k and thickness Δk there are

$$\Delta M = 2\,\frac{4\pi k^2 \Delta k}{\delta^3 k} = \frac{k^2 \Delta k\, V}{\pi^2} \tag{1.4}$$

modes. The extra-factor 2 accounts for the two possible polarizations for each $\vec{k}$ vector. Hence the mode density in frequency is

$$\frac{dM}{d\omega} = \omega^2 V/\pi^2$$

or

$$\frac{dM}{d\nu} = \frac{8\pi\nu^2}{c^3} V \tag{1.5}$$

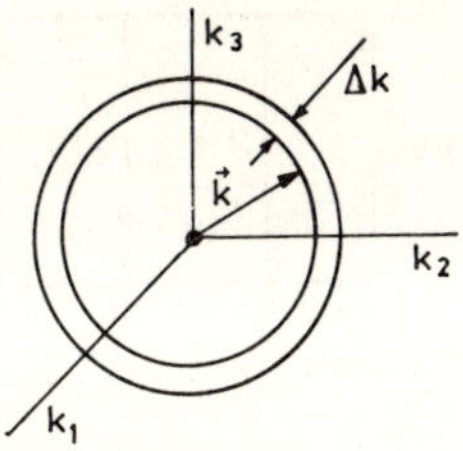

FIG.1.1. Spherical shell in k-space.

If the cavity contains radiators (atoms on the walls or inside) in thermal equilibrium at a temperature T, then the electromagnetic energy density in the cavity is given by Planck's blackbody formula (1900)

$$\frac{dW}{d\nu} = \frac{8\pi\nu^2}{c^3} V \cdot h\nu \cdot \frac{1}{e^{h\nu/k_BT} - 1}$$

$$= \frac{dM}{d\nu} \cdot h\nu \cdot \bar{n}_1(\nu) \qquad (1.6)$$

Here

$$\bar{n}_1(\nu) = (e^{h\nu/k_BT} - 1)^{-1}$$

is the average photon number for each mode whose $\vec{k}$ vector lies on the spherical surface of radius $k = 2\pi\nu/c$.

The distinction between *spontaneous* and *stimulated* emission came in Einstein's derivation of Eq. (1.6) of 1917 as follows: Consider two relevant levels of an atom coupled by an optical transition. Each time the atom goes up *(absorption)* or down *(emission)*, this is a one-photon exchange process.

The emission or decay process can be spontaneous (i.e. not triggered by photons) as well as stimulated (i.e. proportional to the photon number $\bar{n}$ at the frequency $\nu = \Delta E/h$) (see Fig. 1.2).

If there is an ensemble of N atoms in thermal equilibrium, with N_2 in the upper state and $N_1 = N - N_2$ in the lower state, then we have

$$N_1/N_2 = e^{\Delta E/k_BT} \qquad (1.7)$$

$$N_1 B' \bar{n} = N_2 B \bar{n} + N_2 A \qquad (1.8)$$

From the latter equation

$$\bar{n} = \frac{A/B}{\frac{B'}{B}\frac{N_1}{N_2} - 1}$$

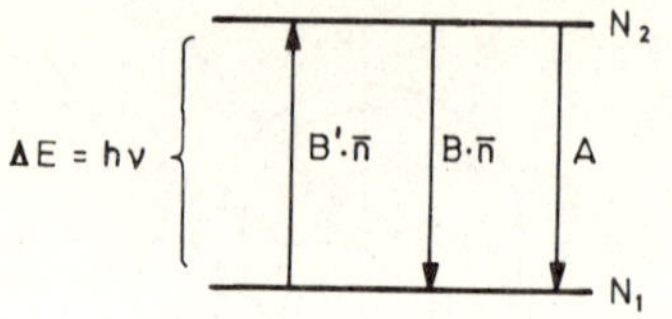

FIG.1.2. Radiative transitions in a two-level atom.

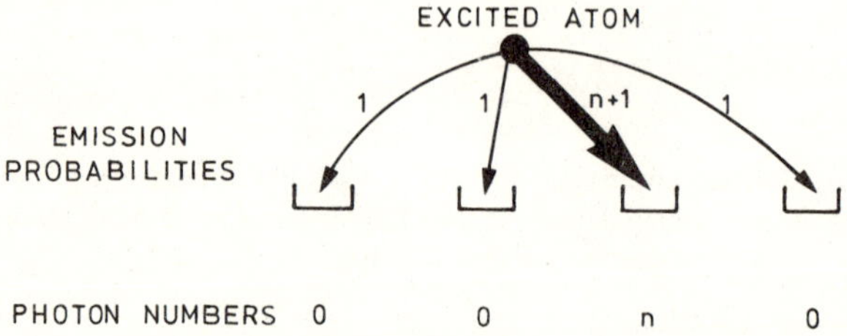

FIG.1.3. Decay channels of an excited atom into different field modes.

By use of Eq. (1.7) and comparison with Eq. (1.6) we get

$$\frac{A}{B} = \frac{8\pi\nu^2}{c^3} V \tag{1.9}$$

$$B' = B \tag{1.10}$$

This can be interpreted by representing the degrees of freedom of the e.m. field as boxes and the excited atom as linked to all of them, as in Fig. 1.3.

With the probabilities there indicated, the stimulated emission probability into one mode is larger than the total spontaneous emission over ΔM modes (all those within the line width $\Delta\nu$ of the atomic emission) when

$$n > \Delta M \tag{1.11}$$

In other words, it is not enough to have a mode with a large population ($n \gg 1$) but it is necessary to obey the specific condition (1.11) in order to observe stimulated effects. Hitherto we have referred to the *whole* photon number in the cavity. It is more convenient to refer to the photon flux, given by $\phi_n = cn/V$, or in terms of the field amplitude and the frequency ω by

$$\phi_n = \frac{c\epsilon_0 E^2}{\hbar\omega} \quad (\text{photons/cm}^2\cdot\text{s}) \tag{1.12}$$

This becomes smaller for high frequencies ω, unless we simultaneously increase the E field strength.

The stimulated transition probability per second per atom is given by

$$p_s = \sigma\phi_n \tag{1.13}$$

where the cross-section σ for the process can be evaluated on a purely classical basis. For a *free* electron it has the Thomson value

$$\sigma_{Th} = \frac{8}{3}\pi r_0^2 \sim 0.6 \times 10^{-24}\ cm^2 \qquad (1.14)$$

where

$r_0 = e^2/(4\pi\epsilon_0 mc^2)$ is the classical electron radius.

For a *bound* electron, if the field frequency is resonant with the atomic transition, the cross-section is

$$\sigma_{res} = \sigma_{Th}\left(\frac{\omega_0}{\gamma}\right)^2 \qquad (1.15)$$

At optical frequencies the decay rate γ can be as small as $10^{-5}\ \omega$, hence σ_{res} can be as large as $10^{10}\ \sigma_{Th}$ (Fig. 1.4). Condition (1.11), with ΔM given by Eq. (1.4), is unfair. We try either to privilege some modes at constant σ, by a tuned cavity, or to privilege a frequency by using the strong frequency dependence of σ_{res}. The former is done at low frequencies (microwaves), the latter at high frequencies (from the i.r. region upwards) (see Fig. 1.5).

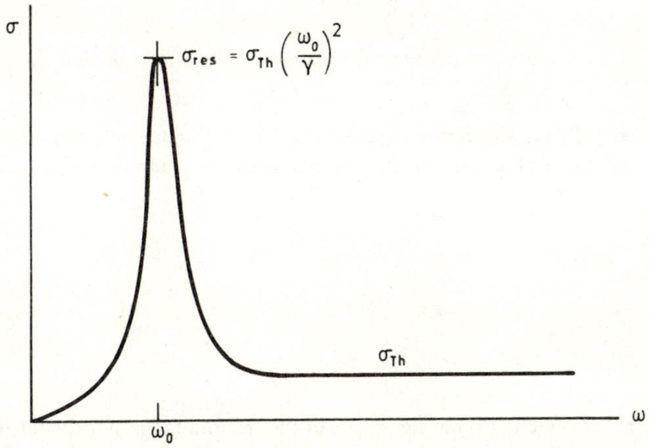

FIG.1.4. Radiative electron cross-section versus frequency.

By Eqs (1.12) and (1.13) the condition (1.11) can be rephrased in terms of σ as $\sigma > \Delta M/\phi \propto \omega^3/E^2$. This shows that for high frequencies and constant σ, stimulated emission can be obtained only by very high fields. Hence above 10^{12} Hz it is no longer convenient to extract energy from free electrons (with $\sigma = \sigma_{Th}$), even though outstanding examples are offered of e.m. fields in the visible and u.v. regions generated by free electrons (Cherenkov radiation, synchrotron radiation). In Fig. 1.5 the regions where most lasers are available are shown. It is useful to compare the λ scale with the eV, Hz and K scales.

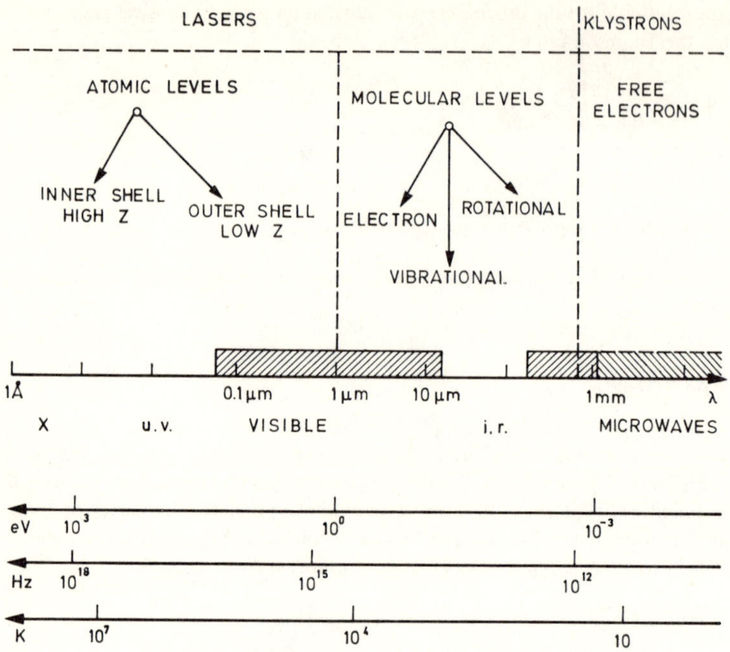

FIG.1.5. Map of coherent radiation emission mechanisms.

To put Eq. (1.15) in a different way, we recall that if γ is due only to radiation damping (no collision broadening or other events), then from classical electrodynamics we have

$$\frac{\omega}{\gamma} \sim \frac{\lambda}{r_0}$$

hence

$$\sigma_{res} \sim \lambda^2/4\pi \tag{1.15'}$$

Notice that the cross-section is not the square of the electron radius nor the square of Bohr's orbit but the squared wavelength which in the visible and u.v. regions is much bigger.

One arrives at the same numerical value by a semi-classical approach. Take the interaction Hamiltonian

$$H' = -\vec{d}\cdot\vec{E} \tag{1.16}$$

with a classical E field and a dipole operator whose matrix element between the two atomic states is $\langle 1|d|2\rangle = \mu$ ($\mu \sim 10^{-27}$ C·cm for an allowed transition). Then by Fermi's golden rule the transition we have

$$w = \frac{2\pi}{\hbar^2}\left(\frac{dn}{d\omega}\right)_{max} \mu^2 E^2 \tag{1.17}$$

Putting $(dn/d\omega)_{max} = 2/(\pi\Delta\omega)$ (Lorentzian curve) and dividing by the photon flux $\phi_n = c\epsilon_0 E^2/(\hbar\omega)$, we have

$$\sigma = \frac{w}{\phi_n} = \frac{4\omega}{\hbar c\epsilon_0 \Delta\omega} \mu^2 \quad (\mathrm{cm}^2) \tag{1.18}$$

This value is numerically given as $\lambda^2/4\pi$.

Dividing w by the photon number in the cavity,

$$n = \epsilon_0 E^2 V/(\hbar\omega)$$

we have the coupling constant per atom and per photon

$$g = \frac{1}{\hbar\epsilon_0 V} \frac{\omega}{\Delta\omega} \mu^2 \tag{1.19}$$

that we shall later use in the fully quantized interaction Hamiltonian (Fig. 1.6)

$$H' = \hbar g(a^+S^- + aS^+) \tag{1.20}$$

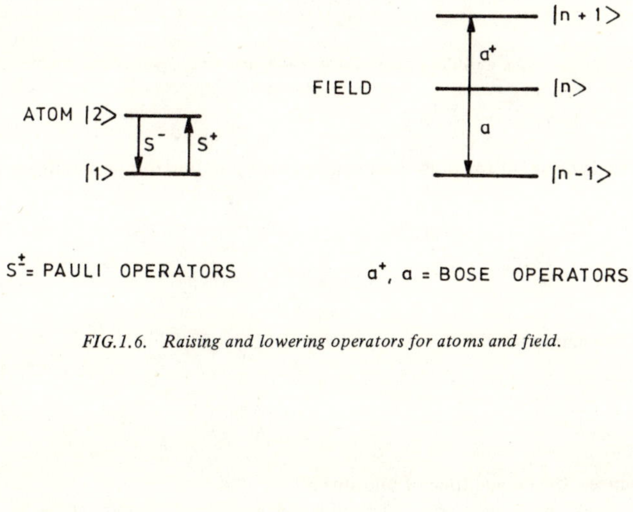

FIG.1.6. Raising and lowering operators for atoms and field.

Let us now go back to condition (1.11). To keep a large photon number within the cavity, we must have small losses. This can be obtained by confining the radiating atoms within two mirrors which act as a photon store.

The 'ingredients' necessary for a laser are:

(a) An *active medium,* i.e. a collection of atoms having a transition at the selected frequency;

(b) A *pump* or excitation mechanism to take the atoms away from thermal equilibrium, up to an excited state;

(c) A *cavity* made of two mirrors facing each other, with high reflectivity (this is called in optics a Fabry-Perot interferometer).

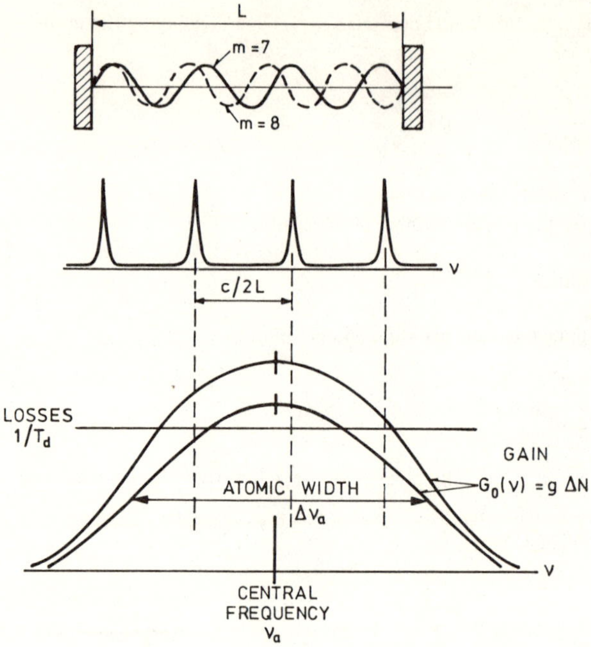

FIG.1.7. Scheme of two standing waves in a Fubry-Perot cavity, and interplay between cavity resonances and atomic gain line.

Such a cavity is resonant for those frequencies corresponding to the standing-wave condition (Fig. 1.7)

$$m \frac{\lambda}{2} = L$$

which amounts to a minimum frequency separation

$$\Delta\nu_{m,m+1} = \frac{c}{2L} \tag{1.21}$$

For these resonances, the escape time of photons is

$$T_d = \frac{L}{c} \frac{1}{1-R} \tag{1.22}$$

that is, a transit time L/c, multiplied by a loss factor, 1/(1–R), R being the mirror reflectivity.

Let us call $N = N_2 - N_1$, the difference of population between upper and lower state in the active medium, *population inversion*. Then, neglecting the spontaneous contributions (i.e. assuming Eq. (1.11) fulfilled), the rate equation for photons in the cavity will be

$$\frac{dn}{dt} = \left(\frac{dn}{dt}\right)_{gain} - \left(\frac{dn}{dt}\right)_{loss}$$

or, by use of relations (1.19) and (1.22),

$$\frac{dn}{dt} = g\,\Delta N \cdot n - \frac{n}{T_d}$$

Hence the *threshold* condition for starting *oscillations* from an initial spontaneous *seed* (in other words: to have a self-sustained oscillator rather than an amplifier; so that we should say LOSER rather than LASER!) will be

$$g\,\Delta N \geqslant \frac{1}{T_d} \tag{1.23}$$

This condition is represented in Fig.1.7 for two different pump values, i.e. for two different ΔN. In the first case, only one mode is above threshold, hence we have a single monochromatic frequency. In the second case we may have emission at three frequencies. Here we must introduce the fundamental difference between *homogeneous* and *inhomogeneous* line width. In the former case a monochromatic transition is broadened by circumstances which are *equal* for all atoms in the cavity (as spontaneous lifetime broadening, atomic collision broadening in a gas, phonon interaction in a solid matrix). All atoms can contribute over the *whole* line width. Hence, once the mode nearest to the peak has been excited, as the associated field 'sweeps' the cavity, it will 'eat' all atomic contributions, forbidding the other modes to go above threshold. In the standing-wave case this frequency picture is not sufficient and one should also consider the space pattern. As sketched in Fig. 1.8, two different modes have modes and maxima in different positions, hence they will 'exploit' different atoms, releasing the competition. The simultaneous laser action over many modes is then possible.

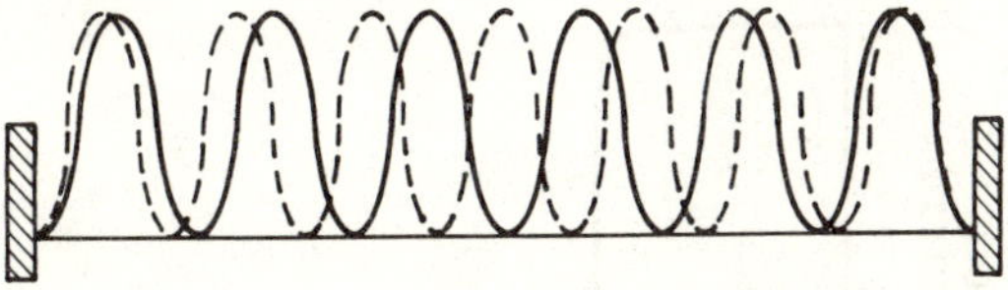

FIG.1.8. Intensity distributions of two standing waves.

The inhomogeneous line broadening corresponds to different frequency locations of different atoms. This can be due, e.g., to Doppler shift in a gas where thermal agitation gives a distribution of velocities to the molecules and hence a distribution of Doppler shifts

$$\Delta\omega = \omega V_k / c$$

where V_k is the velocity component along the $\vec{k}$ vector.

Another inhomogeneity occurs in a crystal where active ions are exposed to a crystal field which changes from site to site, contributing with a spread in the stark shifts. For an inhomogeneous line, different modes can go above threshold, even without a standing-wave pattern.

In general, if c/2L is much smaller than the atomic line width $\Delta\nu_a$, there are many independent laser lines, without phase relations.

In Fig. 1.9 and in the associated caption it is explained how to lock in phase the laser lines, in order to make short pulses (as short as the uncertainty relation permits, i.e. $1/\Delta\nu_a$).

(a)

(b)

(c)

(d)

(e)

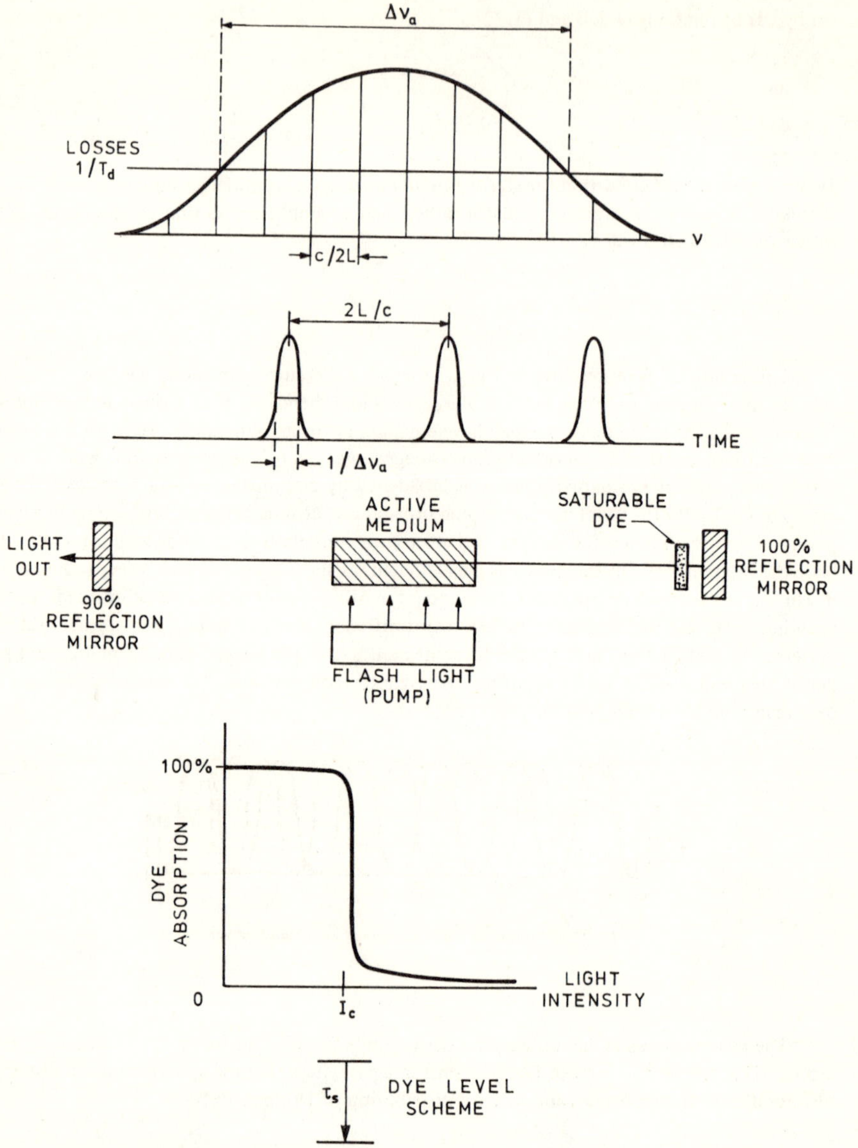

FIG.1.9. Mode locking operation.

(a) Frequency picture of a many-mode laser.

(b) If the different laser fields have fixed phase relations, they act as the different Fourier components of a train of pulses, each lasting $1/\Delta\nu_a$ and separated by $2L/c$. (b) is the Fourier transfer of the amplitude spectrum (a), provided the phases are all equal.

(c) Practical scheme of a mode-locked laser. Besides the three main ingredients (active medium, incoherent light to excite the atoms at the upper level, mirrors) there is also a saturable dye which becomes transparent at a critical light intensity I_c (see (d)). All the standing waves of the different modes will self-adjust their phases to have a maximum when the dye is transparent. Transparency is then lost with a decay time $\tau_s \ll 2L/c$ and then recovered after a transient $2L/c$. This corresponds to having a narrow pulse bouncing forth and back between the two mirrors. Notice that, according to (b), the pulse duration is $1/\Delta\nu_a$.

By using a Doppler broadened atomic line in a gas (like in a He-Ne or in an A^+ laser) we have

$$\Delta\nu \sim \frac{\nu}{c}\sqrt{\frac{k_B T}{M}} \sim 10^9 \text{Hz}$$

hence

$$t_{pulse} \sim 1 \text{ ns}$$

Using ions of a transition element imbedded in a crystal or glass matrix, as Cr^{3+} in Al_2O_3 (ruby), or Nd^{3+} in glass, one may have large $\Delta\nu_a$ because the 3d or 4f electron is strongly affected by the crystal field. A large $\Delta\nu_a$ can also be achieved in the case of complex dye molecules in a liquid solution because of the overlapping among many vibrational and rotational levels. It is nowadays easy to achieve

$$\Delta\nu_a \sim 10^{13} \text{ Hz}$$

and hence

$$t_{pulse} \sim 0.1 \text{ ps}$$

Notice that the range of picosecond times can be attained only by techniques such as in Fig. 1.7, and *not* by electronic shutters.

1.3. STIMULATED EMISSION AND NON-LINEAR OPTICS (NLO)

In Fig. 1.3 we have represented the difference between spontaneous and stimulated emission. With reference to the Hamiltonian (1.20), and considering only the field creation operator, it is well known that it acts on an m-photon state as follows:

$$a^+|n\rangle = \sqrt{n+1}\,|n+1\rangle \tag{1.24}$$

Therefore the transition rate for an emission process will be as follows:

Spontaneous emission: $|\langle 1|a^+|0\rangle|^2 = 1$
Stimulated emission: $|\langle n+1|a^+|n\rangle|^2 = n+1$

Also, in higher-order processes such as those studied in NLO (see Table 1–I) we can have a spontaneous and a stimulated version. Take a parametric process implying three Bose fields as in Fig. 1.10.

The model Hamiltonian for the processes in Fig. 1.10 is

$$H' = \hbar g(a_1 a_2^+ a_3^+ + a_1^+ a_2 a_3)$$

and the transition rate will have matrix elements as follows (see Fig. 1.11):

$|\langle 1|a_2^+|0\rangle|^2 = 1$ Spontaneous emission in field 2

or

$|\langle n_2+1|a_2^+|n_2\rangle|^2 = n_2+1$ Stimulated emission in field 2

TABLE 1-I. EXAMPLES OF NLO PROCESSES

Nature of the quanta		Name of the process
ω_2	ω_3	
Light	Molecular vibrations	Raman
Light	Optical phonons in solids	Raman
Light	Acoustical phonons in solids	Brillouin
Light	Sound waves in liquids	Brillouin
Light	Light	Parametric conversion (sum or difference of frequency, second harmonic generation, etc.)

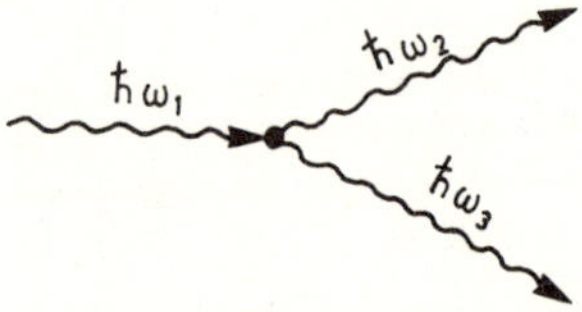

FIG.1.10. Non-linear process.

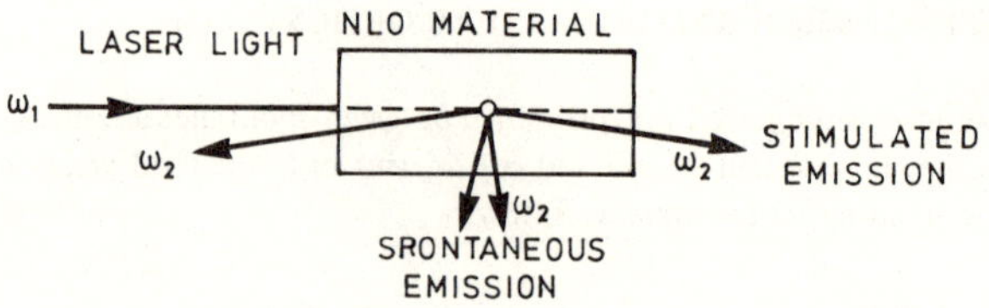

FIG.1.11. Angular relations in a non-linear optical process.

In the first case, we have an angle of 90° and we collect point-like processes, having to satisfy the conservation of energy:

$$\omega_1 = \omega_2 + \omega_3$$

In the second case, we have a direction (forward or backward) almost collinear with the impinging beam. Here, in order to add coherently the field contributions, the momentum matching condition

$$\vec{k}_1 = \vec{k}_2 + \vec{k}_3$$

has also to be satisfied.

In a similar way we may describe the usual light propagation in a transparent medium as an elastic two-photon process. Since the scattered contributions sum in phase, it is more convenient to speak of a linear polarization and a refraction index

$$P_i = \epsilon_0 \chi_{ij}^{(2)} E_j, \qquad n_{ij} = \sqrt{\chi_{ij}^{(2)}} \tag{1.25}$$

rather than stimulated emission in the scattered channel. Similarly, there are four photon processes leading to *'self-actions'* in the propagation of a large e.m. field, i.e. self-focusing, self-defocusing, self-modulation in phase (self-broadening) and amplitude (self-steepening). These non-linearities on the same light beam are described by a non-linear polarization index as

$$P_i = \chi^{(4)}_{ijk\ell} E_j E_k E_\ell \qquad (1.26a)$$

The non-linear refraction index can be written in the isotropic case as

$$n = n_0 + n_2 |E|^2 \qquad (1.26b)$$

In a liquid of anisotropic molecules, self-actions stem from the orientation of the molecules owing to interaction with the induced dipole moments (high-frequency Kerr effect). In a liquid of isotropic molecules, or in solids and gases, self-actions are due to electrostriction. For picosecond pulses, electrostriction is too slow to follow the amplitude change, and the effects are due to distortion of the electron cloud.

1.4. COHERENCE AND COOPERATIVE PHENOMENA

As shown in Fig.1.3, stimulated emission explains the privileged filling of a given mode, i.e. a narrowing in the frequency spectrum and in the spectrum of possible directions (monochromaticity and directionality). This amounts to increasing the *spectral purity* and it can be used in physics and technology (linear spectroscopy, holography, plasma production and compression by powerful laser pulses). But all this has very little to do with *coherence.* Stimulated emission is a first-order effect, i.e. it provides a linear source for Maxwell equations.

Each mode has still harmonic oscillator dynamics, i.e. it is like a particle in a parabolic potential well, with an equilibrium statistical distribution given by Maxwell/Boltzmann. To have a sizable amount of energy $|E_0|^2$, one has to increase the 'temperature', i.e. the excitation, thus broadening the distribution and increasing the entropy as well (Fig.1.12).

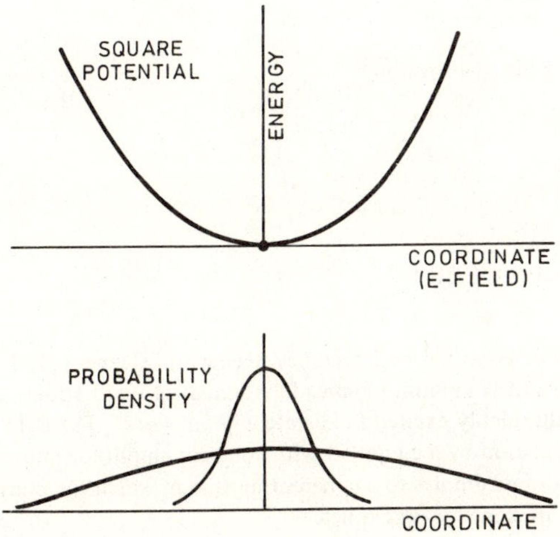

FIG.1.12. Harmonic potential well and equilibrium statistical distribution for a linear field.

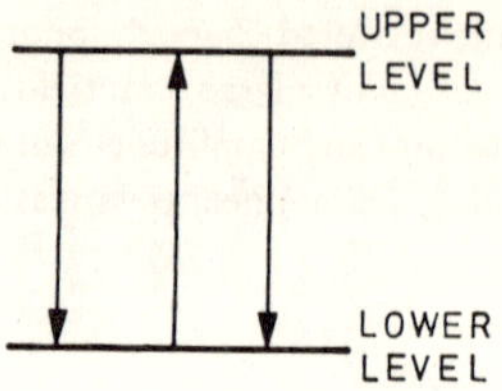

FIG.1.13. Third-order radiative process in a two-level atom.

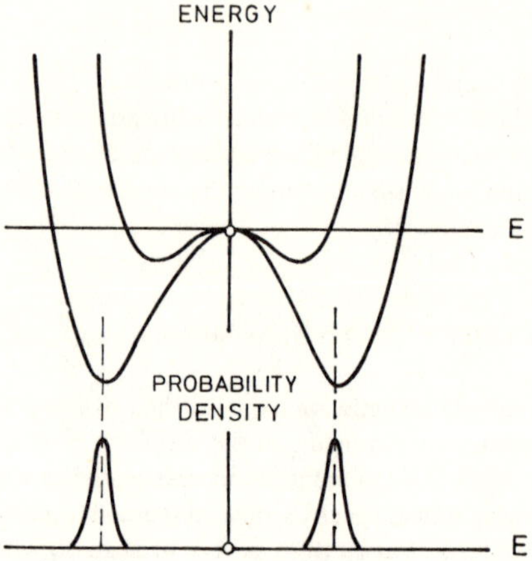

FIG.1.14. Quartic potential well and statistical distribution for a laser field above threshold.

However, as the field E increases, one must consider high-order processes besides the one-photon emission, as e.g. the three-photon process of Fig.1.13, which gives a quadratic field correction in the gain

$$G = G_0 - \beta|E|^2 \tag{1.27}$$

This is equivalent to a cubic polarization

$$P = G_0 E - \tfrac{1}{3}\beta|E|^2 E$$

and to a quartic free energy

$$W = -P \cdot E = -\frac{G_0}{2}|E|^2 + \frac{\beta}{12}|E|^4$$

As E increases, the quartic potential well becomes steeper and steeper (Fig.1.14), so that a useful E_0 can be reached with a little amount of spread, or statistical fluctuations, around it.

We call *coherent* this highly excited field state *without noise*. The field can be described with very good approximation by a c-number with constant amplitude and phase. Such a field can bring the induced atomic dipoles to a coherent motion in which the phase relations among atomic wave functions are kept for long times.

This is the basis for a *coherent non-linear spectroscopy* which sheds information on fine properties of atoms and molecules.

PART 2. PHOTON STATISTICS

2.1. RELEVANCE OF PHOTON STATISTICS (PS)

Consider a photodetector illuminated by a light beam. By an electronic gate lasting for a time T, the number n of photons annihilated at the photosurface in T is counted. The random variable n has a statistical distribution p(n) that can be determined by iterating the above procedure for a large number of samples.

In Fig.2.1 we give an experimental plot of the statistical distribution of photocounts p(n) versus the number of counts.

$$n = \eta \langle I \rangle T$$

$\langle I \rangle$ is the average intensity, T the gating time, η a constant accounting for the quantum efficiency of the detector plus other instrumental factors. The three curves in Fig.2.1 refer to three physical cases which are indistinguishable from the point of view of classical optics. Indeed, in the three cases we have the same average photon number $\langle n \rangle$, the same diffraction-limited plane wave, the same line width $\Delta\omega$ filtered out in such a way that

$$\tau = 1/\Delta\omega \gg T$$

i.e. each sample of the statistical distribution p(n) is collected over a time T during which the field amplitude |E| is practically constant. From the point of view of PS, the three lights are dramatically different, as seen from the figure. The variances associated with distributions (a) and (b) are respectively:

$$\langle \Delta n^2 \rangle = \langle n \rangle$$

$$\langle \Delta n^2 \rangle = \langle n \rangle + \langle n \rangle^2$$

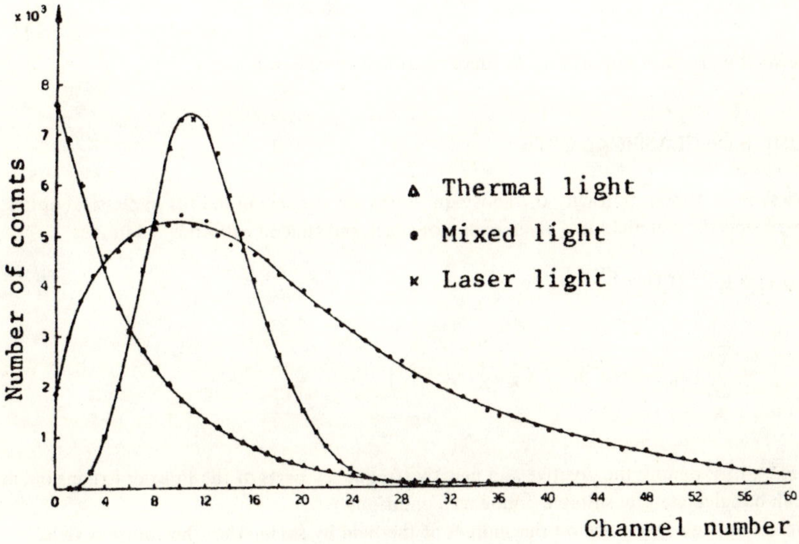

FIG.2.1. Photocount distributions.

In the laser case, p(n) is fitted by the familiar Poisson distribution which describes the number fluctuation in a volume containing a classical gas in equilibrium. The relative rms fluctuation is

$$\frac{\langle \Delta n^2 \rangle^{1/2}}{\langle n \rangle} = \frac{1}{\sqrt{\langle n \rangle}}$$

and for $\langle n \rangle \gg 1$ becomes negligible, justifying a description in terms of averages, as is done in thermodynamics.

In the thermal case (light source in thermal equilibrium as for the black-body) the relative rms fluctuation does not decrease

$$\frac{\langle \Delta n^2 \rangle^{1/2}}{\langle n \rangle} \sim 1$$

hence it is misleading to describe the field only in terms of average values $\langle n \rangle$.

The first contribution of (Δn^2) is a particle-like noise, as in the Poisson case; the second is a wave-like noise. If now one superposes a laser field with average photon number S and a thermal field with average $\langle n \rangle$ onto the same space mode, one obtains the sum of the two variances plus an interference term

$$\langle \Delta n^2 \rangle = S + \langle n \rangle + \langle n^2 \rangle + 2S \langle n \rangle$$

To show the importance of the last term, consider a communication channel with S coherent photons and $\langle n \rangle$ noisy thermal photons. If, e.g., $S = 10^4$, we get for the different $\langle n \rangle$ values:

$\langle n \rangle$	$\langle \Delta n^2 \rangle$
0	10^4
10	20×10^4

This shows the practical importance of these statistical considerations.

2.2. LIMITS OF CLASSICAL OPTICS

We show in this section why the above questions are not accounted for in classical optics. We can expand the free field in a given region of space and time in orthonormal modes

$$\vec{E}(r,t) = \vec{E}^{(+)}(r,t) + \vec{E}^{(-)}(r,t)$$

$$= \sum_K [C_K \vec{u}_K(r) e^{-i_K t} + C_K^* \vec{u}_K^*(r) e^{i\omega_K t}] \quad (2.1)$$

where $E^{(+)}$, $E^{(-)}$ denote the positive and negative frequency parts of the Fourier expansion, and $u_K(r)$ can be calculated by suitable boundary conditions.

We express our ignorance on the sources of the field by saying that the complex field amplitudes C_K are random quantities assigned through a probability distribution

$$p(\{C_K\}) = p(C_1, C_2, \ldots C_K, \ldots) \tag{2.2}$$

so that any field function is given on average by:

$$\langle\langle f(E)\rangle\rangle = \int f[E^{(+)}(\{C_K\})] \, p(\{C_K\}) \prod_K d^2C_K + c \cdot c \tag{2.3}$$

where

$$d^2C_K = d(\mathrm{Re}\, C_K)\, d(\mathrm{Im}\, C_K)$$

We see therefore that a *complete* characterization of the field (2.1) implies a knowledge of the joint probability of *all* the amplitudes C_K or, equivalently, of the correlation functions of the fields

$$G^{(n,m)} = \langle\langle E^{+}(X_1)E^{(+)}(X_2) \ldots E^{(+)}(X_n)E^{(-)}(X_{n+1}) \ldots E^{(-)}(X_{n+m})\rangle\rangle \quad (X_i \equiv r_i, t_i) \tag{2.4}$$

at any order (the ensemble average being performed as in Eq.(2.3)).

Directionality and monochromaticity, the properties associated with classical interferometers, give instead information on the *lowest-order* correlation function, namely

$$G^{(1,1)} = \langle\langle E^{(+)}(r_1 t_1)\, E^{(-)}(r_2 t_2)\rangle\rangle$$

This can be shown with reference to the Young and Michelson interferometers (Figs 2.2 and 2.3).

Let us consider a radiating cavity with an aperture d (Fig.2.2). The expansion (2.1) inside that cavity of each outgoing plane wave is broadened by diffraction by a solid angle $\theta_d^2 \sim (\lambda/d)^2$.

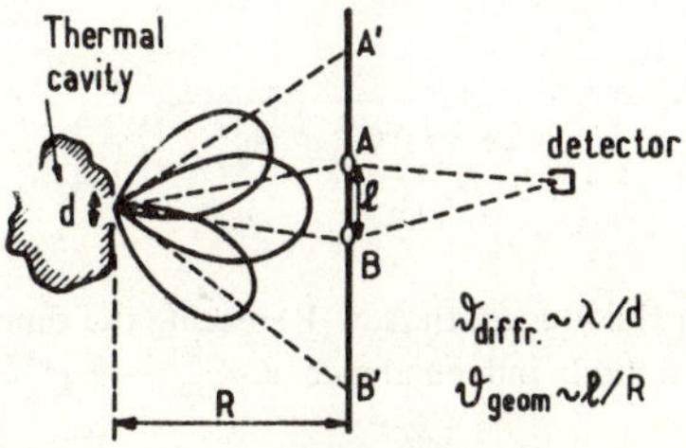

FIG.2.2. Young interferometer and coherence area. To see fringes, the geometrical angle ℓ/R must be smaller than the diffraction angle λ/d, hence (for solid angles) $\ell^2 < (R\lambda/d)^2 = A$ (coherence).

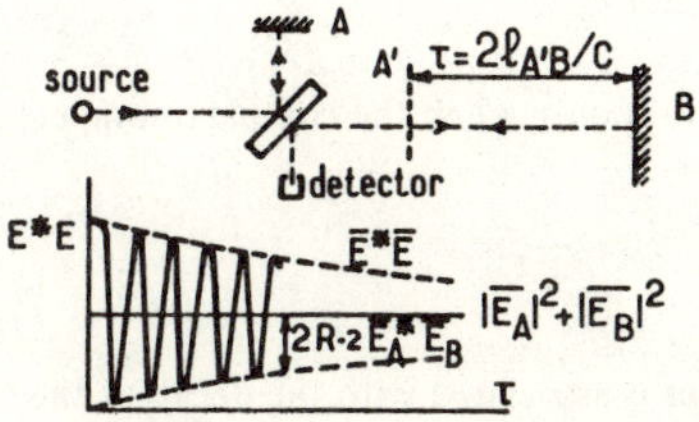

FIG.2.3. Michelson interferometer and coherence time. To see fringes, τ must be smaller than the reciprocal of the source line-width.

In Fig.2.2, a polar diagram for each plane wave is shown, centred at the centre of the aperture, giving the distribution of the intensities around the central propagation vector k for three different plane waves. The Young interferometer is made of a screen with two diffraction holes A, B, plus a square-law detector which time-averages the square of the instantaneous signal

$$|E_A + E_B|^2$$

over a resolving time much longer than an optical period. As the detector moves parallel to the screen, it collects an average contribution $\overline{|E_A|^2} + \overline{|E_B|^2}$ (bar means time average) plus a cross-term contribution $2\mathrm{Re}\,\overline{E_A^* E_B}$ which can be either positive or negative, depending on the phase relationship between E_A and E_B. This extra contribution gives rise to fringes.

If the source is in thermal equilibrium, we expect that the plane waves of the expansion (2.1) are mutually independent. Therefore, the cross-term is non-zero only when the holes A, B pick up fields coming from the same diffracted plane wave, that is, fringes appear only when the distance between A and B is such that the geometrical aperture d/R is smaller than the diffraction aperture λ/d (Fig.2.2).

The fringes are an indication of selection of a single k (directionality). Hence directionality is related to the quantity.

$$G^{(1,1)}(r_A, r_B) = \langle E^{(+)}(r_A, t)\, E^{(-)}(r_B, t)\rangle \tag{2.5}$$

Similarly, in a Michelson interferometer (Fig.2.3), the monochromaticity of a source of band width $\Delta\omega$ is investigated by introducing a delay

$$\tau = t_B - t_A$$

between the two mirrors B and A, and considering the fringe decay as τ is increased. Here light is correlated at the *same space* point but at different times, and the cross-term will be proportional to

$$\overline{E^{(+)}(r, t_A) E^{(-)}(r, t_B)} = \sum_{\ell m} \overline{C_\ell e^{-i\omega_\ell t} C_m^* e^{i\omega_m (t-\tau)}}$$

We have limited the sum over the line width $\Delta\omega$. Replacing the time average with an ensemble average and using the assumed mode independence, $\langle C_\ell C_m^* \rangle = |C_\ell|^2 \delta_{\ell m}$. Hence the sum reduces to

$$\sum_\ell |C_\ell|^2 e^{-i\omega_\ell \tau} \sim |C_0|^2 \sum_{\Delta\omega} e^{-i\omega_\ell \tau}$$

and it gives a non-zero contribution only when the complex numbers are almost phased, i.e. up to $\tau|\omega_\ell - \omega_{\ell'}| \lesssim 1$ or

$$\tau \approx 1/\Delta\omega$$

To conclude, the decay of fringes is associated with the decay of the term

$$G^{(1,1)}(t_A, t_B) = \langle E^{(+)}(r, t_A)\, E^{(-)}(r, t_B)\rangle \tag{2.6}$$

For the Young and the Michelson interferometers, Eqs (2.5) and (2.6) give the number of k values or ω values within the resolution range of the interferometer, i.e. they show how to select a single mode from the mode expansion (2.1). We still have to solve the problem "how is each mode statistically distributed?". We must go over to higher-order correlations such as Eq.(2.4), with $n, m > 1$.

2.3. CHARACTERIZATION OF RANDOM PROCESSES

A random process is described by a random function y(t) of time, which in realistic experiments we take as proceeding by quantized steps τ_R (τ_R = resolving time of the measuring set-up) and varying over a finite interval $(0, T_0)$. If we consider n values $t_1, t_2 \dots t_n$ of t in $(0, T_0)$, then the values $y(t_1), y(t_2) \dots y(t_n)$ are a set of random variables characterized by the hierarchy of joint probability distributions

$$W_1(y_1) \tag{2.7}$$

probability density of finding $y(t_1)$ between y_1 and $y_1 + dy_1$ normalized as $\int_{-\infty}^{\infty} W_1(y_1)\,dy_1 = 1$,

$$W_2(y_1, y_2) \equiv W_1(y_1) P_c(y_1|y_2) \tag{2.8}$$

joint probability of finding y_1 at t_1 and y_2 at t_2; this defines also a conditional probability of finding y_2 at t_2, given y_1 at t_1 through Eq.(2.8) and so on. One should go up to W_n to fully characterize the process in $(0, T_0)$. However, there are two simple random processes. The first is the fully random process in which no memory is kept of previous events, hence

$$W_n(y_1, y_2 \dots y_n) = W_1(y_1) W_1(y_2) \dots W_1(y_n)$$

The second is the Markoff process, completely characterized by W_2 in the sense that the conditional probability is with 'short memory', as expressed by the self-explanatory relation

$$W_3(y_1, y_2, y_3) = W_2(y_1, y_2) P_c(y_1, y_2|y_3)$$
$$= W_2(y_1, y_2) P_c(y_2|y_3)$$

i.e. P_c has no memory of y_1 but only of y_2.

En equivalent way of describing a random process is through the correlation functions

$$\langle y_1 y_2 \dots y_n \rangle \equiv \int y_1 y_2 \dots y_n \, W_n(y_1 y_2 y_n)\, dy_1 dy_2 \dots dy_n$$

for any set of times $(t_1 t_2 \dots t_n)$ including repeated indices. When the random process is an electric field, the first two correlation functions are of particular importance:

$$G^{(1)}(1,2) \equiv \langle E_1^* E_2 \rangle \tag{2.9a}$$

which, in particular for $t_1 = t_2$, becomes

$$I \equiv \langle E_1^* E_1 \rangle \tag{2.9b}$$

and

$$G^{(2)}(1,2) \equiv \langle I_1 I_2 \rangle = \langle E_1^* E_1 E_2^* E_2 \rangle \quad (2.10)$$

The field correlation function was used in Section 2.2. The intensity correlation function will be used in Section 2.4.

2.4. GAUSSIAN PROCESSES AND THE HANBURY-BROWN AND TWISS EFFECT

Gaussian processes with zero average are those whose W_n are Gaussian:

$$W_1(y) = N\, e^{-y^2/\sigma^2}$$

and so on. An equivalent definition is by saying that the odd correlation functions are zero and the even ones factor out in all products of pairs:

$$\langle y_1 y_2 \ldots y_{2n+1} \rangle = 0$$

$$\langle y_1 y_2 \ldots y_{2n} \rangle = \sum_p \langle y_1 y_2 \rangle \langle y_3 y_4 \rangle \ldots \langle y_{2n-1} y_{2n} \rangle$$

where $\sum_p$ is the sum over all permutations of (1 ... 2n). For example:

$$\begin{aligned} \langle y_1 y_2 y_3 y_4 \rangle &= \langle y_1 y_2 \rangle \langle y_3 y_4 \rangle \\ &+ \langle y_1 y_3 \rangle \langle y_2 y_4 \rangle \\ &+ \langle y_1 y_4 \rangle \langle y_2 y_3 \rangle \end{aligned} \quad (2.11)$$

For a complex field, combining Eqs (2.10) and (2.11), one gets

$$\langle I_1 I_2 \rangle = \langle I_1 \rangle \langle I_2 \rangle + |\langle E_1^* E_2 \rangle|^2 \quad (2.12)$$

This is a very important relation, showing that the Gaussian intensity correlation sheds information on the square of the field correlation (see Fig.2.4). Gaussian processes are important for the following two reasons:

(a) Entropy argument

In a cavity at thermal equilibrium the entropy as a function of the variable E is a maximum S_0; hence, expanding around S_0, one has

$$S(E) = S_0 - \alpha E^2$$

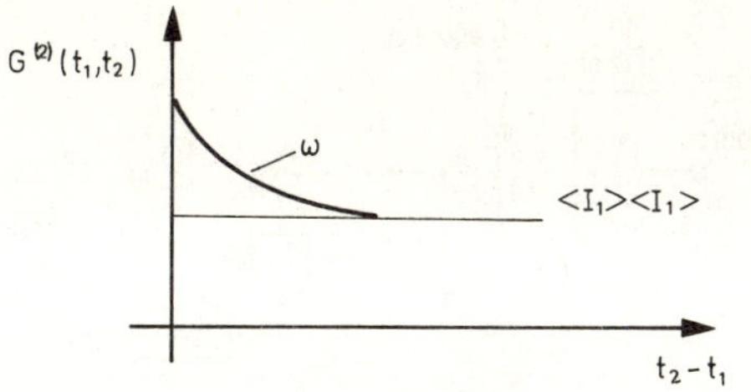

FIG.2.4. Plot of the intensity correlation function.

But the field probability is given by the Boltzmann relation

$$W(E) = e^{S/k} \rightarrow e^{S_0/k} e^{-(\alpha/k)E^2}$$

hence it is Gaussian

(b) Central limit argument

By the central limit theorem, the sum of many uncorrelated events is Gaussian. Such is the case of light generated by the atomic spontaneous emissions in a thermal source, or the light scattered by microscopic bodies in a light scattering experiment. In such a case, the scattered field E_s is proportional to the impinging field E_0 and to the random polarizability $\Delta\epsilon$ of the scattering medium:

$$E_s \propto \Delta\epsilon \cdot E_0$$

If E_0 is free from fluctuations (say, laser light) then the correlations in E_s repeat faithfully the correlations in $\Delta\epsilon$ and hence give information on the medium behaviour. Equation (2.12) says that the *intensity correlation* (as obtained with a photodetector plus an electronic correlator) gives *all* information on the scatterer (except for frequency shifts). Historically, all that was born by an extrapolation of the interferometer idea.

In 1956 a new interferometer was introduced by Hanbury-Brown and Twiss, correlating the outputs of two detectors and therefore correlating the intensities rather than the fields (Fig.2.5). The result of the experiment is as follows:

$$G(2,2) = \langle |E_1|^2 |E_2|^2 \rangle \qquad (2.13)$$

where E_1, E_2 are the fields at the two detectors. Suppose we have already selected by filters a single monochromatic plane wave:

$$E(r,t) = C e^{-i(\omega t - kr)} + C^* e^{i(\omega t - kr)}$$

Then Eq.(2.13) this time implies the statistical distribution of C, *not only* its average intensity, since it is proportional to the fourth moment:

$$\langle |C|^2 |C|^2 \rangle = \langle |C|^4 \rangle$$

Two typical examples are given as follows.

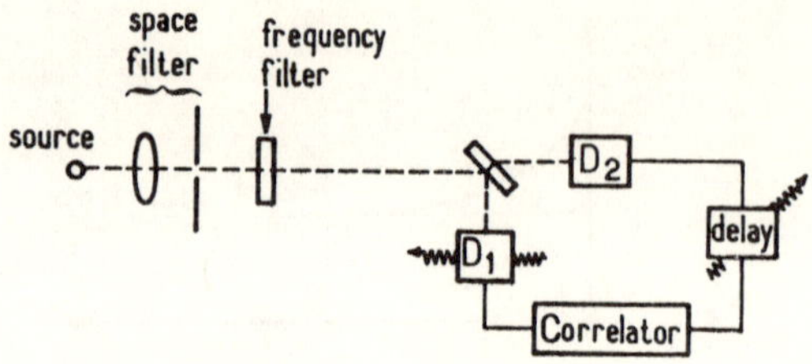

FIG.2.5. Hanbury-Brown/Twiss interferometer. D_1, D_2 = detectors. Motion in D_1 to pick up different angular regions; variable delay between D_1 and D_2.

2.4.1. Gaussian distribution with zero average

$$p(C) = \frac{1}{\pi\langle n\rangle} e^{-|C|^2/\langle n\rangle}$$

This corresponds to thermal equilibrium. Indeed it is a Boltzmann distribution $e^{-W/kT}/Z$ with $W = n\hbar\omega$ (n = photon number $\propto |C|^2$). It is well known that a complex Gaussian distribution has the following relation between fourth and second moment (see Eq.(2.12)):

$$\langle|C|^4\rangle = 2\langle|C|^2\rangle$$

2.4.2. 'Coherent' field without fluctuations

$$p(C) = \delta^{(2)}(C - C_0) \qquad (2.14)$$

($\delta^{(2)}$ is a Dirac δ-function in the complex plane G).

In such a case the intensity correlation is

$$\langle|C|^4\rangle = \langle|C|^2\rangle = |C_0|^4 \qquad (2.15)$$

Equations (2.14) and (2.15) suggest that coherence can be defined as 'δ-like amplitude distribution', or 'correlation functions factorized at any order'.

The above equivalent definitions can be generalized to include a many-mode field. A field made of, say, two modes is coherent when

$$p(C_1, C_2) = \delta^{(2)}(C_1 - C_{10})\, \delta^{(2)}(C_2 - C_{20}) \qquad (2.14)'$$

or

$$\langle E_1^{(+)} E_2^{(+)} \ldots E_{n+m}^{(-)}\rangle = \langle E_1^{(+)}\rangle \langle E_2^{(+)}\rangle \ldots \langle E_{n+m}^{(-)}\rangle \qquad (2.15)'$$

but it may be neither monochromatic nor unidirectional!

2.5. MEASUREMENT OF PHOTON STATISTICS (PS)

We give here a simplified picture based on the argument that photons, being particles with zero mass, cannot be localized when the field is uniform (Fig.2.6(a)).

Hence there is no a-priori correlation between the outputs of two detectors 1 and 2, and the photons, whose average number is proportional to the square field and the measuring time T,

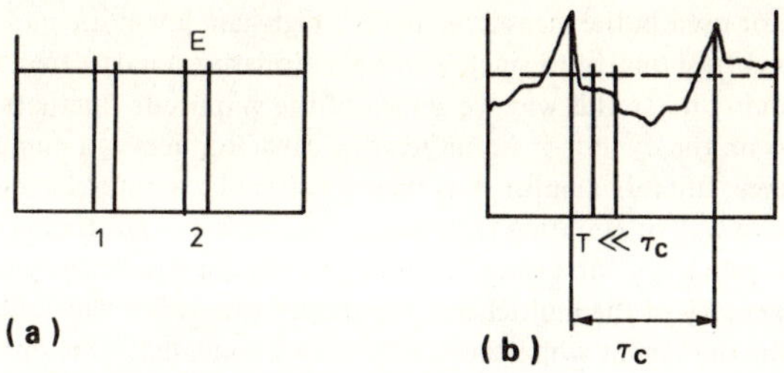

FIG.2.6. Profile of the field amplitude for a coherent case (a) and a fluctuation field case (b). τ_c = coherence time (decay time of the field correlation function); T = observation time.

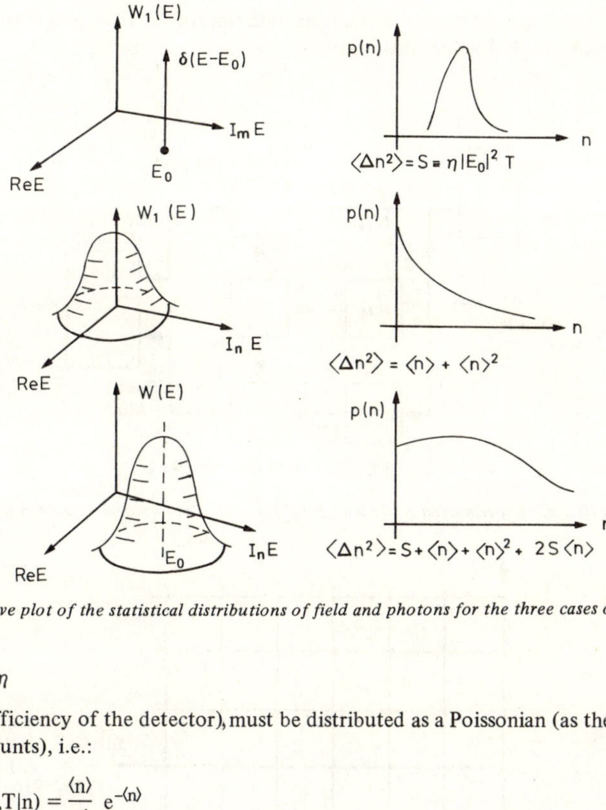

FIG.2.7. Qualitative plot of the statistical distributions of field and photons for the three cases of Fig.2.1.

$$\langle n \rangle = |E|^2 T\eta$$

(η = quantum efficiency of the detector), must be distributed as a Poissonian (as the statistics of radioactive counts), i.e.:

$$p(n) \equiv K(E,T|n) = \frac{\langle n \rangle}{n!} e^{-\langle n \rangle} \qquad (2.16)$$

If now (Fig.2.6(b)) the field is randomly distributed with a statistics $W_1(E,t)$ and each measurement lasts for a time T much smaller than the coherence time τ_c, $T \ll \tau_c$, in order to have a constant field within each sample, then we must average the detector statistics (2.16) over the field statistics

$$p(n,T,t) = \int K(E,T|n) W_1(E)\, dE$$

In Fig.2.7 the results are shown pictorially for the three cases of Fig.2.1.

The photodetector used in the measurements is a high-gain low-noise multiplier phototube. Anode-current pulses corresponding to single photoelectrons are standardized in amplitude and shape by a non-linear circuit. In this way we get rid of the amplitude fluctuations in the multiplication process on the dynodes. An integrating capacitor acts as a number-to-voltage converter and its voltage, suitably amplified, is then classified by a multichannel pulse-height analyser. An alternative way of counting the number of pulses in a given time interval T is to use a fast electronic scale, gated 'on' for a time T, which records the number of pulses and is directly connected with the memory of the multichannel analyser; this avoids the double conversion process, thus increasing the rate at which counts can be accumulated. One obtains directly the distribution p(n,T) of photoelectron numbers. The distributions reported in Fig.2.1 have been obtained by this method.

A single p(n,T) gives only an integrated information on the time evolution of the field. One way of measuring the time evolution would be to measure the PS for increasing time-interval T, up to (or larger than) the characteristic relaxation time of the field, and then to correlate the various shapes of the photocount distributions.

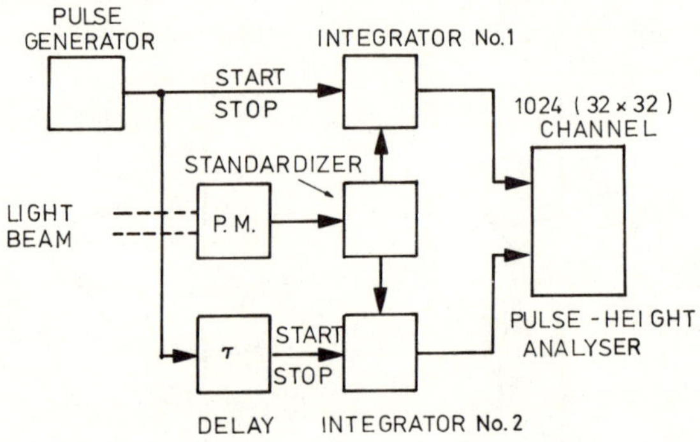

FIG.2.8. Experimental set-up used for measuring joint photocount distributions.

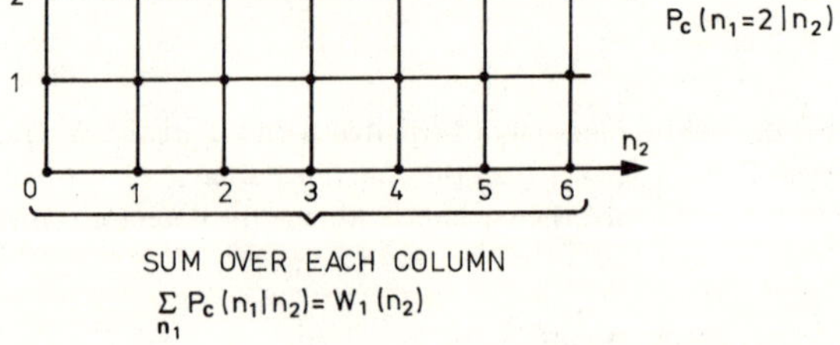

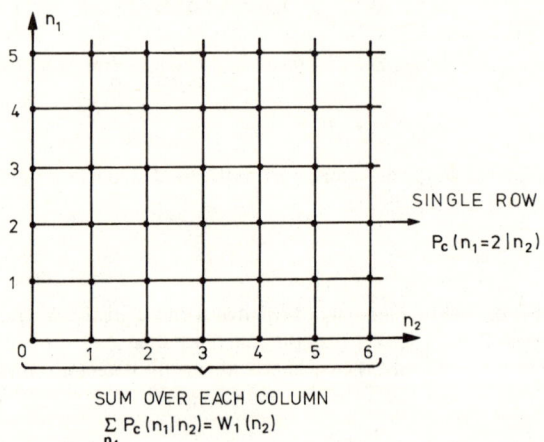

FIG.2.9. Matrix of the output numbers (each represented by a dot) as they are printed at the digital output of the multichannel analyser. The numbers on a given line (e.g. the row $n_1 = 2$) give a conditional distribution.

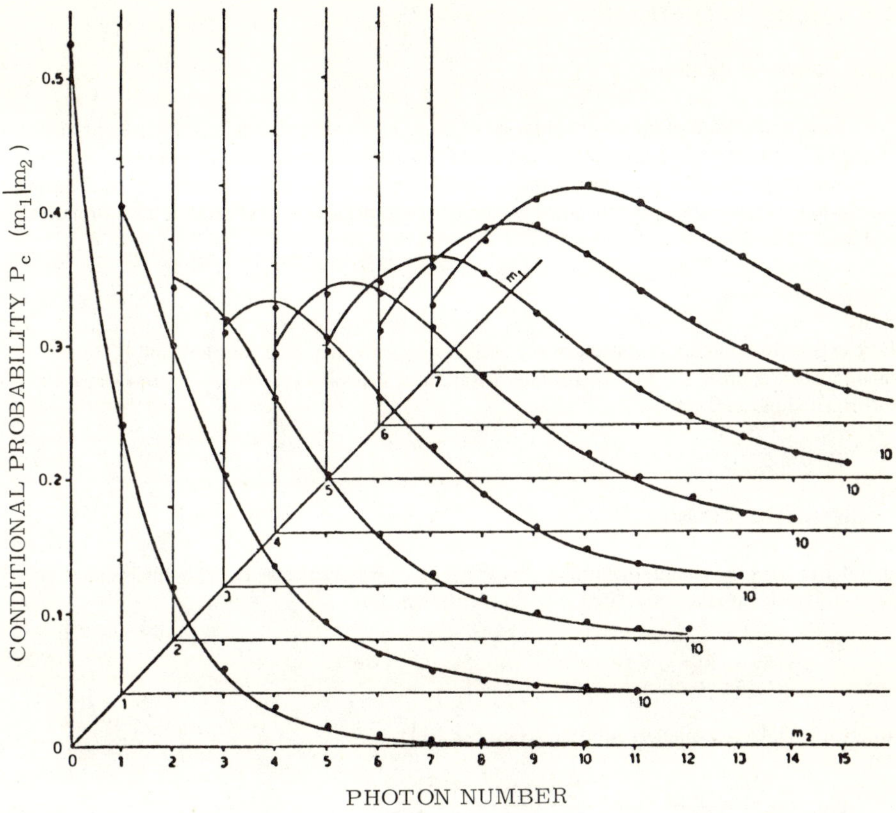

FIG.2.10. Joint photocount distribution for a Gaussian field (laser light scattered by a rotating ground glass disk) with a coherence time of 700 μs. The delay is 150 μs (Arecchi et al. [1]). Total count number per curve = 10^5; delay τ = 150 μs; average photon number m = 2.0; ——— theoretical curves; ● experimental points.

It is better, however, for the interpretation of the results to correlate separate observations, each one made for a time T much shorter than the coherence time, as shown in the experimental set-up (Fig.2.8). Essentially, the operation described before is repeated twice, at times t_1 and t_2, and the two results are sent to a two-dimensional multichannel pulse-height analyser. The results are classified on a two-dimensional matrix, which gives the joint photocount distribution

$$W_2(n_1 t_1, n_2 t_2) = P_c(n_1 t_1 | n_2 t_2)\, W_1(n_1 t_1)$$

One easily realizes that P_c is given by the reading on a row defined by the chosen value of n_1 (Fig.2.9) and is symmetrical with respect to indices 1 and 2 for all stationary processes. The marginal distribution $W_1(n_1) = W_1(n_2)$, corresponding to an uncorrelated experiment of the kind described previously, is obtained by summing for each column (row) the values corresponding to all rows (columns) belonging to that column (row).

Figure 2.10 gives the experimental results for a stationary Gaussian field, together with the theoretical curves.

2.6. LASER FLUCTUATIONS

2.6.1. Review of the theory

As an application of PS we present a set of experiments on both the stationary and transient statistical properties of a laser system. Let us make a heuristic description of a laser.

First let us consider a damped harmonic oscillator. Its statistical amplitude can be described by means of a Langevin equation, which we write in a rotating frame (i.e. after a transformation $\alpha \to \alpha e^{-i\omega t}$) as

$$\dot{\alpha} + \gamma\alpha = \Gamma(t) \tag{2.17}$$

Here $\alpha(t)$ is the complex amplitude of the oscillator, γ is the damping constant, and $\Gamma(t)$ is a random noise source. Let $\Gamma(t)$ be a complex stationary Gaussian process, with zero average and δ-correlated in time:

$$\langle\Gamma(t)\rangle = 0$$

$$\langle\Gamma^*(t)\Gamma(0)\rangle = Q\,\delta(t)$$

In order to have the conditional probability $P(\alpha_0, 0|\alpha,t)$ we must solve the two-dimensional Fokker-Planck equation associated with Eq.(2.17), which is

$$\frac{\partial P}{\partial t} - \gamma\,\mathrm{div}_\alpha(\alpha P) = q\,\nabla_\alpha^2 P$$

where $q = Q/4$. Its complete solution is (see Fig.2.11)

$$P(\alpha_0, 0|\alpha,t) = \frac{1}{\pi\sigma^2(t)}\exp\left[-\frac{|\alpha - \alpha_0 \exp(-\gamma t)|^2}{\sigma^2(t)}\right] \tag{2.18}$$

where

$$\sigma^2(t) = \frac{2q}{\gamma}(1 - e^{-2\gamma t})$$

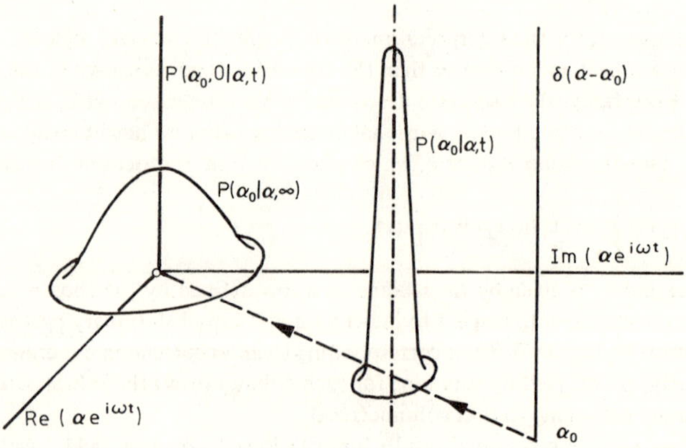

FIG.2.11. Evolution of the conditional probability for a damped harmonic oscillator.

In the limit $t \to \infty$ we obtain the first-order probability density

$$\lim_{t\to\infty} P(\alpha_0 0|\alpha t) = W_1(\alpha) = \frac{1}{\pi 2q/\gamma} \exp\left[-\frac{|\alpha|^2}{2q/\gamma}\right] \tag{2.19}$$

In Eq.(2.18) the oscillator amplitude has been described as a stochastic Markoff process whose ensemble distribution is a Gaussian with variance $\langle n\rangle = 2q/\gamma$ and whose frequency spectrum is Lorentzian with half-width $\Delta\omega = \gamma$ (i.e. whose first-order correlation function has an experimental time decay with a time constant $1/\gamma$).

We may apply this treatment to a single-mode laser field and think of it as an oscillator whose damping is compensated by a gain term G due to the excited atoms in the cavity. The associated Langevin equation is

$$\dot{\alpha} + (\gamma - G)\alpha = \Gamma(t)$$

Generally, G is amplitude dependent, but in first approximation it can be considered as a constant $G = g_0$. In this case the same solutions (Eqs (2.18), (2.19)) still hold, in which we simply have to replace γ by $\gamma - \gamma g_0$. A more realistic picture is obtained considering the dependence of G on $|\alpha|^2$ (cubic non-linearity in the induced dipole moment), i.e.

$$G = g_0 - \beta|\alpha|^2$$

The Langevin equation becomes

$$\dot{\alpha} - \beta(d - |\alpha|^2)\alpha = \Gamma(t) \tag{2.17'}$$

where the linear negative damping $\beta d = g_0 - \gamma$ comes from the linear theory and $\beta|\alpha|^2$ is the first non-linear correction. The Fokker-Planck equation becomes more complicated and we give only the final results. Its stationary solution in the plane $\alpha \equiv re^{i\phi}$ is independent of ϕ and is given by:

$$P(r) = N \exp\left[-\frac{\beta}{4q} r^4 + \frac{\beta d_{r^2}}{2q}\right] \tag{2.19'}$$

It is suitable to introduce the pumping parameter

$$a = \sqrt{\frac{\beta}{q}}\, d \tag{2.20}$$

and to normalize the square modulus as $X^2 = \sqrt{\beta/q}\cdot r^2$. The P(r) distribution is then written as

$$P(X) = N \exp\left[-\frac{1}{4}X^4 + \frac{a}{2}X^2\right] \tag{2.21}$$

We define the threshold point as $\gamma = g_0$, i.e. as that point where the linear gain is equal to the losses. At threshold,

$$d = a = 0$$

Below threshold ($a < 0$) the distribution (2.21) has a peak at $X = 0$, whereas above threshold ($a > 0$) it exhibits a peak at $X = \sqrt{a}$. It is easily shown that, outside the interval around threshold, the following approximations hold:

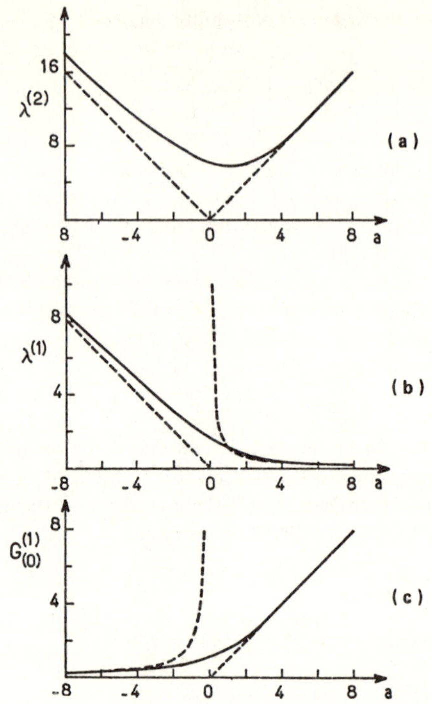

FIG.2.12. Decay constants $\lambda^{(2)}$ of $G^{(2)}(t)$ and $\lambda^{(1)}$ of $G^{(1)}$, and intensity $G^{(1)}_{(0)}$ versus pump parameter a. Dashed lines: linearized theory. Solid lines: exact solution of Fokker-Planck equation.

(a) Laser far below threshold ($a < -10$)

Equation (2.21) reduces to

$$P(r) \sim N \exp\left[-\frac{|a|}{2} X^2\right] = N \exp\left[-\frac{\beta|d|}{2q} r^2\right]$$

which is a Gaussian distribution centred at $r = 0$, with a variance $\langle X^2 \rangle$ equal to $2/|a|$ (Fig.2.12(c) for $a < 0$).

(b) Laser far above threshold ($a > 10$)

The saturation becomes strong and tends to stabilize the field amplitude around its average value $\langle x \rangle = \sqrt{a}$. The field statistics is

$$P(X) = N' \, e^{-a(X-\sqrt{a})^2} \qquad (2.21)'$$

This is a Gaussian distribution centred at $X = \sqrt{a}$. $P(\alpha)$, as given in Eq.(2.21)′, is only a modulus distribution uniform in phase.

We give now the time evolution of the field in the same two limiting cases.

(a) Laser far below threshold

The following formulas hold for the field and intensity correlation functions respectively:

$$G^{(1)}(\tau) = \langle \alpha^*(\tau)\, \alpha(0)\rangle = \frac{2q}{\beta|d|}\, e^{-\beta|d|\tau}$$

$$G^{(2)}(\tau) = \langle|\alpha(\tau)|^2\, |\alpha(0)^2|\rangle + \left(\frac{2q}{\beta|d|}\right)^2 e^{-2\beta|d|\tau}$$

The Fourier transforms of these expressions give the spectra of the field and intensity fluctuations. These spectra have Lorentzian shapes with half-widths $\lambda^{(1)} = \beta|d|$, $\lambda^{(2)} = 2\lambda^{(1)}$, respectively. They are illustrated in Fig.2.12 (a,b) for $a < 0$.

(b) Laser far above threshold

Far above threshold the field distribution is peaked at $\sqrt{d}$. The laser field appears as the linear superposition of a coherent field with amplitude $\sqrt{d}$ plus a Gaussian field with zero average, the photon number being given by

$$\langle n\rangle = \frac{q}{2\beta d}$$

and the decay constant increasing linearly with d, i.e.

$$\Delta\omega = 2\beta d$$

The phase ϕ obeys a diffusion equation whose solution has a diffusion constant q/d proportional to the reciprocal of the output power (Townes formula for a maser oscillator); see Fig.2.12(b) for $a > 0$.

(c) Laser at threshold

So far we have discussed only the cases well above and below threshold. The behaviour in the threshold region cannot be obtained by a simple extrapolation of the previous results and we need a complete solution.

We refer to the theoretical calculations by Risken and Vollmer [2]. Their solutions are plotted as solid lines in Fig.2.12(a, b, c) and for comparison are again reported in the experimental results (Figs 2.13 to 2.16). We see that the threshold point does not show discontinuities as it should appear from a naive linear picture.

2.6.2. Stationary experiments (ensemble distributions and time correlation) [3,4]

In this section we describe the experimental results obtained by means of the PS method in the study of the statistics of the e.m. field of a stabilized laser operating in different conditions. We used a 6328 Å He-Ne laser, single TEM_{00} mode, with one mirror supported by a piezoceramic disc in order to stabilize against fluctuations and to move the mode position with respect to the atomic line.

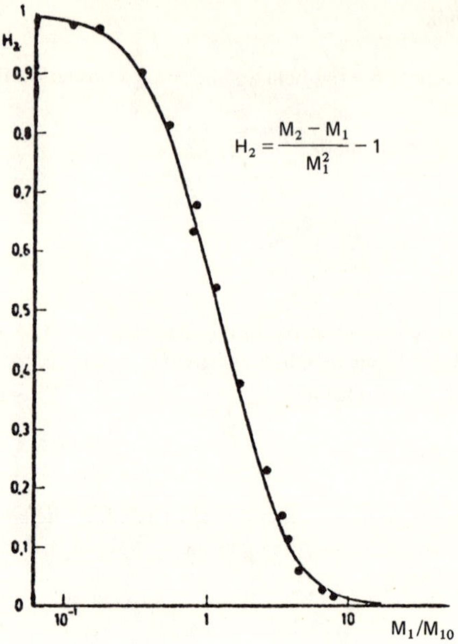

FIG.2.13. *Measured and theoretical values of the reduced second-order factorial moment H_2 as a function of the normalized intensity M_1/M_{10} in the threshold region.* ——— *Theory,* ● *experimental points.*

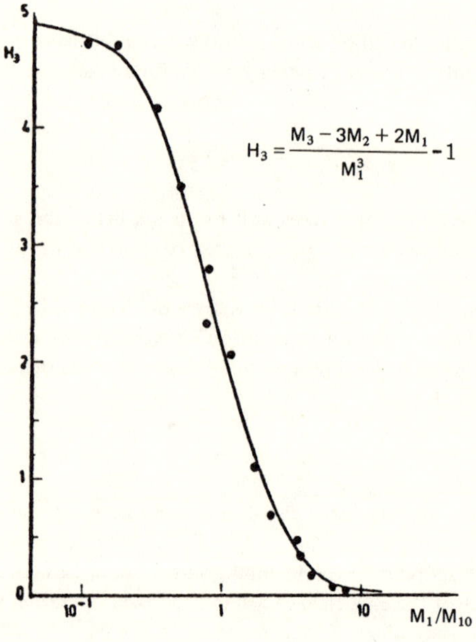

FIG.2.14. *Measured and theoretical values of the reduced third-order factorial moment H_3 as a function of the normalized intensity M_1/M_{10} in the threshold region.* ——— *Theory,* ● *experimental points.*

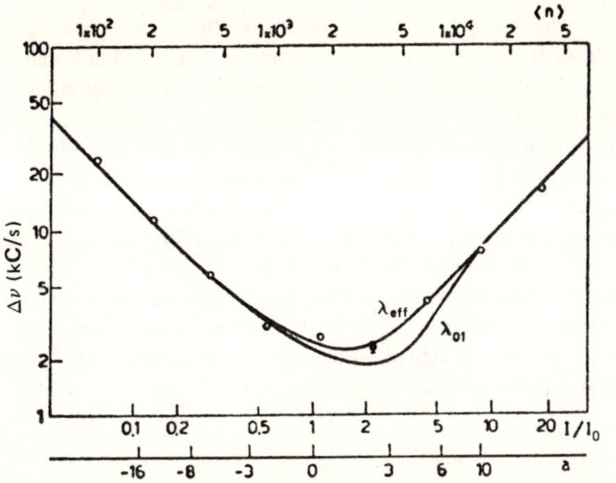

FIG.2.15. Plot of the 'effective' line width $\Delta\nu$ versus the laser intensity I normalized to the threshold values I_0. The horizontal axis is also calibrated in values of the pump parameter a and in average photon number ⟨n⟩ inside the cavity.

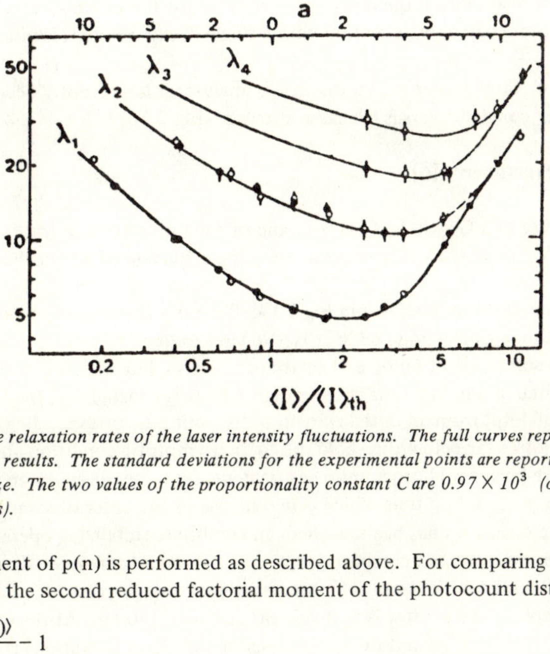

FIG.2.16. Plots of the relaxation rates of the laser intensity fluctuations. The full curves represent the theoretically predicted results. The standard deviations for the experimental points are reported only when they exceed the dot size. The two values of the proportionality constant C are 0.97×10^3 (open circles) and 1.03×10^3 (full circles).

The measurement of p(n) is performed as described above. For comparing experiments and theory we used the second reduced factorial moment of the photocount distribution

$$H_2 = \frac{\langle n(n-1)\rangle}{\langle n\rangle^2} - 1 \tag{2.22}$$

which goes from 1 (Gaussian field distribution; well below threshold) to 0 for an amplitude-stabilized field (well above threshold), and the third one

$$H_3 = \frac{\langle n(n-1)(n-2)\rangle}{\langle n\rangle^3} - 1$$

which goes from 5 (Gaussian distribution) to 0 (amplitude-stabilized field) (Figs 2.13 and 2.14).

From the stationary solution of the Fokker-Planck equation for the statistical distribution of the laser field the distribution of photocounts and the associated factorial moments can be derived. One can see from the figures that the agreement between experiments and theory is very good. Finally we report the frequency spectra of the intensity fluctuations and show that they are consistent with the time-dependent solutions of the same Fokker-Planck equation whose stationary solution is fitted by the ensemble distribution reported above.

The power spectra of the intensity fluctuations measured with a wave analyser showed small deviations from a Lorentzian shape in the threshold region. For any spectral shape we define the 'effective' line width $\Delta\nu$ of an equivalent Lorentzian as the ratio between the total spectral power and the zero frequency value of the spectral density. The measured line width fits the numerical calculations of Risken and Vollmer [2] who gave a dynamical solution for the intensity correlation function as a sum of several exponential terms (with decay constants $\lambda_k^{(2)}$) occurring with weights M_k. This leads to an equivalent decay constant

$$\lambda_{\text{eff}}^{(2)} = \left(\sum_k \frac{M_k}{\lambda_k^{(2)}} \right)^{-1} \qquad \sum_k M_k = 1$$

In Fig.2.16 we have also plotted the main decay constant of the intensity correlation function. It is clear that a Lorentzian approximation with a single decay constant is not an adequate description.

By using a correlator rather than a frequency analyser it has recently been possible to measure the first four decay constants versus the laser intensity (Fig.2.16) [5].

2.6.3. Transient experiments [6]

By the joint use of a Q-switched gas laser and of the linear method for PS a non-stationary statistical ensemble can be studied, measuring the time evolution of a laser field during a fast build-up.

The experimental set-up is shown in Fig.2.17. We put a Kerr cell with end faces at the Brewster angle within a single-mode 6328 Å He-Ne laser cavity.

Starting with some pre-set pump and cavity parameters, but with the optical shutter closed, the Kerr cell is switched 'on' in a time shorter than 5 ns at the instant $t = 0$. The laser field undergoes a transient build-up, from an initial statistical distribution, corresponding to the equilibrium between gain and losses far below threshold, up to an asymptotic condition above threshold. At the instant $t = \tau$ we perform photocount measurements for a measuring interval T of 50 ns, very small compared to the build-up time which is in our case of the order of some microseconds. Once a steady-state condition has been reached, an amplitude-stabilizing operation is performed by sampling the laser output and comparing this with a standard reference signal. This is equivalent to 'preparing' an identical initial state for a successive measuring cycle.

After the sampling, the shutter is switched off for about 10 ms. At the end of this interval the shutter is again switched on and the above-described cycle of operations is repeated. In this way we collect an ensemble of macroscopically identical events. By successively varying τ, we obtain the time evolution of the photocount distribution $p(n,T,\tau)$. A set of experimental results is shown in Fig.2.18. The average photocount number $\langle n \rangle$ and the associated variance $\langle \Delta n^2 \rangle = \langle n^2 \rangle - \langle n \rangle^2$ are reported as a function of the time delay in Figs 2.19 and 2.20 for two different pumping conditions.

Notice that in Fig.2.20 the dashed line just interpolates the experimental points but has no theoretical significance. In Fig.2.21 we correct for attenuation, as described in the next section, and we see that the experimental points agree with the theoretical, solid line.

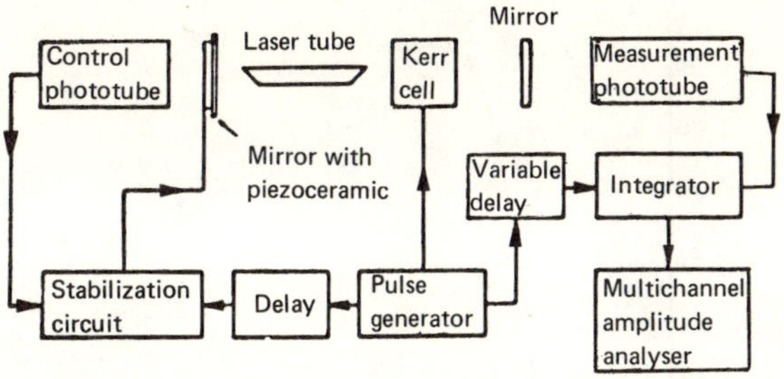

FIG.2.17. Experimental set-up for transient experiments.

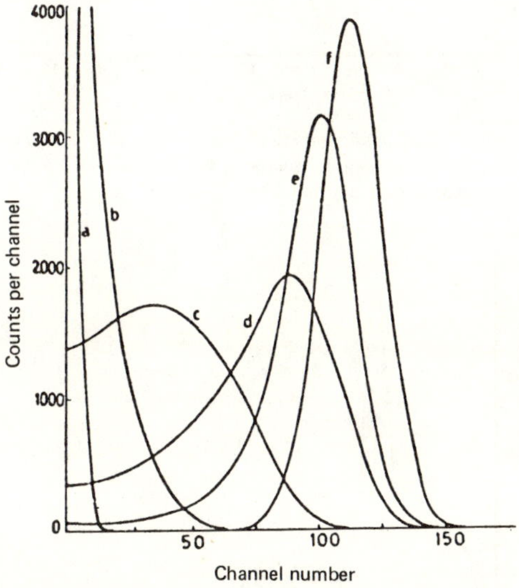

FIG.2.18. Experimental statistical distribution with different time delays obtained on a laser transient. The solid lines connect the experimental points which are not shown, in order to make the figure clearer. All distributions are normalized to the same area: a – 2.6 μs; b – 3.7 μs; c – 4.3 μs; d – 5 μs; e – 5.6 μs; f – 8.8 μs.

2.7. DISTORTION OF PHOTON STATISTICS OWING TO ATTENUATION [7]

The role of attenuation (Fig.2.22) is as follows. For an n-photon field (as described in quantum mechanical books) the attenuation would give rise to a binomial statistics as the partition noise through the grid of an electron vacuum tube. However, for a statistical mixture of coherent fields (each having a Poisson spread in photon number, as described above) the attenuation is a purely classical process

$$E' = \eta E$$

which affects the statistics, leaving unaffected the elementary probability

$$P(E')\, d^2 E' = P(E)\, d^2 E \tag{2.23}$$

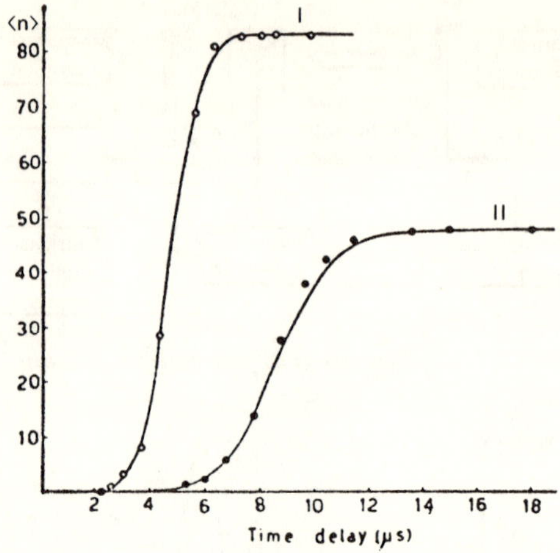

FIG.2.19. Evolution of the mean value ⟨n⟩ of the statistical distribution p(n, T, τ). The solid lines represent the theoretical curves which best fit the experimental points (○, ●). The experimental points related to curve I have been obtained by the statistical distribution of Fig.2.18.

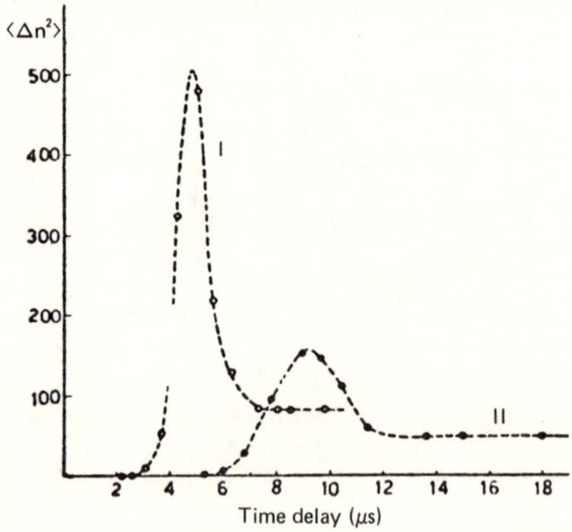

FIG.2.20. Evolution of the variance $\langle \Delta n^2 \rangle = \langle n^2 \rangle - \langle n \rangle^2$ of the statistical distribution p(n, T, τ). Dashed lines represent an interpolation of the experimental points (○, ●).

Using Eq.(2.23) it is easily shown that the factorial moments given by

$$F_k \langle n(n-1) \ldots (n-k+1) \rangle = \int |E|^{2k} \, P(E) \, d^2E$$

change as

$$F'_k = \eta^{2k} F_k \tag{2.24}$$

With Eq.(2.24) one can perform corrections for an attenuation η. This kind of correction has led from Fig.2.20 to Fig.2.21.

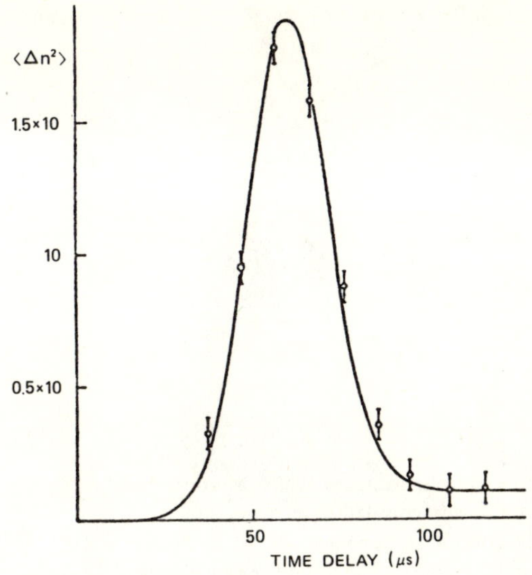

FIG.2.21. Evolution of the variance $\langle \Delta n^2 \rangle$ *of the statistical distribution of photons inside the laser cavity, as a function of the time delay* τ. *The solid line represents theoretical results (Arecchi and Degiorgio, Ref.7).*

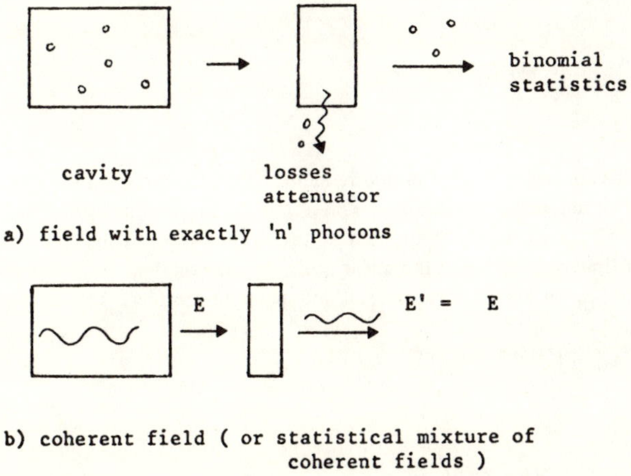

FIG.2.22. Picture of 'particle-like' and 'wave-like' attenuation.

2.8. THE PHOTOMULTIPLIER AS A STATISTICAL DEVICE

The finite response time of the photodetector puts a limit on the use of PS for short times. We show here how intensity correlations can be corrected for the detector resolution. The output i(t) of a photocathode illuminated by a light signal may be represented by a train of photoelectrons localized at random times t_k, i.e.

$$i(t) = \sum_{k=1} \delta(t - t_k)$$

The first correlation function $R_i(\tau) = \langle i(t)\, i(t+\tau)\rangle$ of the random process i(t), which we assume to be stationary, is given by

$$R_i(\tau) = \left\langle \sum_k \sum_j \delta(t-t_k)\,\delta(t+\tau-t_j)\right\rangle$$

$$= \left\langle \sum_k \delta(t-t_k)\,\delta(t+\tau-t_k)\right\rangle + \left\langle \sum_{k\neq j} \sum_j \delta(t-t_k)\,\delta(t+\tau-t_j)\right\rangle \tag{2.25}$$

$$= \mu\delta(\tau) + \mu^2 g^{(2)}(\tau)$$

where μ is the average rate of photo-electrons, which is proportional to the average light intensity, and $g^{(2)}(\tau)$ is the reduced intensity correlation function.

The photomultiplier, considered as a linear stochastic filter (see Ref.[8]) is described by a random response function h(t) which is the output for a single photo-electron emitted at time t = 0. Consequently, the random output current u(t) will be the time convolution of the two random functions h(t) and i(t), i.e.

$$u(t) = \int_{\infty}^{\infty} i(t-t')\,h(t')\,dt' \equiv i(t)*h(t)$$

In performing this convolution, one should remember that the different $\delta(t-t_k)$ of the input are associated with independent realizations $h_{(k)}(t)$ of the random process h(t). The correlation function $R_u(\tau) = \langle u(t)\, u(t+\tau)\rangle$ is computed by taking into account that the statistical properties of the photomultiplier operating in the linear range are independent from the statistics of i(t) and that $h_{(k)}$ and $h_{(j)}$ for $k \neq j$ are statistically independent samples. The correlation function is

$$R_u(\tau) = \mu^2 g^{(2)}(\tau)*G(\tau) + \mu C(\tau) \tag{2.26}$$

where

$$G(\tau) = \int_{\infty}^{\infty} \langle h(t)\rangle\,\langle h(t+\tau)\rangle\,dt$$

and

$$C(\tau) = \int_{\infty}^{\infty} \langle h(t)\,h(t+\tau)\rangle\,dt$$

A full description of the effect of the photomultiplier on the intensity correlation function $R_u(\tau)$ requires therefore two different functions, $C(\tau)$ and $G(\tau)$. $C(\tau)$ modifies the δ-like term of Eq.(2.25), which comes from the discrete character of the detection process, whereas $G(\tau)$ modifies the term describing the joint probability of two different photon events.

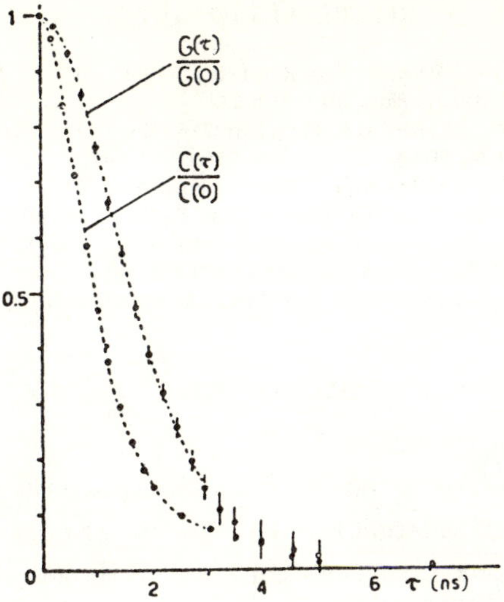

FIG.2.23. Plot of the functions C(τ) and G(τ) for a Philips XP 1210 photomultiplier, operating at 3800 V. The dashed lines interpolate the experimental points (Arecchi et al., Ref.8).

The fluctuations of h(t) are mostly generated by three effects: gain statistics at the dynodes, spread in transit times of secondary electrons from dynode to dynode, and spread in the transit time of photo-electrons from the photocathode to the first dynode. If only the first effects were present, $C(\tau)$ would be proportional to $G(\tau)$, whereas spreads in transit times affect the two functions differently. In particular, the spread associated with the transit time of photo-electrons affects only the location of each h(t) and hence does not affect $C(\tau)$. As an example, we report in Fig.2.23 experimental results on both $C(\tau)$ and $G(\tau)$ (see Ref.[8] and Ch.A5 of Laser Handbook). The two functions in Fig.2.23 are normalized to the same initial value.

To obtain information on the light intensity correlation $g^{(2)}(\tau)$ from the measured correlation $R_u(\tau)$, the term $\mu C(\tau)$ must be subtracted. In this correction experimental errors strongly affect the results, especially at low photon flux. However, the term $\mu C(\tau)$ does not appear at all in the expression for $R_u(\tau)$ when the correlation is performed with two phototubes. In that case, a similar calculation leads to a modified form of Eq.(2.26)

$$R_u(\tau) = \langle u_1(t)\, u_2(t+\tau)\rangle = \mu_1 \mu_2\, g^2(\tau) {*} G_{12}(\tau) \qquad (2.27)$$

where the indices 1 and 2 refer to the two phototubes and

$$G_{12}(\tau) = \int_{-\infty}^{\infty} \langle h_1(t)\rangle \langle h_2(t+\tau)\rangle dt$$

The treatment can easily be generalized to the measurement of higher-order correlation functions.

REFERENCES TO PART 2

[1] ARECCHI, F.T., BERNE, A., SONA, A., Phys. Rev. Lett **17** (1966) 260.
[2] RISKEN, H., VOLLMER, H.D., Z. Phys. **201** (1967) 323.
[3] ARECCHI, F.T., RODARI, G.S., SONA, A., Phys. Lett. **25A** (1967) 59.
[4] ARECCHI, F.T., GIGLIO, M., SONA, A., Phys. Lett. **25A** (1967) 341.
[5] CORTI, M., DEGIORGIO, V., ARECCHI, F.T., Opt. Commun. **8** (1973) 329.
[6] ARECCHI, F.T., DEGIORGIO, V., QUERZOLA, B., Phys. Rev. Lett. **19** (1967) 1168; Phys. Rev. **A3** (1971) 1108.
[7] ARECCHI, F.T., DEGIORGIO, V., Phys. Lett. **27A** (1968) 429.
[8] ARECCHI, F.T., CORTI, M., DEGIORGIO, V., DONATI, S., Opt. Commun. **3** (1971) 284.

BIBLIOGRAPHY

ARECCHI, F.T., Phys. Rev. Lett. **15** (1965) 912.

ARECCHI, F.T., in Quantum Optics (GLAUBER, R.J., Ed.), Academic Press (1967).

ARECCHI, F.T., BERNE, A., BURLAMACCHI, P., Phys. Rev. Lett. **16** (1966) 32.

ARECCHI, F.T., BERNE, A., SONA, A., BURLAMACCHI, P., Proc. 4th Int. Quantum Electronics Conference, Phoenix, 1966; IEEE J. Quant. Electron. **2** (1966) 341.

ARECCHI, F.T., GATTI, E., SONA, A., Phys. Lett. **20** (1966) 27.

ARECCHI, F.T., GIGLIO, M., TARTARI, U., Phys. Rev. **163** (1967).

GLAUBER, R.J., in Quantum Optics and Electronics (Proc. Les Houches 1964 Summer School) (De Witt, C., et al., Eds), New York (1965). (See also references reported therein.)

GLAUBER, R.J., in Physics of Quantum Electronics (Proc. Conf. Puerto Rico, 1965), (KELLEY, P., et al., Eds), New York (1966).

LAMB, W.E., Phys. Rev. **134** (1964) A 1429.

MANDEL, L., WOLF, E., Rev. Mod. Phys. **37** (1965) 213.

RISKEN, H., Z. Phys. **186** (1965) 85; **191** (1966) 302.

WANG, M.C., UHLENBECK, G.E., Rev. Mod. Phys. **17** (1945) 323.

Note: For a more up-to-date reference on photon statistics and laser fluctuations see Part A of Laser Handbook (ARECCHI, F.T., SCHULTZ, E.O., Du BOIS, Eds), North Holland (1972).

PART 3. QUANTUM OPTICS: COHERENT RESONANT SPECTROSCOPY

3.1. INTRODUCTION

In this Part we survey the relevant spectroscopic effects induced by a coherent field near resonance with an atomic or molecular transition. The main results are:

(a) Precise measurements of electric dipole strengths $\mu = \langle 2|e\vec{r}|1\rangle$
(b) Lifetime measurements (spontaneous; collision broadening and its behaviour versus gas pressure which sheds light on interatomic potentials)
(c) Absolute positioning of atomic levels (new wavelength standards, accurate measurements of isotope shifts)

(d) Fine and hyperfine structure (fs, hfs): Landé factors, Lamb shift, tensor polarizability
(e) Metrology: new accurate values for α, Ry, c (finer structure constant, Rydberg constant, light velocity)
(f) Preparation of particular atomic states (coherent excitation) or field states (modulation).

We shall cover the following subjects:
(i) Coherent resonant effects: local effects, propagation effects;
(ii) Non-linear spectroscopy: saturation and two-photon;
(iii) Perturbed fluorescence spectroscopy.

3.2. THE INTERACTION MODEL

Let us consider an e.m. field and a set of atoms confined in a cavity. We discuss how to go from a physical Hamiltonian such as

$$H = H_{\text{field}} + H_{\text{atoms}} + H_{f-a}$$

where

$$H_{f-a} = -\sum_i \frac{e}{m} \vec{A}(x_i)\vec{p}_i$$

x_i being the position of atom i, $\vec{A}(x_i)$ the value of the vector potential at x_i, and p_i the momentum of the i-th electron, to a suitable Hamiltonian.

We expand the field in plane waves (take $\vec{A}$ as a scalar):

$$A(x, t) = \sum_k \sqrt{\frac{\hbar}{2\epsilon_0 \omega_k V}} \, [a_k e^{-i(\omega_k t - kx)} + c \cdot c]$$

(V = cavity volume; mks system used) and consider a_k and the conjugate a_k^+ as Bose operators

$$[a_k, a_{k'}^+] = \delta_{kk'}$$

For simplicity, we skip vector relations and give a scalar theory. ω_k and k are related by the dispersion relation

$$\omega_k = ck$$

We consider the atoms as two-level atoms, so that the Hilbert space of a single atom is fully described by the identity operator I plus the three Pauli operators $\sigma^+, \sigma^-, \sigma_3$

$$[\sigma^+, \sigma^-] = \sigma_3 \qquad [\sigma^\pm, \sigma_3] = \mp 2\sigma^\pm$$

It is then a straightforward matter to obtain the following model Hamiltonian

$$\frac{H}{\hbar} = \sum_k \omega_k a_k^+ a_k + \frac{\omega_0}{2} \sum_{i=1}^{N} \sigma_3(i) + \sum_{k,i} g_k (a_k \sigma_i^+ e^{ikx_i} + a_k^+ \sigma_i^- e^{-ikx_i}) \tag{3.1}$$

We introduce the *collective* operators.

$$J_k^{\pm} = \sum_i \sigma_i^{\pm} e^{\pm ikx_i}$$

$$J_z = \sum \sigma_3(i) \qquad (3.2)$$

When the space dependence can be neglected, i.e. for the three cases
(a) point laser (cavity length $\ll \lambda$),
(b) single mode case (one $k = k_0$ value),
(c) travelling plane wave field (again, one k_0 value),
one has closed commutation relations for $J^{\pm} = \sigma_i^{\pm}$ and J_3, and the Heisenberg equations of motion are obtained.

$$\begin{aligned} \dot{a} &= -i\omega a - igJ^- \\ \dot{J}^- &= -i\omega_0 J^- + igaJ_2 \\ \dot{J}_2 &= -ig(aJ^+ - a^+J^-) \end{aligned} \qquad (3.3)$$

They do not form a closed set, because one would need also an equation for the motion of bilinear operators such as J^+, etc.

We study here the self-consistent field approximation (SCFA).

$$\text{SCFA:} \ \langle aJ \rangle \cong \langle a \rangle \langle J \rangle \qquad (3.4)$$

Classical current $\xrightarrow{\text{Maxwell}}$ field state $\langle a \rangle \neq 0$ ←┐
Coherent states
Classical field $\xrightarrow{\text{Bloch}}$ atomic state $\langle J^{\pm} \rangle \neq 0$ ←┘

For the discussion of coherent states we refer the reader to Part 4. Here we shall discuss Eq.(3.5), analogous to the Bloch equation in Nuclear Magnetic Resonance (NMR), which is the vector form of the atomic equations (3.3) when averaged with the SCFA:

$$\frac{d}{dt} \langle \vec{J} \rangle = \vec{\Omega} \times \langle \vec{J} \rangle \qquad (3.5)$$

$$\Omega \equiv (g\langle a \rangle, 0, \omega_0)$$

3.3. THE TWO-LEVEL ATOM

Under what conditions can a real atom be approximated by a two-level atom? Consider e.g. the allowed electric dipole D-lines of the Na atom: $I = 3/2$, $\vec{F} = \vec{I} + \vec{J}$ (Fig.3.1).

$$3p - 3s \quad \begin{cases} {}^2P_{3/2} - {}^2S_{1/2} & 5890 \text{ Å} \\ {}^2P_{1/2} - {}^2S_{1/2} & 5896 \text{ Å} \end{cases}$$

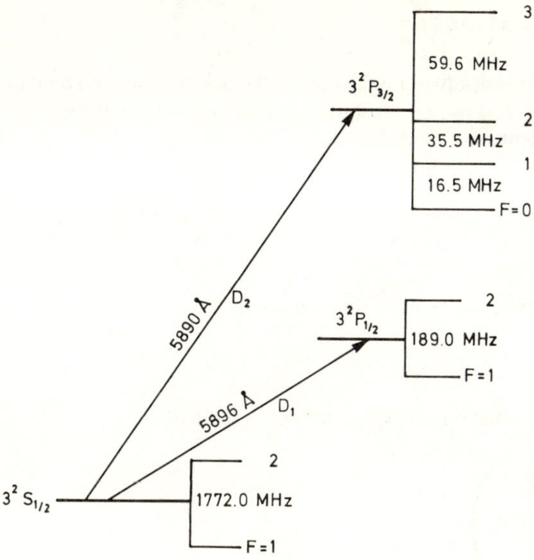

FIG.3.1. Level scheme of Na atom.

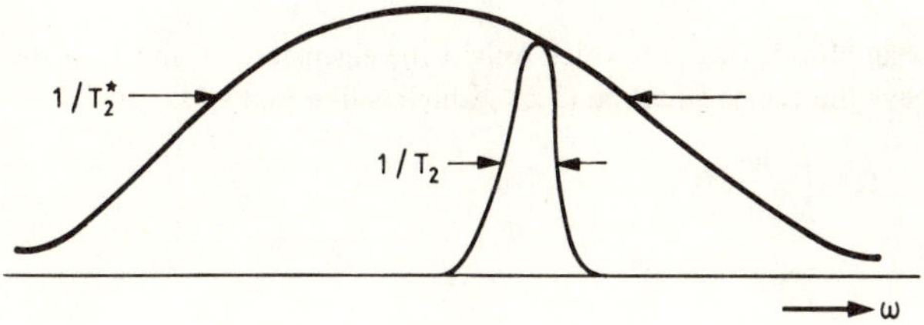

FIG.3.2. Plot of Doppler and natural lines.

In dilute gas (10^{12} atoms/cm^3) the collision time is $\sim 10^{-7}$ s. The associated *homogeneous* line width is

$$\frac{1}{T_2} = \delta\omega_c \sim 10^8 \text{ s}^{-1}$$

Further, there is an *inhomogeneous* Doppler broadening which is typically

$$\frac{1}{T_2^*} = \delta\omega_D \sim \frac{\omega}{c}\sqrt{\frac{kT}{M}} \sim 10^{10} \text{ s}^{-1}$$

(see Fig.3.2). The nuclear spin I = 3/2 combines with J to produce hfs. Each of the F = I + J hfs levels is then magnetically degenerate (2 F + 1) times.

If we want to study coherent excitation on the $^2S_{1/2}$ (F = 1) to $^2P_{3/2}$ (F = 2) transition, avoiding an overlap with F = 1, the pulse must be narrower than 35.5 MHz, i.e. the pulse duration must be longer than 0.5×10^{-8} s.

On the other hand, coherence lasts for less than the collision time, so $t_{pulse} < 10^{-7}$ s. Only under such limits we have a coherent two-level excitation.

3.4. THE BLOCH EQUATIONS

We approach the Bloch equations directly by the Schrödinger method rather than the Heisenberg method. In this way we attribute a physical meaning to the three components of $\vec{J}$.

The time-dependent wave function

$$\psi = a(t)|+\rangle + b(t)|-\rangle$$

$$\mu \equiv \langle +|er|-\rangle \tag{3.6}$$

of a two-level atom, under a coherent field excitation,

$$E = \epsilon \cos \omega t \tag{3.7}$$

obeys the Schrödinger equation. We set up a linear combination

$$\vec{J} \equiv \begin{pmatrix} a^*b + b^*a \\ -i(a^*b - b^*a) \\ |a|^2 - |b|^2 \end{pmatrix} \tag{3.8}$$

of the two complex amplitudes a and b which weight the eigenstates + and − of the free-atom Hamiltonian. It obeys the vector equation (3.6)′, which is like Eq.(3.5)

$$\dot{\vec{J}} = \vec{\Omega} \times \vec{J} \qquad \Omega \equiv \left(-\frac{\mu\epsilon}{\hbar}, 0, \omega_0\right) \tag{3.6$'$}$$

By evaluating

$$\langle \psi | e\vec{r} | \psi \rangle$$

with the help of Eq.(3.6) it is easy to see that the three components of J have the following meaning:

J_1 is proportional to the in-phase polarization (dispersion)
J_2 is proportional to the out-of-phase polarization (absorption or gain)
J_3 is proportional to the population inversion.

We then write

$$\cos \omega t = \frac{e^{i\omega t} + e^{-i\omega t}}{2}$$

go to a frame rotating around z at angular velocity ω_0 (see Fig.3.3),

$$\Omega = \begin{pmatrix} -\frac{\mu}{\hbar}\epsilon \\ 0 \\ \Delta = \omega - \omega_0 \end{pmatrix} \tag{3.9}$$

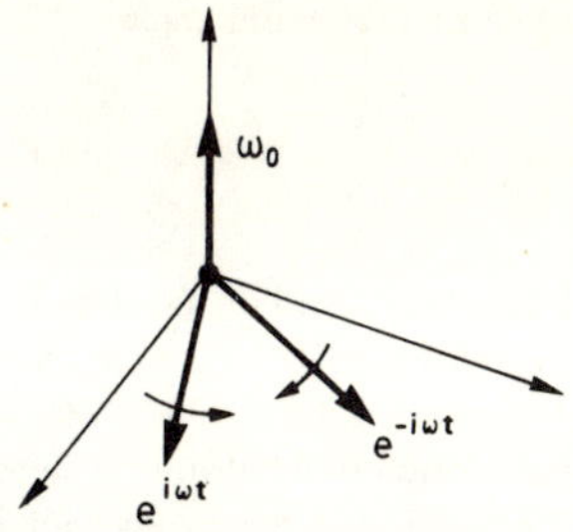

FIG.3.3. Splitting of cos ωt into two counter-rotating components.

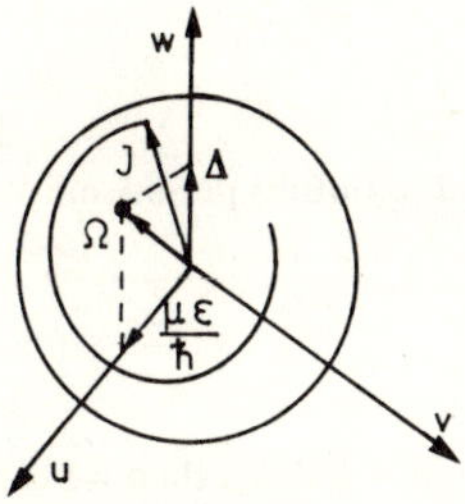

FIG.3.4. Precession of J around Ω. The end-point of J moves on a sphere of radius N/2.

consider $\Delta = \omega - \omega_0 \ll \omega$ and hence neglect the effects of the fast component $e^{-i(\omega+\omega_0)t}$ (rotating wave approximation = RWA). The Bloch equations reduce to

$$\frac{d}{dt}\begin{pmatrix} u \\ v \\ w \end{pmatrix} = \begin{pmatrix} \Delta \cdot v \\ -\Delta \cdot u + \mu\epsilon w \\ -\mu\epsilon v \end{pmatrix} \tag{3.10}$$

This is the motion of a point on a spherical surface, as seen from the conservation relation which stems from Eq.(3.10) (Fig.3.4)

$$u^2 + v^2 + w^2 = \text{constant} = (N/2)^2 \tag{3.11}$$

Indeed, at equilibrium, $u = v = 0$, $w = -N/2$. This character was already implicit in Eq.(3.5). The contribution of the fast component is the Bloch-Siegert shift

$$\delta\omega \sim \frac{1}{4}\frac{(\mu\epsilon/\hbar)^2}{\omega_0} \sim 10^{-10}\,\omega_0 \tag{3.12}$$

It is approximately given by the ratio

$$\frac{(\text{Rabi precession frequency})^2}{\text{transition frequency}}$$

If we consider a 1 ns pulse able to give a precession angle

$$\frac{\mu\epsilon}{\hbar}\cdot t \sim \pi$$

i.e. able to tip the Bloch vector from the South Pole (system in the ground state) to the North Pole (system in the excited state), and if $\mu \sim 10^{-27}$ C X cm (allowed dipole transition), then the Bloch-Siegert shift is only $10^{-10}\ \omega_0$.

The Bloch Eq.(3.10) gives the motion of an atomic system shined by a field (3.7) (local effect). If we want to consider the reaction of the atoms back on a field propagating through the atomic system, we must couple Bloch and Maxwell equations.

For an approximately plane field

$$E = \epsilon(z, t) \cos(\omega t - kz + \varphi(zt)) \tag{3.13}$$

if ϵ and φ are slowly varying compared to optical period and wavelength, i.e. if

$$\frac{1}{\epsilon}\frac{\partial\epsilon}{\partial t} \ll \omega \qquad \frac{1}{\epsilon}\frac{\partial\epsilon}{\partial z} \ll k$$

(slowly varying envelope approximation = SVEA), then we can neglect the second derivatives and have two coupled equations.

$$\begin{aligned} \left(\frac{\partial}{\partial t} + c\frac{\partial}{\partial z}\right)\epsilon &= \frac{\omega N\mu}{2\epsilon} v \\ \epsilon\left(\frac{\partial}{\partial t} + c\frac{\partial}{\partial z}\right)\varphi &= \frac{\omega N\mu}{2\epsilon} u \end{aligned} \tag{3.14}$$

Notice that v is the source term for amplitude and u for phase. (Here N (cm^{-1}) is the density of atoms.)

If we neglect the non-linear coupling $\mu\epsilon w$ in Eq.(3.10), i.e. if we keep constant population inversion $w = -1$ and add phenomenological damping terms $-\gamma u$, $-\gamma v$ to Eq.(3.10), the Fourier transforms of u, v are the usual, linear susceptibilities χ', χ'' (Fig.3.5).

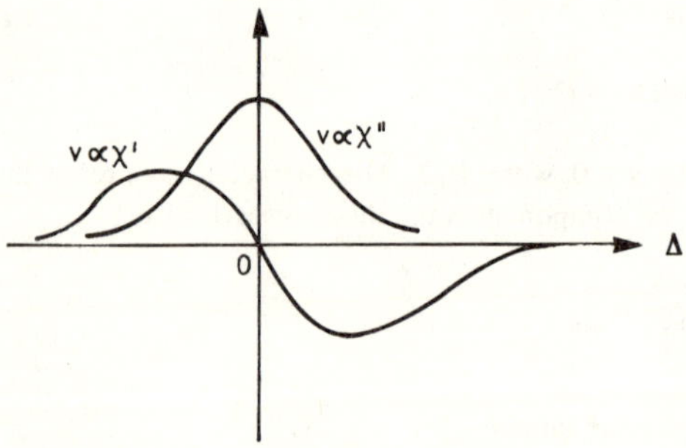

FIG.3.5. Plot of absorption and dispersion versus detuning Δ in the linear case. The two curves are steady-state solutions of Eqs (3.22), with w = − N/2.

At resonance ($\Delta = 0$), we may take $u = 0$, $\varphi = 0$ always and reduce to Eqs (3.10)′, (3.11)′

$$\left.\begin{aligned} \dot{v} &= \mu\epsilon w \\ \dot{w} &= -\mu\epsilon v \end{aligned}\right\} \tag{3.10$'$}$$

$$v^2 + w^2 = 1 \tag{3.11$'$}$$

The motion of $\vec{J}$ on the Bloch sphere is a motion on the (v, w) plane perpendicular to the ϵ direction.

Owing to the conservation law (3.11)′ we can introduce the angle θ as in Eq.(3.15). Its time derivative is the field amplitude (3.16).

$$\left.\begin{aligned} v &= \sin\theta \\ w &= \cos\theta \end{aligned}\right\} \tag{3.15}$$

$$\dot{\theta} = \frac{\mu}{\hbar}\epsilon \tag{3.16}$$

Neglecting $\partial/\partial z$ in Eq.(3.14) (point-like medium) we have the non-linear pendulum

$$\begin{aligned} \ddot{\theta} &= \frac{\omega N\mu^2}{\hbar\epsilon}\sin\theta \\ &= \frac{1}{\tau_c^2}\sin\theta \end{aligned} \tag{3.17}$$

which will yield as solutions Jacobi elliptic functions. In particular, we may have a hyperbolic secant solution. We emphasize the role of the *cooperation time* τ_c

$$\frac{1}{\tau_c^2} = \frac{\omega N\mu^2}{\hbar\epsilon_0} \tag{3.17$'$}$$

It is the characteristic time of radiation/atom interaction. For a linearized problem ($w = -1 = \text{const.}$), Eq.(3.17) reduces to $\ddot{\theta} = -\Omega^2\theta$ where now

$$\Omega^2 = \frac{Ne^2}{m\epsilon_0} \tag{3.17$''$}$$

is the usual plasma frequency for a gas of Lorentz oscillators. Here, m is the electron mass, and μ has been replaced by the classical cross-section for free electrons.

In *propagation,* we transform to a coordinate frame moving at velocity V_0 and look for undistorted pulses or solitons ($\partial/\partial\zeta \equiv 0$)

$$(z, t) \rightarrow \left(\zeta = z;\ \tau = t - \frac{z}{V_0}\right) \tag{3.18}$$

From Eq.(3.14) it is easy to obtain Eq.(3.19) and hence Eq.(3.20).

$$(1 - c/V_0)\frac{d\epsilon}{d\tau} = \frac{\omega N\mu}{2\epsilon} V \tag{3.19}$$

$$\ddot{\theta} = \frac{1}{1 - \dfrac{c}{V_0}} \frac{1}{\tau_c^2} \sin\theta \tag{3.20}$$

The hyperbolic secant solution of the soliton equation (3.20) implies the following constraint between the pulse velocity V_0 and the velocity c in vacuum, the linear attenuation α and the pulse duration τ_p:

$$\frac{1}{V_0} = \frac{1}{c} + \frac{1}{2}\alpha\tau_p \tag{3.21}$$

(α → attenuation; τ_p → pulse duration)

Further, it is easy to prove the following:

(a) The area of the radiation pulse obeys the space propagation equation

$$\Theta \equiv \int_{-\infty}^{+\infty} \frac{\mu}{\hbar}\epsilon\,dt; \qquad \frac{\partial\Theta}{\partial z} = -\frac{\alpha}{2}\sin\Theta(z)$$

which gives *solitons* for 2π, 4π, etc. (Fig.3.6).

(b) The 2π pulse corresponds to an atomic system starting from the South Pole and returning to the South Pole without global exchange of energy with the field (SIT = self-induced transparency). The pulse velocity is given in Eq.(3.21) where α is the linear attenuation

$$\alpha\,(cm^{-1}) = \sigma\,(cm^{-2})\,N\,(cm^{-3})$$

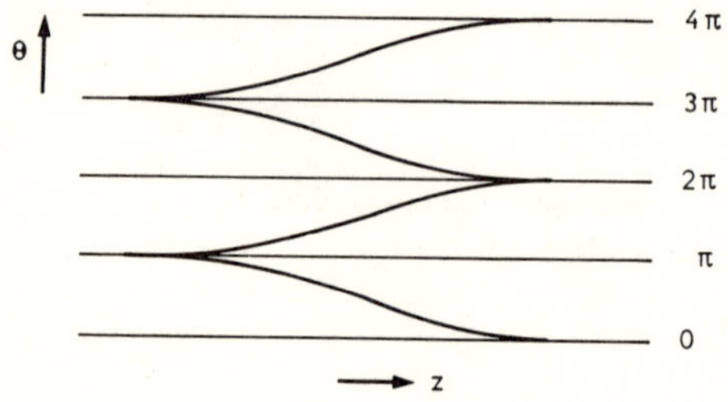

FIG.3.6. Space evolution of an initial Θ towards the asymptotic soliton value.

3.5. IRREVERSIBLE PROCESSES IN THE PRESENCE OF DAMPINGS

Let us introduce phenomenological lifetimes T_2 for polarization components, T_1 for inversion ($T_2 \leqslant T_1$), and 1/k for the field. In case the material is confined in a region of size L without mirrors, 1/k is the transit time L/c.

Further, let us introduce, besides the atoms and field, a third partner, the 'pump', which in the absence of field yields a steady-state value of w

$$w = w_0$$

The Bloch-Maxwell equations become

$$\frac{\partial}{\partial t}\begin{pmatrix} u \\ v \\ w \end{pmatrix} = \begin{pmatrix} \Delta\cdot v & -u/T_2 \\ -\Delta\cdot u + \mu\epsilon w & -v/T_2 \\ -\mu\epsilon v & -\dfrac{w - w_0}{T_1} \end{pmatrix} \tag{3.22}$$

$$\left(\frac{\partial}{\partial t} + \epsilon\frac{\partial}{\partial z}\right)\epsilon = \frac{\omega N\mu}{\hbar\epsilon}\, v - k\epsilon \tag{3.23}$$

In the absence of a pump, we must put

$$w_0 = -1$$

because the atoms will spontaneously decay toward the South Pole of the Bloch sphere, with a lifetime T_1. If $T_2 \ll (T_1, 1/k)$, we can solve at steady state the equations for u, v and have two coupled rate equations for w, ϵ.

$$\dot{w} = \frac{-\mathscr{L}}{T_1} Iw - \frac{w - w_0}{T_1}$$
$$\left(\frac{\partial}{\partial z} + \frac{1}{c}\frac{\partial}{\partial t}\right) I = -\alpha I \tag{3.24}$$

where

$$I \equiv \left(\frac{\mu\epsilon}{\hbar}\right)^2 T_1 T_2 \qquad \text{(normalized intensity)}$$
$$\mathscr{L} = \frac{1}{1 + \Delta^2 T_2^2} \qquad \text{(Lorentzian line shape)} \tag{3.24$'$}$$

Many approaches to lasers are based on Eq.(3.24). On the other hand, if the escape time of the field is faster than T_1, T_2, we solve Eq.(3.23) at steady state, replace in Eq.(3.22), neglect atomic losses and get

$$\dot{\theta} = \frac{1}{k\tau_c^2} \sin\theta \tag{3.25}$$

whose solution is

$$\epsilon \sim \operatorname{sech}\left(\frac{\tau - \tau_M}{\tau_R}\right) \tag{3.26}$$

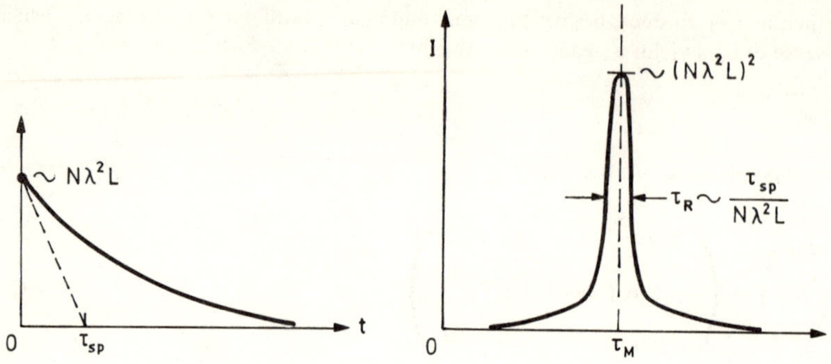

FIG.3.7. Single atom and cooperative spontaneous emissions.

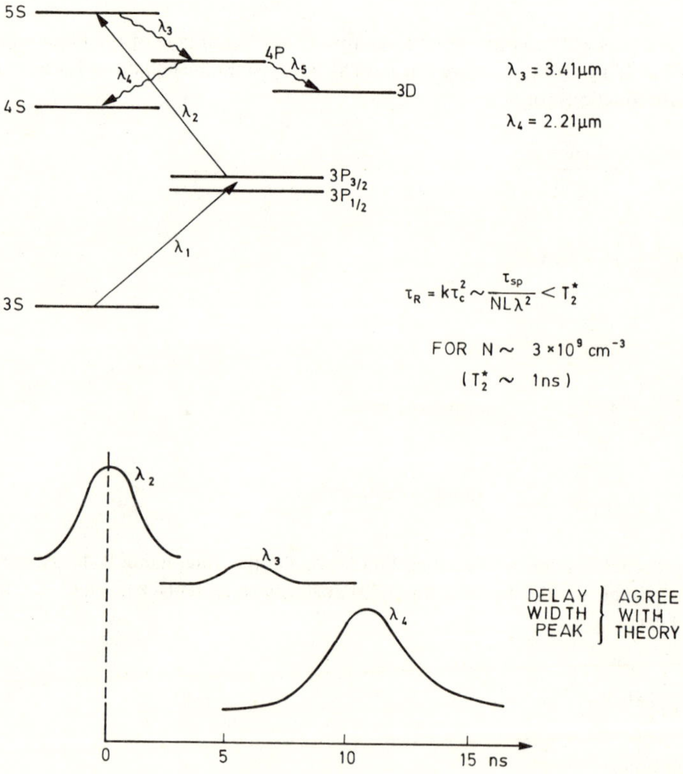

FIG.3.8. Superradiance in Na: level scheme and time behaviour.

with

$$k\tau_c^2 \equiv \tau_R \cong \frac{\tau_{sp}}{N\lambda^2 L} \qquad \text{(pulse width)}^1 \tag{3.27}$$

$$\tau_M = \tau_R \lg N \qquad \text{(pulse delay)} \tag{3.28}$$

Here, time $\tau = t - z/c$ is the *local* time. We have a *soliton* solution or π pulse.

If no applied field is present but if the atoms are prepared in the excited state, we have spontaneous π pulse (superradiance) after a time τ_M.

In Fig.3.7 spontaneous uncorrelated (Weisskopf-Wigner) and correlated emissions are compared. Figure 3.8 shows the experiment by Haroche et al.

3.6. SATURATION AND NON-LINEAR SPECTROSCOPY

The steady-state value of w has an intensity-dependent denominator (see solution (3.29) of Eq.(3.24))

$$w(\Delta) = \frac{w_0}{1+\mathscr{L}I} \xrightarrow{w_0=-1} -\frac{1+(\Delta\cdot T_2)^2}{1+(\Delta\cdot T_2)^2+I} \tag{3.29}$$

where $\mathscr{L}$ and I are as given in Eq.(3.24)′.

As we shine a strong laser field in a Doppler broadened line, we dig a hole into the line whose width depends on T_2 and I, but which is in any way smaller than $1/T_2^*$ (Fig.3.9). By a probe field we can detect the hole, localizing transitions better than in regular linear spectroscopy (limited by the Doppler width).

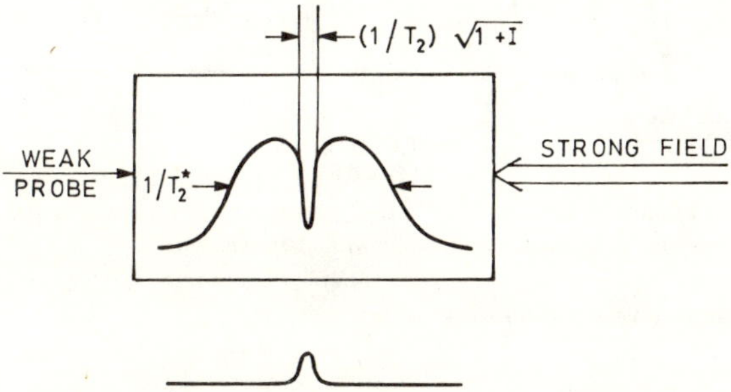

FIG.3.9. Saturation experiment.

3.7. TWO-PHOTON SPECTROSCOPY. COMPARISON WITH SATURATION

A transition between two levels of equal parity, such as 3 s → 4 s in Na, forbidden for one photon, becomes allowed for the simultaneous absorption of two photons. If two lasers are shined in opposite directions into a cell, the Doppler shifts cancel and we have a resonance exactly at the same frequency (Fig.3.10). A comparison between the two methods and a list of the applications is given as follows:

---

[1] $\tau_{sp}^{-1} \sim \mu^2/\lambda^3\hbar\epsilon_0$ is the spontaneous lifetime, τ_c was given in Eq.(3.17)′, k is replaced by c/L.

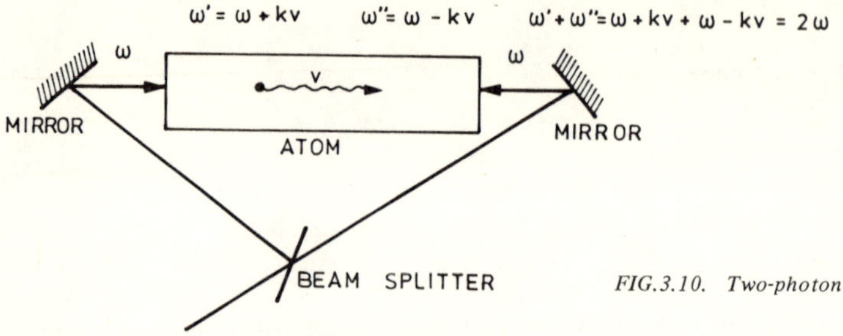

FIG.3.10. Two-photon absorption experiment.

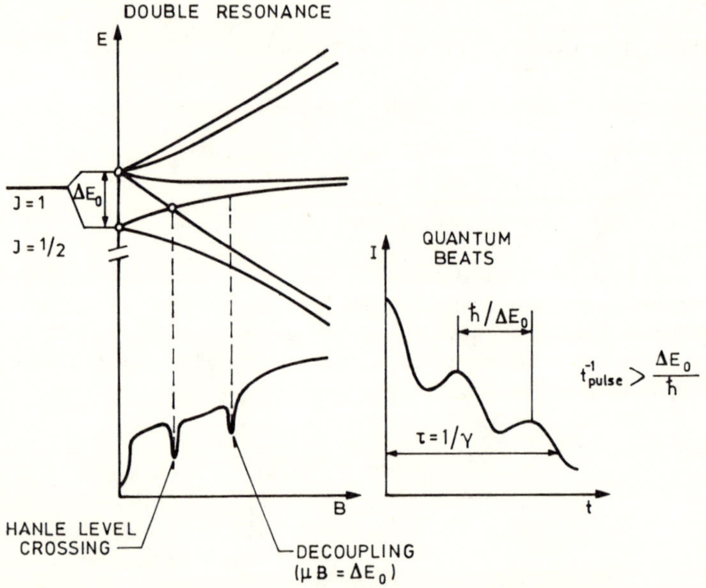

FIG.3.11. Level scheme of the terminal level of an optical transmission versus magnetic field. (Three double resonance effects and time plot of spontaneous intensity with quantum beats.)

Two-photon spectroscopy versus saturation

+ Contrast 100%
+ No saturation broadening
− Level shifts (a.c. Stark)
+ No recoil
− Second-order Doppler (red shift $\propto v^2$)

Applications

fs, hfs
Splitting in low B, E fields (Zeeman, Stark)
Collisional broadening and shift at low pressures
Selective population of single level within Doppler line
Isotope shift
Precise position levels → metrological standards

3.8. PERTURBED FLUORESCENCE SPECTROSCOPY

Hitherto we have described systems which decay after preparation by a laser pulse. Here we consider cases whereby the atoms, after laser preparation, undergo an extra perturbation. This could be an applied B or E field (Zeeman or Stark splitting), internal couplings of the constituent angular momenta of the atom, atomic collisions, etc. We restrict our attention to detection of polarized or anisotropic light.

In particular, for simultaneous excitation of two fs levels, we can consider three techniques: (a) quantum beats, (b) double resonance (light + rf), and (c) modulated light beam. These are illustrated in Figs 3.11 and 3.12.

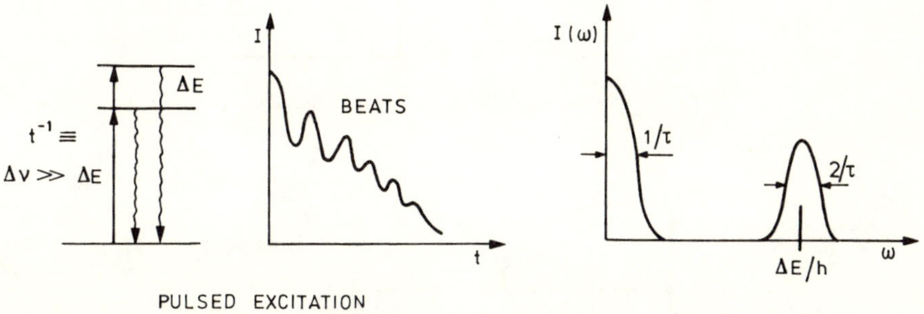

FIG.3.12. Modulated light beam experiment as Fourier transform of the quantum beat experiment.

3.9. DYNAMIC STARK SHIFT

Spontaneous fluorescence after laser excitation is predicted to have a three-peaked spectrum.

Semi-classical ($\langle E^+E^-\rangle \sim \langle E^+\rangle\langle E^-\rangle$!)

$$I(t) = I_0 \left\{ e^{-i\omega_0 t} + \frac{1}{2} \exp\left(-\frac{3}{4}\gamma t\right) \exp\left[-i\left(\omega_0 \pm \frac{2\mu\epsilon}{\hbar} t\right)\right]\right\} \tag{3.30}$$

Quantum. Above a threshold

$$\frac{\mu\epsilon}{\hbar} \geqslant \gamma/4 = 1/T_2 \tag{3.31}$$

there are stationary correlations given by a power spectrum (see Fig.3.13)

$$I(\omega) = \mathscr{F}\left\{G^{(2)}(\tau) \propto 3e^{-(\gamma/2+i\omega_0)\tau} + 1 \exp\left\{-\left[\frac{3}{4}\gamma + i\left(\omega_0 \pm \frac{\mu\epsilon}{\hbar}\right)\right]t\right\}\right\} \tag{3.32}$$

This equation suggests correlation spectroscopy as an alternative to ordinary spectroscopy. The corresponding outcomes are depicted in Fig.3.14.

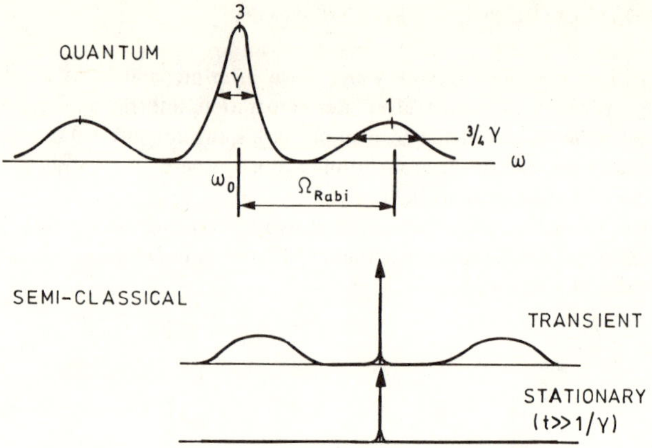

FIG.3.13. Frequency spectrum of dynamic Stark shift.

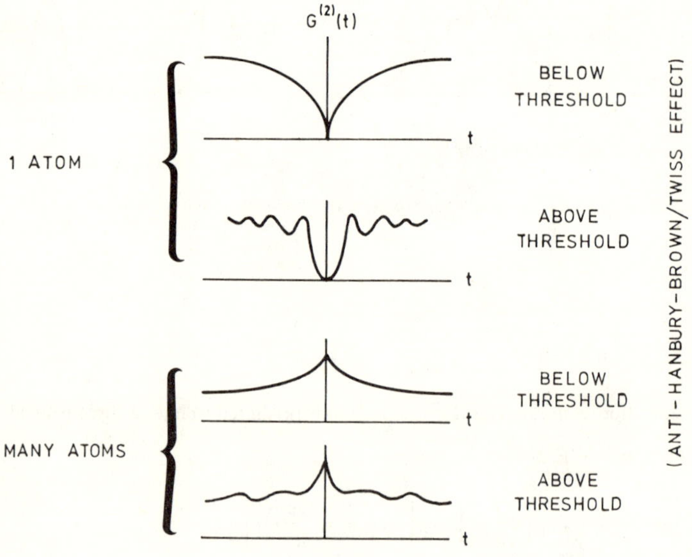

FIG.3.14. Intensity correlations in dynamic Stark shift.

PART 4. FIELD AND ATOMIC COHERENT STATES

4.1. INTRODUCTION

The central problem of quantum optics (laser theory, super-radiance, resonant propagation, etc.) is the description of the interaction between N atoms and an electromagnetic field confined in a cavity of finite volume. A suitable model Hamiltonian for this problem is the following ($\hbar = 1$):

$$H = \sum_k \omega_k a_k^+ a_k + \frac{\omega_0}{2} \sum_{i=1}^{N} S_{3(i)} + \sum_{k,i} g_k (a_k S_i^+ e^{i\vec{k}\cdot\vec{x}_i} + a_k^+ S_i^- e^{-i\vec{k}\cdot\vec{x}_i})$$

where a_k, a_k^+ are Bose operators describing the k-th field mode and $S_i^{\pm}$, S_{3i} are Pauli operators describing the atom located at position $\vec{x}_i$ as a two-level system.

We can introduce the collective operators [1]

$$J_k^{\pm} = \sum_i S_i^{\pm} e^{\pm i\vec{k}\cdot\vec{x}_i}$$

$$J_z = \sum_i S_{3i}$$

They obey the commutation rule

$$[J_k^+, J_{k'}^-] = \sum_i S_{3i} e^{i(\vec{k}-\vec{k}')\cdot\vec{x}_i}$$

This reduces to $\delta_{kk'}$, J_z in the following particular cases which are of extreme physical importance:

(i) point laser (cavity volume $\ll \lambda^3$)
(ii) single-mode laser (travelling wave)
(iii) travelling wave field in an amplifying or absorbing medium

In such cases the above operators obey standard angular momentum commutation rules. The associated Heisenberg equations of motion become (leaving out for simplicity the k index)

$$\dot{a} = -i\omega a - igJ^-$$
$$\dot{J}^- = -i\omega_0 J^- + igaJ_z$$
$$\dot{J}_z = -ig(aJ^+ - a^+J^-)$$

plus similar equations for a^+ and J^+. This set of five equations is not closed. For instance, to solve the second equation, we must know the evolution of the binary operator aJ_z whose equation of motion will imply ternary operators, and so on.

In the self-consistent approximation (SCA), or semi-classical approach, we introduce the approximation

$$\langle aJ^+ \rangle \approx \langle a \rangle \langle J^+ \rangle$$

i.e. we consider only the interaction among the mean fields. The three equations for $J^{\pm}$, J_z can be summarized in the vector equation

$$\langle \dot{\vec{J}} \rangle = \vec{\Omega} \times \langle \vec{J} \rangle$$

where $\vec{\Omega} \equiv (g\langle a\rangle, 0, \omega_0)$. This is a Bloch equation [2] for an angular momentum $\vec{J}$ precessing around a classical field $\langle a\rangle$.

The field equation becomes

$$\langle \dot{a}\rangle = -i\omega\langle a\rangle - ig\langle J^-\rangle$$

This is the equation for a field acted upon by a classical current $\langle\vec{J}\rangle$. Starting with a field in the vacuum state, it leads to a particular field state with $\langle a\rangle \neq 0$, which was introduced by Bloch and Nordsieck [3] to deal with the 'infra-red catastrophe', then formalized by Schwinger [4], and later used by Glauber [5] under the name of coherent state. We shall similarly call a coherent atomic state, or Bloch state, that state with zero-induced dipole ($\langle J^-\rangle \neq 0$) generated from the ground-state by a classical field. We want to show that these states, besides being generated by classical sources, give expectation values of quantum operators whose limits, for large excitations, are the classical values (see Table 4-I). For a more detailed treatment see Ref. [6].

4.2. DESCRIPTION OF THE FREE FIELD

4.2.1. The harmonic oscillator states

To point out with maximum clarity the analogies between the free-field description and the free-atom description we start by listing here, in simple terms, the properties of the single harmonic oscillator. The equation numbering in this section and in Section 4.3 is done in parallel.

The single harmonic oscillator is described by its canonically conjugated coordinates q, p with the commutation relation

$$[q, p] = i\hbar \tag{4.1}$$

The usual lowering and raising operators are formed:

$$a = (2\hbar\omega m)^{-\frac{1}{2}} (\omega mq + ip) \tag{4.2a}$$

$$a^+ = (2\hbar\omega m)^{-\frac{1}{2}} (\omega mq - ip) \tag{4.2b}$$

TABLE 4-I. CLASSICAL EXCITATION AND COHERENT STATES

	Atomic states	Field states
Classical states	Classical current	Classical field
	↕	↕
Quantum states	Coherent atomic state	Coherent field state

Note. The single arrows indicate the direction of production of coherent states starting from classical states. The double arrows indicate states connected by the correspondence principle.

where $\omega m > 0$ is characteristic of the oscillator. These operators satisfy

$$[a, a^+] = 1 \tag{4.3a}$$

from which one obtains

$$[a, a^+a] = a \tag{4.3b}$$

$$[a^+, a^+a] = -a^+ \tag{4.3c}$$

The harmonic oscillator states, or Fock states, are the eigenstates of

$$N = a^+a \tag{4.4}$$

and are given by (see Ref. [7])

$$|n\rangle = \frac{(a^+)^n}{\sqrt{n!}} |0\rangle \qquad (n = 0, 1, 2 \ldots) \tag{4.5}$$

with eigenvalue n. The vacuum state $|0\rangle$ is the harmonic oscillator ground-state defined by

$$a|0\rangle = 0 \tag{4.6}$$

4.2.2. Coherent states of the field

Let us consider the translation operator which produces a shift ξ in q and η in p:

$$T_\alpha = \exp -(i/\hbar)(\xi p - \eta q) = \exp(\alpha a^+ - \alpha^* a) \tag{4.7a}$$

where

$$\alpha = (2\hbar\omega m)^{-\frac{1}{2}}(\omega m\xi + i\eta) \tag{4.7b}$$

A coherent state $|\alpha\rangle$ is obtained by translation of the ground-state [4,5]

$$|\alpha\rangle \equiv T_\alpha |0\rangle \tag{4.8}$$

We shall name these states Glauber states as they have been used extensively by Glauber in quantum optics [5]. Since

$$T_\alpha a T_\alpha^{-1} = a - \alpha \tag{4.9}$$

the state $|\alpha\rangle$ satisfies the eigenvalue equation

$$(a - \alpha)|\alpha\rangle = 0 \tag{4.10}$$

Using the Baker-Campbell-Hausdorf theorem [8] or Feynman disentangling techniques [9] the translation operator can be written in the following forms:

$$T_\alpha = e^{|\alpha|^2/2} e^{-\alpha^* a} e^{\alpha a^+} = e^{-|\alpha|^2/2} e^{\alpha a^+} e^{-\alpha^* a} \tag{4.11}$$

The second of these forms, which is known as the normally ordered form, gives immediately the expansion of $|\alpha\rangle$ in terms of Fock states

$$|\alpha\rangle = T_\alpha|0\rangle = e^{-|\alpha|^2/2} e^{\alpha a^+}|0\rangle \tag{4.12}$$

from which, expanding the exponential and using Eq. (4.5), one obtains

$$\langle n|\alpha\rangle = e^{-|\alpha|^2/2} \frac{\alpha^n}{\sqrt{n!}} \tag{4.13}$$

The scalar product of Glauber states can be obtained either from Eq. (2.12), using the disentangling theorem (4.11), or from Eq. (4.13), using the completeness property of Fock states, $\sum|n\rangle\langle n| = 1$. One gets

$$\langle\alpha|\beta\rangle = \exp[-(1/2)(|\alpha|^2 - 2\alpha^*\beta + |\beta|^2)] \tag{4.14a}$$

from which one obtains

$$|\langle\alpha|\beta\rangle|^2 = \exp(-|\alpha-\beta|^2) \tag{4.14b}$$

The coherent states are minimum uncertainty packets. For three observables A, B, C, which obey a commutation relation $[A, B] = iC$, it is easy to show [7] that $\langle A^2\rangle\langle B^2\rangle \geqslant \langle C\rangle^2/4'$. In particular, with $A = q - \xi$, $B = p - \eta$ and $C = \hbar$, one has

$$\langle(q-\xi)^2\rangle\langle(p-\eta)^2\rangle \geqslant \hbar^2/4 \tag{4.15}$$

for any state. It is easy to show [5] that the equality sign holds for the coherent state $|\alpha\rangle$, where α is related to ξ and η by Eq. (4.7b). This establishes the minimum uncertainty property.

4.2.3. The coherent states as a basis

We now consider the completeness properties of the coherent states. Using Eq. (4.13) and the completeness of Fock states $\sum_n n\rangle\langle n| = 1$, one obtains straightforwardly

$$\frac{1}{\pi}\int d^2\alpha|\alpha\rangle\langle\alpha| = 1 \tag{4.16}$$

The expansion of an arbitrary state in Glauber states follows

$$\begin{aligned} c \equiv \sum_n c_n|n\rangle &= \int \frac{d^2\alpha}{\pi} \sum_n c_n|\alpha\rangle\langle\alpha|n\rangle \\ &= \int \frac{d^2\alpha}{\pi} \exp\left[-\frac{1}{2}|\alpha|^2\right] f(\alpha^*)|\alpha\rangle \end{aligned} \tag{4.17a}$$

where

$$f(\alpha^*) \equiv \sum_n c_n \frac{(\alpha^*)^n}{\sqrt{n!}} = e^{-|\alpha|^2/2} \langle\alpha|c\rangle \tag{4.17b}$$

Using Eq. (4.5) one sees that $|c\rangle$ can also be written as

$$|c\rangle = f(a^+)|0\rangle \tag{4.18}$$

where $f(a^+)$ is defined by its expansion (4.17b). The scalar product of any two states $|c'\rangle$ and $|c\rangle$ is obtained from Eqs (4.16) and (4.17b)

$$\begin{aligned}\langle c'|c\rangle &= \int \frac{d^2\alpha}{\pi} \langle c'|\alpha\rangle\langle\alpha|c\rangle \\ &= \int \frac{d^2\alpha}{\pi} e^{-|\alpha|^2} [f(\alpha^*)]^* f(\alpha^*)\end{aligned} \tag{4.19}$$

In view of the completeness relations, the operators F acting on this Hilbert space can be expanded as

$$F = \sum_{m,n} |m\rangle\langle m|F|n\rangle\langle n| \tag{4.20a}$$

or

$$F = \iint \frac{d^2\alpha\, d^2\beta}{\pi^2} |\beta\rangle\langle\beta|F|\alpha\rangle\langle\alpha| \tag{4.20b}$$

Owing to the overcompleteness of the $|\alpha\rangle$ states, the expansion (4.20b) is in general not unique. This expansion is especially useful if it can be written in the diagonal form

$$F = \int d^2\alpha\, f(\alpha)|\alpha\rangle\langle\alpha| \tag{4.20c}$$

This will be further discussed for the case of the density matrix.

4.2.4. Statistical operator for the field

Up to now we have considered pure quantum states. Since a field in thermal equilibrium with matter at ordinary temperatures is essentially in the ground-state ($\hbar\omega \gg kT$), this is an adequate description for any field obtained from thermal equilibrium in response to a classical current. However, the field radiated by an incoherently pumped medium is a statistical mixture described by a statistical operator ρ, which we assume normalized to unity,

$$\mathrm{tr}\,\rho = 1 \tag{4.21}$$

With the help of this operator, the statistical average of any observable $F(a, a^+)$ is obtained as

$$\langle F\rangle = \mathrm{tr}\,\rho F \tag{4.22}$$

Of particular interest are statistical ensembles described by a statistical operator which is diagonal in the Glauber representation [5]

$$\rho = \int P(\alpha)|\alpha\rangle\langle\alpha|\,d^2\alpha \tag{4.23}$$

where the normalization (4.21) requires

$$\int P(\alpha)d^2\alpha = 1 \tag{4.24}$$

The statistical average of an observable F is then given by an average over the diagonal elements $\langle\alpha|F|\alpha\rangle$:

$$\langle F\rangle = \int P(\alpha)\langle\alpha|F|\alpha\rangle d^2\alpha \tag{4.25}$$

The weight function $P(\alpha)$ has thus the properties of a distribution function in α-space, except that it is not necessarily positive.

Let us define a set of operators $\hat{X}(\lambda)$ such that their expectation values for coherent states

$$b^\alpha(\lambda) = \langle\alpha|\hat{X}(\lambda)|\alpha\rangle \tag{4.26}$$

form a basis in the function space of functions of α. If the statistical ensemble has a diagonal representation (4.23), then the statistical averages of the operators form a kind of 'characteristic function' of $P(\alpha)$:

$$X(\lambda) \equiv \langle\hat{X}(\lambda)\rangle = \int d^2\alpha\, P(\alpha)\, b^\alpha(\lambda) \tag{4.27}$$

The weight function $P(\alpha)$ can be expressed in terms of $X(\lambda)$ with the help of the reciprocal basis $\overline{b}^\lambda(\alpha)$

$$P(\alpha) = \int d^2\lambda\, X(\lambda)\, \overline{b}^\lambda(\alpha) \tag{4.28}$$

A convenient basis is the Fourier basis

$$b^\alpha(\lambda) = \exp(\lambda\alpha^* - \lambda^*\alpha) \tag{4.29a}$$

$$\overline{b}^\lambda(\alpha) = \frac{1}{(2\pi)^2} \exp(-\lambda\alpha^* + \lambda^*\alpha) \tag{4.29b}$$

which is generated by the normally ordered operators

$$\hat{X}_N(\lambda) = \exp(\lambda a^+) \exp(-\lambda^* a) \qquad (4.29c)$$

The question of the existence of the P representation is a complicated one [5, 7]. Using the Fourier basis (4.29) it can be shown, however, that the mere existence of the inverse transformation (4.28) guarantees that the resulting function $P(\alpha)$ can be used to calculate the statistical average of any moment $\langle a^{+m} a^n \rangle$ as if $P(\alpha)$ was the weight function defined in Eq. (4.23). This is due to the fact that the characteristic function $X_N(\lambda)$ plays the role of a generating function for $\langle a^m a^n \rangle$:

$$\langle a^m a^n \rangle = \left(\frac{\partial}{\partial \lambda}\right)^n \left(-\frac{\partial}{\partial \lambda^*}\right)^m X_n(\lambda) \Big|_{\lambda = 0}$$

From this, using Eq. (4.27), one obtains

$$\langle a^{+m} a^n \rangle = \int d^2\alpha \; P(\alpha) \langle \alpha | a^{+m} a^n | \alpha \rangle$$

which is a particular case of Eq. (4.25) and proves the above statement. One could moreover introduce, in addition to Eq. (4.29c), symmetrically ordered $\hat{X}_S(\lambda)$, and antinormally ordered $\hat{X}_A(\lambda)$, exponential operators [5, 8]. The Fourier transforms of their statistical averages are the Wigner distribution and the matrix element $\langle \alpha | \rho | \alpha \rangle / \pi$, respectively. We shall not develop these aspects further as the corresponding expression for atomic coherent states are rather involved and of no clear use as yet.

4.3. DESCRIPTION OF THE FREE ATOMS

4.3.1. The angular momentum states

Angular momentum operators can be defined which act on the N-atom Hilbert space. In particular we can consider a subspace of degenerate eigenstates of J^2 with eigenvalues $J(J+1)$. Since J^2 commutes with J_x, J_y, J_z, these operators only connect states within the same subspace. In general, J^2 and J_z do not form a complete set of commuting observables. As explained in Ref. [5], such a complete set is formed by adding to J^2 and J_z some operators of the permutation group of N objects P_N. These operators play the same role with respect to P_N that J^2 and J_z have with respect to the three-dimensional rotation group. We shall assume that the subspace considered here has also been made invariant under these permutation operations, but for simplicity we shall not indicate this, for the time being, in the labelling of the states. The subspace we are dealing with is identical to a constant angular momentum Hilbert space. The Dicke states, which are the analog of the Fock states (4.5), and the Bloch states, which correspond to the Glauber state (4.8), are most easily defined within such a subspace. The equation numbering is in parallel with that of Section 4.2. From the angular momentum operators J_x and J_y, which satisfy the commutation relation

$$[J_x, J_y] = iJ_z \qquad (4.1)'$$

the lowering and raising operators are formed

$$J_- = J_x - iJ_y \qquad (4.2a)'$$

$$J_+ = J_x + iJ_y \qquad (4.2b)'$$

which obey

$$[J_-, J_+] = -2J_z \tag{4.3a)'}$$

$$[J_-, J_z] = J_- \tag{4.3b)'}$$

$$[J_+, J_z] = -J_+ \tag{4.3c)'}$$

The Dicke states, which are simply the usual angular momentum states, are defined as the eigenstates of

$$J_z = \tfrac{1}{2}(J_+J_- - J_-J_+) \tag{4.4)'}$$

They are given by (Ref. [10])

$$|M\rangle = \frac{1}{(M+J)!}\binom{2J}{M+J}^{-\frac{1}{2}} J_+^{M+J}\,|-J\rangle \qquad (M = -J, -J+1, \ldots, J) \tag{4.5)'}$$

with eigenvalue M. They span the space of angular momentum quantum number J. The ground-state $|-J\rangle$ is defined by

$$J_-|-J\rangle = 0 \tag{4.6)'}$$

4.3.2. Coherent atomic states

Let us consider the rotation operator which produces a rotation through an angle ϑ about an axis $\hat{n} \equiv (\sin\varphi, \cos\varphi, 0)$:

$$\begin{aligned} R_{\vartheta,\varphi} &= e^{-i\vartheta}\,J_n = \exp[-i\vartheta(J_x \sin\varphi - J_y \cos\varphi)] \\ &= \exp[\zeta J_+ - \zeta^* J_-] \end{aligned} \tag{4.7a)'}$$

where

$$\zeta = \frac{\vartheta}{2}\,e^{-i\varphi} \tag{4.7b)'}$$

A coherent atomic state, or Bloch state, $|\vartheta, \varphi\rangle$, is obtained by rotation of the ground-state $|-J\rangle$:

$$|\vartheta, \varphi\rangle \equiv R_{\vartheta,\varphi}|-J\rangle \tag{4.8)'}$$

Furthermore,

$$R_{\vartheta,\varphi} J_n R_{\vartheta,\varphi}^{-1} = J_n$$

$$R_{\vartheta,\varphi} J_k R_{\vartheta,\varphi}^{-1} = J_k \cos\vartheta + J_z \sin\vartheta$$

$$R_{\vartheta,\varphi} J_z R_{\vartheta,\varphi}^{-1} = -J_k \cos\vartheta + J_z \sin\vartheta$$

where

$$J_n = J_x \sin\varphi - J_y \cos\varphi$$

$$J_k = J_x \cos\varphi + J_y \sin\varphi$$

which gives

$$J_+ = (J_k - iJ_n)e^{i\varphi}$$

$$J_- = (J_k + iJ_n)e^{-i\varphi}$$

Using these relations one obtains

$$R_{\vartheta,\varphi} J_- R^{-1}_{\vartheta,\varphi} = e^{-i\varphi}\left[J_- e^{i\varphi} \cos^2\frac{\vartheta}{2} - J_+ e^{-i\varphi} \sin^2\frac{\vartheta}{2} + J_z \sin\vartheta\right] \tag{4.9a)'}$$

and similar relations for J_+ and J_z:

$$R_{\vartheta,\varphi} J_+ R^{-1}_{\vartheta,\varphi} = e^{i\varphi}\left[J_+ e^{i\varphi} \cos^2\frac{\vartheta}{2} - J_- e^{-i\varphi} \sin^2\frac{\vartheta}{2} + J_z \sin\vartheta\right] \tag{4.9b)'}$$

$$R_{\vartheta,\varphi} J_z R^{-1}_{\vartheta,\varphi} = J_z \cos\vartheta - J_- e^{i\varphi} \sin\frac{\vartheta}{2} \cos\frac{\vartheta}{2} - J_+ e^{-i\varphi} \tag{4.9c)'}$$

From Eq. (4.9a)′, and definition (4.8)′, one obtains the eigenvalue equation

$$\left[J_- e^{i\varphi} \cos^2\frac{\vartheta}{2} - J_+ e^{-i\varphi} \sin^2\frac{\vartheta}{2} + J_z \sin\vartheta\right] |\vartheta,\varphi\rangle = 0 \tag{4.10a)'}$$

This equation, together with

$$J^2 |\vartheta,\varphi\rangle = J(J+1) |\vartheta,\varphi\rangle \tag{4.10b)'}$$

specifies uniquely the Bloch state $|\vartheta,\varphi\rangle$. Note that the harmonic oscillator analog of Eq. (3.10b) would have been the trivial relation $(a^+ - \alpha^*)(a - \alpha)|\alpha\rangle = 0$.

Other forms of the eigenvalue equation can be obtained using the relation

$$R_{\vartheta,\varphi} J_z R^{-1}_{\vartheta,\varphi} |\vartheta,\varphi\rangle = -J |\vartheta,\varphi\rangle$$

and Eq. (4.9c)′. The resulting equation can be combined with Eq. (4.10a)′ to eliminate one of the operators J_z, J^+, or J^-, giving

$$\left[J_- e^{i\varphi} \cos^2\frac{\vartheta}{2} + J_+ e^{-i\varphi} \sin^2\frac{\vartheta}{2}\right] |\vartheta,\varphi\rangle = J \sin\vartheta |\vartheta,\varphi\rangle \tag{4.10c)'}$$

$$\left[J_- e^{i\varphi} \cos\frac{\vartheta}{2} + J_z \sin\frac{\vartheta}{2}\right] |\vartheta,\varphi\rangle = J \sin\frac{\vartheta}{2} |\vartheta,\varphi\rangle \tag{4.10d)'}$$

$$\left[J_+ e^{-i\varphi} \sin\frac{\vartheta}{2} - J_z \cos\frac{\vartheta}{2}\right] |\vartheta,\varphi\rangle = J \cos\frac{\vartheta}{2} |\vartheta,\varphi\rangle \tag{4.10e)'}$$

These additional relations are not independent of Eqs (4.10a)′ and (4.10b)′. One notes that these eigenvalue equations are more complicated than their counterpart (4.10). In particular they involve at least two of the three operators J_-, J_+, J_z. This feature is required by the more complicated commutation relation (4.1)′ which applies here. Using the disentangling theorem for angular momentum operators [Ref. 6], the rotation $R_{\vartheta,\varphi}$ given by Eq. (4.7a)′ becomes

$$
\begin{aligned}
R_{\vartheta,\varphi} &= e^{-\tau^* J_-}\, e^{-\ln(1+|\tau|^2) J_z}\, e^{\tau J_+} \\
&= e^{\tau J_+}\, e^{\ln(1+|\tau|^2) J_z}\, e^{-\tau^* J_-}
\end{aligned}
\qquad (4.11a)'
$$

where

$$
\tau \equiv e^{-i\varphi} \tan(\vartheta/2) \qquad (4.11b)'
$$

Let us point out that these expressions are singular for $\vartheta = \pi$, i.e. for the uppermost state. We may have to exclude from some of the following considerations the states contained within an infinitesimally small circle around $\vartheta = \pi$. The validity of expressions such as (4.13)′ for $\vartheta = \pi$ is usually not affected and can be checked directly. The last form of Eq. (4.11a)′, which we call the normally ordered form, gives immediately the expansion of $|\vartheta,\varphi\rangle$ in terms of Dicke states:

$$
|\vartheta,\varphi\rangle = R_{\vartheta,\varphi}|-J\rangle = \left(\frac{1}{1+|\tau|^2}\right)^J e^{\tau J_+}|-J\rangle \qquad (4.12)'
$$

from which, expanding the exponential and using Eq.(3.5), one obtains

$$
\begin{aligned}
\langle M|\vartheta,\varphi\rangle &= \binom{2J}{M+J}^{\frac{1}{2}} \frac{\tau^{M+J}}{(1+|\tau|^2)^J} \\
&= \binom{2J}{M+J}^{\frac{1}{2}} \sin^{J+M}\frac{\vartheta}{2} \cos^{J-M}\frac{\vartheta}{2}\, e^{-i(J+M)\varphi}
\end{aligned}
\qquad (4.13)'
$$

Since the Dicke states form a basis for a well-known irreducible representation of the rotation group, these results could have been derived using the appropriate Wigner $\mathscr{D}^{(J)}$ matrix [10]. The same applies to Eqs (4.12) and (4.13). These could have been obtained, without using the Baker-Campbell-Hausdorf formula, from the transformation properties of an irreducible representation of the group of operations T_α.

The overlap of two Bloch states is obtained either from Eq. (4.12)′, using the disentangling theorem for exponential angular momentum operators, or from Eq. (4.13)′, using the completeness property of Dicke states $\sum_M |M\rangle\langle M| = 1$. One obtains

$$
\begin{aligned}
\langle\vartheta\varphi|\vartheta'\varphi'\rangle &= \left[\frac{(1+\tau^*\tau')^2}{(1+|\tau|^2)(1+|\tau'|^2)}\right]^J \\
&= e^{-iJ(\varphi-\varphi')}\left(\cos\frac{\vartheta-\vartheta'}{2}\cos\frac{\varphi-\varphi'}{2} - i\cos\frac{\vartheta+\vartheta'}{2}\sin\frac{\varphi-\varphi'}{2}\right)
\end{aligned}
\qquad (4.14a)'
$$

from which one obtains

$$|\langle\vartheta\varphi|\vartheta'\varphi'\rangle|^2 = \cos^{4J}\frac{\Theta}{2} \tag{4.14b)'$$

where τ is given by Eq. (4.11b)', τ' is given by the same equation written with the primed quantities, and Θ is the angle between the (ϑ,φ) and (ϑ',φ') directions, as given by

$$\cos\Theta = \cos\vartheta\cos\vartheta' + \sin\vartheta\sin\vartheta'\cos(\varphi-\varphi')$$

The Bloch states form minimum uncertainty packets. The uncertainty relation can be defined in terms of the set of rotated operators $(J_\xi, J_\eta, J_\zeta) = R_{\vartheta,\varphi}(J_x, J_y, J_z)R^{-1}_{\vartheta,\varphi}$. These three observables obey a commutation relation of the type $[A, B] = iC$ with $A = J_\xi$, $B = J_\eta$, $C = J_\zeta$, from which they have the uncertainty property

$$\langle J_\xi^2\rangle\langle J_\eta^2\rangle \geqslant \frac{1}{4}\langle J_\zeta^2\rangle \tag{4.15)'$$

for any states. It is easy to show that the equality sign holds for the Bloch state $|\vartheta,\varphi\rangle$, which is therefore a minimum uncertainty state.

4.3.3. The Bloch states as a basis

Let us now consider the completeness properties of the Bloch states. Using Eq. (4.13)' and the completeness of Dicke states $\sum_M |M\rangle\langle M| = 1$, one obtains

$$\begin{aligned}
(2J+1)\int\frac{d\Omega}{4\pi}\,|\vartheta,\varphi\rangle\langle\vartheta,\varphi|
&= (2J+1)\int\frac{d\Omega}{4\pi}\sum_{M,M'}\binom{2J}{M+J}^{\frac{1}{2}}\binom{2J}{M'+J}^{\frac{1}{2}} e^{i(M-M')\varphi}\\
&\quad\times\left(\cos\frac{\vartheta}{2}\right)^{2J-M-M'}\left(\sin\frac{\vartheta}{2}\right)^{2J+M+M'}|M\rangle\langle M'|\\
&= (2J+1)\int_0^\pi d\vartheta\,\frac{\sin\vartheta}{2}\sum_M\binom{2J}{M+J}\left(\cos\frac{\vartheta}{2}\right)^{2J-2M}\left(\sin\frac{\vartheta}{2}\right)^{2J+2M}|M\rangle\langle M|\\
&= \sum_M |M\rangle\langle M| = 1
\end{aligned} \tag{4.16)'$$

The expansion of an arbitrary state in Bloch states follows:

$$\begin{aligned}
|c\rangle = \sum_M C_M|M\rangle &= (2J+1)\int\frac{d\Omega}{4\pi}\sum_M C_M|\vartheta,\varphi\rangle\langle\vartheta,\varphi|M\rangle\\
&= (2J+1)\int\frac{d\Omega}{4\pi}\,\frac{f(\tau^*)}{(1+|\tau|^2)^J}\,|\vartheta,\varphi\rangle
\end{aligned} \tag{4.17a)'$$

where

$$f(\tau^*) \equiv \sum_M C_M \binom{2J}{J+M}^{\frac{1}{2}} (\tau^*)^{J+M} = (1 + |\tau|^2)^J \langle \vartheta,\varphi|c\rangle \tag{4.17b)$'$}$$

Using Eq. (4.5), one sees that $|c\rangle$ can also be written as

$$|c\rangle = f\left(\frac{1}{J+1-J_z} J_+\right) |-J\rangle \tag{4.18)$'$}$$

The amplitude function $f(\tau^*)$ is, by its definition (4.17b)′, a polynomial of degree 2J. However, any function which has a MacLaurin expansion can be taken as a suitable amplitude function in Eq. (4.17a)′ or Eq. (4.18). Indeed, the powers of τ^* higher than 2J give zero contribution in Eqs (4.17a)′ and (4.18)′. The coefficients C_M are then obtained from the first (2J + 1) terms of the MacLaurin series, using Eq. (4.17b)′.

The scalar product of two states characterized by their amplitude function is, from Eqs (4.16)′ and (4.17b)′,

$$\begin{aligned} \langle c'|c\rangle &= (2J+1) \int \frac{d\Omega}{4\pi} \langle c'|\vartheta,\varphi\rangle\langle\vartheta,\varphi|c\rangle \\ &= (2J+1) \int \frac{d\Omega}{4\pi} \frac{1}{(1+|\tau|^2)^{2J}} [f'(\tau^*)]^* f(\tau^*) \end{aligned} \tag{4.19)$'$}$$

Since Eq. (4.17b)′ was used to derive this equation, its validity is restricted to amplitude functions which are polynomials of degree 2J.

In view of the completeness relations, operators G acting on this Hilbert space can be expanded as

$$G = \sum_{M,M'} |M\rangle\langle M|G|M'\rangle\langle M'| \tag{4.20a)$'$}$$

or

$$G = \frac{(2J+1)^2}{(4\pi)^2} \iint d\Omega\, d\Omega' |\vartheta,\varphi\rangle\langle\vartheta,\varphi|G|\vartheta',\varphi'\rangle\langle\vartheta',\varphi'| \tag{4.20b)$'$}$$

However, G is completely defined by the $(2J + 1)^2$ matrix elements $\langle M|G|M'\rangle$. It results that, except for pathological cases, an operator can always be written in the diagonal form

$$G = \int d\Omega\, g(\vartheta,\varphi) |\vartheta,\varphi\rangle\langle\vartheta,\varphi| \tag{4.20c)$'$}$$

where $g(\vartheta,\varphi)$ is given by a series expansion

$$g(\vartheta,\varphi) = \sum_{\ell,m} G_{\ell,m} \, Y_\ell^m(\vartheta,\varphi)$$

In accordance with Appendix IV of Ref. [6], only the $(2J + 1)^2$ first terms of this sum contribute to Eq. (4.20c)′. These are the terms for which $0 \leqslant \ell \leqslant 2J$. The corresponding coefficients $G_{\ell,m}$ can be expressed as a function of the matrix element $\langle M|G|M'\rangle$.

4.3.4. Statistical operators for the atoms

In order to describe an incoherently pumped system of atoms we introduce a statistical operator ρ with the properties

$$\mathrm{Tr}\,\rho = 1 \tag{4.21$'$}$$

$$\langle G\rangle = \mathrm{Tr}\,\rho G \tag{4.22$'$}$$

As before, the considerations are restricted to states belonging to a single constant angular momentum subspace, and therefore the statistical operator described here does not allow for the most general mixing of atomic states. Of particular interest is the expression of ρ in a diagonal Bloch representation

$$\rho = \int P(\vartheta,\varphi)|\vartheta,\varphi\rangle\langle\vartheta,\varphi|\,d\Omega \tag{4.23$'$}$$

with the normalization

$$\int P(\vartheta,\varphi)\,d\Omega = 1 \tag{4.24$'$}$$

The statistical average of an observable G is then given by

$$\langle G\rangle = \int P(\vartheta,\varphi)\langle\vartheta,\varphi|G|\vartheta,\varphi\rangle \tag{4.25$'$}$$

The weight function $P(\vartheta,\varphi)$ has thus the properties of a distribution function on the unit sphere, except that it is not necessarily positive.

Let us define a set of operators $\hat{X}(\lambda)$ such that their expectation values for Bloch states

$$b_\lambda(\vartheta,\varphi) = \langle\vartheta,\varphi|\hat{X}_\lambda|\vartheta,\varphi\rangle \tag{4.26$'$}$$

form a basis in the space of functions on the unit sphere. Since in this space a discrete basis can be chosen, the parameter λ can be restricted to discrete values $\lambda = 1, 2, \ldots$ For a statistical ensemble described by Eq. (4.23)′ the statistical average of the operators $\hat{X}_\lambda$ forms a set of 'characteristic coefficients' of $P(\vartheta,\varphi)$

$$X_\lambda \equiv \mathrm{Tr}\,\rho\hat{X}_\lambda = \int d\Omega\, P(\vartheta,\varphi)\, b_\lambda(\vartheta,\varphi) \tag{4.27$'$}$$

The weight function can be expressed as a series with the help of the reciprocal basis $\bar{b}^\lambda(\vartheta,\varphi)$,

$$P(\vartheta,\varphi) = \sum_\lambda X_\lambda \bar{b}^\lambda(\vartheta,\varphi) \tag{4.28$'$}$$

A convenient basis is given by the spherical harmonics

$$b_\lambda(\vartheta,\varphi) = Y_\ell^m(\vartheta,\varphi) \qquad (\lambda \equiv \ell, m) \tag{4.29a$'$}$$

$$\bar{b}^\lambda(\vartheta,\varphi) = Y_\ell^{-m}(\vartheta,\varphi) \tag{4.29b$'$}$$

which are generated by the spherical harmonic operators [11]

$$\hat{X}_\lambda = Y_\ell^m(\vec{J}) \tag{4.29c$'$}$$

The already mentioned fact that a diagonal representation always exists in the atomic case also corresponds to the fact that for a given J only the $(2J+1)^2$ operators Y_ℓ^m with $\ell \leqslant 2J$ are different from zero. The finite dimensionality of the basis is required, since ρ is completely determined by its $(2J+1)^2$ matrix elements $\langle M|\rho|M'\rangle$ in the Dicke representation.

Other differences with respect to the field case should also be noted. First, the spherical harmonic operators are usually written in a fully symmetrized form, whereas the operators (4.29c)′ are normally ordered. This is only a formal difficulty as it should be possible to write normally ordered and antinormally ordered multipole operators with properties similar to the $Y_\ell^m(\vec{J})$. A second and more fundamental difference is that the expectation values X_λ are not generating functions for products of the type $\langle J_+^m J_z^n J_-^p\rangle$, in view of the discreteness of the set. This does not cause much difficulty, as generating functions can be defined from exponential operators whose expectation values can be calculated with the help of the disentangling theorem. It is tempting to use for the $\hat{X}_\lambda$ of Eq. (4.29c)′ these exponential operators themselves. Though the case parallel with the field then seems more transparent, the use of the discrete set $\hat{X}_\lambda$ may be of more fundamental significance as it takes into account symmetry properties of the states.

A final comment should be made about the difficulty of dealing with creation and annihilation operations in a finite Hilbert space. The existence of two terminal states, $|J\rangle$ and $|-J\rangle$, requires the presence of a third operator with the properties of J_z and prevents the writing of an eigenvalue equation in terms of one compound operator alone. For instance, the comparison of Eqs (4.18)′ and (4.18) suggests that $J_-(J+1-J_z)^{-1}$ could be a 'good' annihilation operator. Using Eq. (4.13)′ one finds immediately

$$J_-(J+1-J_z)^{-1}|\vartheta,\varphi\rangle = \tau|\vartheta,\varphi\rangle - \tau \sin^{2J}\frac{\vartheta}{2}\, e^{-2iJ\varphi}|J\rangle$$

which for small and large J is almost an eigenvalue equation; so the application of the operator reproduces $|\vartheta,\varphi\rangle$, except for the uppermost Dicke state. There is no doubt that a theory could be developed in terms of more complicated annihilation and creation operators of such type, but the advantages are not clear.

REFERENCES TO PART IV

[1] DICKE, R.H., Phys. Rev. **93** (1954) 99.
[2] BLOCH, F., Phys. Rev. **70** (1946) 460.
[3] BLOCH, F., NORDSIECK, A., Phys. Rev. **52** (1937) 54.
[4] SCHWINGER, J., Phys. Rev. **91** (1953) 728.
[5] GLAUBER, R.J., Phys. Rev. **131** (1963) 2766; GLAUBER, R.J., in Quantum Optics and Electronics, (De WITT, C., et al., Eds), Gordon and Breach, New York (1965).
[6] ARECCHI, F.T., COURTENS, E., GILMORE, R., THOMAS, H., Phys. Rev. A **6** (1972) 2211.
[7] GLAUBER, R.J., in Physics of Quantum Electronics (KELLEY, P.L., LAX, B., TANNENWALD, P.E., Eds), Columbia, New York (1966).
KLAUDER, J.R., SUDARSHAN, E.C.G., Fundamentals of Quantum Optics, Benjamin, New York (1968).
[8] HAKEN, H., in Encyclopedia of Physics (FLÜGGE, S., Ed.), Springer-Verlag, Berlin (1970), Vol. XXV, Ch. 2.
[9] FEYNMAN, R.P., Phys. Rev. **84** (1951) 108.
[10] WIGNER, E.P., Group Theory and its Application to the Quantum Mechanics of Atomic Spectra, Academic Press, New York (1959).
[11] CALLEN, E., CALLEN, H., Phys. Rev. **129** (1963) 578.

EXCITATIONS

IAEA-SMR-20/8

PHONONS AND POLARITONS

R.F. WALLIS*
Laboratoire de Physique des Solides,
Université Pierre et Marie Curie,
Paris,
France

Abstract

PHONONS AND POLARITONS.

The subject of phonons and polaritons is introduced by a discussion of the formalism of lattice dynamics based on the Born-Oppenheimer approximation. The classical equations of motion for the displacements of the atoms of a crystal from their equilibrium positions are set up. Invariance conditions on the coupling constants are derived, and the equations of motion are solved in the harmonic approximation using a normal coordinate transformation. The procedures are illustrated by means of three simple examples. The lattice specific heat is then examined using the Debye elastic continuum model. This leads to a discussion of the frequency distribution function for the normal modes of vibration. Attention is then turned to the calculation of optical properties due to the photon-phonon interaction. A simple model is used to derive expressions for the dielectric constant and the absorption and reflectivity coefficients. These results are then applied to the discussion of poiaritons. The dispersion curves for polaritons are derived from a macroscopic approach based on Maxwell's equations. The use of Raman scattering in the study of polaritons is examined in detail. The effect of a free surface is then considered, and the theory of surface polaritons is developed. The determination of surface polariton dispersion curves by means of attenuated total reflection is discussed. The presentation is concluded by a brief survey of effects due to anharmonicity. A calculation of the linear thermal expansion coefficient is presented for a simple model.

1. Formalism of Lattice Dynamics

We consider a crystal to be made up of a neutral collection of electrons and nuclei. The wave function, $\Psi(\underset{\sim}{r}, \underset{\sim}{R})$, of the crystal is a function of the coordinates, $\underset{\sim}{r}$, of all the electrons and the coordinates, $\underset{\sim}{R}$, of all the nuclei. The Schrödinger equation satisfied by $\Psi(\underset{\sim}{r}, \underset{\sim}{R})$ can be written

$$[T_e + T_n + V(\underset{\sim}{r}, \underset{\sim}{R})] \; \Psi(\underset{\sim}{r}, \underset{\sim}{R}) = E \; \Psi(\underset{\sim}{r}, \underset{\sim}{R}) \tag{1.1}$$

where T_e is the kinetic energy operator for the electrons, T_n is the kinetic energy operator for the nuclei, $V(\underset{\sim}{r}, \underset{\sim}{R})$ is the total potential energy of the electrons and nuclei, and E is the energy eigenvalue.

An approximate separation of the electronic and nuclear coordinates may be achieved following Born and Oppenheimer[1] who

* Permanent address: Department of Physics, University of California, Irvine, California, United States of America.

exploited the large difference in mass of the electrons and nuclei. The electrons are considered to move in the field of the nuclei regarded as instantaneously fixed. The electronic energy eigenvalue so determined then serves as the potential energy for the nuclear motion. Specifically, the wave function $\Psi(\underset{\sim}{r}, \underset{\sim}{R})$ is taken to have the product form

$$\Psi(\underset{\sim}{r}, \underset{\sim}{R}) = \psi(\underset{\sim}{r}, \underset{\sim}{R})\, \chi(\underset{\sim}{R}) \tag{1.2}$$

where the electronic wave function, $\psi(\underset{\sim}{r}, \underset{\sim}{R})$, satisfies the electronic Schrödinger equation

$$[T_e + V(\underset{\sim}{r}, \underset{\sim}{R})]\, \psi(\underset{\sim}{r}, \underset{\sim}{R}) = \Phi(\underset{\sim}{R})\, \psi(\underset{\sim}{r}, \underset{\sim}{R}) \tag{1.3}$$

In Eq.(1.3), the nuclear coordinates $\underset{\sim}{R}$ are assumed to be fixed as parameters and $\Phi(\underset{\sim}{R})$ is the electronic energy eigenvalue which is a function of the nuclear coordinates.

The function $\chi(\underset{\sim}{R})$ in Eq.(1.2) is the nuclear wave function which satisfies the nuclear Schrödinger equation

$$[T_n + \Phi(\underset{\sim}{R})]\, \chi(\underset{\sim}{R}) = E\, \chi(\underset{\sim}{R}) \tag{1.4}$$

In the following we shall assume that the electronic Schrödinger equation, Eq.(1.3), has been solved and that $\Phi(\underset{\sim}{R})$ is a known function of the nuclear coordinates $\underset{\sim}{R}$.

We restrict our attention to stable crystals in which the nuclei vibrate about equilibrium positions that form a lattice. It is convenient to expand the nuclear potential energy in powers of displacements of the nuclei from their equilibrium positions. Let

$$\underset{\sim}{R}(\ell\kappa) = \underset{\sim}{R}^{(o)}(\ell\kappa) + \underset{\sim}{u}(\ell\kappa) \tag{1.5}$$

be the position vector of the nth atom in the ℓth unit cell, where $\underset{\sim}{R}^{(o)}(\ell\kappa)$ is the equilibrium position vector and $\underset{\sim}{u}(\ell\kappa)$ is the displacement vector from equilibrium. Then

$$\Phi(R) = \Phi^{(o)} + \sum_{\ell\kappa\alpha} \Phi_\alpha(\ell\kappa) u_\alpha(\ell\kappa)$$
$$+ \tfrac{1}{2}\sum_{\ell\kappa\alpha} \sum_{\ell'\kappa'\beta} \Phi_{\alpha\beta}(\ell\ell', \kappa\kappa') u_\alpha(\ell\kappa) u_\beta(\ell'\kappa')$$

$$+ \frac{1}{6} \sum_{\ell\varkappa\alpha} \sum_{\ell'\varkappa'\beta} \sum_{\ell''\varkappa''\gamma} \Phi_{\alpha\beta\gamma}(\ell\ell'\ell'';\varkappa\varkappa'\varkappa'')u_\alpha(\ell\varkappa)u_\beta(\ell'\varkappa')u_\gamma(\ell''\varkappa'')$$

$$+ \ldots\ldots \qquad (1.6)$$

where

$$\Phi_\alpha(\ell\varkappa) = \left.\frac{\partial\Phi(\underset{\sim}{R})}{\partial u_\alpha(\ell\varkappa)}\right|_{\underset{\sim}{u}=0}, \quad \Phi_{\alpha\beta}(\ell\ell',\varkappa\varkappa') = \frac{\partial^2\Phi(\underset{\sim}{R})}{\partial u_\alpha(\ell\varkappa)\partial u_\beta(\ell'\varkappa')},$$

$$\text{etc.} \qquad (1.7)$$

The motion of the nuclei described by these equations can be regarded as that of a set of coupled anharmonic oscillators. The terms in the potential energy expansion through quadratic are the harmonic terms and the higher order terms are the anharmonic terms.

The classical force acting on nucleus ($\ell''\varkappa''$) in the γ-direction is given by

$$F_\gamma(\ell''\varkappa'') = -\frac{\partial\Phi(\underset{\sim}{R})}{\partial u_\gamma(\ell''\varkappa'')} \qquad (1.8a)$$

$$= -\{\Phi_\gamma(\ell''\varkappa'') + \tfrac{1}{2}\sum_{\ell'\varkappa'\beta} \Phi_{\gamma\beta}(\ell''\ell',\varkappa''\varkappa')u_\beta(\ell'\varkappa') + \tfrac{1}{2}\sum_{\ell\varkappa\alpha}\Phi_{\alpha\gamma}(\ell\ell'',\varkappa\varkappa'')u_\alpha(\ell\varkappa)$$
$$+ \text{ anharmonic terms } \} \qquad (1.8b)$$

$$= -\{\Phi_\gamma(\ell''\varkappa'') + \tfrac{1}{2}\sum_{\ell\varkappa\alpha} \Phi_{\gamma\alpha}(\ell''\ell,\varkappa''\varkappa)u_\alpha(\ell\varkappa) + \tfrac{1}{2}\sum_{\ell\varkappa\alpha}\Phi_{\alpha\gamma}(\ell\ell'',\varkappa\varkappa'')u_\alpha(\ell\varkappa)$$
$$+ \text{ anharmonic terms } \} \qquad (1.8c)$$

Since $\Phi_{\alpha\gamma}(\ell\ell'',\varkappa\varkappa'')$ is a second partial derivative,

$$\Phi_{\alpha\gamma}(\ell\ell'',\varkappa\varkappa'') = \Phi_{\gamma\alpha}(\ell''\ell,\varkappa''\varkappa) \qquad (1.9)$$

and we can re-express the force as

$$F_\gamma(\ell''\varkappa'') = -\{\Phi_\gamma(\ell''\varkappa'') + \sum_{\ell\varkappa\alpha}\Phi_{\gamma\alpha}(\ell''\ell,\varkappa''\varkappa)u_\alpha(\ell\varkappa)$$
$$+ \text{ anharmonic terms}) \} \qquad (1.10)$$

Now if all the $u_\alpha(\ell\kappa) = 0$, the nuclei are in equilibrium and the force must be zero; hence

$$\Phi_\gamma(\ell''\kappa'') = 0 \quad \text{for all } \gamma, \ell'', \kappa'' \tag{1.11}$$

If all of the displacements are sufficiently small, the anharmonic terms are small compared to the harmonic terms and can be neglected. Then the forces are Hooke's law forces which are linear in the displacements from equilibrium.

One can simplify the notation somewhat by relabeling the indices in the force equation to give

$$F_\alpha(\ell\kappa) = -\sum_{\ell'\kappa'\beta} \Phi_{\alpha\beta}(\ell\ell', \kappa\kappa')u_\beta(\ell'\kappa') + \text{anharmonic terms}. \tag{1.12}$$

The Hamiltonian can be written as

$$H = \sum_{\ell\kappa\alpha} \tfrac{1}{2}M_\kappa \dot{u}_\alpha^2(\ell\kappa) + \tfrac{1}{2}\sum_{\ell\kappa\alpha}\sum_{\ell'\kappa'\beta} \Phi_{\alpha\beta}(\ell\ell', \kappa\kappa')u_\alpha(\ell\kappa)u_\beta(\ell'\kappa') + \frac{1}{6}\sum_{\ell\kappa\alpha}\sum_{\ell'\kappa'\beta}\sum_{\ell''\kappa''\gamma} \Phi_{\alpha\beta\gamma}(\ell\ell'\ell'', \kappa\kappa'\kappa'')u_\alpha(\ell\kappa)u_\beta(\ell'\kappa')u_\gamma(\ell''\kappa'') + \ldots \tag{1.13}$$

The coupling coefficients $\Phi_{\alpha\beta}(\ell\ell', \kappa\kappa')$, $\Phi_{\alpha\beta\gamma}(\ell\ell'\ell'', \kappa\kappa'\kappa'')$, etc., must satisfy several conditions[1]. We shall discuss explicitly two of these conditions -- namely, those that arise from the fact that no interatomic force can be produced if a crystal is subjected to a rigid body translation or to a rigid body rotation, since the relative configuration of the atoms remains unchanged under both of these operations.

Let us consider first infinitesimal translational invariance. Referring to the force equation, Eq.(1.12), if the $u_\beta(\ell'\kappa')$ for all $\ell'\kappa'$ are the same for a particular component γ and are zero for the other components,

$$u_\beta(\ell'\kappa') = \epsilon\delta_{\beta\gamma}, \tag{1.14}$$

then we have a rigid body translation in the direction γ, and the force must be zero:

$$F_{\alpha}(\ell\varkappa) = 0 \tag{1.15}$$

Hence

$$\sum_{\ell'\varkappa'} \Phi_{\alpha\gamma}(\ell\ell',\varkappa\varkappa') = 0 \tag{1.16}$$

or equivalently

$$\sum_{\ell'\varkappa'} \Phi_{\alpha\beta}(\ell\ell',\varkappa\varkappa') = 0 \tag{1.17}$$

Using this last equation, the force equation can be rewritten in the harmonic approximation as

$$F_{\alpha}(\ell\varkappa) = \sum_{\ell'\varkappa'\beta} \Phi_{\alpha\beta}(\ell\ell',\varkappa\varkappa')[u_{\beta}(\ell\varkappa)-u_{\beta}(\ell'\varkappa')] \tag{1.18}$$

that is, the force on a given atom is proportional to the relative displacements of it and the other atoms with which it interacts.

The displacements associated with an infinitesimal rigid body rotation are specified by

$$\begin{aligned} u_{\beta}(\ell'\varkappa') &= \sum_{\gamma} \omega_{\beta\gamma}[R_{\gamma}^{(o)}(\ell'\varkappa') - R_{\gamma}^{(o)}(\ell\varkappa)] \\ &= \sum_{\gamma} \omega_{\beta\gamma}\, X_{\gamma}(\ell'\varkappa',\ell\varkappa) \end{aligned} \tag{1.19}$$

where $X_{\gamma}(\ell'\varkappa',\ell\varkappa) = R_{\gamma}^{(o)}(\ell'\varkappa') - R_{\gamma}^{(o)}(\ell\varkappa)$ and $\omega_{\beta\gamma}$ is an element of an infinitesimal antisymmetric matrix $\omega_{\beta\gamma} = -\omega_{\gamma\beta}$. As a simple example, we consider a rigid rotation in a plane described by the transformation

$$\begin{aligned} x' &= \cos\theta\ x - \sin\theta\ y \\ y' &= \sin\theta\ x + \cos\theta\ y \end{aligned}$$

$$\begin{aligned} x'-x &= (\cos\theta-1)x - \sin\theta\, y \simeq \tfrac{1}{2}\theta^2 x - \theta\, y \simeq 0\, x - \theta\, y \\ y'-y &= \sin\theta\ x + (\cos\theta-1)y \simeq \theta\, x + \tfrac{1}{2}\theta^2 y \simeq \theta\, x + 0\, y \end{aligned}$$

We see that for an infinitesimal rotation, $\theta \to 0$, the transformation matrix is given by

$$\omega = \begin{pmatrix} 0 & -\theta \\ \theta & 0 \end{pmatrix} \tag{1.20}$$

Restricting our attention to the harmonic approximation, we substitute Eq.(1.19) into Eq.(1.12) with the anharmonic terms omitted and set the result equal to zero. We obtain

$$\sum_{\ell'\kappa'\beta\gamma} \Phi_{\alpha\beta}(\ell\ell',\kappa\kappa')\omega_{\beta\gamma}X_{\gamma}(\ell'\kappa',\ell\kappa) = 0 \tag{1.21}$$

This equation must be true for arbitrary values of any element $\omega_{\mu\nu}$. Hence, equating the coefficient of $\omega_{\mu\nu}$ to zero, we get

$$\sum_{\ell'\kappa'} \{\Phi_{\alpha\mu}(\ell\ell',\kappa\kappa')X_{\nu}(\ell'\kappa',\ell\kappa) - \Phi_{\alpha\nu}(\ell\ell',\kappa\kappa')X_{\mu}(\ell'\kappa',\ell\kappa)\} = 0 \tag{1.22}$$

We have thus obtained two conditions, expressed by Eqs.(1.19) and (1.22), on the harmonic coupling constants $\Phi_{\alpha\beta}(\ell\ell',\kappa\kappa')$. Corresponding conditions may be obtained for the anharmonic coupling constants.

Further restrictions on the coupling coefficients are imposed by the symmetry of the crystal[(2)]. Of particular importance are the point group symmetry and the translational symmetry. If a proper or improper rotation of the point group is described by a 3×3 real orthogonal matrix S, the coupling coefficients transform as

$$\Phi_{\alpha\beta}(LL',KK') = \sum_{\lambda\mu} S_{\alpha\lambda}S_{\beta\mu}\Phi_{\lambda\mu}(\ell\ell',\kappa\kappa') \tag{1.23}$$

where the atomic sites, $\ell\kappa, \ell'\kappa'$, are transformed into the sites LK, L'K' by the symmetry operation. Applying this equation to the subgroup of operations leaving $\ell\kappa$ and $\ell'\kappa'$ invariant gives the independent, nonzero elements of $\Phi_{\alpha\beta}(\ell\ell',\kappa\kappa')$.

Under a translational symmetry operation, each atomic site $\ell\kappa$ is carried into another atomic site $L\kappa$ according to the relation

$$\ell\kappa \to L\kappa = \ell + m,\kappa \tag{1.24}$$

where m represents a translation vector of the lattice that is characteristic of the particular operation and the same for all atomic sites $\ell\kappa$. Since the crystal after the translation is indistinguishable from what it was before, we must have

$$\Phi_{\alpha\beta}(\ell + m, \ell' + m; \kappa\kappa') = \Phi_{\alpha\beta}(\ell\ell', \kappa\kappa') \tag{1.25}$$

If we take the special case $m = -\ell$, we find

$$\Phi_{\alpha\beta}(0, \ell'-\ell; \kappa\kappa') = \Phi_{\alpha\beta}(\ell\ell', \kappa\kappa') \tag{1.26}$$

This last relation shows that $\Phi_{\alpha\beta}(\ell\ell', \kappa\kappa')$ is a function only of $\ell' - \ell$ if the crystal is periodic. Similar relations can be established for the anharmonic coefficients.

2. Normal Coordinate Transformation

We now address the problem of solving the classical equations of motion for the nuclei which are given by

$$M_\kappa \frac{\partial^2 u_\alpha(\ell\kappa)}{\partial t^2} = F_\alpha(\ell\kappa) = -\frac{\partial \Phi(\underset{\sim}{R})}{\partial u_\alpha(\ell\kappa)} \tag{2.1}$$

At this point we restrict our attention to the harmonic approximation, so that we only have to deal with linear equations of the form

$$M_\kappa \frac{\partial^2 u_\alpha(\ell\kappa)}{\partial t^2} = \sum_{\ell'\kappa'\beta} \Phi_{\alpha\beta}(\ell\ell', \kappa\kappa')[u_\beta(\ell\kappa) - u_\beta(\ell'\kappa')] \tag{2.2}$$

The corresponding Hamiltonian can be written as

$$H = \sum_{\ell\kappa\alpha} \tfrac{1}{2} M_\kappa \dot{u}_\alpha^2(\ell\kappa) + \tfrac{1}{2} \sum_{\ell\kappa\alpha} \sum_{\ell'\kappa'\beta} \Phi_{\alpha\beta}(\ell\ell', \kappa\kappa') u_\alpha(\ell\kappa) u_\beta(\ell'\kappa') \tag{2.3}$$

Thus, we are faced with the problem of many coupled harmonic oscillators.

This problem may be solved by means of a coordinate transformation to new coordinates called normal coordinates[1,2].

This transformation diagonalizes the Hamiltonian and reduces the problem to one of uncoupled oscillators.

The first step is to get rid of the complicating mass factors in the kinetic energy by means of the transformation

$$w_\alpha(\ell\kappa) = M_\kappa^{\frac{1}{2}} u_\alpha(\ell\kappa) \tag{2.4}$$

Then the Hamiltonian becomes

$$H = \tfrac{1}{2} \sum_{\ell\kappa\alpha} \dot{w}_\alpha^2(\ell\kappa) + \tfrac{1}{2} \sum_{\ell\kappa\alpha} \sum_{\ell'\kappa'\beta} D_{\alpha\beta}(\ell\ell',\kappa\kappa') w_\alpha(\ell\kappa) w_\beta(\ell'\kappa') \tag{2.5}$$

where

$$D_{\alpha\beta}(\ell\ell',\kappa\kappa') = \frac{\Phi_{\alpha\beta}(\ell\ell',\kappa\kappa')}{(M_\kappa M_{\kappa'})^{\frac{1}{2}}} \tag{2.6}$$

is the <u>dynamical matrix</u>. The equations of motion become

$$\ddot{w}_\alpha(\ell\kappa) = - \sum_{\ell'\kappa'\beta} D_{\alpha\beta}(\ell\ell',\kappa\kappa') w_\beta(\ell'\kappa') \tag{2.7}$$

The time dependence may be eliminated by the ansatz

$$w_\alpha(\ell\kappa) = w_\alpha^{(o)}(\ell\kappa) e^{-i\omega t} \tag{2.8}$$

We then obtain

$$\omega^2 w_\alpha(\ell\kappa) = \sum_{\ell'\kappa'\beta} D_{\alpha\beta}(\ell\ell',\kappa\kappa') w_\beta(\ell'\kappa') \tag{2.9}$$

Our problem is thus equivalent to the diagonalization of the dynamical matrix. In general, this is enormously difficult because the dynamical matrix is very large.

We can drastically simplify the calculation by using periodic boundary conditions which exploit the fact that $\Phi_{\alpha\beta}(\ell\ell',\kappa\kappa')$ and $D_{\alpha\beta}(\ell\ell',\kappa\kappa')$ depend only on the difference $\ell - \ell'$ for a periodic crystal. This suggests that we seek a wave-like solution to Eq.(2.7) of the form

$$w_\alpha(\ell\kappa) = W_{\alpha\kappa}(\underset{\sim}{k}) e^{i[\underset{\sim}{k}\cdot\underset{\sim}{R}^{(o)}(\ell) - \omega t]} \tag{2.10}$$

Substituting into Eq.(2.7), we get

$$\omega^2 W_{\alpha\varkappa}(\underset{\sim}{k}) = \sum_{\ell'\varkappa'\beta} D_{\alpha\beta}(\ell\ell',\varkappa\varkappa') e^{i\underset{\sim}{k}\cdot[\underset{\sim}{R}^{(o)}(\ell')-\underset{\sim}{R}^{(o)}(\ell)]} W_{\beta\varkappa'}(\underset{\sim}{k}) \tag{2.11}$$

If we introduce a new summation variable $\bar{\ell} = \ell' - \ell$, we find

$$\omega^2 W_{\alpha\varkappa}(\underset{\sim}{k}) = \sum_{\varkappa'\beta} D_{\alpha\beta}(k,\varkappa\varkappa') W_{\beta\varkappa'}(\underset{\sim}{k}) \tag{2.12}$$

where

$$D_{\alpha\beta}(\underset{\sim}{k},\varkappa\varkappa') = \sum_{\bar{\ell}} D_{\alpha\beta}(0,\bar{\ell};\varkappa\varkappa') e^{i\underset{\sim}{k}\cdot\underset{\sim}{R}^{(o)}(\bar{\ell})} \tag{2.13}$$

is independent of $\bar{\ell}$.

For a given wave vector $\underset{\sim}{k}$, Eqs. (2.12) constitute a set of $3r$ linear homogeneous algebraic equations where r is the number of atoms per unit cell. This is a manageable number of equations. For a nontrivial solution we must have

$$|D_{\alpha\beta}(\underset{\sim}{k},\varkappa\varkappa') - \omega^2 \delta_{\alpha\beta}\delta_{\varkappa\varkappa'}| = 0 \tag{2.14}$$

The $3r$ solutions to Eq.(2.14) for given $\underset{\sim}{k}$ we label by $\omega(\underset{\sim}{k}j)$, $1 \le j \le 3r$. The quantity j is called the branch index. The quantities $\omega^2(\underset{\sim}{k}j)$ are the normal mode frequencies. For a stable crystal it is necessary that $\omega^2(\underset{\sim}{k}j)$ be non-negative for every normal mode.

Let us denote the eigenvectors of the matrix $\underset{\sim}{D}(\underset{\sim}{k},\varkappa\varkappa')$ by $\underset{\sim}{e}_{\varkappa}(\underset{\sim}{k}j)$. The components of the eigenvectors can be chosen to be orthonormal:

$$\sum_{\varkappa\alpha} e^*_{\alpha\varkappa}(\underset{\sim}{k}j) e_{\alpha\varkappa}(\underset{\sim}{k}j') = \delta_{jj'} \tag{2.15}$$

They also satisfy the closure relation

$$\sum_j e^*_{\alpha\varkappa}(kj) e_{\beta\varkappa'}(\underset{\sim}{k}j) = \delta_{\varkappa\varkappa'}\delta_{\alpha\beta} \tag{2.16}$$

and the equation

$$\sum_{\beta\varkappa'} D_{\alpha\beta}(\underset{\sim}{k},\varkappa\varkappa') e_{\beta\varkappa'}(\underset{\sim}{k}j) = \omega^2(\underset{\sim}{k}j) e_{\alpha\varkappa}(\underset{\sim}{k}j) \tag{2.17}$$

The eigenvector components $e_{\alpha\varkappa}(\underset{\sim}{k}j)$ serve as a unitary transformation from the quantities $W_{\alpha\varkappa}(\underset{\sim}{k})$ to the normal coordinates $Q(\underset{\sim}{k}j)$:

$$W_{\alpha\varkappa}(\underset{\sim}{k}) = \sum_j e_{\alpha\varkappa}(\underset{\sim}{k}j)Q(\underset{\sim}{k}j) \tag{2.18}$$

The normal coordinate transformation from the original displacements $u_\alpha(\ell\varkappa)$ to the normal coordinates is

$$u_\alpha(\ell\varkappa) = \frac{1}{(NM_\varkappa)^{\frac{1}{2}}} \sum_{\underset{\sim}{k}j} e_{\alpha\varkappa}(\underset{\sim}{k}j)e^{i\underset{\sim}{k}\cdot\underset{\sim}{R}^{(o)}(\ell)}Q(\underset{\sim}{k}j) \tag{2.19}$$

From Eq.(2.13) we see that $D^*_{\alpha\beta}(-\underset{\sim}{k},\varkappa\varkappa') = D_{\alpha\beta}(\underset{\sim}{k},nn')$. It follows that $\underset{\sim}{e}^*_\varkappa(-\underset{\sim}{k}j)$ is also an eigenvector of $\underset{\sim}{D}(\underset{\sim}{k},\varkappa\varkappa')$ with the same eigenvalue $\omega^2(\underset{\sim}{k}j)$ as $\underset{\sim}{e}_\varkappa(\underset{\sim}{k}j)$. If $\underset{\sim}{k}$ is not a point of degeneracy, $\underset{\sim}{e}_\varkappa(\underset{\sim}{k}j)$ and $\underset{\sim}{e}_\varkappa(-\underset{\sim}{k}j)$ differ at most by a phase factor which we take to be +1 so that

$$\underset{\sim}{e}^*_\varkappa(-\underset{\sim}{k}j) = \underset{\sim}{e}_\varkappa(\underset{\sim}{k}j) \tag{2.20}$$

Since the displacements $\underset{\sim}{u}(\ell\varkappa)$ are real, it then follows from Eqs.(2.19) and (2.20) that

$$Q^*(-\underset{\sim}{k}j) = Q(\underset{\sim}{k}j) \tag{2.21}$$

If the normal coordinate transformation, Eq.(2.19), is substituted in the Hamiltonian, Eq.(2.3), one obtains the result

$$H = \tfrac{1}{2}\sum_{\underset{\sim}{k}j}\{\dot{Q}^*(\underset{\sim}{k}j)\dot{Q}(\underset{\sim}{k}j) + \omega^2(\underset{\sim}{k}j)Q^*(\underset{\sim}{k}j)Q(\underset{\sim}{k}j)\} \tag{2.22}$$

where use has been made of the orthonormality of the eigenvectors and the result

$$\sum_\ell e^{i(\underset{\sim}{k}'-\underset{\sim}{k})\cdot\underset{\sim}{R}^{(o)}(\ell)} = N\delta_{\underset{\sim}{k}',\underset{\sim}{k}} \tag{2.23}$$

for wave vectors $\underset{\sim}{k}$ and $\underset{\sim}{k}'$ both in the first Brillouin zone. From the Lagrangian $L = T_n - \Phi(\underset{\sim}{R})$ we obtain the momentum conjugate to $Q^*(\underset{\sim}{k}j)$:

$$P(\underset{\sim}{k}j) = \frac{\partial L}{\partial\dot{Q}^*(\underset{\sim}{k}j)} = \dot{Q}(\underset{\sim}{k}j) \tag{2.24}$$

The equations of motion are

$$\dot{P}(\underset{\sim}{k}j) = -\frac{\partial H}{\partial Q^*(\underset{\sim}{k}j)} = -\omega^2(\underset{\sim}{k}j)Q(\underset{\sim}{k}j) \tag{2.25}$$

or

$$\ddot{Q}(\underset{\sim}{k}j) = -\omega^2(\underset{\sim}{k}j)Q(\underset{\sim}{k}j) \tag{2.26}$$

The solution to Eq.(2.26) can be written as

$$Q(\underset{\sim}{k}j,t) = Q_+(\underset{\sim}{k}j)e^{i\omega(\underset{\sim}{k}j)t} + Q_-(\underset{\sim}{k}j)e^{-i\omega(\underset{\sim}{k}j)t} \tag{2.27}$$

We note that the solution of the equations of motion in terms of the normal coordinates $Q(\underset{\sim}{k}j)$ is very simple because the Hamiltonian is diagonal in terms of these coordinates and the equations of motion are uncoupled.

In studying the quantum mechanics of lattice vibrations, we need real normal coordinates. They can be obtained by the transformation(1,2)

$$Q(\underset{\sim}{k}j) = \frac{1}{\sqrt{2}}[q_1(\underset{\sim}{k}j) + iq_2(\underset{\sim}{k}j)] \tag{2.28}$$

$$Q(-\underset{\sim}{k}j) = \frac{1}{\sqrt{2}}[q_1(\underset{\sim}{k}j) - iq_2(\underset{\sim}{k}j)] \tag{2.29}$$

We see that $q_1(\underset{\sim}{k}j) = q_1(-\underset{\sim}{k}j)$ and $q_2(\underset{\sim}{k}j) = -q_2(-\underset{\sim}{k}j)$, so only one-half of the quantities $q_1(\underset{\sim}{k}j)$ and $q_2(\underset{\sim}{k}j)$ are independent. The independent coordinates may be chosen by taking only $\underset{\sim}{k}$ vectors lying on one side of a plane through the origin of the Brillouin zone. The Hamiltonian now becomes

$$H = \frac{1}{2}\sum_{\underset{\sim}{k}}{}' \sum_j \sum_{\lambda=1,2} \{\dot{q}_\lambda^2(\underset{\sim}{k}j) + \omega_\lambda^2(\underset{\sim}{k}j)q_\lambda^2(\underset{\sim}{k}j)\} \tag{2.30}$$

and the Schrödinger equation is

$$\frac{1}{2}\sum_{\underset{\sim}{k}}{}' \sum_j \sum_\lambda \{-\hbar^2 \frac{\partial^2}{\partial q_\lambda^2(\underset{\sim}{k}j)} + \omega_\lambda^2(\underset{\sim}{k}j)q_\lambda^2(\underset{\sim}{k}j)\}\chi = E\chi \tag{2.31}$$

The variables are separable, so

$$\chi = \prod_{\underset{\sim}{k}j\lambda}{}' \varphi_{n_\lambda(\underset{\sim}{k}j)}(q_\lambda(\underset{\sim}{k}j)) \tag{2.32}$$

where $\varphi_{n_\lambda(\underset{\sim}{k}j)}(q_\lambda(\underset{\sim}{k}j))$ is the wave function for the normal coordinate of mode $\underset{\sim}{k}j\lambda$, and

$$E = \sum_{\underset{\sim}{k}j\lambda} \hbar\omega_\lambda(\underset{\sim}{k}j)[n_\lambda(\underset{\sim}{k}j) + \tfrac{1}{2}] \tag{2.33}$$

where $n_\lambda(\underset{\sim}{k}j)$ is the quantum number for mode $\underset{\sim}{k}j\,\lambda$.

Another useful procedure is to introduce creation and destruction operators through the transformations

$$Q(\underset{\sim}{k}j) = (\hbar/2\,\omega(\underset{\sim}{k}j))^{\frac{1}{2}}(a_{\underset{\sim}{k}j} + a^+_{-\underset{\sim}{k}j}) \tag{2.34}$$

$$P(\underset{\sim}{k}j) = -i(\hbar\omega\ (\underset{\sim}{k}j)/2)^{\frac{1}{2}}(a_{\underset{\sim}{k}j} - a^+_{-\underset{\sim}{k}j}) \tag{2.35}$$

From the commutation relation

$$[u_\alpha(\ell\varkappa), p_\beta(\ell'\varkappa')] = i\hbar\delta_{\ell\ell'}\delta_{\varkappa\varkappa'}\delta_{\alpha\beta} \tag{2.36}$$

one can show that

$$[Q^*\underset{\sim}{k}j), P(\underset{\sim}{k}'j')] = i\hbar\delta_{\underset{\sim}{k},\underset{\sim}{k}'}\delta_{jj'} \tag{2.37}$$

$$[Q(\underset{\sim}{k}j), P^*(\underset{\sim}{k}'j')] = i\hbar\delta_{\underset{\sim}{k},\underset{\sim}{k}'}\delta_{jj'} \tag{2.38}$$

and thence that

$$[a_{\underset{\sim}{k}j}, a^*_{\underset{\sim}{k}'j'}] = \delta_{\underset{\sim}{k},\underset{\sim}{k}'}\delta_{jj'} \tag{2.39}$$

$$[a_{\underset{\sim}{k}j}, a_{\underset{\sim}{k}'j'}] = [a^+_{\underset{\sim}{k}j}, a^+_{\underset{\sim}{k}'j'}] = 0 \tag{2.40}$$

The Hamiltonian becomes

$$H = \sum_{\underset{\sim}{k}j} \hbar\omega(\underset{\sim}{k}j)[a^+_{\underset{\sim}{k}j}a_{\underset{\sim}{k}j} + \tfrac{1}{2}]$$

which is the standard form for a collection of independent harmonic oscillators. The results of the operations of a^+ and a on an eigenstate of quantum number n are

$$a^+|n\rangle = (n+1)^{\frac{1}{2}}|n+1\rangle \tag{2.41}$$

$$a|n\rangle = n^{\frac{1}{2}}|n-1\rangle \tag{2.42}$$

Also, one sees that

$$a^+a|n\rangle = n|n\rangle \tag{2.43}$$

so a^+a is the number operator. The energy eigenvalue is then given by

$$E = \sum_{\underset{\sim}{k}j} \hbar\omega(\underset{\sim}{k}j)[n_{\underset{\sim}{k}j} + \tfrac{1}{2}] \tag{2.44}$$

where each $n_{\underset{\sim}{k}j}$ can take on any integral value from zero to infinity. The energy eigenvalues are therefore quantized. The quantum of excitation energy of a lattice vibration is called a phonon.

3. Simple Examples

a. Linear monatomic chain

Let us consider a one-dimensional array of particles of equal mass which interact with one another through nearest-neighbour Hooke's law forces of force constant (Fig. 1).

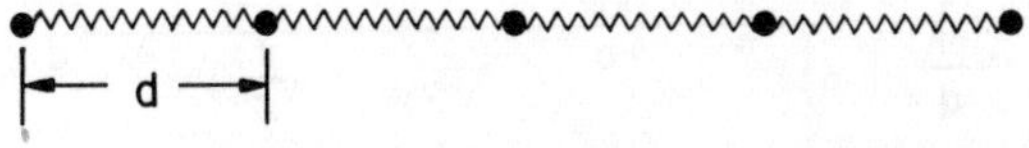

FIG.1. Linear monatomic chain with nearest neighbour interactions.

For this case there is one atom per unit cell, and the indices $\varkappa, \varkappa'$ are superfluous. The coupling coefficients which are non-zero are

$$\Phi_{xx}(\ell,\ell-1) = \Phi_{xx}(\ell,\ell+1) = -\varphi'' = -\sigma \qquad (3.1)$$

$$\Phi_{xx}(\ell,\ell) = 2\varphi'' = 2\sigma \qquad (3.2)$$

where φ'' is the second derivative of the nearest-neighbour interaction potential $\varphi(\underset{\sim}{r})$ evaluated at the equilibrium distance d.

We assume periodic boundary conditions so each particle satisfies an equation of motion of the form

$$M\ddot{u}(\ell) = \sigma[u(\ell+1) + u(\ell-1) - 2u(\ell)] \qquad (3.3)$$

The displacement $u(\ell)$ can be assumed to have the form

$$u(\ell) = Ue^{iqx(\ell)}e^{-i\omega t} \qquad (3.4)$$

where $x(\ell) = d\ell$ and the periodic boundary conditions require that $q = 2\pi n/Nd$ with n an integer and N the number of unit cells. If we substitute Eq.(3.4) into Eq.(3.3), we obtain a nontrivial solution only if the frequency ω is related to the wave vector q by

$$\omega^2 = \frac{4\sigma}{M}\sin^2\frac{dq}{2} \qquad (3.5)$$

In the limit $N \to \infty$, the frequencies of the normal modes form a quasi-continuum as shown in Fig. 2.

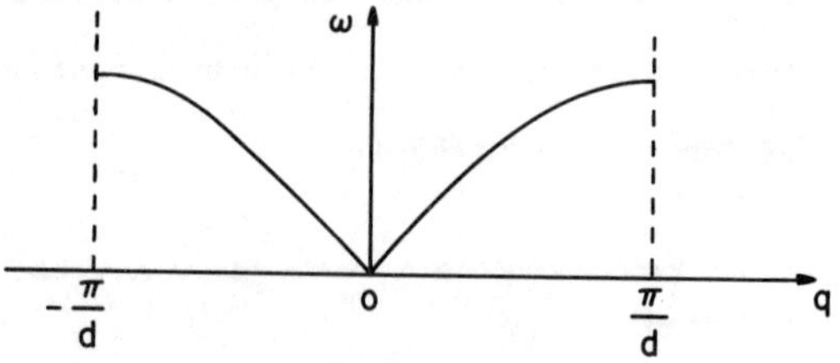

FIG.2. Frequency plotted against wave vector for the normal modes of the linear monatomic chain.

We notice that the allowed frequencies lie in a band from zero to the maximum value $(4\sigma/M)^{\frac{1}{2}}$. At small q, $\omega \simeq (\sigma/M)^{\frac{1}{2}}\,dq$. The wave vectors corresponding to the independent normal modes lie in the first Brillouin zone as shown in the figure; the pattern of displacements of the particles for a wave vector outside the first Brillouin zone is exactly the same as that for some wave vector inside the first Brillouin zone. From Eq.(3.4) we see that the independent normal modes correspond to running waves with a wavelength $\lambda = 2\pi/q$.

b. Linear diatomic chain

We now consider a linear chain whose particles alternately take the value for their masses of M_1 and M_2 and interact with nearest-neighbour Hooke's law forces (Fig.3).

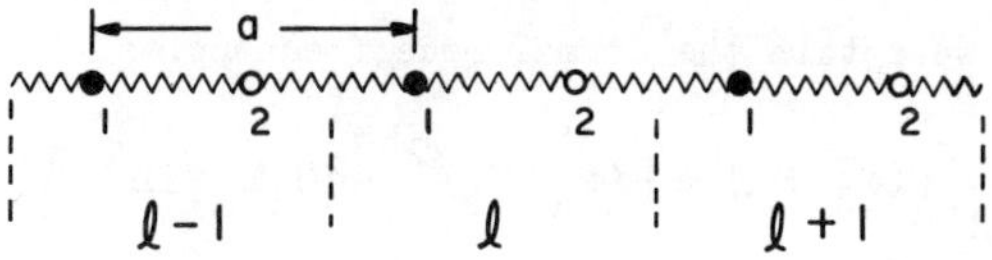

FIG.3. Linear diatomic chain with nearest neighbour interactions.

There are now two particles per unit cell as shown in the figure. The lattice constant is denoted by a. The non-zero coupling coefficients are

$$\Phi_{xx}(\ell,\ell;12) = \Phi_{xx}(\ell,\ell-1;12)$$

$$= \Phi_{xx}(\ell,\ell+1;21) = \Phi_{xx}(\ell,\ell;21) = -\varphi'' = -\sigma \qquad (3.6)$$

$$\Phi_{xx}(\ell,\ell;11) = \Phi_{xx}(\ell,\ell;22) = 2\varphi'' = 2\sigma \qquad (3.7)$$

If we consider a periodic chain, then the equations of motion of each type of particle can be written

$$M_1\ddot{u}(\ell 1) = \sigma[u(\ell 2) + u(\ell-1,2) - 2u(\ell 1)] \qquad (3.8)$$

$$M_2\ddot{u}(\ell 2) = \sigma[u(\ell+1,1) + u(\ell 1) - 2u(\ell 2)] \qquad (3.9)$$

Again assuming a running wave solution,

$$u(\ell\kappa) = U_\kappa e^{iqx(\ell)}e^{-i\omega t}, \qquad \kappa = 1,2 \qquad (3.10)$$

where $x(\ell) = \ell a$, we substitute into Eqs.(3.8) and (3.9) and obtain the following algebraic equations

$$- \omega^2 M_1 U_1 = \sigma[U_2 + U_2 e^{-iqa} - 2U_1] \tag{3.11}$$

$$- \omega^2 M_2 U_2 = \sigma[U_1 e^{iqa} + U_1 - 2U_2] \tag{3.12}$$

which have a non-trivial solution only if the determinant of the coefficients of U_1 and U_2 is zero:

$$\begin{vmatrix} \omega^2 M_1 - 2\sigma & \sigma(1+e^{-iqa}) \\ \sigma(e^{iqa}+1) & \omega^2 M_2 - 2\sigma \end{vmatrix} = 0 \tag{3.13}$$

Solving Eq.(3.13) we obtain the normal mode frequencies

$$\omega^2(q,\pm) = \frac{\sigma}{M_1 M_2} \{(M_1+M_2) \pm [(M_1+M_2)^2 - 4M_1M_2 \sin^2 \frac{qa}{2}]^{\frac{1}{2}}\}. \tag{3.14}$$

The periodic boundary conditions specify that $q = 2\pi n/Na$ with n an integer in the range $-\frac{N}{2} + 1 \le n \le +\frac{N}{2}$ and we have assumed N to be even. The first non-zero reciprocal lattice vector is $2\pi/a$, so that the boundaries of the first Brillouin zone are at $\pm \pi/a$.

For the special q values 0 and $\pm \pi/a$, simple expressions for ω^2 can be obtained:

$$q = 0 \; : \; \omega^2 = 0, \frac{2\sigma}{\mu} \; ; \; \frac{1}{\mu} = \frac{1}{M_1} + \frac{1}{M_2} \tag{3.15}$$

$$q = \pm \frac{\pi}{a} : \omega^2 = \frac{2\sigma}{M_1}, \frac{2\sigma}{M_2} \tag{3.16}$$

In the limit $N \to \infty$, the normal mode frequencies become quasicontinuous and have the general form shown in Fig.4.

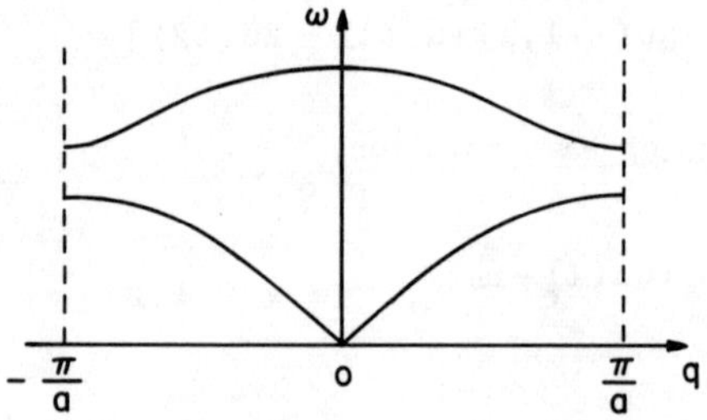

FIG.4. Frequency plotted against wave vector for the normal modes of the linear diatomic chain in the reduced zone scheme.

We see for the diatomic case that the allowed frequencies lie within two bands rather than one. Between the two bands is a region of forbidden frequencies or a frequency "gap". The lower frequency band is associated with the so-called "acoustical" branch, while the higher frequency band is associated with the "optical" branch. The acoustical branch, of course, includes the ordinary long-wavelength acoustical waves, while the optical branch includes those normal modes primarily responsible for the absorption of electromagnetic radiation in the infrared.

The presentation of the normal mode frequencies in the above figure corresponds to the so-called "reduced" zone scheme. The frequencies are restricted to the first Brillouin zone. However, one sees from Eq.(3.14) that the frequency is a periodic function of the wave vector, so it is possible, for example, to choose the optical branch frequencies to lie in the second Brillouin zone rather than the first. This is known as the "extended" zone scheme and is shown in Fig. 5.

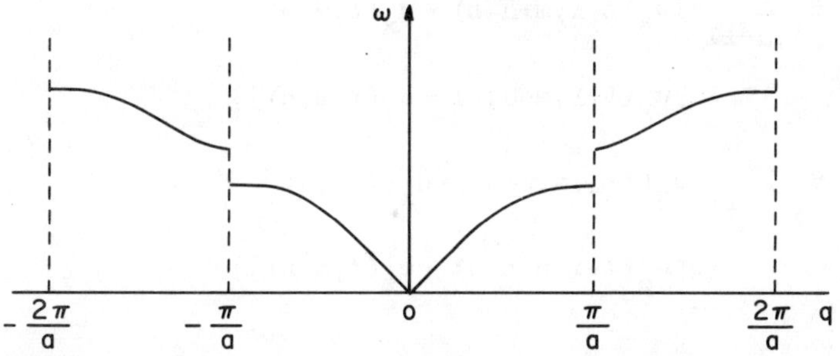

FIG.5. Frequency plotted against wave vector for the normal modes of the linear diatomic chain in the extended zone scheme.

The extended zone scheme is useful in establishing the relation of the diatomic to the monatomic case. If $M_2 \rightarrow M_1$, the gap disappears, the curves become continuous, and the size of the first Brillouin zone doubles.

The amplitudes U_1 and U_2 are important to know in various calculations. From Eqs.(3.11) and (3.12) we can show that they can be written as

$$U_1 = U(\omega^2 M_2 - 2\sigma) \tag{3.17}$$

$$U_2 = -U\sigma(e^{iqa} + 1) \tag{3.18}$$

where U is an arbitrary constant. Let us consider the limit $q \to 0$. For the acoustical branch one finds that $U_1 = U_2$, so all the particles move in phase. For the optical branch, one finds that $M_1 U_1 = - M_2 U_2$, so particles of different mass move out of phase with one another. If the particles of different mass also have different electrical charge, then a fluctuating electrical dipole moment is set up during an optical mode vibration.

c. Monatomic simple cubic lattice

The cases discussed so far have been one-dimensional. To see what happens for a three-dimensional case, we examine a monatomic simple cubic lattice with nearest and next-nearest central force interactions. The equations of motion can be written as

$$\begin{aligned} M\ddot{u}_x(\ell,m,n) &= \alpha \sum_{\lambda=\pm 1} [u_x(\ell+\lambda,m,n) - u_x(\ell,m,n)] \\ &+ \beta \sum_{\lambda,\mu=\pm 1} \{u_x(\ell+\lambda,m+\mu,n) - u_x(\ell,m,n) \\ &\qquad + \lambda\mu[u_y(\ell+\lambda,m+\mu,n) - u_y(\ell,m,n)]\} \\ &+ \beta \sum_{\lambda,\nu=\pm 1} \{u_x(\ell+\lambda,m,n+\nu) - u_x(\ell,m,n) \\ &\qquad + \lambda\nu[u_z(\ell+\lambda,m,n+\nu) - u_z(\ell,m,n)]\} \end{aligned} \tag{3.19}$$

where α and β are the nearest and next-nearest neighbour force constants and the lattice sites are specified by the three integers ℓ, m, n. There are two additional equations which can be obtained by a cyclic permutation of x, y, z and the increments of ℓ, m, n.

We seek a solution to the equations of motion of the form

$$u_j(\ell,m,n) = U_j e^{i\underset{\sim}{q}\cdot\underset{\sim}{x}(\ell,m,n)} e^{-i\omega t}, \quad j=x,y,z \tag{3.20}$$

where $\underset{\sim}{x}(\ell,m,n) = \ell d\hat{x} + md\hat{y} + nd\hat{z}$ and d is the lattice spacing. Substituting Eq.(3.20) into the equations of motion, we obtain the linear algebraic equations

$$\omega^2 MU_x = [2\alpha(1-C_x) + 4\beta(2-C_xC_y-C_xC_z)]U_x + 4\beta S_xS_yU_y \quad + 4\beta S_xS_zU_z$$

$$\omega^2 MU_y = 4\beta S_xS_yU_x + [2\alpha(1-C_y) + 4\alpha(2-C_xC_y-C_yC_z)]U_y + 4\beta S_yS_zU_z$$

$$\omega^2 MU_z = 4\beta S_xS_zU_x + 4\beta S_yS_zU_y + [2\alpha(1-C_z) + 4\beta(2-C_xC_z-C_yC_z)]U_z \tag{3.21}$$

where $C_j = \cos q_jd$ and $S_j = \sin q_jd$ with $j = x, y, z$. For a non-trivial solution, the determinant of the coefficients of U_x, U_y, U_z must be zero:

$$\begin{vmatrix} 2\alpha(1-C_x) + 4\beta(2-C_xC_y-C_xC_z) - \omega^2 M & 4\beta S_xS_y & 4\beta S_xS_z \\ 4\beta S_xS_y & 2\alpha(1-C_y) + 4\beta(2-C_xC_y-C_yC_z) - \omega^2 M & 4\beta S_yS_z \\ 4\beta S_xS_z & 4\beta S_yS_z & 2\alpha(1-C_z) + 4\beta(2-C_xC_z-C_yC_z) - \omega^2 M \end{vmatrix} = 0 \tag{3.22}$$

In general, Eq.(3.22) is a cubic equation in ω^2. However, for directions of propagation of high symmetry, simple expressions can be obtained for ω^2. Thus, for propagation along the x-axis, $\underset{\sim}{q} = (q, o, o)$ and $S_y = S_z = 0$. Then we obtain

$$\omega^2 = \frac{4\alpha}{M}(1 + 4\,\frac{\beta}{\alpha}\,\sin^2\frac{qd}{2} \qquad \text{(one solution)} \tag{3.23}$$

$$\omega^2 = \frac{8\beta}{M}\,\sin^2\frac{qd}{2} \qquad \text{(two solutions)} \tag{3.24}$$

The case specified by Eq.(3.23) corresponds to longitudinal waves where the displacements are parallel to the direction of propagation (here, the x-direction). The solutions specified by Eq.(3.24) correspond to two degenerate transverse waves where the displacements are perpendicular to the direction of propagation. For our case, one solution has displacements in the y-direction and the other in the z-direction. The form of the dispersion curves is shown in Fig. 6.

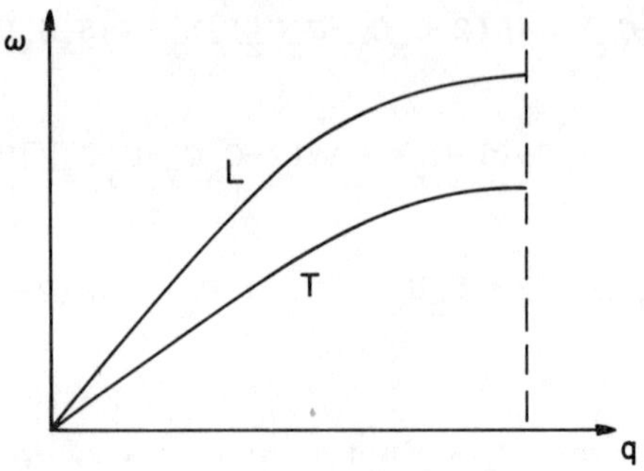

FIG.6. Frequency plotted against wave vector in the (100) direction for a monatomic simple cubic lattice.

For a general direction of propagation, the normal modes of vibration are mixtures of longitudinal and transverse components. Only in directions of high symmetry can the modes be classified as purely longitudinal or purely transverse.

4. Specific Heat

In the harmonic approximation the nuclear motion of a crystal is described by a set of independent harmonic oscillators corresponding to the normal coordinates. Thermodynamic quantities such as the mean energy and the specific heat are then given by sums of the corresponding quantities for the individual oscillators. Thus, the mean energy can be written as

$$\langle E \rangle = \sum_s \bar{\epsilon}_s \tag{4.1}$$

where $\bar{\epsilon}_s$ is the mean energy of oscillator s.

To calculate $\bar{\epsilon}_s$, we start from the energy ϵ_s of oscillator s given by

$$\epsilon_s = \hbar\omega_s(n_s + \tfrac{1}{2}), \quad n_s = 0,1,2,\ldots, \tag{4.2}$$

where ω_s is the frequency of the oscillator. The mean energy is specified by

$$\bar{\epsilon}_s = \frac{\sum_{n_s=0}^{\infty} \epsilon_s e^{-\beta\epsilon_s}}{\sum_{n_s=0}^{\infty} e^{-\beta\epsilon_s}} \tag{4.3}$$

$$= \hbar\omega_s[\bar{n}(\omega_s)+\tfrac{1}{2}] \tag{4.4}$$

where

$$\bar{n}(\omega_s) = [e^{\beta\hbar\omega_s} - 1]^{-1} \tag{4.5}$$

and $\beta = 1/k_BT$. The quantity $\bar{n}(\omega_s)$ is the mean occupation number of oscillator s and is described by the Bose factor.

In a crystal of macroscopic size, the frequencies are very closely spaced and one may replace the sum over s in Eq.(4.1) by an integral over ω,

$$\langle E\rangle = \int_0^{\omega_m} g(\omega)\bar{\epsilon}(\omega)\,d\omega , \tag{4.6}$$

where ω_m is the maximum normal mode frequency and $g(\omega)d\omega$ is the number of normal mode frequencies in the range ω to $\omega + d\omega$. In general, the frequency distribution function $g(\omega)$ is a complicated function of ω. At low temperatures, however, the important frequencies are the low frequencies and for them it is possible to obtain a simple expression for $g(\omega)$.

For low frequencies, the wavelength is large compared to a lattice spacing and one can use continuum theory following Debye. Then the frequency of branch j is given by

$$\omega = c_j|\underset{\sim}{k}| = c_jk \tag{4.7}$$

where c_j is the speed of the wave and the components of the wave vector $\underset{\sim}{k}$ may be specified by periodic boundary conditions as

$$k_i = \frac{2\pi}{L}n_i \quad , \quad n_i = \ldots.-2,-1,0,+1,+2,\ldots. \tag{4.8}$$

We have assumed that the crystal has the shape of a cube of edge L, but this is not a restrictive assumption. Let us consider a spherical shell in $\underset{\sim}{k}$-space of radius k and thickness dk. The volume of this shell is $4\pi k^2dk$. Since the density of normal modes in $\underset{\sim}{k}$-space according to Eq.(4.8) is $(2\pi/L)^3$, the number of normal modes in the spherical shell is

$$dN = (\frac{L}{2\pi})^3 4\pi k^2dk \tag{4.9}$$

If we eliminate k in favour of ω using Eq.(4.7) we can rewrite dN in the form

$$dN = \frac{\Omega}{2\pi^2 c_j^3} \omega^2 \, d\omega \tag{4.10}$$

where $\Omega = L^3$ is the volume of the crystal. For a three-dimensional crystal, there are two transverse acoustic branches and one longitudinal acoustic branch. Consequently, the total number of modes in the shell is given by

$$dN = \frac{\Omega}{2\pi^2} \left(\frac{2}{c_t^3} + \frac{1}{c_\ell^3} \right) \omega^2 d\omega \tag{4.11}$$

and the frequency distribution function is

$$g(\omega) = \frac{\Omega}{2\pi^2} \left(\frac{2}{c_t^3} + \frac{1}{c_\ell^3} \right) \omega^2 \tag{4.12}$$

If the crystal has N atoms, it has 3N normal modes, and the frequency distribution function must satisfy a constraint of the form

$$\int_0^{\omega_D} g(\omega) \, d\omega = 3N \tag{4.13}$$

where ω_D is a "cutoff" frequency that arises because the crystal is not really a continuum. If one substitutes Eq.(4.12) into Eq.(4.13) and carries out the integration, one obtains

$$\omega_D = 2\pi \left(\frac{9N}{4\pi\Omega} \right)^{\frac{1}{3}} \left(\frac{2}{c_t^3} + \frac{1}{c_\ell^3} \right)^{-\frac{1}{3}} \tag{4.14}$$

and one can rewrite Eq.(4.12) in the form

$$g(\omega) = 9N \frac{\omega^2}{\omega_D^3} \tag{4.15}$$

The specific heat is defined by

$$C_v = \frac{\partial \langle E \rangle}{\partial T} \tag{4.16}$$

Using Eqs.(4.4), (4.6), and (4.16) and identifying ω_m with ω_D, we can write the specific heat in the form

$$C_v = \int_0^{\omega_D} g(\omega) \frac{\frac{\hbar^2\omega^2}{k_B T^2} e^{\beta\hbar\omega}}{\left(e^{\beta\hbar\omega}-1\right)^2} d\omega \qquad (4.17)$$

Introducing the dimensionless variable $x = \beta\hbar\omega$ and the Debye temperature $\Theta_D = \hbar\omega_D/k_B$, we can rewrite Eq.(4.17) as

$$C_v = 9Nk_B\left(\frac{T}{\Theta_D}\right)^3 \int_0^{\frac{\Theta_D}{T}} \frac{x^4 e^x}{(e^x-1)^2} dx \qquad (4.18)$$

where we have used the expression for $g(\omega)$ given by Eq.(4.15). At low temperatures, $T \ll \Theta_D$, the upper limit of the integral in Eq.(4.18) can be taken to be infinity. The value of the integral is found to be $4\pi^4/15$, so the specific heat becomes

$$C_v = 3Nk_B \cdot \frac{4\pi^4}{5}\left(\frac{T}{\Theta_D}\right)^3 \qquad (4.19)$$

which is the famous "Debye T^3 law".

At high temperatures, $T \gg \Theta_D$, we can expand the exponentials in Eq.(4.18) and retain only the lowest order non-vanishing terms. Performing the integral, we get

$$C_v = 3Nk_B \qquad (4.20)$$

which is the classical value. This value is independent of the frequency distribution function, $g(\omega)$, because each normal mode contributes the same amount, k_B, to the specific heat.

At intermediate temperatures, the frequency distribution function specified by Eq.(4.15) is inadequate to describe the specific heat of real crystals. One must use the correct distribution function for the whole range of frequencies, not just low frequencies.

We now consider the nature of the frequency distribution function as given by lattice dynamical calculations. We first discuss the monatomic linear chain which can be treated analytically.

If the length of the chain is L, the number of normal modes dN in the wave vector range dk is given by

$$dN = \frac{L}{2\pi}\,dk \tag{4.21}$$

The normal mode frequencies are specified by Eq.(3.5). In order to avoid confusion in notation we relabel the lattice spacing d by a and write

$$\omega = \sqrt{\frac{4\sigma}{M}}\,\sin\frac{|ak|}{2} \tag{4.22}$$

Differentiating Eq.(4.22), we obtain, for $k > 0$,

$$d\omega = a\sqrt{\frac{\sigma}{M}}\,\cos\frac{ak}{2}\;dk \tag{4.23}$$

We solve Eq.(4.23) for dk and substitute into Eq.(4.21) to give

$$dN = \frac{L}{\pi a}\cdot\frac{d\omega}{\sqrt{\omega_m^2 - \omega^2}} \tag{4.24}$$

The frequency distribution function is then given by

$$g(\omega) = \frac{2L}{\pi a}\cdot\frac{1}{\sqrt{\omega_m^2 - \omega^2}} \tag{4.25}$$

where the factor 2 arises from the degeneracy of the modes for k and −k. In Eqs.(4.24) and (4.25), we have denoted the maximum normal mode frequency by $\omega_m = \sqrt{4\sigma/M}$. The dependence of $g(\omega)$ on ω is shown in Fig. 7.

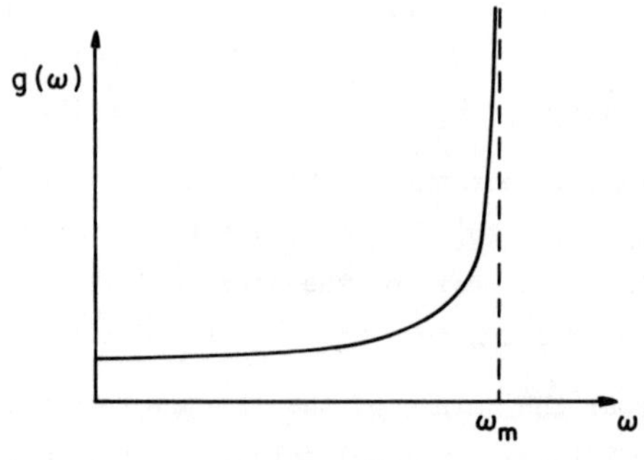

FIG. 7. Frequency distribution for the linear monatomic chain.

We notice immediately that $g(\omega)$ has a singularity at the maximum frequency ω_m. This is just the point at which the dispersion curve has zero slope — i.e., $\frac{d\omega}{dk} = 0$. Such a point is called a <u>critical</u> point. If we rewrite Eq.(4.21) in the form

$$dN = \frac{L}{d\pi} \cdot \frac{d\omega}{(\frac{d\omega}{dk})} \tag{4.26}$$

then we see that $g(\omega)$ is simply

$$g(\omega) = \frac{L}{2\pi} \cdot \frac{1}{(\frac{d\omega}{dk})} \tag{4.27}$$

and the divergence of $g(\omega)$ at the critical point is immediately apparent.

In two- and three-dimensions, the frequency distribution function can be written as

$$g(\omega) = (\frac{L}{2\pi})^{\ell} \int_{S(\omega)} \frac{dS}{|\nabla_k \omega|} \tag{4.28}$$

where S is a surface in $\underset{\sim}{k}$-space of constant ω and ℓ is the dimensionality of the crystal. When $|\nabla_k \omega|$ vanishes, we have a critical point. Especially important critical points are saddle points. In two dimensions, one finds that $g(\omega)$ has logarithmic singularities at the critical points. In three dimensions, the derivative of $g(\omega)$ has singularities. A typical frequency distribution for the three-dimensional case is shown in Fig.8.

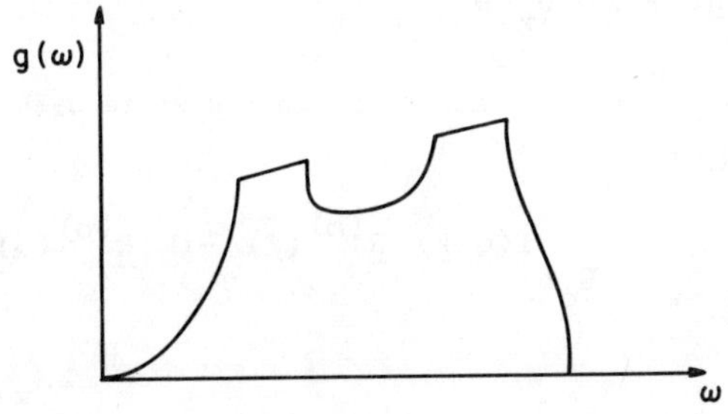

FIG. 8. Typical frequency distribution for a three-dimensional crystal with one atom per unit cell.

The existence of singularities in the distribution function has been shown by Van Hove[3] to be a consequence of the periodicity of $\omega(\underset{\sim}{k})$. According to a topological theorem of M. Morse, any function of more than one variable which is periodic in all its variables has at least a certain number of saddle points.

5. Optical Absorption Due to Phonons

Let us consider a crystal with two different kinds of atoms with charges e_1 and e_2. From electroneutrality we can write $e_1 + e_2 = 0$ or alternatively,

$$e_1 = e^* \tag{5.1}$$

$$e_2 = -e^* \tag{5.2}$$

We now assume that an electromagnetic field is present with an electric vector of the form

$$\underset{\sim}{E}(\underset{\sim}{x},t) = \underset{\sim}{E}_o e^{-i\underset{\sim}{k}\cdot\underset{\sim}{x}} e^{i\omega t} \tag{5.3}$$

The coupling of the crystal to the electromagnetic field is described by the interaction Hamiltonian

$$H' = \sum_{\ell\varkappa} e_\varkappa \, \underset{\sim}{u}(\ell\varkappa)\cdot\underset{\sim}{E}(\underset{\sim}{R}(\ell\varkappa),t) \tag{5.4}$$

If we assume periodic boundary conditions and take the displacement $\underset{\sim}{u}(\ell\varkappa)$ to have the form

$$\underset{\sim}{u}(\ell\varkappa) = \underset{\sim}{U}_\varkappa \, e^{i\underset{\sim}{q}\cdot\underset{\sim}{R}^{(o)}(\ell)} \tag{5.5}$$

then to first order in the displacements, we can write with the aid of Eqs. (5.3) and (5.5)

$$H' = \sum_{\ell\varkappa} e_\varkappa \underset{\sim}{U}_\varkappa \cdot \underset{\sim}{E}_o e^{i(\underset{\sim}{q}-\underset{\sim}{k})\cdot\underset{\sim}{R}^{(o)}(\ell)} e^{-i\underset{\sim}{k}\cdot\underset{\sim}{R}^{(o)}(\varkappa)} e^{i\omega t} \tag{5.6}$$

where we have used the relation $\underset{\sim}{R}^{(o)}(\ell\varkappa) = \underset{\sim}{R}^{(o)}(\ell) + \underset{\sim}{R}^{(o)}(\varkappa)$. Carrying out the sum over ℓ, we obtain the result

$$H' = N\Delta(\underset{\sim}{q} - \underset{\sim}{k}) \sum_{\varkappa} e_{\varkappa} \underset{\sim}{U}_{\varkappa} \cdot \underset{\sim}{E}_{o} \, e^{-i\underset{\sim}{k} \cdot \underset{\sim}{R}^{(o)}(\varkappa)} \, e^{i\omega t} \tag{5.7}$$

where $\Delta(\underset{\sim}{q} - \underset{\sim}{k}) = 1$ if $\underset{\sim}{q} - \underset{\sim}{k}$ is a reciprocal lattice vector and is zero otherwise. Now for the infrared radiation of interest in the optical absorption by phonons, the wave vector $\underset{\sim}{k}$ is in magnitude much smaller than any non-zero reciprocal lattice vector and can be taken to be zero as a good approximation. If we use the reduced zone scheme, the factor $\Delta(\underset{\sim}{q})$ in H' requires that $\underset{\sim}{q}$ also be zero. Thus, we reach the important conclusion that only the normal modes of vibration of wave vector zero interact strongly with electromagnetic waves. Furthermore, with $\underset{\sim}{k} \simeq 0$, we see from Eq.(5.7) that H' is proportional to $\sum_{\varkappa} e_{\varkappa} \underset{\sim}{U}_{\varkappa}$. For acoustical modes we have seen that $U_1 = U_2$ in the limit $\underset{\sim}{q} \to 0$, so with the aid of Eqs. (5.1) and (5.2), we find that H' = 0 for acoustical modes. Hence, only <u>optical</u> modes give strong optical absorption.

If we now take into account the fact that $\underset{\sim}{k}$ is not quite zero, but finite, we can reach another important conclusion. We know that the electromagnetic wave is transverse, so $\underset{\sim}{E}_o \perp \underset{\sim}{k}$. For a longitudinal optical mode, $\underset{\sim}{U}_{\varkappa} \parallel \underset{\sim}{q}$ since the displacements are parallel to the direction of propagation. The factor $\Delta(\underset{\sim}{q} - \underset{\sim}{k})$ in H' tells us that $\underset{\sim}{q} = \underset{\sim}{k}$, so $\underset{\sim}{U}_{\varkappa} \parallel \underset{\sim}{k}$ for longitudinal modes. This means that $\underset{\sim}{U}_{\varkappa} \cdot \underset{\sim}{E}_o = 0$ for these modes and hence H' = 0. Thus, only <u>transverse</u> optical modes interact strongly with the electromagnetic field.

The optical properties of a crystal can be discussed in terms of the dielectric constant $\epsilon(\omega)$ which we now calculate. We need consider only the $\underset{\sim}{q} = 0$ optical modes and for these the displacement pattern in a given unit cell is the same as in every other unit cell. Hence, the indices ℓ and $\varkappa$ are superfluous. For NaCl type crystals, the equations of motion can then be written as[1]

$$M_1 \ddot{\underset{\sim}{u}}_1 = 2\sigma(\underset{\sim}{u}_2 - \underset{\sim}{u}_1) + e^* \underset{\sim}{E} \tag{5.8}$$

$$M_2 \ddot{\underset{\sim}{u}}_2 = 2\sigma(\underset{\sim}{u}_1 - \underset{\sim}{u}_2) - e^* \underset{\sim}{E} \tag{5.9}$$

where

$$\sigma = \sum_{\bar{\ell}} \Phi_{\alpha\alpha}(o,\bar{\ell};11) \tag{5.10}$$

and $\underset{\sim}{E}$ is the macroscopic electric field of the radiation. Short range interactions are included in σ, while long-range Coulomb interactions are included in $\underset{\sim}{E}$.

If we transform to center of mass coordinate $\underset{\sim}{u}$ and difference coordinate $\underset{\sim}{w}$ given by

$$\underset{\sim}{u} = \frac{M_1 \underset{\sim}{u}_1 + M_2 \underset{\sim}{u}_2}{M_1+M_2} \tag{5.11}$$

$$\underset{\sim}{w} = \underset{\sim}{u}_1 - \underset{\sim}{u}_2 \tag{5.12}$$

The equation for $\underset{\sim}{w}$ becomes

$$\mu \ddot{\underset{\sim}{w}} + 2\sigma \underset{\sim}{w} = e^* \underset{\sim}{E} \tag{5.13}$$

or

$$\ddot{\underset{\sim}{w}} + \omega_o^2 \underset{\sim}{w} = (e^*/\mu) \underset{\sim}{E} \tag{5.14}$$

where μ is the reduced mass and $\omega_o^2 = 2\sigma/\mu$. We can include damping in a phenomenological way by generalizing Eq.(5.14) to read

$$\ddot{\underset{\sim}{w}} + \gamma \dot{\underset{\sim}{w}} + \omega_o^2 \underset{\sim}{w} = (e^*/\mu) \underset{\sim}{E} \tag{5.15}$$

where γ is the damping constant. If we assume

$$\underset{\sim}{w} = \underset{\sim}{W}\, e^{i\omega t} \tag{5.16}$$

$$\underset{\sim}{E} = \underset{\sim}{E}_o e^{i\omega t} \tag{5.17}$$

then we find after substituting into Eq.(5.15) that

$$\underset{\sim}{W} = \frac{(e^*/\mu)\underset{\sim}{E}_o}{\omega_o^2 - \omega^2 + i\gamma\omega} \tag{5.18}$$

The dielectric constant may be calculated from the polarization which is the electric dipole moment per unit volume and is given by

$$\underset{\sim}{P} = \frac{1}{V} \sum_{\ell\kappa} e_\kappa \underset{\sim}{u}(\ell\kappa) \tag{5.19}$$

$$= \frac{e^* N}{V} (\underset{\sim}{u}_1 - \underset{\sim}{u}_2) = \frac{e^*}{v_a} \underset{\sim}{w} \tag{5.20}$$

$$= \left(\frac{e^{*2}}{\mu v_a}\right) \frac{1}{\omega_o^2 - \omega^2 + i\gamma\omega} \underset{\sim}{E} \tag{5.21}$$

where V is the volume of the crystal and v_a is the volume of a unit cell. The dielectric susceptibility $\chi(\omega)$ is defined by

$$\underset{\sim}{P} = \chi(\omega) \underset{\sim}{E} \tag{5.22}$$

so that

$$\chi(\omega) = \left(\frac{e^{*2}}{\mu v_a}\right) \frac{1}{\omega_o^2 - \omega^2 + i\gamma\omega} \tag{5.23}$$

The dielectric constant $\epsilon(\omega)$ is given by

$$\epsilon(\omega) = 1 + 4\pi\chi(\omega) + 4\pi\chi_e \tag{5.24}$$

where χ_e is the electronic contribution to the dielectric susceptibility. If we define $\epsilon_e = 1 + 4\pi\chi_e$ to be the electronic contribution to the dielectric constant and use Eq.(5.23), we obtain

$$\epsilon(\omega) = \epsilon_e \left(1 + \frac{\Omega^2}{\omega_o^2 - \omega^2 + i\gamma\omega}\right) \tag{5.25}$$

where $\Omega^2 = 4\pi e^{*2}/\mu v_a \epsilon_e$. We see that the dielectric constant is complex,

$$\epsilon(\omega) = \epsilon_1(\omega) + i\epsilon_2(\omega) \tag{5.26}$$

where

$$\epsilon_1(\omega) = \epsilon_e \left(1 + \frac{\Omega^2(\omega_o^2-\omega^2)}{(\omega_o^2-\omega^2) + \gamma^2\omega^2}\right) \tag{5.27}$$

$$\epsilon_2(\omega) = -\frac{\epsilon_e \gamma\omega\Omega^2}{(\omega_o^2 - \omega^2) + \gamma^2\omega^2} \tag{5.28}$$

We regard ϵ_e as arising from interband electronic transitions which occur at much higher frequencies than the phonon frequencies.

Let us consider some limiting cases.

For $\omega \to \infty$,

$$\epsilon(\omega) \to \epsilon_e \equiv \epsilon_\infty \tag{5.29}$$

while for $\omega \to 0$,

$$\epsilon(\omega) \to \epsilon_e\left(1 + \frac{\Omega^2}{\omega_o^2}\right) \equiv \epsilon_o \tag{5.30}$$

Then

$$\frac{\epsilon_o}{\epsilon_\infty} = 1 + \frac{\Omega^2}{\omega_o^2} = \frac{\omega_{LO}^2}{\omega_o^2} \tag{5.31}$$

where

$$\omega_{LO}^2 = \omega_o^2 + \Omega^2 \tag{5.32}$$

It turns out that ω_o is the frequency of long wavelength transverse optical phonons and that ω_{LO} is the frequency of long wavelength longitudinal optical phonons. Thus we can write

$$\frac{\epsilon_o}{\epsilon_\infty} = \frac{\omega_{LO}^2}{\omega_{TO}^2} \tag{5.33}$$

This is the well-known Lyddane-Sachs-Teller relation. If we neglect damping, the real part of the dielectric constant can be written in the useful form

$$\epsilon_1(\omega) = \epsilon_e \frac{\omega_{LO}^2 - \omega^2}{\omega_{TO}^2 - \omega^2} \tag{5.34}$$

A plot of $\epsilon_1(\omega)$ versus ω is given in Fig. 9.

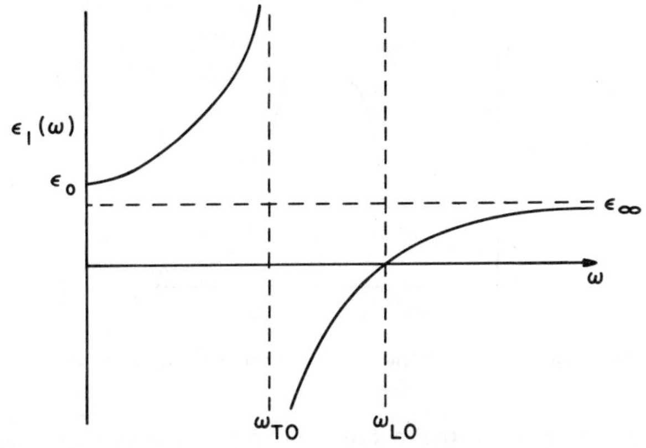

FIG.9. Dielectric constant plotted against frequency for a crystal with two atoms per unit cell.

We see that $\epsilon_1(\omega)$ has a pole at ω_{TO} and a zero at ω_{LO}.

To discuss the optical properties it is convenient to introduce the complex index of refraction η by

$$\eta = \sqrt{\epsilon(\omega)} \tag{5.35}$$

If we set $\eta = n - i\varkappa$ and use Eq.(5.26), we find that

$$\epsilon_1(\omega) = n^2 - \varkappa^2 \tag{5.36}$$

$$\epsilon_2(\omega) = -2n\varkappa \tag{5.37}$$

If we wish, we can solve Eqs.(5.36) and (5.37) for n and $\varkappa$ in terms of ϵ_1 and ϵ_2.

From the study of optics, one finds that the reflectivity R for normal incidence is given by

$$R = \frac{(n-1)^2+\varkappa^2}{(n+1)^2+\varkappa^2} \tag{5.38}$$

One can calculate R as a function of ω using Eqs.(5.27), (5.28), (5.36), and (5.37). A typical result is shown in Fig.10.

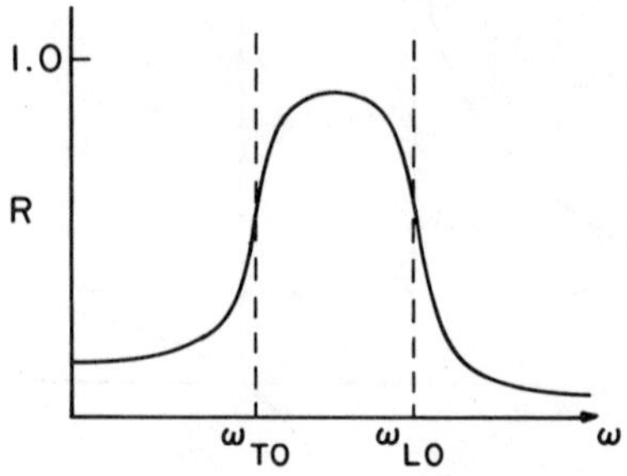

FIG.10. Reflectivity versus frequency for a crystal with two atoms per unit cell.

The important point is that the reflectivity is high between ω_{TO} and ω_{LO}. This is the region where, if one neglects damping, $\epsilon(\omega)$ is negative and η is imaginary. An electromagnetic wave does not propagate in such a region.

The absorption coefficient a is given by

$$a = -2\frac{\omega}{c}\epsilon_2(\omega) \tag{5.39}$$

$$= 4\frac{\omega}{c}n\varkappa \tag{5.40}$$

We notice from Eq.(5.28) that $\epsilon_2(\omega)$ is sharply peaked at $\omega = \omega_o = \omega_{TO}$ if γ is not too large. A typical plot of the absorption coefficient versus frequency is shown in Fig.11.

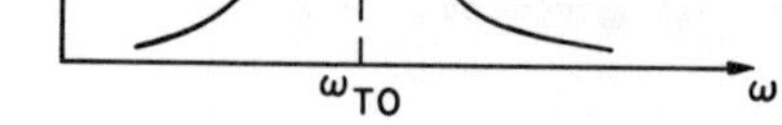

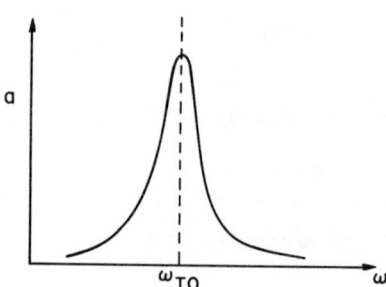

FIG.11. Absorption coefficient versus frequency for a crystal with two atoms per unit cell.

From experimental measurements of a versus ω, one can obtain values of ω_{TO} and γ. From experimental measurements of R versus ω, one can obtain values of ω_{TO}, ω_{LO} and γ. From $\Omega^2 = \omega_{LO}^2 - \omega_{TO}^2$ one can get a value of e^*.

6. Polaritons

We shall define a polariton as an excitation which arises from the coupling of an electromagnetic wave with an elementary excitation of a crystal(4). The elementary excitation may be a phonon, a magnon, a plasmon, an exciton, etc., or it may be a coupled excitation such as a phonon-plasmon, etc. Polaritons may be discussed from either a microscopic or macroscopic point of view. In our presentation we shall use a macroscopic approach which has the advantages of mathematical simplicity and ease of comparison with experiment.

Let us start our discussion with Maxwell's equations which describe the propagation of electromagnetic waves. The particular equations of interest are

$$\nabla \times \underset{\sim}{E} = -\frac{1}{c}\frac{\partial \underset{\sim}{B}}{\partial t} \tag{6.1}$$

$$\nabla \times \underset{\sim}{H} = \frac{1}{c}\frac{\partial \underset{\sim}{D}}{\partial t} \tag{6.2}$$

where $\underset{\sim}{E}$, $\underset{\sim}{H}$, $\underset{\sim}{D}$, and $\underset{\sim}{B}$ are the electric field, the magnetic field, the electric displacement, and the magnetic induction, respectively. We shall assume that all four of these vectors have the form of plane waves characterized by a wave vector $\underset{\sim}{k}$ and a circular frequency ω. Thus, for $\underset{\sim}{E}$ we have

$$\underset{\sim}{E} = \underset{\sim}{E}_o\, e^{i(\underset{\sim}{k}\cdot\underset{\sim}{r} - \omega t)} \tag{6.3}$$

where $\underset{\sim}{E}_o$ is the amplitude of the electric field. In general, the dielectric tensor $\epsilon_{\alpha\beta}$ and the magnetic permeability tensor $\mu_{\alpha\beta}$ are functions of both the wave vector and frequency, and one can write the constitutive relations in the form

$$D_\alpha = \sum_\beta \epsilon_{\alpha\beta}(\underset{\sim}{k},\omega)\, E_\beta \tag{6.4}$$

$$B_\alpha = \sum_\beta \mu_{\alpha\beta}(\underset{\sim}{k},\omega)\, H_\beta \tag{6.5}$$

If one substitutes the plane wave form for the fields into Maxwell's equations, Eqs.(6.1) and (6.2), one obtains

$$\underset{\sim}{k} \times \underset{\sim}{E} = \frac{\omega}{c} \underset{\sim}{B} \tag{6.6}$$

$$\underset{\sim}{k} \times \underset{\sim}{H} = -\frac{\omega}{c} \underset{\sim}{D} \tag{6.7}$$

Equations (6.4), (6.5), (6.6), and (6.7) permit one to eliminate three of the fields and obtain an equation for one field alone. In the case of the electric field, this equation can be written as

$$\underset{\sim}{k} \times [\mu^{-1}(\underset{\sim}{k},\omega)\cdot(\underset{\sim}{k}\times\underset{\sim}{E})] + \frac{\omega^2}{c^2}\in(\underset{\sim}{k},\omega)\cdot\underset{\sim}{E} = 0 \tag{6.8}$$

where $\mu^{-1}(\underset{\sim}{k},\omega)$ is the inverse of the magnetic permeability tensor.

Equation (6.8) in fact represents three linear homogeneous algebraic equations in the components E_α, $\alpha = x, y, z$. To obtain a nontrivial solution, the determinant of coefficients of the E_α must be zero. This determinantal equation constitutes the dispersion relation for the electromagnetic wave in the crystal and is, in fact, the dispersion relation for the polaritons.

We shall simplify the discussion by treating only dielectrically active crystals -- i.e., we assume that the magnetic permeability tensor is the unit tensor. Then Eq.(6.8) becomes

$$-k^2\underset{\sim}{E} + \underset{\sim}{k}(\underset{\sim}{k}\cdot\underset{\sim}{E}) + \frac{\omega^2}{c^2}\in(\underset{\sim}{k},\omega)\cdot\underset{\sim}{E} = 0 \tag{6.9}$$

If we now restrict our attention to isotropic crystals, the electromagnetic wave is purely transverse, so $\underset{\sim}{k}\cdot\underset{\sim}{E} = 0$. Also, the dielectric tensor is proportional to the unit tensor, and we obtain for the dispersion relation

$$\frac{c^2k^2}{\omega^2} = \in(\underset{\sim}{k},\omega) \tag{6.10}$$

To elucidate the dependence of ω on k, we consider the case of optical phonons in a cubic crystal with two atoms per unit cell such as NaCl. For the moment, we shall also neglect the wave vector dependence of $\epsilon(\underset{\sim}{k}, \omega)$. Then $\epsilon(\omega)$ has the form given in Eq.(5.34). Substituting this form into Eq.(6.10) and solving for ω^2, we get

$$\omega_{\pm}^2(k) = \frac{1}{2}\left\{\omega_{LO}^2 + \frac{c^2k^2}{\epsilon_\infty} \pm \left[\left(\omega_{LO}^2 - \frac{c^2k^2}{\epsilon_\infty}\right)^2 + \frac{4c^2k^2(\omega_{LO}^2-\omega_{TO}^2)}{\epsilon_\infty}\right]^{\frac{1}{2}}\right\} \tag{6.11}$$

Consider the limiting cases $k \to 0$ and $k \to \infty$. For $k \to 0$, we obtain $\omega_+(0) = \omega_{LO}$ and $\omega_-(0) = 0$, while for $k \to \infty$, we get $\omega_+(\infty) = ck/\sqrt{\epsilon_\infty}$ and $\omega_-(\infty) = \omega_{TO}$. A plot of $\omega_+(k)$ and $\omega_-(k)$ versus k is given in Fig.12.

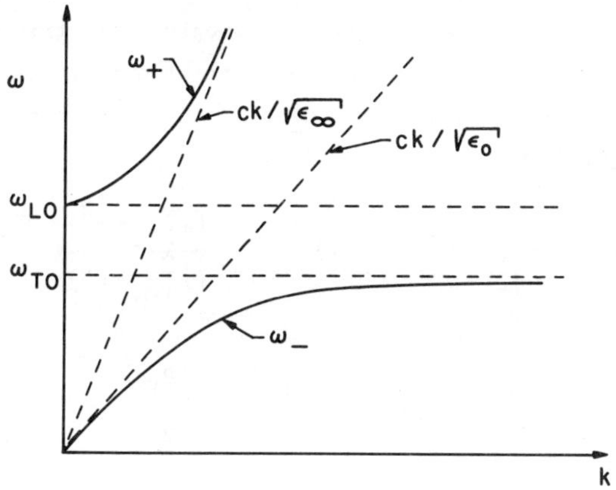

FIG.12. Frequency versus wave vector for polaritons in a crystal with two atoms per unit cell.

We observe from the figure that the limiting slope of $\omega_-(k)$ as $k \to 0$ is $c/\sqrt{\epsilon_0}$, while the limiting slope of $\omega_+(k)$ as $k \to \infty$ is $c/\sqrt{\epsilon_\infty}$.

From Eq.(5.34) we see that ω_{LO} is the frequency for which $\epsilon(\omega) = 0$. We can show that ω_{LO} is the frequency of the limiting longitudinal optical mode of long wavelength by the following argument. Since the crystal has no free charges, $\nabla \cdot D = 0$. But $\nabla \cdot D = \nabla \cdot [\epsilon(\omega)\underset{\sim}{E}] = i\epsilon(\omega)(\underset{\sim}{k}\cdot\underset{\sim}{E})$. For a longitudinal wave, $\underset{\sim}{k}$ and $\underset{\sim}{E}$ are parallel, so $\underset{\sim}{k}\cdot\underset{\sim}{E}$ does not vanish for $\underset{\sim}{k}$ small but finite. Thus, for $\nabla \cdot D$ to vanish, one must have $\epsilon(\omega) = 0$. However, it must be emphasized that for an isotropic crystal, the modes corresponding to both ω_+ and ω_- are transverse. We note also that there is no propagating wave with frequency between ω_{LO} and ω_{TO}. This is reasonable since the crystal is highly reflecting in this frequency range.

The form of the dispersion curves shown in the figure is typical of those for interacting modes. In particular, the crossing of the curves for the noninteracting modes is eliminated. For a given branch of the interacting system, the character of the mode changes as one passes from $k = 0$ to the limit $k \to \infty$. Thus, the mode associated with the lower branch acquires increasing phonon content and decreasing photon content as k increases, until in the limit $k \to \infty$, the mode is entirely phonon-like. The opposite is true of the mode associated with the upper branch.

We now consider the case of an anisotropic dielectric medium. In order to keep the discussion simple we restrict our attention to uniaxial crystals where the dielectric tensor can be taken to be diagonal with diagonal elements of the form

$$\epsilon_{xx}(\omega) = \epsilon_{yy}(\omega) = \epsilon_{\perp}(\omega) = \epsilon_{\infty\perp} \frac{(\omega_{LO\perp}^2 - \omega^2)}{(\omega_{TO\perp}^2 - \omega^2)} \tag{6.12}$$

$$\epsilon_{zz}(\omega) = \epsilon_{\parallel}(\omega) = \epsilon_{\infty\parallel} \frac{(\omega_{LO\parallel}^2 - \omega^2)}{(\omega_{TO\parallel}^2 - \omega^2)} \tag{6.13}$$

where the notation is an obvious generalization of that used for the isotropic case. The problem possesses cylindrical symmetry about the z-axis (the optical axis), so the results must be independent of the orientation of $\underset{\sim}{k}$ relative to the x- and y-axes. We can therefore choose our axes so $\underset{\sim}{k}$ lies in the xz-plane with components

$$k_x = k \sin\theta \quad , \quad k_z = k \cos\theta \tag{6.14}$$

where θ is the angle between $\underset{\sim}{k}$ and the z-axis.

If Eqs.(6.12), (6.13), and (6.14) are substituted into Eq.(6.9) for the electric field, the individual component equations can be written as

$$\left(\frac{\omega^2}{c^2 k^2} \epsilon_{\perp}(\omega) - \cos^2\theta \right) E_x + \sin\theta\cos\theta E_z = 0 \tag{6.15}$$

$$\left(\frac{\omega^2}{c^2k^2}\,\epsilon_\perp(\omega) - 1\right)E_y = 0 \tag{6.16}$$

$$\left(\frac{\omega^2}{c^2k^2}\,\epsilon_\parallel(\omega) - \sin^2\theta\right)E_z + \sin\theta\cos\theta E_x = 0 \tag{6.17}$$

We see immediately that the y-component of the electric field is decoupled from the x- and z-components. A purely transverse polariton exists for any direction of propagation with the electric field perpendicular to the plane formed by the wave vector and the optical axis. The dispersion relation is independent of θ and is given by

$$\frac{c^2k^2}{\omega^2} = \epsilon_\perp(\omega) \tag{6.18}$$

We see that this dispersion relation is identical in form to that for the isotropic case. One sometimes refers to this polariton as the ordinary polariton for the uniaxial crystal.

We now turn to the coupled equations involving E_x and E_z. Setting the determinant of coefficients equal to zero gives the dispersion relation

$$\frac{c^2k^2}{\omega^2}\left(\sin^2\theta\epsilon_\perp(\omega) + \cos^2\theta\epsilon_\parallel(\omega)\right) = \epsilon_\parallel(\omega)\epsilon_\perp(\omega) \tag{6.19}$$

If Eqs.(6.12) and (6.13) are substituted into this dispersion relation, one obtains a cubic equation in ω^2 which may be solved for ω in terms of k and θ. Two simple cases which may be easily discussed are $\theta = 0^\circ$ and $\theta = 90^\circ$.

For $\theta = 0^\circ$, the dispersion reduces to

$$\epsilon_\parallel(\omega)\left(\frac{c^2k^2}{\omega^2} - \epsilon_\perp(\omega)\right) = 0 \tag{6.20}$$

There are two distinct normal modes. The first corresponds to

$$\epsilon_\parallel(\omega) = 0 \tag{6.21}$$

and is purely longitudinal in character with the electric field parallel to the optical axis and the frequency independent of wave vector. The second corresponds to

$$\frac{c^2k^2}{\omega^2} = \epsilon_\perp(\omega) \tag{6.22}$$

and is purely transverse with the electric vector perpendicular to the optical axis. This mode has the same dispersion relation as the ordinary polariton and differs from it only in polarization.

For $\theta = 90^\circ$, we again have a purely longitudinal mode, but with the electric vector normal to the optical axis and the frequency specified by

$$\epsilon_\perp(\omega) = 0 \tag{6.23}$$

There is also a purely transverse mode with electric vector parallel to the optical axis and dispersion relation given by

$$\frac{c^2k^2}{\omega^2} = \epsilon_\parallel(\omega) \tag{6.24}$$

Plots of the dispersion relations for these two cases are shown in Fig.13.

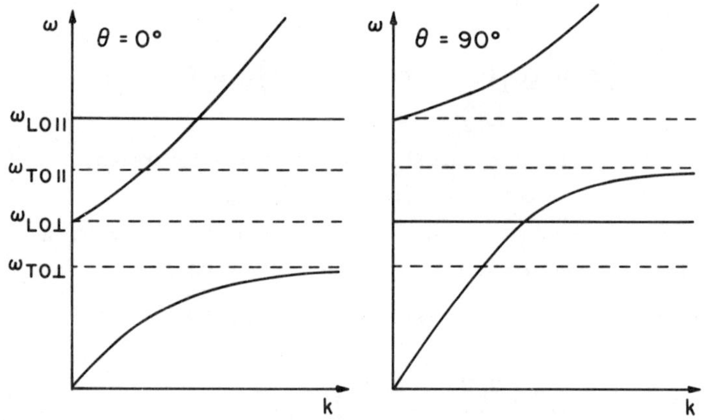

FIG.13. Frequency versus wave vector for polaritons in an anisotropic crystal for θ = 0° and θ = 90°.

When we consider general values of θ, we find that the two normal modes are coupled together and that neither is purely longitudinal or purely transverse. The crossings of the dispersion curves are eliminated with the result that the curves have the qualitative form shown in Fig. 14.

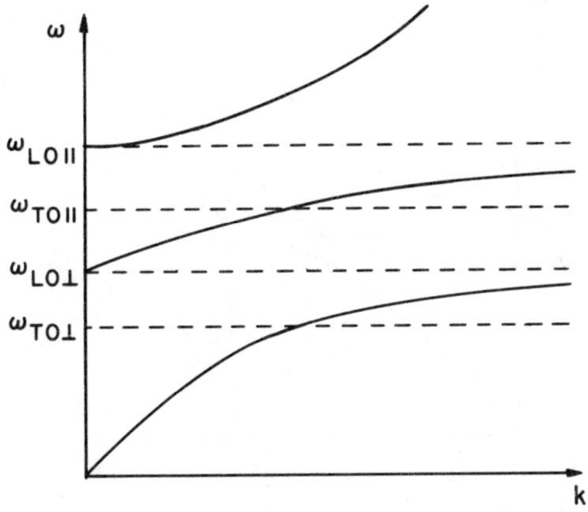

FIG.14. Frequency versus wave vector for polaritons in an anisotropic crystal for a general value of θ.

We note that two modes have frequencies that remain finite as $k \to \infty$. These limiting frequencies are specified by the equation

$$\epsilon_{\perp}(\omega)\sin^2\theta + \epsilon_{\parallel}(\omega)\cos^2\theta = 0 \qquad (6.25)$$

The third mode has a limiting frequency given by

$$\omega = ck\left(\frac{\sin^2\theta}{\epsilon_{\infty\parallel}} + \frac{\cos^2\theta}{\epsilon_{\infty\perp}}\right)^{\frac{1}{2}} \qquad (6.26)$$

We now examine the effect of spatial dispersion or the wave vector dependence of the dielectric tensor on the properties of polaritons. Spatial dispersion is a manifestation of a nonlocal relation between the electric displacement $\tilde{D}$ and the electric field $\tilde{E}$. A dependence on wave vector may enter the dielectric tensor through a k-dependence of the effective charge, the effective mass, or the transverse optical phonon frequency. In our discussion we shall ignore the k-dependence of the first two quantities and consider only the k-dependence of ω_{TO} for crystals with a center of inversion. For small k, one can write

$$\omega_{TO}^2(k) \simeq \omega_{TO}^2 + Ak^2 \qquad (6.27)$$

and calculate from the dispersion relation, Eq.(6.10), the normal mode frequencies which are plotted in a qualitative manner in Fig. 15 for the case $A > 0$.

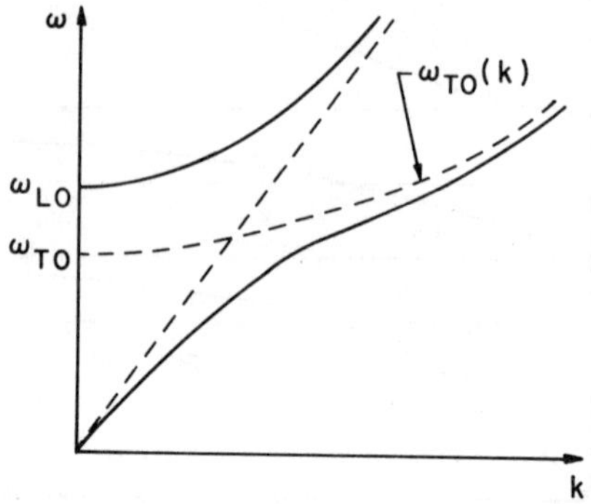

FIG.15. Frequency versus wave vector in the presence of spatial dispersion.

In contrast to the case without spatial dispersion, there is now a normal mode with frequency between ω_{LO} and ω_{TO} Also, in the limit $k \to \infty$, , there are now two modes with group velocity greater than zero instead of one.

For the case of optical phonons, the constant A may be positive or negative, but its magnitude is typically quite small. For the case of excitons, however, A is positive and, in a material such as CdS, is fairly sizeable. Various features in the reflectivity of CdS near the exciton lines can be attributed to spatial dispersion effects.

In our discussion of polaritons so far, we have focused on those associated with optical phonons. The same sort of treatment can be applied to polaritons associated with excitons, plasmons, magnons, etc. In the case of excitons, ω_{TO} is replaced by the resonant frequency of the exciton. In the case of plasmons, ω_{TO} is zero and ω_{LO} is the plasma frequency. In the case of magnons, one considers elements of the magnetic permeability tensor with frequency dependence similar to that of Eq.(5.25). For anti-ferromagnetic crystals, ω_{TO} is replaced by the antiferromagnetic resonance frequency.

One of the most fruitful experimental tools for the study of polaritons is Raman scattering. Indeed, using this technique, one is able to obtain information concerning the dispersion curves of polaritons.

Raman scattering is the inelastic scattering of light by a crystal or other material. Consider a light beam, typically from a laser, of frequency ω_o and wave vector $\underset{\sim}{q}_o$ incident on a crystal. In addition to the usual Rayleigh process, the light may be scattered by the Raman process in which both the frequency and the wave vector are changed. If the frequency of the scattered light is less than that of the incident light, we have Stokes scattering, whereas, if the frequency of the scattered light is greater than that of the incident light, we have antiStokes scattering. In Stokes scattering, a polariton is created in the crystal; in antiStokes scattering, a polariton is destroyed.

We shall focus our attention on the Stokes process. Let $\underset{\sim}{q}_s$ and ω_s be the wave vector and frequency of the scattered light and let $\underset{\sim}{k}$ and ω_π be the wave vector and frequency of the polariton created, respectively. Then conservation of wave vector and conservation of energy in the scattering process require that

$$\underset{\sim}{k} = \underset{\sim}{q}_o - \underset{\sim}{q}_s \tag{6.28}$$

$$\omega_\pi = \omega_o - \omega_s \tag{6.29}$$

Let θ be the angle between the direction of the scattered beam and the direction of the incident beam. Then after squaring Eq.(6.28), we obtain

$$k^2 = (q_o - q_s)^2 + 2q_o q_s(1 - \cos\theta) \tag{6.30}$$

We shall be interested in small values of θ — i.e., in scattering near the forward direction. We can then expand $\cos\theta$ in powers of θ and obtain approximately

$$k^2 \simeq (q_o - q_s)^2 + q_o q_s \theta^2 \tag{6.31}$$

Typical Raman experiments are done with incident light in the visible, so $\omega_o \gg \omega_\pi$. Consequently, the wave vector magnitudes q_o and q_s are very close to one another. If

n_o and n_s are the refractive indices of the crystal at ω_o and ω_s, respectively, we can then write

$$q_o = \frac{n_o \omega_s}{c} \qquad (6.32)$$

$$q_s = \frac{n_s \omega_s}{c} \simeq q_o - \frac{\omega_\pi}{c}\left(n_o + \omega_o \frac{\partial n}{\partial \omega}\right) \qquad (6.33)$$

Substitution of these relations into Eq.(6.31) gives the result

$$c^2 k^2 = \omega_\pi^2 \left(n_o + \omega_o \frac{\partial n}{\partial \omega}\right)^2 + \omega_o^2 n_o^2 \theta^2 \qquad (6.34)$$

From this equation one can calculate the wave vector k of the created polariton from measured values of the frequency shift ω_π and the scattering angle θ. Since ω_π is the frequency of the created polariton, we thus have obtained a point on the polariton dispersion curve. By considering different angles of scattering, one can then trace out the dispersion curve.

There are, however, some restrictions on what one can do. We note that $c/[n_o + \omega_o \frac{\partial n}{\partial \omega}]$ is the group velocity, v_{go}, of the incident light. Furthermore, from Eq.(6.34), we see that

$$k \geq \frac{\omega_\pi}{v_{go}} \qquad (6.35)$$

Also, $\frac{\partial n}{\partial \omega}$ is typically positive and $n_o \simeq \sqrt{\epsilon_\infty}$, so v_{go} is generally smaller than the phase velocity v_{po} given by

$$v_{po} = \frac{c}{n_o} \simeq \frac{c}{\sqrt{\epsilon_\infty}} \qquad (6.36)$$

Thus,

$$\frac{\omega_\pi}{v_{go}} > \frac{\omega_\pi}{v_{po}} = \frac{\omega_\pi}{c}\sqrt{\epsilon_\infty} \qquad (6.37)$$

and we must have

$$k \geq \frac{\omega_\pi}{c}\sqrt{\epsilon_\infty} \qquad (6.38)$$

This last relation means that one cannot observe the upper branch of the polariton dispersion curve by Raman scattering since for the upper branch

$$k < \frac{\omega_\pi}{c} \sqrt{\epsilon_\infty} \tag{6.39}$$

However, for anisotropic crystals, it turns out that information about the upper branch can be obtained[5].

The intensity of Raman scattering can be investigated if one realizes that the scattering is produced by the modulation of the dielectric susceptibility by the optical phonons. Indeed, both the atomic displacement $\underset{\sim}{w}$ and the macroscopic electric field $\underset{\sim}{E}$ contribute. One can write the change in the dielectric susceptibility tensor element $\chi_{\mu\nu}$ as

$$\delta\chi_{\mu\nu} = \sum_\lambda \frac{\partial\chi_{\mu\nu}}{\partial w_\lambda} w_\lambda + \sum_\lambda \frac{\partial\chi_{\mu\nu}}{\partial E_\lambda} E_\lambda \tag{6.40}$$

$$= \sum_\lambda a_{\mu\nu\lambda} w_\lambda + \sum_\lambda b_{\mu\nu\lambda} E_\lambda \tag{6.41}$$

where $a_{\mu\nu\lambda}$ is a tensor describing the modulation of $\chi_{\mu\nu}$ by the atomic displacements of optical character and $b_{\mu\nu\lambda}$ is the electro-optic tensor.

If the incident photon travels a distance L in the crystal, then one can show that the scattering efficiency per unit solid angle associated with transitions between polariton states of quantum numbers n and n + 1 is given by

$$\frac{dS}{d\Omega} = \left(\frac{\omega_s}{c}\right)^4 VL \left|\langle n+1| \sum_{\mu\nu} \hat{e}_\mu(s)\,\delta\chi_{\mu\nu}\,\hat{e}_\nu(I)\,|n\rangle\right|^2 \tag{6.42}$$

where $\hat{e}(I)$ and $\hat{e}(s)$ are unit vectors in the directions of polarization of the incident and scattered photons, respectively, and V is the volume of the crystal. If we use Eqs.(5.16) - (5.18), we can eliminate $\underset{\sim}{E}$ in favour of $\underset{\sim}{w}$ and write Eq.(6.42) in the alternative form

$$\frac{dS}{d\Omega} = \left(\frac{\omega_s}{c}\right)^4 VL \left|\sum_{\mu\nu\lambda} \hat{e}_\mu(s)\hat{e}_\nu(I)\left[a_{\mu\nu\lambda} + \frac{\mu}{e^*}(\omega_{TO}^2 - \omega^2)\, b_{\mu\nu\lambda}\right]\right|^2 |\langle n+1|w_\lambda|n\rangle|^2 \tag{6.43}$$

where ω is the polariton frequency.

Depending on the relative signs of $a_{\mu\nu\lambda}$ and $b_{\mu\nu\lambda}/e^*$, the atomic displacement and electro-optic contributions to the scattering may interfere either constructively or destructively. We also note from Eq.(6.43) that for $\omega = \omega_{TO}$ the electro-optic contribution vanishes. This occurs for large scattering angles where the change in wave vector during the scattering is large.

7. Surface Polaritons

If a crystal is not infinite but possesses a free surface, then it is possible under proper conditions to have a surface polariton in which the associated electromagnetic field is localized at the surface of the medium. We shall restrict our attention to surface polaritons which are associated with a planar interface between two media.

We present an elementary theoretical treatment of surface polaritons for the case where both media are isotropic(6). The interface separating the two media is assumed to be specified by $x = 0$. The frequency-dependent dielectric constant for the material in the half space $x > 0$ is $\epsilon_a(\omega)$, while that for the material in the half space $x < 0$ is $\epsilon_b(\omega)$. We assume that losses are sufficiently small to permit us to neglect the imaginary parts of the dielectric constants. We also neglect any wave vector dependence of the dielectric constants and assume the magnetic permeability to be unity everywhere.

Our starting point is Maxwell's equations given by Eqs.(6.1) and (6.2). We eliminate the magnetic vector and obtain the equation

$$\nabla \times \nabla \times \underset{\sim}{E} + \frac{1}{c^2}\frac{\partial^2 \underset{\sim}{D}}{\partial t^2} = 0 \qquad (7.1)$$

where $\underset{\sim}{D} = \epsilon(\omega)\underset{\sim}{E}$. Consider a solution to Eq.(7.1) corresponding to a surface polariton propagating in the z-direction. The electric field varies in a wavelike manner in the z-direction but decreases exponentially away from the interface. Thus,

$$\underset{\sim}{E}_a(\underset{\sim}{r},t) = \underset{\sim}{E}_a^o\, e^{-\alpha_a x}\, e^{i(kz-\omega t)}, \quad x > 0 \qquad (7.2a)$$

$$\underset{\sim}{E}_b(\underset{\sim}{r}, t) = \underset{\sim}{E}_b^o \, e^{\alpha_b x} \, e^{i(kz-\omega t)} , \quad x < 0 \tag{7.2b}$$

where k is the wave vector describing the propagation parallel to the surface and the decay constants α_a and α_b must have positive real parts in order to have a bona fide surface polariton.

We substitute Eqs.(7.2) into Eq.(7.1) and obtain a set of algebraic equations whose solution yields the following expressions for the electric fields:

$$\underset{\sim}{E}_a(r,t) = \left(\frac{ik}{\alpha_a} E_{az}^o, E_{ay}^o, E_{az}^o \right) e^{-\alpha_a x + ikz - i\omega t} , \quad x>0 \tag{7.3a}$$

$$\underset{\sim}{E}_b(r,t) = \left(- \frac{ik}{\alpha_b} E_{bz}^o, E_{by}^o, E_{bz}^o \right) e^{\alpha_b x + ikz - i\omega t} , \quad x<0 \tag{7.3b}$$

The decay constants are given by

$$\alpha_a^2 = k^2 - \epsilon_a(\omega) \frac{\omega^2}{c^2} \tag{7.4a}$$

$$\alpha_b^2 = k^2 - \epsilon_b(\omega) \frac{\omega^2}{c^2} \tag{7.4b}$$

Using Maxwell's equations, we find the magnetic fields to be

$$\underset{\sim}{B}_b(\underset{\sim}{r}, t) = \frac{ic}{\omega} \left(ikE_{ay}^o, \frac{\omega^2}{\alpha_a c^2} \epsilon_a(\omega) E_{az}^o, \alpha_a E_{ay}^o \right) e^{-\alpha_a x + i(kz-\omega t)} \quad x > 0 \tag{7.5a}$$

$$\underset{\sim}{B}_b(\underset{\sim}{r}, t) = \frac{ic}{\omega} \left(ikE_{by}^o, - \frac{\omega^2}{\alpha_b c^2} \epsilon_b(\omega) E_{bz}^o, -\alpha_b E_{by}^o \right) e^{\alpha_b x + i(kz-\omega t)} \quad x < 0 \tag{7.5b}$$

where we have assumed that $\underset{\sim}{B}(\underset{\sim}{r}, t) \sim e^{-i\omega t}$.

We now take into consideration the boundary conditions which state that the tangential components $\underset{\sim}{E}$ and the normal components of $\underset{\sim}{D}$ are continuous at the interface $x = 0$. From the continuity of the tangential components of $\underset{\sim}{E}$, we find

$$E_{ay}^o = E_{by}^o , \quad E_{az}^o = E_{bz}^o \tag{7.6}$$

From the continuity of the normal component of $\underset{\sim}{D}$, we obtain

$$\frac{\epsilon_a(\omega)}{\epsilon_b(\omega)} = - \frac{\alpha_a}{\alpha_b} \tag{7.7}$$

We also have the boundary condition that the tangential components of $\underset{\sim}{B}$ are continuous at the interface. Using this condition and Eqs.(7.5), we see that

$$(\alpha_a + \alpha_b) E^o_{ay} = 0 \tag{7.8}$$

Since α_a and α_b must have positive real parts for a surface polariton, Eq.(7.8) can be satisfied only if $E^o_{ay} = E^o_{by} = 0$. Accordingly, the electric field of the surface polariton lies in the sagittal plane which is the plane defined by the direction of propagation and the surface normal. Such a mode is known as a TM mode.

Equation (7.7) constitutes the dispersion relation for surface polaritons. By squaring Eq.(7.7) and using Eqs.(7.4) we can re-express the dispersion relation in the form

$$\frac{c^2 k^2}{\omega^2} = \frac{\epsilon_a(\omega)\epsilon_b(\omega)}{\epsilon_a(\omega)+\epsilon_b(\omega)} \tag{7.9}$$

which may be compared with that for bulk polaritons given by Eq.(6.10). A word of caution must be presented in using Eq.(7.9). Some of the solutions of this equation will be spurious since it was obtained by squaring Eq.(7.7).

Consider a surface polariton which is characterized by real and positive α_a and α_b. At the frequency ω corresponding to the surface polariton, we see from Eq.(7.7) that $\epsilon_a(\omega)$ and $\epsilon_b(\omega)$ must have opposite signs. The medium having the negative value of $\epsilon(\omega)$ is termed the <u>surface active</u> medium, and that having the positive value of $\epsilon(\omega)$ is termed the <u>surface inactive</u> medium.

As a specific example, we consider the case of surface polaritons associated with surface optical phonons in a cubic ionic crystal with two atoms per unit cell. The dielectric constant is given by Eq.(5.34) for medium a which we take to be the surface active medium. We take medium b to be vacuum, so $\epsilon_b(\omega) = 1$. If we substitute these values of the dielectric constant into the dispersion relation, Eq.(7.9), and solve for the surface polariton frequency, $\omega_{so}(k)$, we obtain

$$\omega_{so}^2(k) = \frac{1}{2\epsilon_\infty}\{(1+\epsilon_\infty)c^2k^2 + \epsilon_\infty\omega_{LO}^2 - [((1+\epsilon_\infty)c^2k^2 + \epsilon_\infty\omega_{LO}^2)^2 - 4\epsilon_\infty c^2k^2(\omega_{TO}^2+\epsilon_\infty\omega_{LO}^2)]^{\frac{1}{2}}\} \quad (7.10)$$

where we have replaced ϵ_e by ϵ_∞ in accordance with Eq.(5.29). A typical plot of $\omega_{so}(k)$ versus k is given in Fig.16.

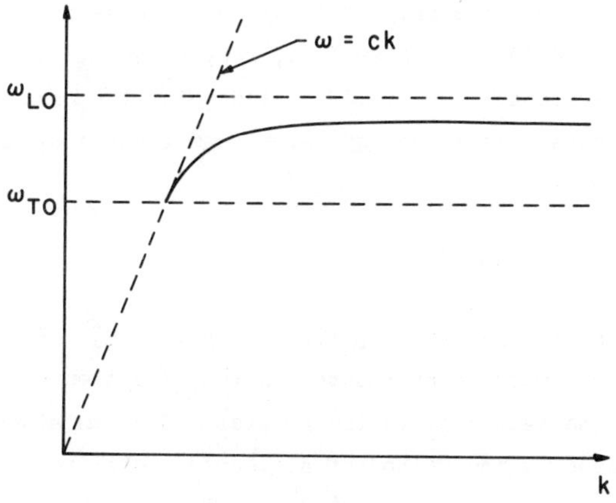

FIG.16. Frequency versus wave vector for surface polaritons on the surface of an isotropic crystal with two atoms per unit cell.

We comment on several features of the dispersion curve. First, we observe that it lies entirely to the right of the light line in vacuum, $\omega = ck$. This has the consequence that no simple resonant interaction between the surface polariton and light incident from the vacuum is possible. Second, the dispersion curve starts at the light line at $\omega = \omega_{TO}$, rises and bends over, and approaches an asymptote characterized by the frequency

$$\omega_{so}(\infty) = \left\{\omega_{TO}^2 + \frac{\omega_{LO}^2 - \omega_{TO}^2}{1 + \frac{1}{\epsilon_\infty}}\right\}^{\frac{1}{2}} \quad (7.11)$$

at large values of k (the unretarded limit). Near the light line the surface polariton is primarily photon-like, whereas at large values of k it is primarily phonon-like.

We now make some remarks concerning the experimental observation of surface polaritons. We have already noted that the use of conventional optical absorption measurements to observe surface polaritons is not possible. The frequency of the incident light must equal the frequency of the surface polariton,

$$c(k_{\parallel}^2 + k_{\perp}^2)^{\frac{1}{2}} = \omega_{so}(k_{\parallel}) \tag{7.12}$$

where $k_{\parallel}$ and $k_{\perp}$ are the components of the wave vector parallel and perpendicular to the surface, respectively. On the other hand, the electromagnetic fields of the polariton must decay with increasing distance from the crystal-vacuum interface, and this imposes the condition

$$ck_{\parallel} > \omega_{so}(k_{\parallel}) \tag{7.13}$$

Equations (7.12) and (7.13) are compatible only if $k_{\perp}^2 < 0$, so the incident light must be attenuated in the direction normal to the surface in the vacuum above the crystal. Such an attenuated field can be achieved by the method of attenuated total reflection (ATR) developed by Otto. In this method one places a prism of dielectric constant ϵ_p above the crystal so that it is separated from the crystal by a gap of thickness d. Light with its electric vector in the xz-plane is incident on the interface between the prism and the vacuum and makes an angle θ with the x-axis (normal to the interface). If $\theta > \sin^{-1}(1/\epsilon_p)$, the light undergoes total internal reflection within the prism and gives rise to an electric field in the gap which varies exponentially with x — i.e., a field for which $k_{\perp}^2 < 0$. We also note that the value for $k_{\parallel}$ is given by

$$k_{\parallel} = \sqrt{\epsilon_p}\ \frac{\omega}{c} \sin\theta \tag{7.14}$$

where ω is the frequency of the incident light, and that for given $\epsilon_p > 1$, there is a range of θ values for which $k_{\parallel} > \omega/c$. It is therefore possible to have the frequency of the incident light equal to the frequency of the surface polariton for some value of $k_{\parallel}$. When this occurs, there is a dip in the reflectivity of the system as a function ω corresponding to the excitation of surface polaritons. By varying the angle of incidence θ, one can trace out the dispersion curve.

8. Anharmonic Effects

The terms in the expansion of the potential energy, Eq.(1.6), which are cubic, quartic, ..., in the displacements are known collectively as anharmonic terms. They lead to a number of interesting physical phenomena including

a) thermal expansion
b) finite thermal conductivity
c) attenuation of phonons
d) broadening of optical absorption lines.

The lattice dynamical theory of the last three phenomena is rather complicated mathematically and will not be dealt with here. We present a calculation of the thermal expansion of a face-centered cubic lattice with nearest neighbour interactions due to Maradudin and Flinn[7].

It is convenient to use the Helmholtz free energy F which can be written as

$$F = U + F_v \tag{8.1}$$

where U is the static lattice energy in the equilibrium configuration and F_v is the vibrational contribution to the free energy. Let $\varphi(r)$ be the potential energy of interaction of two nearest neighbour atoms separated by a distance r. Then

$$U = 6N\varphi(r_o) \tag{8.2}$$

where r_o is the equilibrium separation and N is the number of atoms in the crystal.

The vibrational free energy F_v is given in general by

$$F_v = - k_B T \log Z \tag{8.3}$$

where the partition function Z is

$$Z = \mathrm{Tr}\, e^{-\beta H} \tag{8.4}$$

In Eq.(8.4), H is the Hamiltonian including anharmonic terms and $\beta = 1/k_B T$. We write

$$H = H_o + H_A \tag{8.5}$$

where H_o is the harmonic Hamiltonian and H_A is the anharmonic Hamiltonian. For a single harmonic oscillator the partition function is given by

$$Z = \sum_{n=o}^{\infty} e^{-\beta\hbar\omega(n+\frac{1}{2})} = \frac{1}{2\sinh\frac{\beta\hbar\omega}{2}} \tag{8.6}$$

For the entire crystal in the harmonic approximation the partition function is

$$Z = \sum_{n_1=o}^{\infty} \sum_{n_2=o}^{\infty} \ldots\ldots e^{-\beta[\hbar\omega_1(n_1+\frac{1}{2})+\hbar\omega_2(n_2+\frac{1}{2})+\ldots\ldots]}$$

$$= \prod_s \left\{ \sum_{n_s=o}^{\infty} e^{-\beta\hbar\omega_s(n_s+\frac{1}{2})} \right\}$$

$$= \prod_s \frac{1}{2\sinh\frac{\beta\hbar\omega_s}{2}} \tag{8.7}$$

where s specifies the various normal modes. The harmonic contribution to the vibrational free energy can be obtained by substituting Eq.(8.7) into Eq.(8.3). The result is

$$F_{vo} = k_B T \sum_s \log\left[2\sinh\frac{\hbar\omega_s}{2k_B T}\right] \tag{8.8}$$

We shall restrict our discussion to the high temperature limit where $\hbar\omega_s << 2k_B T$ for all s. Then we can expand the hyperbolic sine in power series in its argument and obtain

$$F_{vo} = k_B T \sum_s \log\left[\frac{\hbar\omega_s}{k_B T} + \frac{1}{3}\left(\frac{\hbar\omega_s}{2k_B T}\right)^3 + \ldots\right]$$

$$= k_B T \sum_s \log\left\{\frac{\hbar\omega_s}{k_B T}\left[1 + \frac{1}{24}\left(\frac{\hbar\omega_s}{k_B T}\right)^2 + \ldots\right]\right\}$$

$$= k_B T \sum_s \log\frac{\hbar\omega_s}{k_B T} + \frac{k_B T}{24}\sum_s\left(\frac{\hbar\omega_s}{k_B T}\right)^2 + \ldots$$

$$= k_B T \sum_s \log\frac{\hbar\omega_s}{k_B T} + 0(1/T) \tag{8.9}$$

For our model of a face-centered cubic crystal with nearest neighbour interactions, one can calculate the normal mode frequencies ω_s and then evaluate the sum over s in Eq.(8.9) numerically. The result is

$$F_{vo} = 3Nk_BT \log[\, 0.6505\, (\hbar\omega_m/k_BT)\,] \tag{8.10}$$

where ω_m is the maximum normal mode frequency specified by

$$\omega_m^2 = \frac{8\varphi''(r_o)}{M} \tag{8.11}$$

and $\varphi''(r_o)$ is the second derivative of $\varphi(r)$ evaluated at $r = r_o$.

The evaluation of the anharmonic contribution to the free energy is in general rather complicated. However, we can obtain this contribution correct to lowest order in the cubic anharmonic coefficient by the following procedure. We treat the frequency ω_m in Eq.(8.10) as temperature dependent. The temperature dependence arises from the dependence of r_o on temperature. We write

$$r_o = (1 + \epsilon)\,\bar{r}_o \tag{8.12}$$

where ϵ is the linear expansivity and $\bar{r}_o$ is independent of temperature and specifies the nearest neighbour separation corresponding to the minimum static energy. Thus, we have

$$\frac{\partial U}{\partial r_o} = 6N\varphi'(r_o) = 0 \qquad \text{for } r_o = \bar{r}_o \tag{8.13}$$

so $\varphi'(\bar{r}_o) = 0$.

We can expand U in powers of ϵ and obtain

$$U = 6N\,\{\varphi(\bar{r}_o) + \tfrac{1}{2}\epsilon^2\bar{r}_o^2\varphi''(\bar{r}_o) + \ldots\} \tag{8.14}$$

where we have used the fact that $\varphi'(\bar{r}_o) = 0$. We can also expand $\varphi''(r_o)$, which appears in Eq.(8.11), and obtain the result

$$\varphi''(r_o) = \varphi''(\bar{r}_o) + \epsilon\bar{r}_o\varphi'''(\bar{r}_o) + \ldots \tag{8.15}$$

Substituting Eq.(8.15) into Eq.(8.11) gives

$$\omega_m^2 = \frac{8\varphi''(\bar{r}_o)}{M}\left[1 + \epsilon\bar{r}_o \frac{\varphi'''(\bar{r}_o)}{\varphi''(\bar{r}_o)} + \ldots\right] \tag{8.16}$$

If we now eliminate ω_m from Eq.(8.10) with the aid of Eq.(8.16), we obtain

$$F_v = 3Nk_BT \log\left[0.6505 \frac{\hbar}{k_BT}\left(\frac{8\varphi''(\bar{r}_o)}{M}\right)^{\frac{1}{2}}\right] + \epsilon\bar{r}_o \frac{3Nk_BT}{2} \frac{\varphi'''(\bar{r}_o)}{\varphi''(\bar{r}_o)} + 0(\epsilon^2) \tag{8.17}$$

The equilibrium value of ϵ is determined by minimizing the total free energy $F = U + F_v$ with respect to ϵ:

$$\frac{\partial F}{\partial \epsilon} = 0 \tag{8.18}$$

Using Eqs.(8.14) and (8.17), we find that

$$\epsilon = -\frac{k_BT}{4\bar{r}_o} \frac{\varphi'''(\bar{r}_o)}{[\varphi''(\bar{r}_o)]^2} \tag{8.19}$$

Since $\varphi'''(r_o)$ is typically negative, the thermal expansivity is ordinarily positive. The linear thermal expansion coefficient, α, is given by

$$\alpha = \frac{\partial r_o}{\partial T} = -\frac{k_B\varphi'''(\bar{r}_o)}{4[\varphi''(\bar{r}_o)]^2} \tag{8.20}$$

This equation exhibits explicitly the dependence of the thermal expansion coefficient on the anharmonic coefficient $\varphi'''(\bar{r}_o)$.

References

1. Born, M., Huang, K., Dynamical Theory of Crystal Lattices, Oxford University Press, Oxford (1954).

2. Maradudin, A.A., Montroll, E.W., Weiss, F.H., Ipatova, I.P., Theory of Lattice Dynamics in the Harmonic Approximation, 2nd Edn., Academic Press, New York (1972).

3, Van Hove, L., Phys. Rev. 84 (1953) 1189.

4. For a discussion of the definition of the term polariton, see Mills, D.L., Burstein, E., Repts. on Prog. Phys. 37 (1974) 817.

5. Nicola, J.H., Freitus, J.A., Leite, R.C.C., Proceedings of the Third International Conference on Light Scattering in Solids, Ed. Balkanski, M., Leite, R.C.C., Porto, S.P.S., Flammarion Sciences, Paris (1976) 497.

6. Wallis, R.F., Prog. in Surf. Sci. 4 (1974) 233.

7. Maradudin, A.A., Flinn, P.A., Phys. Rev. 129 (1963) 2529.

MANY-PHONON PROCESSES INDUCED BY LIGHT ABSORPTION AT LOCALIZED CENTRES IN NON-CONDUCTING CRYSTALS

N. TERZI
Instituto di Fisica dell' Università di Milano,
and
Gruppo Nazionale di Struttura della Materia del CNR,
Milan,
Italy

Abstract

MANY-PHONON PROCESSES INDUCED BY LIGHT ABSORPTION AT LOCALIZED CENTRES IN NON CONDUCTING CRYSTALS.

A theoretical analysis of the phonon processes involved in the optical transitions of a bound electron in non-conducting crystals is presented. The absorption coefficient is deduced from the response function and its Fourier transform is analysed using the technique of the many-body theory. The diagrammatic expansion, representing the phonon processes in all orders, is resummed and the final expression is evaluated for weak and strong electron-phonon coupling.

1. INTRODUCTION

The investigation of the processes involved in the absorption of light in matter usually begins with the study of the simplest case, i.e., when a free elementary excitation is created and a photon is destroyed.

The main features of the absorption spectrum can be interpreted on this basis, and different elementary excitations (electrons,excitons, plasmons, phonons, etc.) are in turn involved, depending on the energy of the absorbed photons.

The finer details of the spectra, however, are often related to more entangled but more interesting processes. The elementary excitations are actually not independent quasi-particles inside the matter, but interact and their mutual action can be detected and evaluated by a closer analysis of the spectrum. Among the processes which involve two elementary excitations of the matter at the same time, I think that the light-induced excitation of an electron bound to a localised centre,such as a defect in a non-conducting crystal, has a very interesting feature. Infact, this process involves the photon, the electron and the crystal phonons, but only two interactions play an important role in determining the absorption coefficient:

i) the electron-radiation interaction, from which depends the intensity of the band at the excitation energy in the absorption spectrum;

ii) the electron-phonon interaction (i.e. the interaction of the electron with the surrounding lattice), whose intensity determines the shape of the band.

Furthermore the electron-phonon interaction has the very interesting characteristic that it is not stationary, but is switched by the electron-radiation interaction. I explain in other words what I mean.

Owing to the photon absorption process, the optic electron localized at the centre goes to the excited state, leaving a hole in its ground state: a bound hole-electron pair is created inside the crystal. (Since the electron is by assumption bound at the defect and its lowest excited state has an energy which lies inside the forbidden gap, neither the electron nor the hole can propagate). The presence of such a hole-electron pair modifies locally both the equilibrium conditions and the dynamics of the crystal ions, which before the excitation vibrate around an equilibrium position determined by the electron charge distribution relative to the ground state of the optic electron. When the adiabatic approximation is used to separate the coupled motion of the electron and ions in presence of the hole-electron excitation, in the total Hamiltonian a new term appears which corresponds to an interaction energy between the hole-electron pair and the phonon field (1). In what follows I call it the electron-phonon (EP) interaction and indicate it by H_{EP}. H_{EP} can be developed in series of the vibrational displacements $\underline{u}_1$ of the 1-th ion around its equilibrium position $\underline{R}_1$ (without excitations) or better in series of the operators of creation and annihilation of phonons (2).

Therefore, when acting on the ground state of the system (remember that the lattice is at the equilibrium before the absorption takes place), H_{EP} creates or destroys phonons i.e. is responsible for a many-phonon process. Briefly, the photon is eventually able to create or destroy phonons, through both the adiabatic approximation and the electron-photon interaction, even if there is no direct photon-phonon interaction. Note that this effect is rather peculiar to the localized centres, because a change in the electronic structure can modify sensibly, but locally, the forces between the centre and the surrounding ions.

The most interesting feature of the EP interaction is that it is transient, because it acts only during the time the pair exists. This characteristic plays a fundamental role in the theoretical evaluation of the shape of the absorption band.

I remember that the absorption coefficient $R(\omega)$ is related to the imaginary part of the dialectric function $\varepsilon(k,\omega)$, and that R(t) the Fourier transform of R(ω), is a function of the electron-hole pair autocorrelation function between the time $t_o = 0$ the pair is created and a photon is absorbed, and the time t the pair is destroyed and a photon is emitted. The EP interaction then acts only during the time interval t and it is not at all stationary. As I'll show in the next paragraph, the EP interaction acts on the hole-electron propagator rather like an external field, which is able to create or destroy phonons during the correlation time t. Depending on the strength of the EP interaction, the number of phonons involved can vary from zero (no interaction at all; the absorption spectrum shows a single line at the hole-electron formation energy in the rigid lattice) to even several tens. In this last case, the line corresponding to the pure electronic transition broadens, is often structured, and takes a shape (hereafter called the lineshape) which depends on the details of the EP interaction.

Such effect is well known for molecules, whose electronic spectrum shows almost always vibrational and rotational structure(3). It is also present in the spectrum of localized elctron systems in non-conducting crystals, such as those due to imperfections or impurities, and has been often interpreted on the basis of molecular models, even if in this case the physical situation is much more complex than for molecules. The experimental shape of the absorption bands has been shown and discussed for several centres in the lectures of Prof. Chiarotti and an interpretation has been presented in the framework of the configurational coordinate model(4).

Furthermore, a great number of spectra are presented and commented upon in the book "Physics of color centres" where the reader can also find a rich reference to the literature.(5)

Here I spend only a few words on the very interesting spectrum of the excitons. Their absorption bands show often the same type of vibrational shape as the localized centres. One must consider that the exciton is a complex elementary excitation, which has got essentially two degrees of freedom, the first associated to the translation of the centre of mass of the hole-electron pair, and the second associated to the internal motion. The photons can interact with the exciton through both of them. Depending on what interaction dominates, one can have either the phonon assisted processes with the $\underline{K} = \underset{\sim}{q}$ wavevector conservation law (as it occurs mainly in the semi-conductors) or trapped excitons whose hole-electron pair interacts with the phonons as local centre via the process just described (this can be the most usual case in alkali halides).(6)

Note that the present problem seems to be similar to the polaron problem, because in both cases there is an electron interacting with the phonons.[2]

However, the likeness is only apparent for the following two reasons:

i) Here the optic electron (the electron excited by the light) is localised and its energy levels are discrete, the lowest ones falling inside the forbidden energy gap of the host crystal. It has not got a quasimomentum $\underset{\sim}{k}$ which is related to the translational motion. Therefore, when the electron-hole pair is created and its evolution in time is considered, the graphs due to the EP interaction can be resummed exactly, as it will be shown in the next sections. Conversely, only special classes of graphs can be resummed in the propagation of the polaron, just because in that case the selection rule about the conservation of the quasimomentum holds at the interaction vertices, and the higher order graphs cannot be simply written as powers of a fundamental graph, as it occurs in the present situation.

ii) The EP interaction is transient, but it is stationary in the case of the polaron.

In what follows I present the theoretical evaluation of the absorption coefficient for local centres. The second quantization formalism and the diagram technique is used. The lineshape function is discussed for the special case, which corresponds to the configurational model, when the excitation interacts with only one normal mode. The results are then extended to include the interaction with a continuum of normal modes, as it happens for centres in crystals. Finally, it is discussed how the electronic degeneracy in the excited state can be taken into account (dynamical Jahn-Teller centres).

2. THE MODEL HAMILTONIAN

In this section I present and discuss the assumptions constituting the theoretical model under which the absorption coefficient is evaluated.

a) The photon-electron interaction

As already said in the introduction, during the absorption process, an electron of a localized centre is excited and a photon is destroyed. The interaction is considered in the dipole approximation, and the final electronic excitation is then dipole allowed.

b) The electron Hamiltonian

We consider that only the excitations of one electron are involved in the light-absorption process. The level structure of the electron consists of a nondegenerate ground state and in a degenerate excited state. This model, which does not give any attention to other excited electronic states or to their mixing, may nevertheless work for a great variety of impurities in polar crystals. We indicate by E_1 the energy of the excited state; by a_i^+ the creation operator of the electron in the excited state $|i>$ (i=1,.....,n where n is the degeneracy of the level) by a_o the annihilation operator of the electron in the ground state $|g>$, whose energy is E_o, and by H_{el} the electron Hamiltonian:

$$H_{el} = E_1 \Sigma_i a_i^+ a_i + E_o a_o^+ a_o \qquad (1)$$

The electric-dipole-moment operator of the allowed transition from the ground state to the excited state is:

$$d = \Sigma_i (M_{oi} a_i^+ a_o + M_{io} a_i a_o^+) \qquad (2)$$

where M_{oi} is the transition matrix element

$$M_{oi} = \langle g| e \underset{\sim}{r} |i\rangle \qquad (3)$$

Note that the same Hamiltonian holds if we are considering a molecule instead of a localized centre in crystal.

c) The Phonon Hamiltonian

The lattice dynamics is that of a non-conducting crystal, perturbed by the presence of localised centres. As well as known, the adiabatic approximation holds also for perturbed crystals and the motion of the ions can be separated by the motion of the optic electron.(2)
Let us assume that the ions in the crystal vibrate around the equilibrium position when the electron is in the ground state and the lattice dynamics are harmonic. Labelling by λ the normal modes (both continuum and local modes) the phonon Hamiltonian is:

$$H_P = \Sigma_\lambda b_\lambda^+ b_\lambda \qquad (4)$$

where b_λ^+ , b_λ are the phonon operators and ω_λ is their frequency ($\hbar = 1$). Note that λ is a quasi-continuum index, which is identical to the quasi-momentum $\underset{\sim}{q}$ when the crystal is perfect. ω_λ belongs to a spectrum which goes from zero to a maximum value. Conversely, λ is a discrete index if we are dealing with a molecule and, in particular, $\lambda = 1$ for a diatomic molecule, which has only one normal mode. The relation between the operators b_λ^+ and b_λ and the cartesian components α of the dynamical displacements $u_{1\alpha}$ of the 1-th ion from its equilibrium position can be found for instance in Refs (7) and (8).

d) The EP interaction

When the electron is excited, leaving a hole in the ground state, the phonon Hamiltonian H_p for the excited state can be written:

$$H_p^{ex} = H_p + H_{EP} \tag{5}$$

H_{EP} is the EP interaction term induced by the excitation. It can be developed either in series of the lattice displacements $u_{1\alpha}$ or in series of the phonon operators b_λ^+ and b_λ which are related to $u_{1\alpha}$ by a linear transformation. H_{EP} is given then by

$$H_{EP} = \sum_{ij \neq 0} h_{ij}\, a_i^+ a_j \tag{6}$$

where

$$h_{ij} = \sum_\lambda h_{ij}(\lambda)\ (b_\lambda^+ + b_\lambda) + \tag{7a}$$

$$+ \tfrac{1}{2}\sum_{\lambda,\lambda'} h_{ij}(\lambda,\lambda')(b_\lambda^+ + b_\lambda)(b_{\lambda'}^+ + b_{\lambda'}) + \ldots \tag{7b}$$

Note that $h_{oo} = 0$, owing to the assumption made about the groundstate dynamics. Furthermore, $h_{oi} = 0$ because we assumed that the electronic excitations are induced only by interaction with the light and not by phonons. Eqs. (7a) and (7b) give the linear and the quadratic contributions, respectively, to the EP interaction Hamiltonian.

In what follows, we will take into account only the effect coming from the linear EP interaction. Each term in (7a) corresponds to an interaction in which the electron undergoes a transition from the state i to the state j, while one phonon λ is created or destroyed. In the many-body treatment of the absorption problem and by using the diagram technique, such an interaction gives rise, as will be explained in the next paragraph, to graphs where the interaction vertices are of the type shown in Fig.1. Here the dot represents the interaction coefficient $h_{ij}(\lambda)$, the full lines represent the electron propagators and the dotted line represents the phonon propagator. No arrow has been drawn on the phonon line, because the phonon can be either destroyed or created in the interaction. For a fixed λ, $h_{ij}(\lambda)$ are the elements of a matrix n x n, n being the degeneracy of the excited states interacting with the phonons.

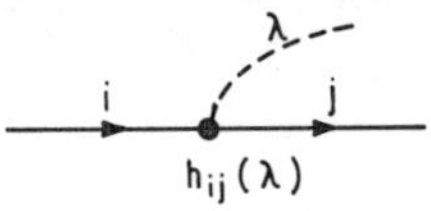

FIG.1.

e) Total Hamiltonian

The total hamiltonian of the electron-phonon interacting system is therefore the following:

$$H = H_{el} + H_P + H_{EP} = H_o + H_{EP} \tag{8}$$

The ground state of the system, indicated by $|0\rangle$, corresponds to the electron in its ground state $|g\rangle$ at thermal equilibrium with the surrounding lattice. In what follows, the thermal averaging operation on the ground state phonon thermal population will be indicated by $\langle\cdots\rangle_\beta$ where $\beta = 1/KT$.

3. THE RESPONSE FUNCTION

The absorption coefficient $I(\omega)$ of light of frequency ω is related to the line-shape function $R(\omega)$ by the relation

$$I(\omega) = (4\pi^2\omega/3\hbar cn)\ R(\omega) \tag{9}$$

where n is the refraction index at the frequency ω, c is the speed of light. $R(\omega)$ is the Fourier transform of the trace of the dipole-moment autocorrelation function R(t) which in the present scheme has the form:

$$R(\omega) = \frac{1}{2\pi}\int_{-\infty}^{+\infty} dt\, R(t)\, e^{i\omega t} \tag{10}$$

$$R(t) = \mathrm{Tr}\,\langle 0|\, T\{d^{+}(t)\, d(0)\}\,|0\rangle \tag{11a}$$

$$= \sum_i M_{oi}\, \langle 0|\, T\{a_i(t)a_o^{+}(t)a_o(0)a_i^{+}(0)\}\,|0\rangle_\beta\, M_{io} \tag{11b}$$

where d and M_{oi} are given in Eqs (2) and (3) and the operators a and a^{+} are in the Heisenberg representation

$$a(t) = e^{iHt}\, a\, e^{-iHt}$$

H is the total Hamiltonian of Eq. (8); T{• • •} is the time ordering operator. Note that the expression in brackets in Eq. (11b) is actually an autocorrelation function of the hole (operators a_o and a_o^{+}) -electron pair.

I remember that the hole in the ground state is assumed to be structureless and interacting neither with the electron nor with the phonons. Therefore it contributes with a pure phase factor $e^{iE_o t}$ to R(t). Furthermore, because all the energy will be measured from E_o, from now on I'll put $E_o = 0$.

One obtains for R(t) of Eq. (11b):

$$R(t) = \sum_i |M_{oi}|^2\, \langle 0|\, T\{a_i(t)\, a_i^{+}(0)\}\,|0\rangle \tag{11c}$$

$$= \sum_i |M_{oi}|^2\, r_i(t) \tag{11d}$$

So far I have used the linear response theory in second quantization language. For completeness I have to remember here also the semiclassical expression of R(ω), which is that usually employed in studying the vibrational shape of the absorption bands of molecules and localized centres in crystals. The absorption coefficient for a dipole allowed transition is proportional to the thermal average of the squared dipole moment of the optical transition, when the optic electron goes from the ground (electron plus vibrational) state $|g,\chi_n\rangle$ and an excited state $|1,\chi_{n'}\rangle$. $R(\omega)$ is given by:

$$R(\omega) = Av_n \sum_{n'} |\langle g,\chi_n| e\underset{\sim}{r} |1,\chi'_{n'}\rangle|^2 \delta(E_{1n'} - E_{gn} - \omega)$$

$$= \sum_{nn'} \frac{e^{-\beta\omega_n}}{Z} |M_{01}|^2 |\langle\chi_n|\chi'_{n'}\rangle|^2 \delta(E_{1n'} - E_{gn} - \omega) \qquad (12)$$

Here $|\chi_n\rangle$ and $|\chi'_{n'}\rangle$ are the adiabatic vibrational wave functions for the ground and the excited state respectively. ω_n is the n-th normal mode frequency for the ground state. E_{gn} and E_{in} are the electronic plus vibrational energy for the two states respectively. The shape of the band is determined by the overlap integral $\langle\chi_n|\chi'_{n'}\rangle$ weighted by the energy-conservation delta function. This integral can be evaluated in an analytical form[9], but the rather involved result does not allow for an easy interpretation of the elementary processes involved, mainly when several normal modes are involved. Furthermore, it is very hard, if not impossible, to handle it when the effect of a possible degeneracy of the electronic levels are important, as for the Jahn-Teller impurities. In what follows, the many-body analysis of $R(\omega)$ for a simple case (Sect. 4) leads to a result which can be easily extended to take into account local centres either interacting with a quasi-continuum of normal modes (Sect.5) or having degenerate electronic states, involved in the optic excitation (Sect.6).

4. NON-DEGENERATE TWO-LEVEL ELECTRON IN INTERACTION WITH A SINGLE NORMAL MODE

I will discuss first the most simple case of a localised electron, with two non-degenerate levels and interacting with only one normal mode. This is actually the case of a diatomic molecule whose possible degeneration can be neglected, but it has been widely adopted under the name of configurational coordinate model to describe the many-phonon processes of localized centres in crystals. Let ω be the normal mode frequency. H_{EP} becomes:

$$H_{EP} = h_{11}\, a_1^+ a_1 = \gamma\,(b+b^+)\, a_1^+ a_1 \qquad (13)$$

where γ is the EP coupling constant.

Let us now go from the Heisenberg to the interaction representation, in which the unperturbed Hamiltonian is H_o of Eq.(8). One has (see for instance Ref.10)

$$r_1(t)=\langle\tilde{a}_1(t)\tilde{a}_1^+(0)T\{\exp[-i\int_0^t dsh_{11}(s)a_1^+(s)a_1(s)]\}\rangle_{\beta,conn} \tag{14}$$

where

$$a_1(s) = e^{iH_o s}\, a_1\, e^{-iH_o s} \tag{15}$$

s is the time inside the interval 0-t at which the interaction $h_{11}(s)$ takes place and a phonon is either absorbed or destroyed. As was said in the introduction, $h_{11}(s)$ depends on the time s only because it acts during the interval 0-t. $\langle\cdots\cdots\rangle_{conn}$ indicates that only the terms corresponding to the diagrams with connected electron lines have to be considered. The connected diagrams of lowest order, contributing to the expression (13), can be then obtained by interpreting with the usual rules the devolpment of Eq. (13) as follows:

$$\begin{aligned} R(t) &= \langle\tilde{a}_1(t)\tilde{a}_1^+(0)\rangle \qquad \text{(zero-order term)} \\ &+ (i\gamma)^2\langle\tilde{a}_1(t)\int_0^t ds\int_0^s ds'\tilde{a}_1^+(s')\tilde{a}_1(s')[b(s')+b^+(s')] \\ &\quad [b(s)+b^+(s)]\tilde{a}^+(s)\tilde{a}(s)\tilde{a}(0)\rangle_\beta \qquad \text{(1-st-order term)} \\ &+ \cdots\cdots \qquad \text{(second and higher order terms)} \end{aligned} \tag{16}$$

This expression can be written as the sum of graphs of increasing order, as is shown in Fig. 2.

Note that the phonon (dotted line in Fig. 2) created or absorbed by the EP interaction (dot in Fig.2) exists only inside the time interval 0-t: these are not phonons from the outside which go in or out. Furthermore, the electron (full lines) propagates from 0 to t without changing structure, because by assumption its structure consists of one single level. It can be then evaluated at once, being a simple phase factor $\exp(-iE_1t)$. In the graph language it means that the graphs in Fig. 2 can be redrawn as in Fig. 3, where $o < s''' < s'' < s' < s < t$. As one can see from Fig. 3, the EP interaction has the same effect of an external field,(10) which is able to create or destroy independent phonons, so each graph in curly brackets is formed by a certain number of unlinked one-phonon propagators. I remember that

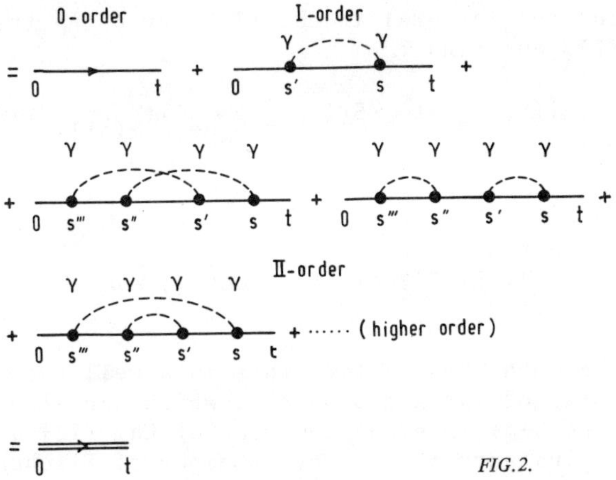

FIG.2.

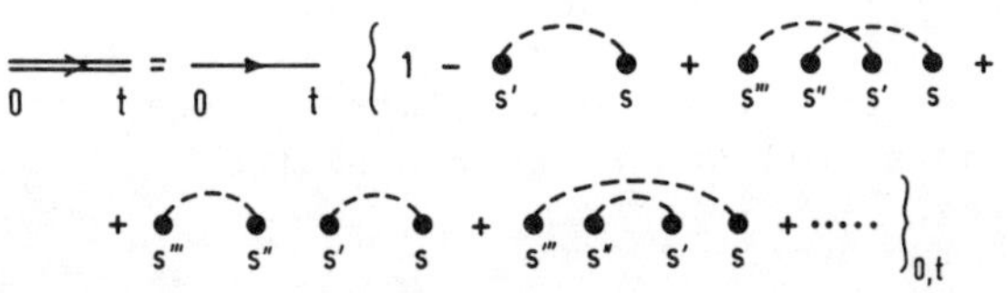

FIG.3.

in Eq. (14) and (15) the thermal average $\langle\cdots\rangle_\beta$ works only on the phonons, because the electron energies are too high. It follows that each phonon line corresponds to an independent thermal average given by the expression:

$$D(t,\bar{\omega}) = \langle [b^+(t)+b(t)]\,[b^+(0)+b(0)]\rangle_\beta$$

$$= \bar{n}\, e^{i\bar{\omega}|t|} + (\bar{n}+1)\, e^{-i\bar{\omega}|t|} \qquad (17)$$

where

$$\bar{n} = (\exp\beta\bar{\omega} - 1)^{-1} \qquad (18)$$

The "linked cluster theorem", usually applied to evaluate the vacuum fluctuations, can then be applied. The theorem states that the unlinked graphs in Fig. 3 form an exponential series. The argument of the exponential is the first-order term in Eq. (16) or in Fig. 3.[11] One has

$$r_1(t) = [0 \longrightarrow t]\ \exp - [\overset{s'}{\bullet} - - - - \overset{s}{\bullet}] \qquad (19a)$$

$$= e^{-iE_1 t} \exp\left\{-\gamma^2 \int_0^t ds \int_0^s ds'\ D(s-s';\bar{\omega})\right\} \qquad (19b)$$

By performing the integral in Eq. (19b) by using the relation (17), one finds:

$$r_1(t) = e^{-i(E_1 - S_o\bar{\omega})t}\, e^{-S(T)} \exp\left\{S_o\left[\bar{n}e^{i\bar{\omega}t} + (\bar{n}+1)e^{-i\bar{\omega}t}\right]\right\} \quad (20)$$

where $S_o = (\gamma/\bar{\omega})^2$ and

$$S(T) = (\gamma/\bar{\omega})^2(2\bar{n}+1) = S_o \coth(\tfrac{1}{2}\beta\bar{\omega}) \quad (21)$$

is called the Huang-Rys factor. This is a well known result since the works of Lax and O'Rourke, which can also be deduced by the semiclassical approach of Eq. (12) and can be found for instance in the review paper of Fitchen, ref.(4). Here I shall discuss only the main features of $r_1(\omega)$, the Fourier transform of $r_1(t)$ in Eq.(20).

The possible singularities of $r_1(\omega)$ are related to the $t\to\infty$ behavior of $r_1(t)$, as a well known theorem on the Fourier transform states.

In this limit ($t \gg 1/\bar{\omega}$) the oscillating factors in the exponents inside the curly brackets in Eq. (20) go to zero, so that

$$\lim_{t\to\infty} r_1(t) = e^{-i(E_1-\gamma^2/\bar{\omega})t}\, e^{-S(T)} \quad (22)$$

and

$$r_1(\omega) = \frac{1}{2\pi}\, e^{-S(T)}\, \delta(\omega - E_1 + S_o\bar{\omega}) \quad (23)$$

$r_1(\omega)$ shows a δ-like singularity at the energy

$$\omega = E_1 - S_o\bar{\omega} = \Omega \quad (24)$$

called zero-phonon energy. I remember that E_1 is a pure electronic energy (see the Hamiltonian of Eq.(1)) which corresponds to the excitation energy of the electron-hole pair when the ions are not allowed to vibrate. Therefore, in absence of vibrations the absorption coefficient should be a δ-like function centred at $\omega = E_1$. By letting the ions move and interact with the electron, the singularity shifts to the lower energy Ω, and the difference is equal to $S_o\bar{\omega}$.

We can then say that E_1 is a static energy while Ω is a relaxed energy. However, the δ-like peak at Ω can be seen only when S_o is small ($S_o \lesssim 4$) and at low temperatures ($\bar{n}\sim 0$)

because its intensity is ruled out by the exp(-S(T)) factor in front of the delta (Eq.(23)): the stronger is the coupling constant γ, the greater is $S_o = (\gamma/\bar{\omega})^2$ and the lower is the zero-phonon intensity.

To what characteristics of the spectrum is E_1 now related? It is the average energy of the absorption band, as one can see using a well-known relation between the moments of a distribution $r(\omega)$ and the derivative of its Fourier transform $r(t)$, which states that

$$\int \omega^K r(\omega)\, d\omega = i^K \left.\frac{d^K}{dt^K} r(t)\right|_0 \tag{25}$$

For the first moment (K = 1) one finds

$$\int \omega\, r(\omega)\, d\omega = i \left.\frac{dr(t)}{dt}\right|_o = E_1 \tag{26}$$

Note that $r(\omega)$ is normalised to unity, because when K=0 in Eq. (25), one finds

$$\int r(\omega)\, d\omega = 1$$

By recalling Eqs. (11d) and (9), this means that while the band shape is determined by the strength of the EP interaction, its integrated intensity is ruled only by the dipole moment of the transition.

Let us now see what is the shape of the band which is described by Eq. (20).

By developing in series the exponential with oscillating factors in Eq. (20), one has

$$\begin{aligned} r_1(t) &= e^{-S(T)}\, e^{-i\Omega t} \sum_k \frac{1}{k!} S_o^k \left[\bar{n} e^{i\bar{\omega} t} + (\bar{n}+1) e^{-i\bar{\omega} t}\right]^k \\ &= e^{-S(T)}\, e^{-i\Omega t} \left\{1 + S_o\left[\bar{n}\, e^{i\bar{\omega} t} + (\bar{n}+1)\, e^{-i\bar{\omega} t}\right] + \dots\right\} \end{aligned} \tag{27}$$

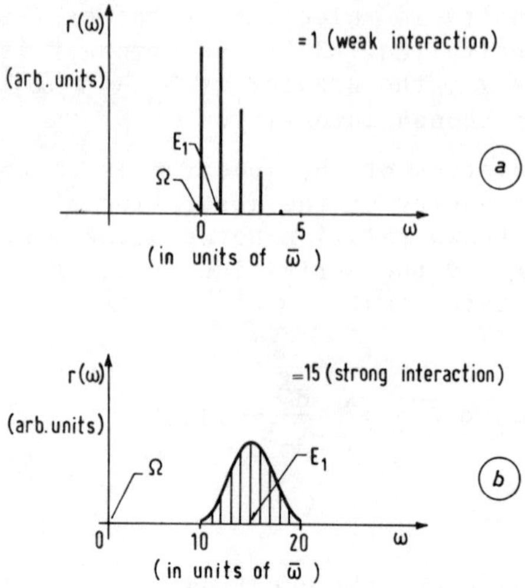

FIG.4. Schematic absorption coefficient at zero temperature ($\bar{n} = 0$) for a centre with one normal mode (frequency $\bar{\omega}$) interacting with the electron for: (a) weak coupling ($S_0 = 1$); (b) strong coupling ($S_0 = 15$). Ω is the zero-phonon energy and E_1 is the baricentre of the band.

Then the Fourier transform of $r_1(t)$ is

$$r(\omega) = \frac{e^{S(T)}}{2\pi} \Big\{ \underbrace{\delta(\omega - \Omega)}_{\text{zero-phonon}} + \underbrace{S_0 \bar{n}\, \delta(\omega - \Omega + \bar{\omega})}_{\text{1-phonon (absorbed)}}$$

$$+ \underbrace{S_0 (\bar{n}+1)\, \delta(\omega - \Omega - \bar{\omega})}_{\text{1-phonon (emitted)}} + \ldots \Big\} \qquad (28)$$

The spectrum is composed of δ - lines at distance $\bar{\omega}$. At T=0°K the intensity of the k-th line, corresponding to the creation of k phonons, is $e^{-S_0}(S_0^{\,k}/k!)$, which has its maximum value when $k \sim S_0$. By comparing the graphs of Fig. 2 with Eq. (28), it follows that the most probable process is represented by the graph with S_0 phonons. S_0 has, therefore, the meaning of the most probable number of phonons involved in the many-phonon process associated with the electronic excitation.

In Fig. 4, $r_1(\omega)$ is reported for two extreme cases:

i) S_0=1, weak EP interaction;

ii) S_0=15, strong EP interaction.

In the first case (Fig. 4a) the lines are few and form an asymmetric spectrum, whose lowest energy line is the zero-phonon line. In the second case (Fig. 4b) several lines appear in the spectrum whose envelop has a fairly gaussian shape, whose maximum corresponds to E_1 and the zero-phonon line has zero intensity ($e^{-15} = 3.06\ 10^{-7}$).

Note that, since the two spectra in Fig. 4 have the same integrated intensity, the smaller S_o is the more intense are the lines corresponding to a specific process (with one phonon, with two phonons, etc.). The spectra reported in Fig.4 are typical of the diatomic molecules (see the book of Herzberg) where the anharmonicity can be neglected. Their normal mode frequency is usually of the order of 0.1 eV, so that on one hand the lines are well separated and measurable, but on the other hand $n \sim 0$ up to very high temperatures, so that the temperature effects are difficult to be detected.

In the $S_o \gg 1$ case, it could be interesting to evaluate the half width W (i.e. the width at the half height) of the envelop curve in Fig. 4b. W is related to the second moment of the distribution, which in turn is related to the $t \to 0$ limit of the second derivative of $r_1(t)$. We evaluate then $r_1(t \to 0)$ from Eq.(20) and develop the argument of the exponent up to t^2. One has

$$\lim_{t\to 0} r_1(t) = \exp\Big\{-S_o(2\bar{n}+1) - i(E_1 - S_o\bar{\omega})t + S_o\bar{n}(1 + i\bar{\omega}t - \tfrac{1}{2}\bar{\omega}^2 t^2 + \ldots)$$
$$+ S_o(\bar{n}+1)\ (1 - i\bar{\omega}t - \tfrac{1}{2}\bar{\omega}^2 t^2 + \ldots)\Big\} =$$

$$= \exp\left\{-iE_1 t - \tfrac{1}{2}S_o\ (2\bar{n}+1)\bar{\omega}^2\ t^2\right\} \qquad (29)$$

Its Fourier transform gives

$$r_1(\omega) = \left[2\pi S(T)\bar{\omega}^2\right]^{\frac{1}{2}} \exp - \left[(\omega - E_1)^2 / 2\bar{\omega}^2 S(T)\right] \qquad (30)$$

This is the envelope function which is a gaussian function centred at $\omega = E_1$ (see Fig.4b) and whose halfwidth W is related to the variance as follows:

$$W = 4\ \ln 2 \left[2\bar{\omega}^2\ S(T)\right] = 8\ \ln 2\ S_o\,\bar{\omega}^2\ \coth\tfrac{1}{2}\beta\bar{\omega} \qquad (31)$$

5. NON-DEGENERATE TWO-LEVEL ELECTRON INTERACTING WITH SEVERAL NORMAL MODES

The model can be applied to localised excitations in molecules or in crystals. The theoretical framework used in the previous section can still be straightforwardly applied (harmonic dynamics, two-level electron, linear EP interaction), the only difference being that in the present case there are phonons involved with different frequency ω_λ and different symmetry. Figs. 2 and 3 still hold, because on one hand the linear EP interaction H_{EP} of Eq. (7a) is linear as before and, when put in Eq (14), gives rise to processes where only one phonon is involved at the same time. On the other hand, the electronic excited state is still assumed structureless and localised, so that the hole-electron propagator can be taken in front of the series (see Eq. 19a) and the "linked graph theorem", used to sum the series, holds again.

However the dotted lines (the phonon propagators) must be now labelled because they can correspond to different phonons. The phonon-propagator will be indicated by $D(s,\omega_\lambda)$, where λ runs from 1 to N, where N is the number of normal modes of the system.

One finds (see Eq. (19b):

$$r_1(t) = \langle {}^{0}\!\!\longrightarrow^{t} \rangle \exp\left\{ -\sum_\lambda s' \overset{\lambda}{\text{- - - - - -}} s \right\}$$

$$= e^{-iE_1 t} \exp\left\{ -\sum_\lambda (\gamma_\lambda/\omega_\lambda)^2 \int_0^t ds \int_0^s ds' \, D(s-s',\omega_\lambda) \right\} \quad (32)$$

where $D(t,\omega_\lambda)$ is given by Eq.(17) and γ_λ is identical with $h_{11}(\lambda)$ in Eq.(7b).

Let us now discuss in detail Eq.(32) for impurities in crystals. In such a case the index λ labels both the continuum and the local modes of a perturbed lattice, so that the sum over λ in Eq. (32) is better replaced by an integration as follows

$$\sum_\lambda \longrightarrow \int d\omega \left[\sum_\lambda \delta(\omega - \omega_\lambda) \right] \quad (33)$$

In Eq.(32) we have two quantities: the first $D(t,\omega_\lambda)/\omega_\lambda$ is a bulk quantity, because it is related to the frequencies

ω_λ of the whole crystal, when the electron is in its ground state (Eq.4). The second γ_λ is a local quantity, because it is the λ-th component of the perturbation switched by the electronic excitation in the perturbation space. Such a local perturbation is then better classified on the basis of the local symmetries of the centre, instead of on the basis of the crystal normal modes [7,12] If by α we indicate the irreducible representations (irr. rep.) of the point group at the impurity site, γ_α will be the α -symmetry component of the EP interaction at such a point. The relation between γ_α and γ_λ is a simple linear combination:

$$\gamma_\alpha = \Sigma_\lambda \, c_{\alpha\lambda} \, \gamma_\lambda \tag{34}$$

where $c_{\alpha\lambda}$ are projection coefficients. The sum in Eq.(32) can be then written using Eqs (33) and (34) as follows:

$$\begin{aligned} \Sigma_\lambda(\gamma_\lambda/\omega_\lambda)^2 \, D(t,\omega_\lambda) &= \\ &= \sum_\alpha \gamma_\alpha^2 \int d\omega \; \omega^{-2} D(t,\omega) \, \Sigma_\lambda |c_{\alpha\lambda}|^2 \, \delta(\omega-\omega_\lambda) \\ &= \sum_\alpha \gamma_\alpha^2 \int d\omega \; \omega^{-2} D(t,\omega) \, \rho_\alpha(\omega) \\ &= \sum_\alpha \gamma_\alpha^2 \, D_\alpha(t) \end{aligned} \tag{35}$$

The quantity

$$\rho_\alpha(\omega) = \sum_\lambda |c_{\alpha\lambda}|^2 \, \delta(\omega-\omega_\lambda) \tag{36}$$

is usually called the α - symmetry projected density of phonon states, because it is a spectral density of the crystal phonons but is projected into the symmetry α of the perturbation space around the centre, owing to the coefficients $c_{\alpha\lambda}$ in Eq. (36). It may be evaluated by means of a computer from the host crystal dynamics (by taking into account the perturbation on the dynamical matrix introduced by the imperfection) and the point symmetries of the hole-electron excitation (see Refs 7, 12). Not all the irr. rep.'s relative to the impurity site group are involved in the sum in Eq. (35), because the symmetry β of the wave function of the electronic excitation imposes that[13]:

$$[\beta^2] \times [\alpha] \in I$$

i.e. the identity representation must be contained in the symmetrical product of the square of the irr. rep. β with α.

By putting Eq.(35) in (32) and by evaluating the integration on the time, one obtains an expression very similar to Eq.(20), where the substantial difference lies in the integration over the density of phonon states.

For instance, the Huang-Rys factor and the zero-phonon energy are now (compare them with the corresponding quantities in Eqs (21) and (24)):

$$S(T) = \Sigma_\lambda (\gamma_\lambda/\omega_\lambda)^2(2n_\lambda+1) = \sum_\lambda S_\lambda(2n_\lambda+1)$$

$$= \sum_\alpha \gamma_\alpha^2 \int d\omega\, \omega^{-2} \rho_\alpha(\omega) \coth\tfrac{1}{2}\beta\omega \tag{37}$$

$$\Omega = E_1 - \Sigma_\lambda S_\lambda \omega_\lambda = E_1 - \Sigma_\alpha \gamma_\alpha^2 \int d\omega\, \omega^{-1} \rho_\alpha(\omega) \tag{38}$$

Because the density of the phonon states appears as an integrated quantity in (37) and (38), S(T) and Ω are not sensible to the details of the lattice dynamics. On the contrary, the structures supported by the time-dependent term in $r_1(t)$ change completely in going from a single frequency model to a frequency distribution situation.

i) $S \sim 1$)

Let us consider the weak interaction limit ($S \sim 1$) and develop $r_1(t)$ as in Eq.(27). By evaluating the Fourier transform as in Eq.(28), one finds at T=0°K (n=0):

$$r(\omega) = \frac{e^{-S(T)}}{2\pi} \Big\{\delta(\omega-\Omega) + \Sigma_\lambda S_\lambda\, \delta(\omega-\Omega-\omega_\lambda)$$

$$+ \Sigma_{\lambda\lambda'} S_\lambda S_{\lambda'}\, \delta(\omega-\Omega-\omega_\lambda-\omega_{\lambda'}) +$$

$$+ \text{(three-phonon processes)} + \ldots\Big\} \tag{39a}$$

$$= \frac{1}{2\pi} e^{-S(T)} \Big\{ \delta(\omega-\Omega) + \sum_\alpha \gamma_\alpha^2 (\omega-\Omega)^{-2} \rho_\alpha(\omega-\Omega)$$

$$+ \text{(convolution of two projected densities)}$$

$$+ \ldots \Big\} \tag{39b}$$

Therefore at the low-energy side of the zero phonon the shape of the absorption spectrum is that of the one-phonon

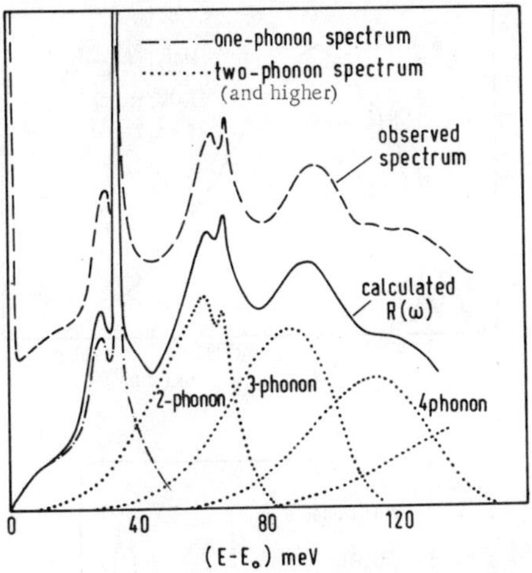

FIG.5. Experimental absorption spectrum (dashed line) of the M′ centre in LiF, showing a multi-phonon side-band at the high-energy side of the zero-phonon line. The continuum line shows the calculated absorption lineshape, while the one-phonon contribution is that shown by the _ . _ . _ line (from Ref. (5)).

projected density itself. The term $(\omega-\Omega)^{-2}$ in Eq.(39b), does not introduce divergencies because $\rho_\alpha(\omega-\Omega)$ goes to zero at least as $(\omega-\Omega)^3$. Instead of the δ-like spikes as in Fig.4, the spectrum forms a continuum due to the superposition of the one-phonon, two-phonon, etc. terms in Eq.(39b), and its structures and singularities reproduce those of the involved $\rho_\alpha(\omega)$ and of their convolutions. In Fig.5 the spectra corresponding to two, three and four phonon convolution processes are shown. Note that the higher the order of the convolution, the smoother the corresponding spectrum. Therefore only when S is small has the absorption spectrum got spikes and narrow structures. The relation between the integrated intensities of different order is the same as in the previous section: the spectrum of the S-th order still has the greatest intensity.

Fig.6b shows the result of an inverse process, i.e. the numerical deconvolution of an experimental spectrum (Fig.6a) from which it was possible to deduce the one-phonon spectral density (the quantity $\sum \gamma_\alpha^2(\omega-\Omega)^{-2}\rho_\alpha(\omega-\Omega)$ in Eq.(39b)) from the total spectrum.

An interesting lineshape is shown by those centres that induce local modes in the α-perturbed dynamics (δ-like lines in the $\rho_\alpha(\omega)$ spectra) and by substitutional molecules or impurity aggregates, whose internal modes transform like the α - symmetry irr. rep. and have frequencies in the continuum of $\rho_\alpha(\omega)$. In that case the δ-spikes such as those of Fig.4 are superimposed to a continuum spectrum.(14)

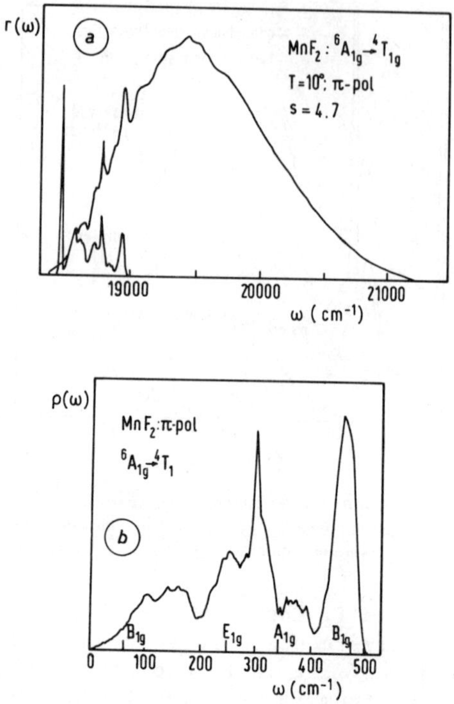

FIG.6. (a) The experimental absorption coefficient for the transition $^6A_{1g} \rightarrow {}^4T_{1g}$ in MnF_2 (from Ref. (17). (b) The one-phonon part (Eq. (39b)) of $r(\omega)$ deduced from the experimental spectrum of Fig. 6a; the $\alpha = A_{1g}$, B_{1g} and E_{1g}-symmetry contributions are shown.

ii) $S \gg 1$

At the other limit of strong interactions ($S \gg 1$), or at $t \rightarrow 0$ limit of $r_1(t)$, Eqs. (29), (30) and (31) still hold. However in this case $r_1(\omega)$ of Eq. (30) does not describe only the envelope function of the spectrum, but gives the true lineshape. In fact the main contribution to the band comes from the high-order (S-order) phonon processes, which correspond to the S-th order convolution processes of the $\rho_\alpha(\omega)$'s (see Eq. (39b)). Because the result of a convolution process between the $\rho_\alpha(\omega)$ is a smoother function that the $\rho_\alpha(\omega)$ themselves, the lineshape function turns out to be a structureless function, when $S \gg 1$.

See in Fig.5 how the three-phonon process contribution to $r_1(\omega)$ is a smoother function than that coming from the first order process (the $\rho_\alpha(\omega)$ itself).
Such a band is therefore well described by the short time limit of $r_1(t)$; the higher the power in t considered in Eq. (29), the higher the order of the moment of the band one must take into account in $r_1(t)$.

Therefore, in a first approximation the band shows a gaussian shape, the possible asymmetry coming from the third moment of $r_1(\omega)$, i.e. the t^3 term in the argument of the exponential of Eq.(29).

6. ELECTRON WITH A DEGENERATE LEVEL INTERACTING WITH SEVERAL NORMAL MODES

This situation occurs when the defect is in a site of high symmetry in the static lattice, so that its energy levels can be degenerate. Such a degeneracy is partially removed by the vibrations via the Jahn-Teller effect, which is due to the interaction of the electron with the phonons.(15)

Here I consider the case when only the excited state is degenerate. Therefore the Jahn-Teller interaction is exactly the same EP interaction responsible for the shape of the band, which has been discussed in the previous sections. Furthermore the stronger the EP coupling the greater the action of the phonons in removing the degeneracy. In the experimental absorption spectrum the effect of the Jahn-Teller effect can be detected on broad bands (strong EP interaction), which show large shoulders or splittings, (in two or three sub-bands). Such structures have been interpreted in the configurational coordinate model as due to dynamical splitting of the adiabatic potentials in the excited state, induced by local vibrations whose symmetry does not have the total point symmetry of the defect.(15)

Such a theory, while able to justify why the Jahn-Teller bands are structured, does not give results which agree completely with the experimental spectra. It does not explain for instance the asymmetry of the sub-bands, their asymmetric dependence on the temperature and for some centres, the division in two instead of three sub-bands. In order to justify such features, a more careful analysis of the EP interaction on the Jahn-Teller absorption bands must be performed. The method of treating the EP interaction used in the previous sections seems to afford the required framework.(16)

Here I explain only the main line of the analysis, because it can be explained as exactly the same process (a many-phonon process) responsible for the line-shape of the bands (see Fig.3), once the degeneracy of the level has been taken into account in the evaluation of its interaction (Fig.2) with the phonons.(16)

Of course the electron has now got an internal structure, due to the degeneracy, and the interaction vertex $h_{ij}(\lambda)$ is no more a simple coefficient but must be represented by a matrix, whose dimensions are the order of the degeneracy. At each interaction the electron scatters from the state i to the state j of the degenerate level. The high order graphs of Fig.3 can be reduced to a power of a fundamental graph as before, only if the time-ordering (indicated below by $T\{..\}$) of the many phonon processes is taken into account. The final expression, which is equivalent to that in Eq.(32) with Eq.(35) is then:

$$r_1(t) = \left({}^{0}\!\!\longrightarrow\!\!{}^{t} \right) T\left\{\exp -\Sigma_\lambda \; {}^{s'}\!\bullet\!\overset{}{\underset{\lambda}{- - - - - - -}}\!\bullet^{s} \right\}$$

$$= e^{-iE_1 t}\; T\left\{\exp -\Sigma_\alpha \int_0^t ds \int_0^s ds' \; h_\alpha(s)\; D_\alpha(s-s') h_\alpha(s')\right\}_{11} \qquad (40)$$

where $h_\alpha(s)$ means that the EP interaction of α-symmetry acts at the time s.
(The matrix indices in $h_\alpha(s)$ have been omitted for sake of brevity).

Eq. (40) differs from Eq. (32) essentially for two reasons: the former is that it is an exponential, whose argument is a quadratic form of the matrices $h_\alpha(s)$ (i.e. it is a gaussian functional); the latter is that the matrices $h_\alpha(s)$ must be ordered in time. Therefore the Fourier transform of Eq. (40) is not an easy task to perform.

But it can be done in the short-time limit (strong EP interaction), because in that case the time-ordering operator does not effect the evaluation of the exponential in Eq. (40).

The more interesting results of such treatment are that on one hand, at the lowest order of approximation (semiclassical approximation, i.e. when the matrices are treated as parameters) Eq.(40) gives the same result as the configurational coordinate model. On the other hand, the other characteristics of the band, such as the temperature dependence of the intensity and of the splitting can be found in the further order of approximation.[16]

REFERENCES

(1) BORN, M., HUANG, K., in Dynamical Theory of Crystal Lattices, Oxford Press, London (1954). A review paper of the adiabatic approximation applied to local centres can be found in PERLIN, Y., Sov. Phys.-Usp. 6 (1964) 542 and MARADUDIN, A.A., in Solid State Physics (SEITZ, F., TURNBULL, D., Eds), vol. 18, Academic Press, New York (1966).

(2) WALLIS, R.F., these Proceedings, Vol. 1, Paper IAEA-SMR-20/8.

(3) HERZBERG, G., Molecular spectra and Molecular Structure, Van Nostrand, New York (1966).
(4) CHIAROTTI, G., lectures on the optical properties of solids given at the Trieste Winter College on Interaction of Radiation with Condensed Matter, January–March 1976 (not included in the present Proceedings).
(5) FITCHEN, D.B., in Physics of Color Centres (FOWLER, W.B., Ed.), Academic Press, New York (1968).
(6) For a general review of the excitons see: KNOX, R.S., Theory of excitons, Suppl. 5 of Solid State Physics (SEITZ, F., TURNBULL, D., Eds), Academic Press, New York (1963).
(7) See MARADUDIN, A.A., in Solid State Physics (SEITZ, F., TURNBULL, D., Eds), Vol. 19, Academic Press, New York (1966).
(8) KLEIN, M.V., in same book as Ref. (5) above.
(9) LAX, M., J. Chem. Phys. **20** (1952) 1752; O'ROURKE, R.C., Phys. Rev. **91** (1953) 265.
(10) ABRIKOSOV, A.A., GORKOV, L.P., DZYALOSHINSKI, I.E., Methods of Quantum Field Theory in Statistical Physics, Prentice Hall, Englewood Cliff (1963).
(11) RAIMES, S., Many-electron Theory, North-Holland, Amsterdam (1972). See also NOZIERES, P., de DOMINICIS, C.T., Phys. Rev. **178**, (1969) 1097.
(12) See Ref. (8). See also MULAZZI, E., NARDELLI, G.F., TERZI, N., Phys. Rev. **172** (1968) 847.
(13) See for instance HEINE, W., Group Theory in Quantum Mechanics, Pergamon Press, London (1960).
(14) See for instance the spectra of Sm^{2+} in alkali halides: WAGNER, M., BRON, W.E., Phys. Rev. **139** (1965) A223; BRON, W.E., Phys. Rev. **140** (1965) A2005.
(15) For a general account of the Jahn-Teller Effect see the recent book of ENGLMAN, R., The Jahn-Teller Effect in Molecules and crystals, Wiley, London (1972). For the interpretation of the dynamic Jahn-Teller structure in the configurational coordinate model see TOYOZAWA, Y., INOUE, M., J. Phys. Soc. Japan **21** (1966) 1663.
(16) MULAZZI, E., TERZI, N., Phys. Rev. B **10** (1974) 3552.
(17) RODER, U., et al., Solid State Commun. **9** (1971) 733.

IAEA-SMR-20/14

MAGNONS

S.M. REZENDE
Departamento de Fisica,
Universidade Federal de Pernambuco,
Recife,
Brazil

Abstract

MAGNONS.

This course is devoted to the study of the properties of spin waves, or magnons, in magnetic insulators. The study is directed to simple systems in which the physical concepts can be understood without extensive calculations. Representative experimental results are presented and compared with the theoretical interpretation. The course is divided into six main sections, namely: 1. Spin waves; 2. Statistical and thermodynamic properties of magnons; 3. Magnon interactions; 4. Anti-ferromagnetic magnons; 5. Mixed excitations (magnetoelastic waves and nuclear magnons); 6. Magnons in disordered systems.

1. SPIN WAVES

A. Introduction: Spin waves, or magnons, are the elementary excitations of coupled spin systems in condensed matter. The necessary condition for the existence of magnons is that the spins be interacting. Thus magnons are found in a variety of magnetic systems, such as ferromagnetic, ferrimagnetic and antiferromagnetic insulators [1,2], semiconductors or metals, in crystalline or amorphous form. The term magnon is usually employed to characterize excitations of electronic spins, but it can also be used for collective nuclear spin excitations. The frequency or energy of these excitations depends on the detailed form of the interactions among the spins in each material and on the spatial distribution of the spin disturbances. Typically the frequency falls in the range of a few hundred MHz ($\sim 10^9 sec^{-1}$) to a few hundred cm^{-1} ($\sim 10^{13} sec^{-1}$).

Since the early work of Bloch [3], spin waves have been extensively studied both theoretically and experimentally [2]. The interest in them stems from several factors. First, like other excitations in matter, they play an important role in determining the thermodynamic properties of a system. Hence, on one hand the knowledge of the properties of magnons at temperatures where they are not too densely populated allows one to predict the thermodynamic properties. On the other hand, measurements of thermodynamic quantities yield information on the properties of magnons. Second, spin waves are driven by external stimuli of different kinds such as electromagnetic radiation. Thus, they have a direct influence on the dynamic response of a magnetic system. Magnons have been excited and detected in a large number of materials under different circumstances by several techniques, such as radio-frequency, microwave, infrared and optical spectroscopy, and inelastic light and neutron scattering. Their experimental excitation serves not only to give better understanding of the collective vibrations of matter but also to yield measurements of material parameters. In addition, since magnons can be generated at a given time in a region of a sample and due to propagation can be detected at a later time at a different point, they are potentially useful for technological applications in signal processing. What makes them most attractive in these applications is the fact that their properties can be electrically adjusted by means of an external applied magnetic field, which can easily be produced by a current through a coil.

In this course we shall study the basic properties of magnons and their role on the thermodynamics of a system. Due to limitations we shall restrict ourselves to simple ferro- and antiferromagnetic insulators.

B. Ferromagnetic Magnons: In this course we shall take the Heisenberg model as the ordering spin interaction. In addition we shall always assume that the system is subjected to an external magnetic field, so that the Zeeman interaction is also present. Assume a ferromagnetic system with N spins S, at lattice sites i, interacting with an exchange constant J (>0) which depends only on the difference (i-j), i.e., on the distance between spins. The external field is taken to lie along the z-axis of a cartesian coordinate system, which becomes the direction of the quantization of the spins. The Hamiltonian of the system is then [2],

$$\begin{aligned}\mathcal{H} = \mathcal{H}_z + \mathcal{H}_{exc} &= -g\mu_B \sum_i H_0 S_i^z - \sum_{i\neq j} J_{ij}\, \vec{S}_i\cdot\vec{S}_j \\ &= -g\mu_B H_0 \sum_i S_i^z - \sum_{i\neq j} J_{ij}[S_i^z S_j^z + \tfrac{1}{2}(S_i^- S_j^+ + S_i^- S_j^+)]\end{aligned} \tag{1.1}$$

where H_0 is the applied field, g is the spectroscopy g-factor, μ_B is the Bohr magneton, and S^+ and S^- are the raising and lowering operators. The Zeeman term is also commonly written $\gamma\hbar$ in the place of $g\mu_B$. Here $\gamma = g\mu_B/\hbar$ is the electron gyromagnetic ratio. It can be easily seen that the ground state of a system described by (1.1) has all the spins pointing in the z-direction, i.e., $|0\rangle = |S_1^z = S, \ldots S_N^z = S\rangle$. The low lying energy excitations above the ground state are formed by spin deviations which propagate the spin waves. These collective excitations are quantized in energy. A quantum of spin wave is called a magnon. In the study of collective excitations of the spin system the most commonly used technique is the so-called second quantization. The idea is to look for canonical transformations which cast the Hamiltonian in the form [4]

$$\mathcal{H} = E_0 + \sum_k \hbar\omega_k a_k^\dagger a_k + \mathcal{H}^{int} \tag{1.2}$$

i.e., the sum of a ground state energy, a term which represents a collection of independent harmonic-oscillator type of excitations and terms which represent the interactions among these excitations. Provided that the relaxation times due to the interactions are much longer than the periods of oscillation, the concept of excitation is meaningful and the form (1.2) is very useful. The statistics of the excitations is important in determining the properties of the spin wave operators. Magnons are found to behave as bosons. Therefore the operators we want must obey the boson commutation relations

$$[a_k, a_{k'}^\dagger] = \delta_{kk'} \quad , \quad [a_k, a_{k'}] = [a_k^\dagger, a_{k'}^\dagger] = 0 \tag{1.3}$$

The starting point to find the spin wave operators is the first Holstein-Primakoff transformation [5]. It introduces the localized spin-deviation operators. The operator for the number of spin deviations at site i is

$$n_i = S - S_i^z \tag{1.4}$$

Clearly the eigenstates of this operator are the states $|S_i^z\rangle$. The spin-deviation creation operator $a_i^\dagger$ is the boson operator which creates a quantum of spin-deviation; i.e., which reduces S_i^z by one unit. Analogously the spin-deviation annihilation operator a_i increases S_i^z by one unit. Denoting by $|n_i\rangle \equiv |S_i^z\rangle$ the eigenstates of n_i, we require that $a_i^\dagger$ and a_i have the harmonic oscillator properties

$$a_i^\dagger |n_i\rangle = (n_i + 1)^{1/2} |n_i + 1\rangle \tag{1.5a}$$

$$a_i |n_i\rangle = (n_i)^{1/2} |n_i - 1\rangle \tag{1.5b}$$

It can be shown that the transformations that satisfy (1.4) - (1.6) and the spin commutation rules are [2,5,6,8]

$$S_i^+ = (2S)^{1/2} \left(1 - \frac{a_i^\dagger a_i}{2S}\right)^{1/2} a_i \tag{1.6a}$$

$$S_i^- = (2S)^{1/2} a_i^\dagger \left(1 - \frac{a_i^\dagger a_i}{2S}\right)^{1/2} \tag{1.6b}$$

$$S_i^z = S - a_i^\dagger a_i = S - n_i \tag{1.6c}$$

For these transformations to be useful we expand the operator in the square root with the usual binomial expansion. Clearly we want to approximate the expansion by the lowest order terms. This is valid only if $\langle n_i\rangle /2S \ll 1$, i.e., if the expectation value of the number of deviations in the spin S_i is much smaller than the maximum possible value, which is 2S. Nevertheless, the change from the operators S_i to the boson operators $a_i^\dagger$ and a_i introduces an error because the former operate in a finite space and the latter operate in an infinite metric space proper for bosons. This is the kinematical error first studied by Dyson [7]. This error is small when the number of magnons in the system is much less than 2SN. This condition is comfortably satisfied in the low-temperature regime $T \ll T_c$. The first terms in the expansions of (1.6) will give rise to the quadratic part of the Hamiltonian. The others will result in terms with three or more operators, which represent the interaction between the harmonic oscillator type states. Let us consider first the terms

$$S_i^+ = \sqrt{2S}\, a_i \tag{1.7a}$$

$$S_i^- = \sqrt{2S}\, a_i^\dagger \tag{1.7b}$$

$$S_i^z = S - a_i^\dagger a_i \tag{1.7c}$$

Using (1.7) in (1.1) the quadratic part of the Hamiltonian becomes of the form

$$\mathcal{H}^{(2)} = \sum_i A_{ij} a_i^\dagger a_j \tag{1.8}$$

Eq. (1.8) is still not in the diagonal form (1.2). This can be obtained with another transformation from the field operators a_i and $a_i^\dagger$ to the normal mode collective operators a_k and $a_k^\dagger$. In general this transformation is

$$a_i = \sum_k \phi_k^i\, a_k \tag{1.9a}$$

$$a_i^\dagger = \sum_k \phi_k^{i*}\, a_k \tag{1.9b}$$

In the case of translational symmetry $\phi_k^i = \exp(i\vec{k}\cdot\vec{r}_i)$ is a solution of the eigenvalue problem and (1.9) reduces to the usual Fourier transformation.

$$a_i = \frac{1}{N^{1/2}} \sum_k e^{i\vec{k}\cdot\vec{r}_i} a_k \tag{1.10a}$$

$$a_i^\dagger = \frac{1}{N^{1/2}} \sum_k e^{-i\vec{k}\cdot\vec{r}_i} a_k^\dagger \tag{1.10b}$$

Recall the completeness and orthogonality relations of the eigenfunctions ϕ_k^i

$$\sum_i e^{i(\vec{k}-\vec{k}')\cdot\vec{r}_i} = N\delta_{kk'} \tag{1.11a}$$

$$\sum_i e^{i\vec{k}\cdot(\vec{r}_i-\vec{r}_j)} = N\delta_{ij} \tag{1.11b}$$

With the transformations (1.10) in (1.8) and using (1.11) one can easily cast the quadratic part of the Hamiltonian in the form

$$\mathcal{H}^{(2)} = \sum_k \hbar\,\omega_k a_k^\dagger a_k \tag{1.12}$$

where the energy is

$$\hbar\,\omega_k = g\mu_B H_0 + 2zJS(1-\gamma_k) \tag{1.13}$$

We have assumed only nearest neighbor exchange interaction and have used the convenient parameter

$$\gamma_k = \frac{1}{2} \sum_{\vec{\delta}} e^{i\vec{k}\cdot\vec{\delta}} \tag{1.14}$$

where $\vec{\delta}$ is the vector that connects a site with any of its z nearest neighbors. In the case of a simple cubic lattice with parameter <u>a</u>

$$\gamma_k = \frac{1}{3}(\cos k_x a + \cos k_y a + \cos k_z a) \tag{1.15}$$

Fig. 1 shows a plot of the dispersion relation for $\vec{k}$ along the (100) direction in a s.c. lattice. Notice that at the edge of the zone ($k_x a = \pi$, $k_y a = k_z a = 0$), $\gamma_k = -1$. If the wavelength of the spin waves is much larger than the lattice spacing (in other words, if $ka \ll 1$), γ_k can be approximated to

$$\gamma_k = 1 - \frac{1}{z} k^2 a^2$$

and the dispersion relation can be written approximately as

$$E_k = \hbar\omega_k = g\mu_B H_0 + 2JSa^2k^2 \tag{1.16}$$

or

$$\omega_k = \gamma H_0 + Dk^2 \tag{1.17}$$

where $D = 2JSa^2/\hbar$ and $\gamma = g\mu_B/\hbar$ is the gyromagnetic ratio. For g = 2

$$\gamma = 1.759 \times 10^7\ \text{sec}^{-1}/\text{Gauss or } \gamma = 2.8\ \text{MHz/Gauss} \tag{1.18}$$

Notice that for k = 0 the frequency $\omega_0 = \gamma H_0$ contains no contribution from the exchange interaction. This is so because with k = 0 all the spin deviations are in phase, so that the relative deviation of two neighboring spins is zero. Hence, due to the form of the exchange interaction (1.1),

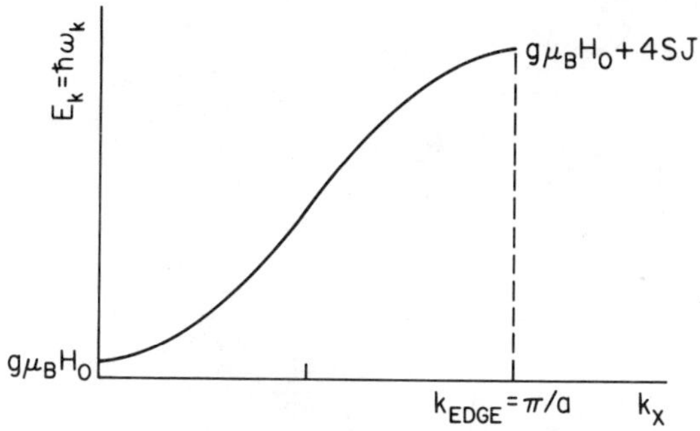

FIG.1. Ferromagnetic magnon dispersion curve.

a k = 0 excitation has the same exchange energy as the ground state. In this situation the frequency is determined by the applied field, as well as by the effective fields experienced by the spins due to other interactions in the crystal. Notice also that for a paramagnetic system, i.e., a system of uncoupled spins, $J = 0$ and $\omega = \gamma H_0$. The measurement of this paramagnetic resonance frequency yields information on the g factors of the ions and on the interaction fields. For this reason electron paramagnetic resonance (EPR) spectroscopy is a very useful tool for investigating substances. For applied fields of the order of a few kilogauss, EPR frequencies fall in the microwave range of a few GHz. The expression for the frequency of ferromagnetic resonance (FMR) (the k = 0 magnon mode) is more complex due to the effect of demagnetizing r.f. frequencies, which we have not considered for simplicity. Only for the case of a spherical sample it is simply γH_0.

Fig. 1 shows that as k increases the magnon energy increases. This is clearly due to the fact that as the relative deviation of neighboring spins gets larger the exchange energy increases. In particular, magnons with $k = k_{edge}$ have the maximum relative deviation between neighbors ($\lambda \sim a$) and consequently the maximum energy. The zone edge magnon energy is approximately 8JS. Since J is typically 10^{-14} - 10^{-15} erg the energy is in the range 10 - 100 meV, which corresponds to a frequency 10^{12} - 10^{13} Hz (see Table I).

The Zeeman and the exchange contributions to the Hamiltonian of a ferromagnetic system are sufficient to give the wavelike nature to its excitations. In real crystals, however, other contributions such as the anisotropy and the dipolar interactions cannot be neglected [2]. Their presence modifies the characteristics of the spin waves. The second quantized form of these interactions can be found by using the magnon transformations (1.10) in the corresponding spin Hamiltonians. In some cases, such as with the dipolar interaction, the quadratic Hamiltonian found is not in the diagonal form (1.2). In these cases one has to find the new normal mode operators by means of additional canonical transformations [2,5,6,8].

2. STATISTICAL AND THERMODYNAMIC PROPERTIES OF MAGNONS

A. Eigenstates of the Hamiltonian: The eigenstates of the Hamiltonian (1.12) are independent harmonic-oscillator-type states, which are denoted by $|n_1, n_2, \ldots n_k \ldots n_N\rangle$ where n_k is the number of magnons with energy ω_k. These states form a complete ortho-normal set. Magnons are bosons and therefore (within our assumptions) n_k can vary from 0 to ∞. $a_k^\dagger$ and a_k are respectively magnon creation and annihilation operators, which have the boson properties, such as in (1.5) (with i replaced by k, of course). The state with zero magnons is the vacuum state, $|0_1 0_2 \ldots 0_k \ldots 0_N\rangle \equiv |0\rangle$.

Table I: Energy/frequency conversion units.

	GHz	Sec^{-1}	cm^{-1}	eV	erg	°K	Oe*
GHz		6.2832×10^{9}	3.3357×10^{-2}	4.1357×10^{-6}	6.6262×10^{-18}	4.7994×10^{-2}	3.5714×10^{2}
Sec^{-1}	1.5915×10^{-10}		5.3052×10^{-12}	6.5824×10^{-16}	1.0546×10^{-27}	7.6385×10^{-12}	5.6840×10^{-8}
cm^{-1}	29.979	1.8836×10^{11}		1.2398×10^{-4}	1.9865×10^{-16}	1.4388	1.0707×10^{4}
eV	2.4180×10^{5}	1.5193×10^{15}	8.0655×10^{3}		1.6022×10^{-12}	1.1605×10^{4}	8.6355×10^{7}
erg	1.5092×10^{17}	9.4824×10^{26}	5.0341×10^{15}	6.2414×10^{11}		7.2431×10^{15}	5.3898×10^{19}
°K	20.836	1.3092×10^{11}	0.69502	8.6170×10^{-5}	1.3806×10^{-16}		7.4413×10^{3}
Oe*	2.80×10^{-3}	1.7593×10^{7}	9.3399×10^{-5}	1.1580×10^{-8}	1.8554×10^{-20}	1.3438×10^{-4}	

*calculated for γ = 2.8 MHz/Oe.

The state with n_k magnons with energy ω_k and zero magnons with other energies is represented briefly as $|n_k\rangle$. We can generate the state $|n_k\rangle$ from the vacuum by the successive applications of the creation operator,

$$|n_k\rangle = \frac{1}{(n_k!)^{\frac{1}{2}}} (a_k^{\dagger})^{n_k} |0\rangle \tag{2.1}$$

This eigenstate of $\mathcal{H}$ has a well-defined number of magnons, each with energy $\hbar\omega_k$ and quasi-momentum $\hbar k$. In order to understand some properties of the magnon states let us relate the operators $a_k^{\dagger}$ and a_k to the original spin variables. With (1.7) and (1.10) we have

$$S_i^x = \frac{1}{2}(S_i^+ + S_i^-) = (\frac{S}{2N})^{\frac{1}{2}} \sum_k (a_k e^{i\vec{k}\cdot\vec{r}_i} + \text{h.c.}) \tag{2.2a}$$

$$S_i^y = \frac{1}{2i}(S_i^+ - S_i^-) = -i(\frac{S}{2N})^{\frac{1}{2}} \sum_k (a_k e^{i\vec{k}\cdot\vec{r}_i} - \text{h.c.}) \tag{2.2b}$$

$$S_i^z = S - \frac{1}{N} \sum_{kk'} a_{k'}^{\dagger} a_k e^{i(\vec{k}-\vec{k}')\cdot\vec{r}_i} \tag{2.3}$$

and also, using (1.11)

$$S_T^z = \sum_i S_i^z = NS - \sum_k a_k^{\dagger} a_k \tag{2.4}$$

Notice that the magnetization vector can be related to these spin operator components through

$$M^{\alpha}(r_i) = \frac{N}{V} g\mu_B S_i^{\alpha} \tag{2.5}$$

Using (1.5) we see that $\langle n_k | a_k | n_k\rangle = \langle n_k | a_k^{\dagger} | n_k\rangle = 0$, due to the orthogonality of states with different number of magnons. Hence

$$\langle n_k | S_i^x(t) | n_k\rangle = \langle n_k | S_i^y(t) | n_k\rangle = 0 \tag{2.6}$$

and there is no component of the spin transverse to the applied field when the spin system is in an eigenstate of $\mathcal{H}$. In other words, the states (2.2) which have a well defined number of magnons have uncertain phase. They do produce a net change in the z-component of the spin, which is simply related to the number of magnons only when the dipolar interaction is neglected by (2.4).

The result (2.6) reveals that the states with a well defined number of magnons do not correspond to the "classical" picture of the spin wave in which the spins precess about the external field. The states which correspond to a classical or macroscopic situation should involve a large and uncertain number of magnons. The coherent magnon states presented in Section 2C are shown to have the desired properties for a macroscopic situation.

B. <u>Temperature Dependence of the Magnetization</u>: At 0°K the spins are fully aligned in a ferromagnetic system resulting in a magnetization given by $Ng\mu_B S/V$. At $T \neq 0$ the temperature dependence of the magnetization is governed by the thermodynamics of the spin excitations. At low-temperatures, i.e., $T \ll T_c$, the number of thermal magnons is small and the magnon interactions can be neglected, so that the magnetization can be calculated with simple spin wave theory. For simplicity let us consider a system only with exchange interaction. With (2.4)

$$\langle S^z \rangle = S - \frac{1}{N}\sum_k \langle n_k \rangle \qquad (2.7)$$

Since magnons are bosons their population in thermal equilibrium is

$$\langle n_k \rangle = \frac{1}{e^{E_k/k_BT} - 1} \qquad (2.8)$$

In order to evaluate (2.7) we replace the sum by an integral over the Brillouin Zone

$$\frac{1}{N}\sum_k \rightarrow \frac{\Omega}{(2\pi)^3} \; d^3k$$

where Ω is the volume of the unit cell. At low temperatures the thermal population decreases rapidly as k increases. Thus most of the contribution to the integral comes from the small k-region and we can extend the upper limit to ∞. The decrease in the magnetization is then $M(0) - M(T) = g\mu_B N/V (S-S^z)$. Using the small k expansion for the energy $E_k = \hbar Dk^2$ and introducing the dimensionless variable $x \equiv \hbar Dk^2/k_BT$, the temperature dependence of the result becomes readily apparent

$$S - \langle S^z \rangle = \frac{\Omega}{(2\pi)^2}\left(\frac{k_BT}{\hbar D}\right)^{3/2} \int_0^\infty \frac{x^{1/2}dx}{e^x - 1} \qquad (2.9)$$

The integral is equal to $\sqrt{\pi}\,\zeta(3/2)/2$, where ζ is the Riemann zeta function (2). For a simple cubic lattice, $\Omega = a^3$ and $\hbar D = 2JSa^2$, so that (2.9) becomes

$$S - \langle S^z \rangle = \zeta(3/2)\left(\frac{k_B}{2\pi JS}\right)^{3/2} T^{3/2} \qquad (2.10)$$

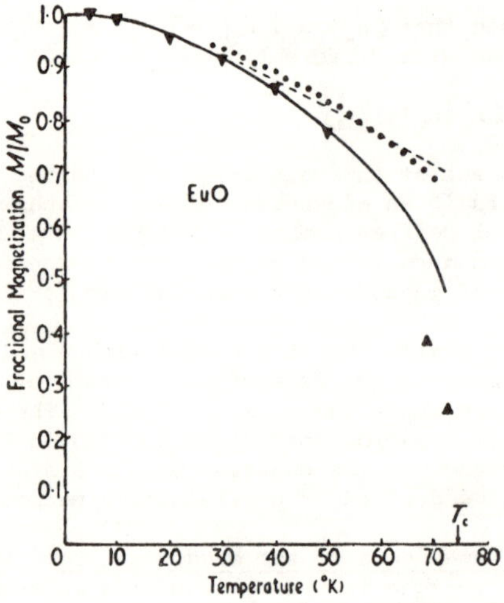

FIG.2. Theoretical and experimental temperature dependence of the magnetization of EuO. The solid curve is calculated from spin wave theory including the effects of four-magnon interactions and subsequent renormalization of the spin wave energies used in evaluating the magnon populations. The broken curve corresponds to non-interacting spin waves and the dotted curve to the usual series expansion in terms dependent on $T^{3/2}$ and $T^{5/2}$. (After Low [9]).

This is the well known Bloch $T^{3/2}$ law [2]. It governs the behavior of the magnetization at low temperatures and has been verified to hold for a wide variety of magnetic materials. At higher temperatures a number of corrections have to be introduced in the calculation so that spin wave theory can describe the behavior of M(T). Using the proper corrections Low [9] was able to show that spin wave theory correctly describes M(T) for the simple ferromagnet EuO for temperatures as high as 3/4 T_c, as shown in Fig. 2.

C. Coherent Magnon States: As we have shown previously the eigenstates of the Hamiltonian (1.12) do not have the properties required to represent a macroscopic spin wave. These properties belong to coherent magnons. In analogy with the coherent photon states used in quantum optics [10] one can define a coherent magnon state [11,12] as the eigenstate of the magnon annihilation operator; i.e.,

$$a_k|\alpha_k\rangle = \alpha_k|\alpha_k\rangle \tag{2.11}$$

where α_k is a complex number which characterizes the state. This state can be expanded in terms of the eigenstates of the Hamiltonian:

$$|\alpha_k\rangle = e^{-\frac{1}{2}|\alpha_k|^2} \sum_{n_k=0}^{\infty} \frac{(\alpha_k)^{n_k}}{(n_k!)^{\frac{1}{2}}}|n_k\rangle \tag{2.12}$$

The properties of the states $|\alpha_k\rangle$ have been extensively studied by Glauber [10]. We review here a few important ones. The probability of finding n_k magnons in the coherent state $|\alpha_k\rangle$ follows immediately from (2.12). It is given by a Poisson distribution

$$|\langle n_k|\alpha_k\rangle|^2 = \frac{|\alpha_k|^{2n_k}}{n_k!} e^{-|\alpha_k|^2} \tag{2.13}$$

The mean value of the distribution is the expectation value of the occupation number operator, which is found from (2.11) and its conjugate:

$$\langle \alpha_k|n_k|\alpha_k\rangle = |\alpha_k|^2 \tag{2.14}$$

It can be shown that coherent states are not orthogonal to one another, but they form a complete set. This property allows the expansion of an arbitrary state in terms of coherent states.

Using the definition of the coherent state (2.11) we can calculate the expectation value of the spin operator components in this state. Eqs. (2.2), and (2.11) give readily

$$\langle \alpha_k|S_i^x(t)|\alpha_k\rangle = \left(\frac{S}{2N}\right)^{1/2}\left[\alpha_k e^{i(\vec{k}\cdot\vec{r}_i-\omega_k t)} + \text{c.c.}\right]$$

$$\langle \alpha_k|S_i^y(t)|\alpha_k\rangle = -i\left(\frac{S}{2N}\right)^{1/2}\left[\alpha_k e^{i(\vec{k}\cdot\vec{r}_i-\omega_k t)} - \text{c.c.}\right]$$

Thus we can write for a coherent magnon state

$$\langle \vec{S}_i\rangle = \hat{z}\langle S_i^z\rangle + \hat{x}S_k\cos(\vec{k}\cdot\vec{r}_i-\omega_k t+\beta_k) + \hat{y}S_k\sin(\vec{k}\cdot\vec{r}_i-\omega_k t+\beta_k) \tag{2.15}$$

where β_k is the argument of α_k, and

$$S_k = \left(\frac{2S}{N}\right)^{1/2}|\alpha_k| \tag{2.16}$$

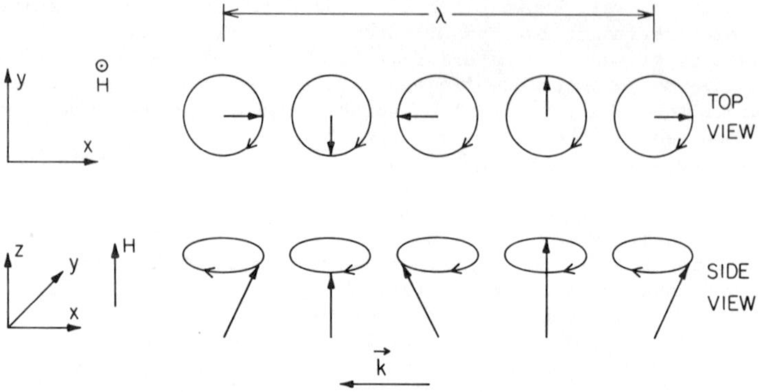

FIG.3. Spin wave propagating perpendicularly to the magnetic field.

The result (2.15) provides the view of a spin wave corresponding to a coherent magnon state. As pictured in Fig. 3, each spin precesses with angular frequency ω_k about the direction of propagation of the wave. The shortest distance between parallel spins in the direction of k is the wavelength λ, which is related to the wavenumber through $\lambda = 2\pi/k$.

The behavior of the expectation value of the spin operator in a coherent magnon state is the same as in a "semi-classical" spin wave, which is usually derived based on the equation of motion

$$\frac{d\vec{S}}{dt} = \gamma\vec{S} \times \vec{H} \tag{2.17}$$

Although the semi-classical and quantum theories of spin waves have been developed quite extensively in the past thirty years, only recently the proper correspondence between these treatments by means of the coherent states was pointed out [11,12].

To conclude this section let us make some remarks on the statistical properties of magnon states. The eigenstates $|n_k\rangle$ of the harmonic oscillator Hamiltonian, studied in (2A), are not the states in which the spin system is likely to be found in any physical situation. They provide, however, a very convenient basis for the expansion of any physical state. The two physical situations of most interest are certainly the one in which the system is excited by a coherent external field, and the thermal equilibrium state. In the first case the system is in a coherent state [12] with frequency ω_k and wavevector $\vec{k}$, determined by the driving excitation. The probability of finding a given n_k number of magnons in this state is, as can be shown with (2.13) and (2.14),

$$\rho_{coh}(n_k) \equiv |\langle n_k|\alpha_k\rangle|^2 = \frac{(\langle n\rangle)^{n_k}}{n_k!} e^{-\langle n\rangle} \tag{2.18}$$

where $\langle n\rangle$ denotes the average number of magnons. This is a Poisson distribution, plotted in Fig. 4a. This distribution is sharply peaked around $\langle n\rangle$, whose square root is proportional to the amplitude of the spin excitation (Eq. (2.16)).

The statistics of the "state" of thermal equilibrium are quite different. First of all, we note that in thermal equilibrium the system cannot be described by simple quantum states. Rather, it is a mixture, which can be specified by the density operator

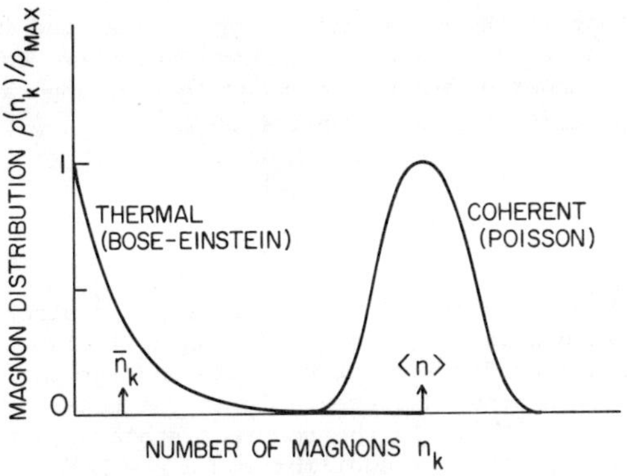

FIG.4. Normalized number distribution of magnon states.

$$\rho_0 = \frac{1}{Z} e^{-\beta \mathcal{H}_0} \tag{2.19}$$

where $\beta = 1/k_B T$, $\mathcal{H}_0$ is the Hamiltonian (1.12) and $Z = \mathrm{Tr}(e^{-\beta\mathcal{H}_0})$ is the partition function. With (2.19) we can calculate the properties of the system in thermal equilibrium. First, let us evaluate the average number of magnons with wavevector k at a given temperature, i.e.,

$$\bar{n}_k \equiv \langle n_k \rangle = \mathrm{Tr}(\rho_0 n_k) \tag{2.20}$$

Since $|n_k\rangle$ is an eigenstate of $\mathcal{H}_0$, the traces are easily calculated on this basis

$$Z = \mathrm{Tr}(e^{-\beta\mathcal{H}_0}) = \sum_{n_k} e^{-\beta\hbar\omega_k n_k} = \frac{1}{1-e^{-\beta\hbar\omega_k}} \tag{2.21}$$

$$\mathrm{Tr}(n_k e^{-\beta\mathcal{H}_0}) = \sum_{n_k} n_k e^{-\beta\mathcal{H}_0} = \frac{\partial}{\partial(-\beta\hbar\omega_k)} \sum_{n_k} e^{-\beta\hbar\omega_k n_k} = \frac{e^{-\beta\hbar\omega_k}}{(1-e^{-\beta\hbar\omega_k})^2}$$

Using these results in (2.20) we obtain

$$\bar{n}_k = \frac{1}{e^{\beta\hbar\omega_k}-1} \tag{2.22}$$

the known result which, of course, is correct only in the absence of interactions. The probability of finding a given n_k number of magnons when the system is in the thermal equilibrium can be calculated expressing ρ_0 as (using the completeness relation $\sum_n |n\rangle\langle n| = 1$)

$$\rho_0 = \frac{1}{Z}\sum_{n_k} e^{-\beta\mathcal{H}_0} |n_k\rangle\langle n_k| = \frac{1}{Z}\sum_{n_k} e^{-\beta\hbar\omega_k n_k} |n_k\rangle\langle n_k|$$

Using (2.21) and (2.22) this can be rewritten in the form

$$\rho_0 = \sum_{n_k} \frac{(\bar{n}_k)^{n_k}}{(1+\bar{n}_k)^{n_k+1}} |n_k\rangle\langle n_k| \tag{2.23}$$

This shows that in thermal equilibrium one can find in the system any number of magnons with any energy (and wavevector). For a given energy the average number of magnons depends on the temperature as in (2.22). The distribution of magnons of a given energy ω_k is

$$\rho_{th}(n_k) = \frac{(\bar{n}_k)^{n_k}}{(1+\bar{n}_k)^{n_k+1}} \tag{2.24}$$

This is the Bose-Einstein distribution which is plotted in Fig. 4. It approaches an exponential function $e^{-\bar{n}_k}$ when $\bar{n}_k$ is large. The differences between the distributions for a coherent state and for thermal equilibrium are clear in Fig. 4. Another difference is in the behavior of the expectation values of $\langle S^x \rangle$ and $\langle S^y \rangle$. In the coherent state we have the precessing spin picture, whereas in thermal equilibrium $\langle S^x \rangle = \langle S^y \rangle = 0$, as can be readily proved with the use of (2.23).

3. MAGNON INTERACTIONS

The magnon states we have been studying do not have infinite lifetimes as implied in (2.15). The absence of a decay rate in our results so far is due to the fact that the quadratic Hamiltonian (2.12) characterizes a system of independent collective modes. Actually the magnon modes interact among themselves and with other elementary excitations in the crystal, such as phonons, electrons, plasmons, etc. Typically a more realistic Hamiltonian could be written as

$$\begin{aligned}\mathcal{H} = {} & \sum_k \hbar\,\omega_k a_k^\dagger a_k + \sum_{k_1k_2k_3} \left[V^{(3)}_{m-m}(k_1,k_2\,k_3)\, a_{k_1} a^\dagger_{k_2} a^\dagger_{k_3}\, \delta(\vec{k}_1-\vec{k}_2-\vec{k}_3) + \text{h.c.} \right] \\ & + \sum_{k_1k_2} V^{(4)}_{m-m}(k_1k_2,k_3k_4)\, a_{k_1} a_{k_2} a^\dagger_{k_3} a^\dagger_{k_4}\, \delta(\vec{k}_1+\vec{k}_2-\vec{k}_3-\vec{k}_4) + \sum_k \left[V_{m-p}(k)\, a_k b_k + \text{h.c.} \right] \\ & + \sum_{k_1k_2k_3} \left[V_{m-p}(k_1k_2k_3)\, a_{k_1} a^\dagger_{k_2} b^\dagger_{k_3}\, \delta(\vec{k}_1-\vec{k}_2-\vec{k}_3) + \text{h.c.} \right] + \ldots\end{aligned} \tag{3.1}$$

where the second term represents 3-magnon processes which are illustrated in Fig. 5a. The delta function appears naturally from Eq. (1.11) and expresses the conservation of quasi-momentum.

The magnon interactions are responsible for the scattering of the energy in a given k-mode and therefore they account for its relaxation, i.e., its finite lifetime. In addition they provide the means for nonlinear dynamic interactions among several excitations leading to very interesting phenomena.

A. Origin of the Magnon Interactions: The 3-magnon interaction of (3.1) arises from the dipolar Hamiltonian. It is obtained with the use of the terms with three operators in the expansions of S_i^+ and S_i^- in Eq. (1.6). It is not difficult to make an order of magnitude estimate of the vertex of this interaction. In the long wavelength limit [2,8,13]

$$V_{m-m}(k_1,k_2\,k_3) \sim \frac{\mu_B M}{\sqrt{N}} \tag{3.2}$$

which is of the order of $(10^{-18} - 10^{-17})/\sqrt{N}$ erg.

The 4-magnon interaction arises from anisotropy, dipolar and exchange interactions between the spins. The contribution from the exchange is larger and is obtained readily from (1.1) with the expansions of (1.6) and the transformations (1.9). One can show that

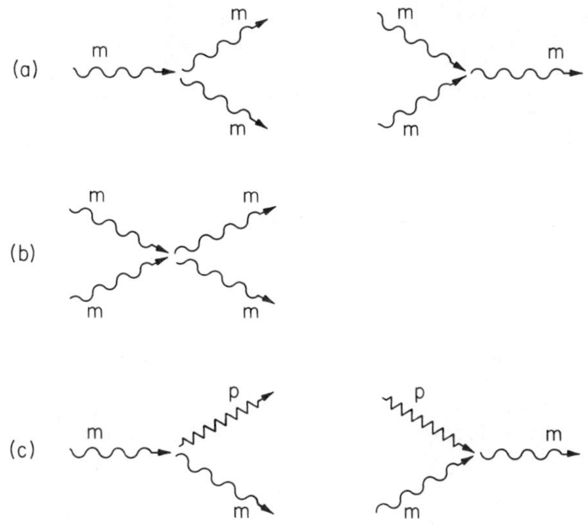

FIG.5. Magnon interaction processes: (a) 3-magnon; (b) 4-magnon; (c) 2-magnon/1-phonon.

$$V_{m-m}(k_1k_2,k_3k_4) = \frac{1}{2N}(J_{k_1} + J_{k_4} - 2J_{k_1-k_3}) \tag{3.3}$$

where $J_k = J\sum_\delta e^{i\vec{k}\cdot\vec{\delta}}$

Notice that this 4-magnon interaction conserves the number of magnons in the approximation of no dipolar interaction. This is a result of the fact that the total z-component of the spin, $\sum_i S_i^z$, commutes with the Hamiltonian (1.1). Therefore $\sum_i S_i^z$, and hence the number of spin deviations, are conserved.

The magnon-phonon interaction results from the dependence on the ion distances of the exchange, dipole-dipole, and the spin-orbit interactions. This is the same interaction which leads to the coupling between the macroscopic magnetization and the strain, resulting in the phenomenon called magnetostriction. The magnetoelastic Hamiltonian can be written in a power series of the spin operators and the lattice displacement operators. In the single-ion approximation [2]

$$\mathcal{H}_{me} = \sum_i b_{\alpha\beta\gamma\delta} S_i^\alpha S_i^\beta \frac{\partial R_i^\gamma}{\partial x_\delta} + \sum_{\substack{i\\ \delta\varepsilon\zeta}} b_{\alpha\beta\gamma} S_i^\alpha S_i^\beta \frac{\partial R_i^\gamma}{\partial x_\delta} \frac{\partial R^\varepsilon}{\partial x_\zeta} + \ldots \tag{3.4}$$

where $b_{\alpha\beta\gamma\delta}$ and $b_{\alpha\beta\gamma\delta\varepsilon\zeta}$ are, respectively, the first and second order magnetoelastic coefficients. The calculation of the bs is very difficult for it involves a detailed knowledge of the atomic orbitals and the crystalline electric fields. The crystal symmetry usually reduces the number of non-zero coefficients to a few ones. In the case of cubic symmetry there are only two non-zero first-order magnetoelastic constants [2]. The Hamiltonian (3.4) can be written in terms of operators of collective excitations using the transformations for the spin and the standard expansion [15] for the displacement operator $\vec{R}$,

$$\vec{R} = \sum_{k,\mu} \left(\frac{\hbar}{2\rho V\omega_{p\mu}}\right)^{1/2} \hat{e}_\mu (b_{k\mu}^\dagger e^{-i\vec{k}\cdot\vec{r}} + b_{k\mu} e^{i\vec{k}\cdot\vec{r}}) \tag{3.5}$$

where ρ is the mass density, $\omega_{p\mu}$ is the frequency of the phonon with polarization μ, and $b^{\dagger}_{k\mu}$ and $b_{k\mu}$ are the corresponding creation and annihilation operators. With (3.5) we see clearly that the first term of the Hamiltonian (3.4) represents the interaction between one phonon and one or more magnons. If we use for one of the spin operators the component S^z and for the others S^x and S^y the lowest order interaction can be arranged approximately as

$$\mathcal{H}^{(2)}_{m-p} = \sum_k i \frac{\hbar}{2} \sigma_k (a^{\dagger}_k b_k - b^{\dagger}_k a_k) \tag{3.6a}$$

where

$$\sigma_k = k \left(\frac{4b^2 S^3 N}{\hbar \omega_p \rho V} \right)^{1/2} \tag{3.6b}$$

where b_k is the operator for circularly polarized transverse phonons and b is a transverse magnetoelastic constant, related to the usual b_2 constant of cubic crystals by $b = (b_2 V/\sqrt{2S^2N})$. The Hamiltonian (3.6) represents a linear coupling between magnons and phonons which is important only in the region of k-space where their dispersion curves cross. In this region the character of the excitation contains an admixture of magnons and phonons. This mixed excitation will be studied later. In the next order approximation the magnetoelastic Hamiltonian contains two magnon operators and one phonon operator, such as the last term in (3.1). This interaction represents a phonon emission process in which the energy of the spin wave is transferred to the lattice.

B. Magnon Energy Renormalization: The magnon energy (1.13) obtained with the spin wave approximation was found to be independent of the temperature (for J is essentially temperature-independent). Actually resonance and neutron scattering experiments show that this is not the case. To a first approximation a Green's function theory [4,14] gives a magnon exchange energy which is proportional to $\langle S^z(T)\rangle$ rather than to S, as in (1.13). Thus the energy decreases as T increases and approaches zero at the transition temperature T_c. In the spin wave theory, corrections to the magnon energy can be introduced by the magnon interactions. The most important of them is the 4-magnon exchange interaction (3.3), which is sometimes called the dynamical interaction. The correct calculation of the effect of $\mathcal{H}^{(4)}$ on the energy is quite involved, for it requires the evaluation of $\mathrm{Tr}(\mathcal{H}\rho_0)$ where ρ_0 now contains $\mathcal{H}^{(2)}$ and $\mathcal{H}^{(4)}$. In a first approximation, however, we can use a Hartree-Fock approach and find the interaction of magnons with the average field of the others. If we calculate the energy of the system by taking the expectation value of the interaction Hamiltonian $\mathcal{H}^{(4)}$ in an unperturbed state, only the terms that return the system to its original state will contribute. Thus we write the interaction (3.3) as

$$\mathcal{H}^{(4)}_{ex} \simeq \frac{1}{2N} \sum_{kk'} 4(J_k + J_{k'} - J_0 - J_{\vec{k}-\vec{k}'}) a^{\dagger}_k a_k a^{\dagger}_{k'} a_{k'} \tag{3.7}$$

and the correction to the energy of the k mode is

$$E^{(4)}_k = \frac{4}{N} \sum_{k'} (J_k + J_{k'} - J_0 - J_{\vec{k}-\vec{k}'}) \langle n_{k'} \rangle \tag{3.8}$$

where the numerical factors account for the various combinations over the k_s. As in the previous section we can find the leading term in the expansion of (3.8). First we expand J_k in powers of k keeping the terms in k^4

$$J_k = J_0 + \alpha k^2 + \beta k^4$$

For instance, for simple cubic lattices we find $\alpha = -Ja^2$ and $\beta = Ja^4/12$. The correction then becomes

$$E_k^{(4)} = \frac{4}{N}\sum_{k'} [\,-\alpha 2\vec{k}\cdot\vec{k}' + \beta(k^4 + k'^4 - |\vec{k}-\vec{k}'|^4)]\langle n_{k'}\rangle = -\frac{8}{N}\sum_{k'} \beta k^2 k'^2 \langle n_{k'}\rangle \tag{3.9}$$

The last result is obtained using the fact that the sum over k' eliminates the terms $\vec{k}\cdot\vec{k}'$. Notice that the terms proportional to α are not present in (3.9). This is a significant result, for if the magnon dispersion relation were exactly quadratic, there would be no correction to the energy in first order. Therefore the non-interacting spin waves provide a good description of a ferromagnet at low temperatures. For simple cubic lattices (3.9) can be written as

$$E_k^{(4)} = -(zJS^2N)^{-1}\,\hbar\omega_k \sum_{k'} \hbar\omega_{k'}\langle n_{k'}\rangle \tag{3.10}$$

This shows that as the temperature increases, the magnon energy $\hbar\omega_k + E_k^{(4)}$ decreases. With the same arguments employed with (2.9) one can show that the leading term in the expansion of the renormalization energy (3.10) varies with $T^{5/2}$.

C. Spin Wave Relaxation: The interactions (3.1) are processes which tend to share the energy of a given magnon mode with all the excitations in the crystal. This means that when the occupation number of a given mode is changed from its thermal equilibrium value by some external excitation, the interactions act as to reestablish the thermalization. This is the same as to say that magnons have a finite lifetime. The relaxation rates of magnons are measured directly through the linewidths in a number of experiments, such as in magnetic resonance, and in neutron and light scattering. Therefore relaxation studies provide a powerful tool for the understanding of microscopic interaction mechanisms [6]. In order to indicate how magnon relaxation phenomena are studied let us consider, for example, 3-magnon interaction processes, represented by

$$\mathcal{H}^{(3m)} = \sum_{k_1k_2k_3} (V_{123}a^{\dagger}_{k_1}a^{\dagger}_{k_2}a_{k_3} + V^{*}_{123}a_{k_1}a_{k_2}a^{\dagger}_{k_3})\,\Delta(\vec{k}_1+\vec{k}_2-\vec{k}_3) \tag{3.11}$$

This interaction can relax a given k-mode through two different processes, illustrated in Fig. 6. In the confluent process the k-magnon absorbs a thermal magnon and emits another magnon. In the splitting process the k-magnon emits two magnons. Clearly the energy and momentum conservation in these two processes are different and lead to different temperature dependences for the two relaxation rates. Let us calculate the relaxation rate for the confluent process. The idea is to try to write the time rate of change of the occupation number for the mode k in the form

$$\frac{dn_k}{dt} = -\frac{1}{T_k}(n_k - \bar{n}_k) \tag{3.12}$$

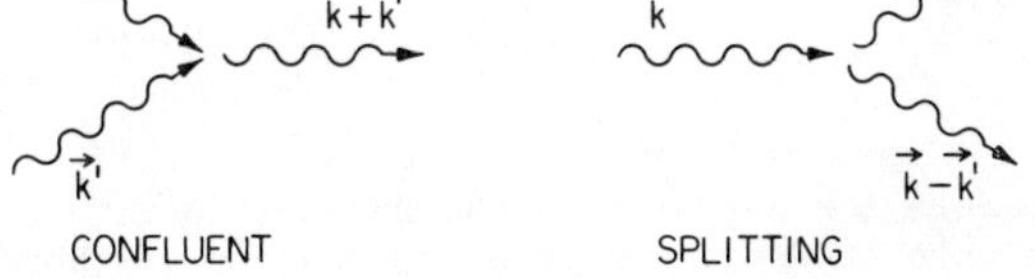

FIG.6. Three-magnon processes that contribute to the decay of magnons.

where $\eta_k = 1/T_k$ is the relaxation rate. dn_k/dt can be calculated with first-order perturbation theory. Due to the form of (3.11) n_k can change by only one unit. Thus

$$\frac{dn_k}{dt} = - [TP(n_k \to n_k - 1) - TP(n_k \to n_k + 1)] \tag{3.13}$$

Using the Golden-rule for the transition probabilities (TP) we have

$$\frac{dn_k}{dt} = -\frac{2\pi}{\hbar^2} \sum_{ff'} \{ |\langle f, n_k - 1| \mathcal{H}^{3m} | n_1 n_2 n_k \rangle|^2 - |\langle f', n_k + 1 | \mathcal{H}^{3m} | n_1 n_2 \ldots n_k \rangle|^2 \} \delta(\omega_f - \omega_{f'}) \tag{3.14}$$

Notice now that we can operate on the state $|n_k\rangle$ with any of the three magnon operators appearing in (3.11), such that the first term in (3.14) becomes for the confluent process

$$|\langle n_k - 1, n_{k'} - 1, n_{k+k'} + 1 | \mathcal{H}^{3m} | n_k n_{k'} n_{k+k'} \rangle|^2 = n_k n_{k'} (n_{k+k'} + 1) | V^*_{kk'(k+k')} + V^*_{k'k(k+k')} |^2$$

Using the same procedure with the second term of (3.14) and dropping (k+k') in V to simplify the notation we obtain [16]

$$\frac{dn_k}{dt} = -\frac{2\pi}{\hbar^2} \sum_{k'} |V_{kk'} + V_{k'k}|^2 \left[n_k n_{k'} (n_{k+k'} + 1) - (n_k + 1)(n_{k'} + 1) n_{k+k'} \right] \delta(\omega_k + \omega_{k'} - \omega_{k+k'}) \tag{3.15}$$

We now assume that all magnon modes are in thermal equilibrium, except the k mode, and set $n_{k'} \to \bar{n}_{k'}$ and $n_{k+k'} \to \bar{n}_{k+k'}$ in (3.15). If we also make $n_k \to \bar{n}_{k'}$ we must get

$$\bar{n}_k \bar{n}_{k'} (\bar{n}_{k+k'} + 1) - (\bar{n}_k + 1)(\bar{n}_{k'} + 1) \bar{n}_{k+k'} = 0 \tag{3.16}$$

since when all modes are in thermal equilibrium $dn_k/dt = 0$. The result (3.16) can be shown to be just a consequence of the form for $\bar{n}$ (2.22) and the energy conservation ($\omega_{k+k'} = \omega_k + \omega_{k'}$). If we now subtract (3.15) with all $n_k \to \bar{n}$ (i.e., zero) from (3.15) with only $n_k \to \bar{n}_{k'}$ and $n_{k+k} \to \bar{n}_{k+k'}$ the term $(n_k - \bar{n}_k)$ can be factorized to give,

$$\frac{1}{T_k} = \frac{2\pi}{\hbar^2} \sum_{k'} |V_{kk'} + V_{k'k}|^2 \left[\bar{n}_{k'} (\bar{n}_{k+k'} + 1) - (\bar{n}_{k'} + 1) n_{k+k'} \right] \delta(\omega)$$

Using (2.22) the decay rate for the confluent process can be finally written in the form

$$\frac{1}{T_k} = \frac{2\pi}{\hbar^2} (e^{\beta\hbar\omega_k} - 1) \sum_{k'} |V_{kk'} + V_{k'k}|^2 e^{\beta\hbar\omega_{k'}} \bar{n}_{k'} \bar{n}_{k+k'} \delta(\omega_k + \omega_{k'} - \omega_{k+k'}) \tag{3.17}$$

This procedure for calculating the relaxation rate can be generalized for a process in which there are N incoming magnons (or bosons in general) and M outcoming magnons (or bosons). In this case, the decay rate of the mode 1 is given by [16]

$$\frac{1}{T_{k_1}} = \frac{2\pi}{\hbar^2}\left(e^{\beta\hbar\omega_1} - 1\right) \sum_{k_1 k_2 \ldots k_{M+N}} |V_{NM}|^2 \, e^{\beta\hbar(\omega_2+\omega_3+\ldots\omega_N)} \bar{n}_2 \bar{n}_3 \ldots \bar{n}_{M+N} \Delta(k)\delta(\omega) \tag{3.18}$$

where V_{NM} is the appropriate coupling coefficient and $\Delta(k)$ and $\delta(\omega)$ are the momentum and energy conservation delta function.

The total relaxation rate for a magnon is the sum of the contributions from all possible interaction processes. The book of Sparks [6] has a detailed account of relaxation calculations in ferromagnets. Usually different processes dominate the decay rate at different temperature or wavevector ranges, depending on the strength of the interactions. In YIG, for example, the k=0 or uniform mode, whose decay rate is measured directly in ferromagnetic resonance experiments, relaxes through the following processes [6,17]: Two-magnon pit scattering, in which the k=0 mode decays into degenerate modes, due to the non-momentum conserving potentials created by the absence of the magnetic moments at the volume and surface pits in the sample; impurity scattering; and at higher temperatures, 2-magnon-1-phonon and 3-magnon processes. For $k\neq 0$ modes, at room temperature, 3-magnon confluence processes due to the dipolar interaction can explain the experimental results with an accuracy better than 10%. Fig. 7 shows the comparison between the theoretical results [18] and the experimental data [19] for the magnon linewidth as measured by parallel pumping techniques. Note that the linewidth is related to the decay rate η_k by $\Delta H_k = \eta_k/\gamma$.

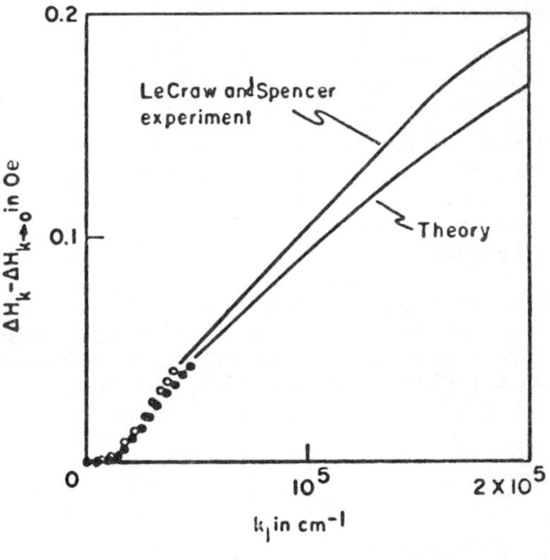

FIG. 7. Comparison between magnon linewidth data [19] *and 3-magnon relaxation calculation* [18] *in YIG.*

4. ANTIFERROMAGNETIC MAGNONS

The simplest type of antiferromagnet has two magnetic sublattices with equal spins. The intersublattice exchange constant is the largest and is negative so that the spins on the two sublattices tend to align oppositely to each other. As in ferromagnets the exchange interaction gives rise to spin excitations with wavelike characteristics. Classically when the applied

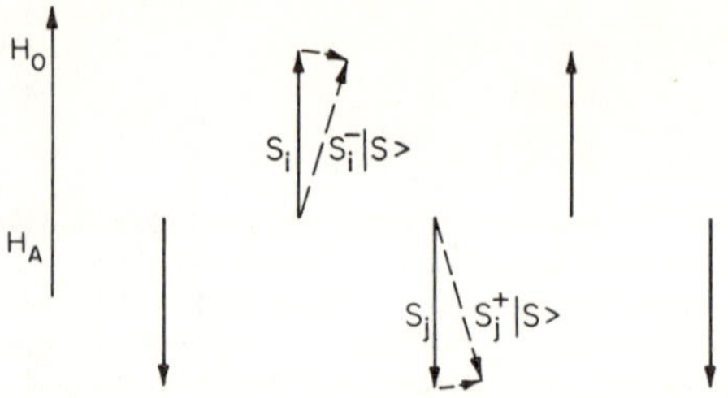

FIG.8. Ground state (classical) and spin derivations in normal antiferromagnet.

field is less than a critical value the state of minimum energy is that shown in Fig. 8. The spins of the sublattices i and j are exactly antiparallel to each other, in the so-called Néel ground state. This is easy to see if we use $J < 0$ in the Heisenberg Hamiltonian (1.1) and assume that $\vec{S}_i$ and $\vec{S}_j$ are classical vectors. Quantum mechanically the situation is not so simple, because to be the ground state of a system it must be an eigenstate of the Hamiltonian besides having the minimum energy. If we operate with the Hamiltonian (1.1) on the state of Fig. 8 we clearly do not obtain the same state, because the term $S_i^- S_j^+$ flips S_j up and S_i down.

The ground state difficulty in antiferromagnets is circumvented with the assumption that the Néel state is an eigenstate of the Hamiltonian. Spin waves are then generated by flipping the spins away from this ordered state. We shall take for the Hamiltonian the form

$$\mathcal{H} = -g\mu_B H_0\left(\sum_i S_i^z + \sum_j S_j^z\right) + 2J\sum_{i,j=i+\delta} \vec{S}_i\cdot\vec{S}_j - K\sum_{ij}\left[(S_i^z)^2 + (S_j^z)\right]^2 \tag{4.1}$$

where the indices i and j refer to the two sublattices, the applied field is along the z-direction, the exchange interaction is non-zero only for nearest neighbors, and the anisotropy interaction has uniaxial nature. In (4.1) we have asssumed that the external field is applied parallel to the symmetry axis of the crystal because this is the simplest configuration to find the ground state.

If we assume in (4.1) that S_i and S_j are classical vectors, the ground state can be found with the minimum energy conditions $(\partial\mathcal{H}/\partial\theta)_{i,j} = (\partial\mathcal{H}/\partial\phi)_{ij} = 0$, where θ and ϕ are the polar and azimuthal angles of the spins. For H_0 parallel to the symmetry axis, the ground state is that shown in Fig. 8 only if H_0 is less than the critical value

$$H_c = \sqrt{H_A\,(2H_E + H_A)} \tag{4.2}$$

where

$$H_A = \frac{(2S-1)K}{\gamma\hbar} \qquad H_E = \frac{2SzJ}{\gamma\hbar} \tag{4.3}$$

are the effective anisotropy and exchange fields seen by the spins. At $H_0 = H_c$ there is a magnetic phase transition to a spin-flop state which will be discussed in the next section. From (4.2) we see that the anisotropy contribution to the Hamiltonian cannot be neglected or else there is no normal state.

The excitations of the Hamiltonian (4.1) can be studied in the spin wave approximation employed in Section 1. We associate a spin-deviation creation operator $a_i^\dagger$ to S_i^-, a destruction operator to S_i^+ and the number of spin deviations to S_i^z. For sublattice j we note that the spins are reversed (Fig. 8).

$$\begin{aligned} S_i^+ &= (2S)^{1/2} a_i & S_j^+ &= (2S)^{1/2} b_j^\dagger \\ S_j^- &= (2S)^{1/2} a_i^\dagger & S_j^- &= (2S)^{1/2} b_j \\ S_i^z &= S - a_i^\dagger a_i & S_j^z &= -S + b_j^\dagger b_j \end{aligned} \tag{4.4}$$

Substitution of the transformations (4.4) in (4.1) will lead to terms which couple the spin deviations at different lattice sites. Collective modes are introduced with the transformations to collective spin variables

$$\begin{aligned} a_i &= \frac{1}{N^{1/2}} \sum_k e^{i\vec{k}\cdot\vec{r}_i} a_k \\ b_j &= \frac{1}{N^{1/2}} \sum_k e^{i\vec{k}\cdot\vec{r}_j} a_k \end{aligned} \tag{4.5}$$

where the new operators satisfy the usual Bose commutation relations. Manipulation of (4.1) and (4.5) in a manner similar to that used in Section 1 leads to the following quadratic Hamiltonian

$$\mathcal{H} = \sum_k (A_k a_k^\dagger a_k + A'_k b_k^\dagger b_k + B_k a_k b_{-k} + B_k^* a_k^\dagger b_{-k}^\dagger) \tag{4.6}$$

where

$$\begin{aligned} A_k &= \gamma\hbar (H_E + H_A + H_0) & A'_k &= \gamma\hbar (H_E + H_A - H_0) \\ B_k &= \gamma\hbar H_E \gamma_k & \gamma_k &= \frac{1}{z} \sum_\delta \exp(i\vec{k}\cdot\vec{\delta}) \end{aligned} \tag{4.7}$$

The Hamiltonian (4.7) is not in the diagonal form. Clearly the diagonalization can be performed with linear combinations of the operators a_k and $b_{-k}^\dagger$. This means that the spin wave excitations will involve admixtures of spin deviations of both sublattices. We can diagonalize (4.6) with the transformations to new normal mode operators α_k and β_k.

$$\begin{aligned} a_k &= u_k \alpha_k - v_k \beta_{-k}^\dagger \\ b_{-k}^\dagger &= - v_k \alpha_k + u_k \beta_{-k}^\dagger \end{aligned} \tag{4.8}$$

Using (4.8) in (4.6) we can write the Hamiltonian for the antiferromagnetic magnons as follows

$$\mathcal{H} = \sum_k (\hbar\omega_{\alpha k} \alpha_k^\dagger \alpha_k + \hbar \omega_{\beta k} \beta_k^\dagger \beta_k) \tag{4.9a}$$

where

$$\begin{aligned} \omega_{\alpha k, \beta k} &= \omega_k \pm \gamma H_0 \\ \omega_k &= \gamma [H_c^2 + H_E^2 (1 - \gamma_k^2)]^{1/2} \end{aligned} \tag{4.9b}$$

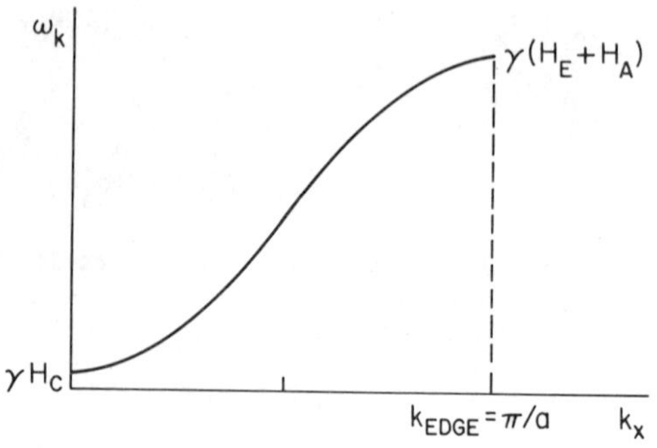

FIG.9. Antiferromagnetic magnon dispersion with $H_0 = 0$.

Table II: T_N - Neel ordering temperature; M - Sublattice magnetization at T = 0; H_A - Anisotropy field (for $RbMnF_3$, which is cubic, this is valid for the [111] direction); H_E - Exchange field; H_c - spin-flop field.

	T_N(°K)	M(Gauss)	H_A(Oe)	H_E(Oe)	H_c(Oe)	H_c(GHz)
$RbMnF_3$	82	300	4.5	830 K	2.7 K	7.6
MnF_2	67	576	7.9 K	550 K	93 K	260
FeF_2	78	560	200 K	580 K	521 K	1.57 K

The conditions on the transformation coefficients for diagonalizing the Hamiltonian are:

$$u_k = \left(\frac{\omega_{ZB} + \omega_k}{2\omega_k}\right)^{1/2} , \quad v_k = \left(\frac{\omega_{ZB} - \omega_k}{2\omega_k}\right)^{1/2} \tag{4.10}$$

where $\omega_{ZB} = \gamma(H_E + H_A)$ is the value of ω_k at the Brillouin zone boundary. Note that the transformation coefficients are independent of the applied field H_0. They are, however, strongly dependent on k. At the zone edge, for example, u_k=1 and v_k=0, so that those magnons involve spin deviations only from each of the sublattices. Notice also that the existence of the two spin wave modes is a consequence of the fact that there are two spins in the magnetic unit cell. The two modes are degenerate in the absence of an applied field. In this case the dispersion curve for both modes has the form shown in Fig. 9. At the center of the zone the frequency is H_c. The energy gap gives rise to several differences between the properties of ferro- and antiferromagnetic materials. For example, antiferromagnetic magnons are thermally excited with appreciable populations only at temperatures $T \gtrsim \hbar H_c/K_B$, leading to different thermodynamic properties than in ferromagnets, where no such gap exists with $H_0 = 0$. Another important difference lies on the (k=0) resonance studies. While FMR frequencies depend mainly on the applied field and usually can be chosen to fall in low microwave ranges, antiferromagnetic resonance (AFMR) frequencies are given by

$$\omega_0 = \gamma(H_c \pm H_0)$$

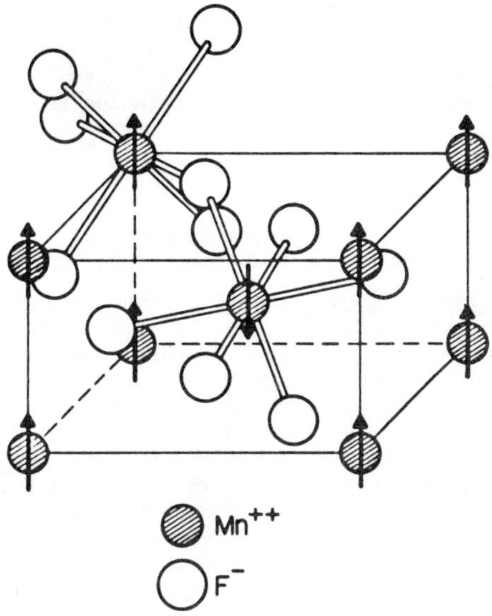

FIG.10. Magnetic structure of MnF_2. The direction of the Mn^{2+} spin ordering is shown. The dimensions of the magnetic unit all are 4.87 × 4.87 × 3.3 A.

Since the applied field cannot have any arbitrary value, the frequency is strongly determined by the size of the gap. Thus AFMR experiments have been carried out with several spectroscopy techniques, such as microwave, millimeter and far infrared wavelengths. In order to illustrate this point better we give in Table II the values of parameters of interest for three simple representative antiferromagnets. $RbMnF_3$ has a cubic structure (perovskite) in which the magnetic ions are Mn^{2+}. Since these are S-state ions, the spin-orbit coupling is negligible and there is no crystalline anisotropy. Another possible source of anisotropy in crystals is the dipole-dipole interaction. In these materials this interaction is also almost cancelled out due to the cubic lattice, so that its anisotropy field is very small (4.5 Oe). For this reason the gap frequency in $RbMnF_3$ falls in a convenient microwave range, and its magnon properties have been widely investigated [20]. Another extensively studied antiferromagnet [2,20-23] is MnF_2, which has a tetragonal structure (rutile). Due to a smaller distance between the spins along the c-axis (see Fig. 10) than along the a-axis, the dipole-dipole field is not cancelled out and the anisotropy field is three orders of magnitude larger than in $RbMnF_3$. MnF_2 is said to be an ideal uniaxial antiferromagnet because of its very simple structure, its S-state ions that are free from the hard-to-calculate electronic orbital effects and because its parameters make it accessible to be studied with a variety of experimental techniques. The other representative antiferromagnet [24-26] is FeF_2, which has the same structure as MnF_2 and several common properties. The main difference between them is due to the very large anisotropy field of FeF_2, which results from the spin-orbit interaction caused by the non S-state of the Fe^{2+} ions. As a consequence, for instance, FeF_2 has a much larger gap than MnF_2 and the temperature dependence of their thermodynamic properties is different.

We shall not go into as much detail on the properties of antiferromagnetic magnons as we did for ferromagnets. It is interesting, however, to review briefly some of their unique properties. For concreteness let us focus attention on the "ideal" system, MnF_2. Its magnetic structure is illustrated in Fig. 10. The F^- ions have overlapping orbitals which lead to superexchange interactions between the Mn^{2+} spins. The interaction between the near neighbors on opposite sublattices is much larger than between the spins on the same sublattice ($J_2 >> J_1$ and $Z_2 = 8$, $Z_1 = 2$). It is easy to show that the γ_k given by (4.7) which enters the dispersion relation is

$$\gamma_k = \cos\frac{k_x a}{2}\cos\frac{k_y a}{2}\cos\frac{k_z c}{2} \tag{4.11}$$

Note that one of the uniform mode (k=0) frequencies vanishes at $H_0=H_c$, in which there is a transition to the spin-flop phase.

Sublattice Magnetization: An antiferromagnet such as the one we have been studying has no net spontaneous magnetization because of the cancellation of the moments of the two sublattices. However, one can define a sublattice magnetization as the magnetic moment per unit volume, per sublattice. This is a useful concept not only from a theoretical point of view, but also of experimental interest, since it can be measured directly. One way of measuring M(T) is by nuclear magnetic resonance (NMR) [24]. As we shall see later in this chapter, the nuclear spins in a magnetic material are under the influence of a hyperfine field created by the electronic spins. This is a large field and usually determines the NMR frequency. In MnF_2, for instance, both Mn^{55} and F^{19} nuclear spins resonate in the fields created by the Mn^{2+} spins. The measurement of the NMR ν^{19}, which is a sharp and easily detectable line, gives direct information on the T dependence of the sublattice magnetization.

The change in the sublattice magnetization can be found as in Sec. 2B. Using the expressions $S_i^z = S - a_i^\dagger a_i$ and $S_j^z = -S + b_j^\dagger b_j$ and the transformations (4.5) and (4.8) one can show that

$$S - \langle S_i^z\rangle = \frac{1}{N}\sum_k (u_k^2\langle n_{\alpha k}\rangle + v_k^2\langle n_{\beta k}\rangle + v_k^2)$$

$$\langle S_j^z\rangle - S = \frac{1}{N}\sum_k (u_k^2\langle n_{\beta k}\rangle + v_k^2\langle n_{\alpha k}\rangle + v_k^2) \tag{4.12}$$

The temperature dependence of the two sublattice magnetizations is contained in the first two terms, which involve the thermal occupation numbers. This temperature dependence is different than in ferromagnets for two reasons, namely, the existence of the gap and the admixtures of the two spin wave modes. One can show [15], for instance, that even in the absence of the gap (H_A=0) M(T) varies with T^2 at low temperatures, rather than $T^{3/2}$ as in a ferromagnet. This is due to the k-dependence of u_k^2 and v_k^2. Spin wave theory can describe very well the behavior of M(T) at temperatures quite close to the Neel temperature. As in the ferromagnetic case, at the high temperature region it is important to use the correct dispersion relation, including the effect of the energy renormalization, and to carry out the integrations in the finite Brillouin zone. The results of this calculation [9] for MnF_2 together with the experimental data are shown in Fig. 11.

The last terms in the two equations (4.12) represent deviations of the z-components of the spins from the fully aligned antiparallel configuration assumed. They are independent of temperature and therefore exist at T=0. They are called the zero-point deviation of the magnetization and are due to the fact that the Neel ground state is not the true quantum ground state. This zero-point derivation is usually a few percents of the value of S and has been measured in several crystals.

Magnon Interactions: When the expansions of the spin operators (4.4) are extended to higher order terms one obtains an interaction Hamiltonian, in the same manner as in ferromagnets. These magnon interactions are responsible for several effects, such as the energy renormalization and the

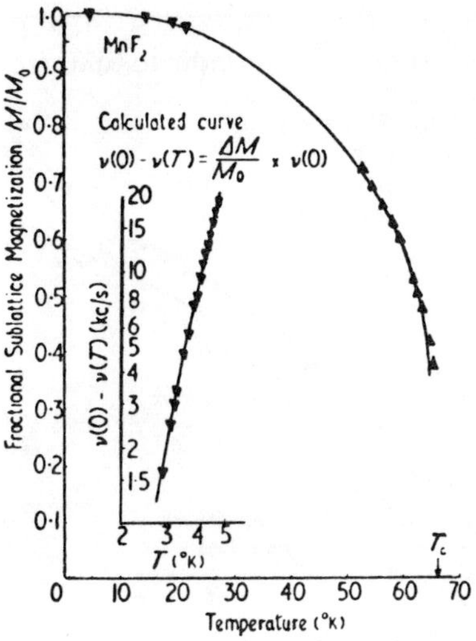

FIG.11. Theoretical and experimental variation with temperature of the sublattice magnetization of MnF_2. The solid curves correspond to calculated values based on a model involving exchange interactions between both nearest and next nearest neighbours and a temperature dependent anisotropy field [9, 24].

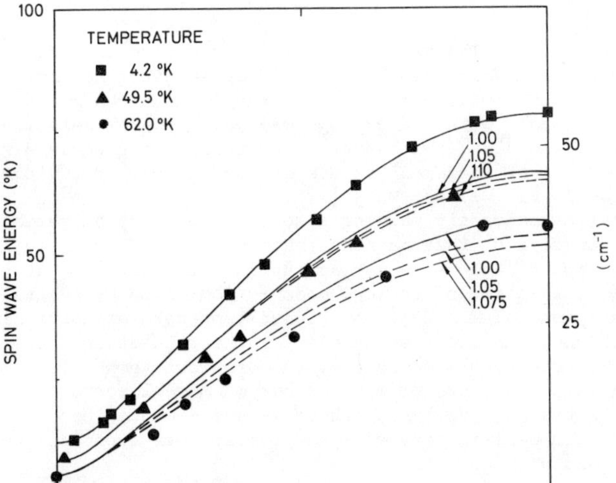

FIG.12. Neutron scattering data for magnons with k along the c-axis in MnF_2 at three temperatures. The curves represent the results of the calculations including energy renormalization, for different values of an adjustable parameter [27].

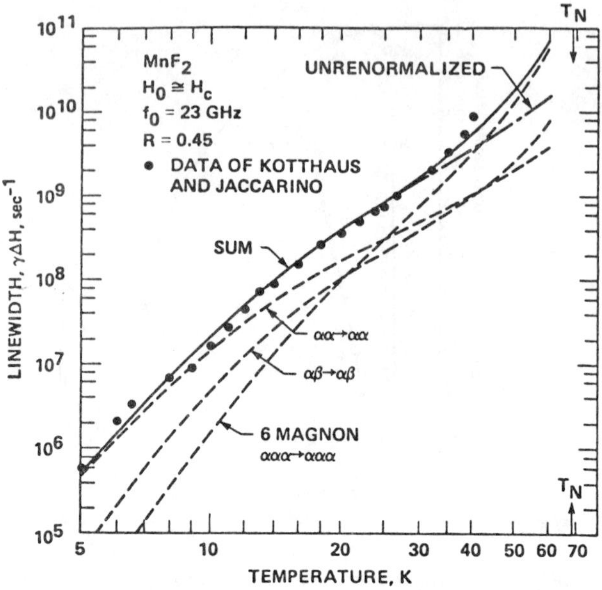

FIG.13. Comparison of theoretical [16] and experimental [22] linewidths in MnF_2. AFMR data taken with a large applied field which reduces the frequency of the k = 0 β mode to 23 GHz (note that at zero field the frequency is 260 GHz).

magnon relaxation. The renormalization gives rise to a temperature dependence in the frequency, which can be measured by means of several techniques. With AFMR at zero applied field, for example, one can measure $\omega_{k=0}(T)$ directly [21]. Crystals with larger gaps may have $\omega_{k\simeq 0}(T)$ measured by one-magnon light scattering [26]. With neutron scattering the whole dispersion curve can be measured at different values of the temperature, yielding data such as is shown in Fig. 12 for MnF_2 [27].

Antiferromagnetic magnons also decay due to the various interaction processes involving other magnons, phonons, etc. The decay rates can be calculated much in the same way as outlined in Sec. 3C. At very low temperatures the relaxation of the k=0 mode is dominated by several temperature independent mechanisms [22]. At higher temperatures intrinsic magnon-magnon interactions are responsible for the relaxation. One then expects that as the temperature approaches the critical temperature, processes involving larger numbers of magnons become more important. This has been shown to be the case for MnF_2, where 4- and 6-magnon processes lead to results that can explain very well the temperature dependent part of the AFMR (k=0) linewidth (Fig. 13).

5. MIXED EXCITATIONS

A. Nuclear Magnons: The nuclei of the atoms are known [24] to possess angular momentum characterized by the operator $\hbar\vec{I}$ and an associated magnetic moment,

$$\vec{\mu}_I = \gamma_N \hbar \vec{I} \tag{5.1}$$

where γ_N is the nuclear gyromagnetic ratio. Its value for a proton is $\gamma_N^1 = 2.68 \times 10^4$ sec^{-1} gauss^{-1}. For Mn^{55}, $\gamma_N^{55} = 6.6 \times 10^3$ sec^{-1} Gauss^{-1}. Notice that

$$\frac{\gamma_N^1}{\gamma_e} = 1.52 \times 10^{-3} \tag{5.2}$$

which indicates that nuclear Zeeman energies are much smaller than the electronic ones. The nuclear angular momentum components obey the same commutation relations as S. Thus clearly many of the spin electronic phenomena in paramagnetic materials have nuclear counterparts. In particular, in an external dc magnetic field H_0 the nuclear spins tend to precess with frequency $\omega_0 = \gamma_N H_0$, which usually falls in the range 1-100 MHz. The nuclear magnetic resonance (NMR) frequency is a result of various interactions, the most important of which is the dipole-dipole coupling between nuclear and electronic spins, which are called the hyperfine interactions. In the simplest form, it can be written as

$$\mathcal{H}_{IS} = \hbar A\, \vec{I}\cdot\vec{S} \tag{5.3}$$

where A is the hyperfine constant. This hyperfine coupling changes the NMR frequency to

$$\omega_N = \gamma_N H_0 + AS \tag{5.4}$$

In paramagnetic materials, in which there is no strong coupling between electronic spins, the nuclear spins are independent of each other. Therefore there is no collective excitation of the nuclear spins. In strongly magnetic materials, however, the nuclear spins can interact indirectly, via the hyperfine coupling with electronic spins and the exchange interaction. This interaction is illustrated in Fig. 14a. It was first proposed by Suhl [28] and Nakamura [29] to explain the linewidth of the NMR of nonmagnetic ions in ordered materials. Later de Gennes et al. [30] showed that this coupling leads to collective nuclear modes with a k-dependent frequency. In order to understand the origin of these nuclear modes let us take the following Hamiltonian for an electron-nuclei spin system,

$$\mathcal{H} = \sum_i \gamma_e \hbar H_0 S_i^z - \sum_i \gamma_N \hbar H_0 I_i^z - \sum_{\substack{i \\ j=i+\delta}} J\vec{S}_i\cdot\vec{S}_j + \sum_i \hbar A \vec{I}_i\cdot\vec{S}_i \tag{5.5}$$

Notice that we have taken the electronic Zeeman term with the positive sign. This is done here to contrast it with the nuclear term, because they do have opposite signs. The ground state of the system is illustrated in Fig. 14b. In the spin wave approximation we can introduce spin-deviation

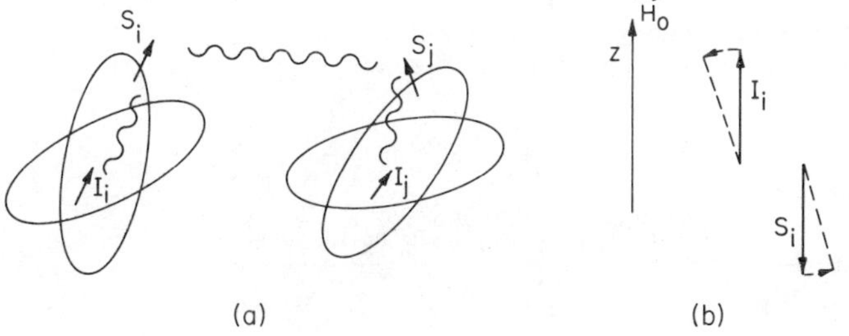

FIG.14. (a) Indirect coupling between nuclear spins. (b) Electron-nucleus spin ground state.

creation and annihilation operators for both electronic and nuclear spins. As usual we set

$$S_i^+ = (2S)^{1/2} a_i^\dagger \qquad I_i^+ = (2\langle I^z\rangle)^{1/2} b_i$$
$$S_i^- = (2S)^{1/2} a_i \qquad I_i^- = (2\langle I^z\rangle)^{1/2} b_i^\dagger$$
$$S_i^z = -S + a_i^\dagger a_i \qquad I_i^z = \langle I^z\rangle - b_i^\dagger b_i \tag{5.6}$$

The use of $\langle I^z\rangle$ instead of I in (5.6) is explained by the fact that, evan at low temperatures, say at liquid helium, the upper Zeeman levels of the nuclear system are reasonably well populated and $\langle I^z\rangle << I$. The use of the average is well justified because it does [30] lead to meaningful collective excitations of the system. Inserting (5.6) in (5.5) and dropping quartic terms gives

$$\mathcal{H} = \gamma_e \hbar H_0 \sum_i a_i^\dagger a_i + \gamma_N \hbar H_0 \sum_i b_i^\dagger b_i + 2SJ \sum_{i,\delta} (a_i^\dagger a_i - a_i^\dagger a_{i+\delta})$$
$$+ \hbar A \sum_i \left[\langle I^z\rangle a_i^\dagger a_i + S b_i^\dagger b_i + (S\langle I^z\rangle)^{1/2} (a_i b_i + a_i^\dagger b_i^\dagger) \right] \tag{5.7}$$

The presence of the hyperfine interaction shows that neither the pure electronic magnons nor the simple nuclear spin excitations can be eigenstates of the Hamiltonian (5.5). To understand the nature of the new normal modes we first introduce the collective spin variables a_k and b_k,

$$a_i = \frac{1}{N^{1/2}} \sum_i e^{i\vec{k}\cdot\vec{r}_i} a_k \qquad b_i = \frac{1}{N^{1/2}} \sum_i e^{i\vec{k}\cdot\vec{r}_i} b_k \tag{5.8}$$

Substitution in (5.7) gives

$$\mathcal{H} = \sum_k \left[\hbar (\omega_k + A\langle I^z\rangle) a_k^\dagger a_k + \hbar \omega_N b_k^\dagger b_k + \hbar A (S\langle I^z\rangle)^{1/2} (a_k b_{-k} + a_k^\dagger b_{-k}^\dagger) \right] \tag{5.9}$$

where ω_k is the frequency of pure ferromagnetic magnons, given in (1.13). If the last term in the Hamiltonian is neglected, the eigenstates of the system are the direct products of the independent electronic and nuclear excitations, $|n_k^e\rangle|n_k^N\rangle$, where

$$|n_k^e\rangle = \frac{(a_k^\dagger)^{n_k^e}}{\sqrt{n_k^e!}} |0\rangle \qquad |n_k^N\rangle = \frac{(b_k^\dagger)^{n_k^N}}{\sqrt{n_k^N!}} |0\rangle$$

Using perturbation theory we can calculate the correction of the energy of the excitations due to the term $\mathcal{H}_{SI} = \hbar A (S\langle I^z\rangle)^{1/2} (a_k b_{-k} + a_k^\dagger b_{-k}^\dagger)$

$$\varepsilon_k^N = \varepsilon_k^{(0)} + \langle n_k^N, n_k^e | \mathcal{H}_{SI} | n_k^N, n_k^e\rangle +$$
$$+ \sum_{\substack{n'^N_k \\ n'^e_k}} \frac{\langle n_k^N, n_k^e | \mathcal{H}_{SI} | n'^N_k, n'^e_k\rangle \langle n'^N_k n'^e_k | \mathcal{H}_{SI} | n_k^N, n_k^e\rangle}{\varepsilon_k^{(0)} - \varepsilon'^{(0)}_k} \tag{5.10}$$

where the zero-order term is $\varepsilon_k^{(0)} = \hbar\omega_N n_k^n + \hbar(\omega_k + A\langle I^z\rangle) n_k^e$, and the first-order correction is zero, because the operators in the Hamiltonian change the numbers by one unit, generating orthogonal states. As we want to calculate the effect of the hyperfine interaction on the nuclear excitation, we consider that the electronic system is in its vacuum state; i.e., $|n_k^e\rangle = |0\rangle$. Next we assume that the nuclear energies in the denominator are negligible compared to the electronic magnon energy, a condition which is comfortably satisfied in usual cases, and note that the only magnon state which contributes to the sum is $|n_k'^e\rangle = |1_{-k}^e\rangle = a_{-k}^\dagger|0\rangle$. The correction to the nuclear energy thus becomes

$$\varepsilon_k^{(2)} = \sum_{n_k'^N} \hbar^2 A^2 S\langle I^z\rangle \frac{\langle n_k^N|b_k|n_k'^N\rangle\langle n_k'^N|b_k^\dagger|n_k^N\rangle}{-\hbar\omega_k} \tag{5.11}$$

which can be reduced further with the use of the completeness relation

$$\sum_{n'^N} |n_k'^N\rangle\langle n_k'^N| = 1$$

We see that the effect of the electronic magnons on the nuclear excitations is to create an effective interaction between nuclear spins with Hamiltonian

$$\mathcal{H}_{SN} = -\hbar\sum_k \frac{A^2 S\langle I^z\rangle}{\omega_k} b_k^\dagger b_k \tag{5.12}$$

The energy eigenvalues of this Hamiltonian depend on the wavevector $\vec{k}$. Therefore the nuclear excitations have a collective nature in the crystal. The Hamiltonian (5.12) is called the Suhl-Nakamura Hamiltonian. As usual $\gamma_N H_0 << AS$, ($\omega_N \simeq AS$), the frequency in (5.4) becomes

$$\omega_N(k) \simeq \omega_N\left(1 - \gamma_e\frac{H_N}{\omega_k}\right) \tag{5.13}$$

where $H_N = A\langle I^z\rangle/\gamma_e$ is the hyperfine field seen by the electronic spins. The second term in (5.12) is the fractional frequency pulling of the NMR frequency due to the Suhl-Nakamura interaction.

We can gain more insight into the nature of the Suhl-Nakamura interaction by rewriting the interaction Hamiltonian (5.12) using (5.6) and (5.8),

$$\mathcal{H}_{SN} = -\sum_{ii'} \frac{\hbar A^2 S}{2N} \sum_k \frac{1}{\omega_k} e^{i\vec{k}\cdot(\vec{r}_{i'}-\vec{r}_i)} I_i^+ I_{i'}^-,$$

which we can write as

$$\mathcal{H}_{SN} = -\frac{1}{2}\sum_{ii'} B_{ii'} I_i^+ I_{i'}^- \tag{5.14}$$

where

$$B_{ii'} = \frac{\hbar A^2 S}{N}\sum_k \frac{e^{i\vec{k}\cdot(\vec{r}_{i'}-\vec{r}_i)}}{\omega k}$$

The Hamiltonian (5.14) represents an effective interaction between nuclear spins, which resembles the exchange interaction between electronic spins. It is an indirect interaction, which takes place via the emission of a virtual electronic magnon by a spin flip I_i^-, and the absorption of this magnon by another spin flip I_i^+. This process is similar to the indirect interaction between nuclear spins via conduction electrons in metals, the Ruderman-Kittel interaction [31]. The distance through which the Suhl-Nakamura interaction is effective is determined by the range function $B_{ii'}$. Under suitable approximations this function can be cast in a convenient form. These approximations are $\omega_k = \gamma H + Dk^2$, $\Sigma \rightarrow (N\Omega/2^3\pi^3) \int d^3k$ and extension of the upper limit of the integral to ∞. This last approximation is worst for the interaction between nearest neighbor spins, because what it essentially does is to assume a continuous medium. Using these in (5.15) and making $\vec{R} = \vec{r}_{i'} - \vec{r}_i$ we obtain

$$B(R) = \frac{\hbar A^2 S}{D} \frac{\Omega}{(2\pi)^3} \int d^3k \frac{e^{ikR}}{\alpha^2+k^2}$$

where $\alpha \equiv (\gamma H/D)^{1/2}$ which is, for a simple cubic lattice,

$$\alpha = \left(\frac{g\mu_B H}{2SJa^2}\right)^{1/2} = \left(\frac{1}{a}\frac{H}{H_E}\right)^{1/2} \qquad (5.16)$$

We can recognize the integrand as the Fourier transform of a Yukawa potential. Performing the integral we obtain (28)

$$B(R) = B\frac{e^{-\alpha R}}{R}$$

$$B = (\hbar^2 A^2 Sa/8\pi SJ) \qquad (5.17)$$

From (5.16) and (5.17) we see that the interaction between nuclear spins is proportional to $(\hbar A)^2/J$ and decays rapidly with distance. Its range is of the order of $a(H_E/H)^{1/2}$, which is typically 10-30 lattice spacings. It is actually a long range interaction when compared with the exchange interaction. This interaction was first proposed by Suhl and Nakamura to explain the broad NMR lines of magnetic materials.

Instead of using the approximation just described, one can also diagonalize (5.9) exactly, using the same formalism employed in previous sections. To apply it here we first rewrite (5.9) as [32,33]

$$\mathcal{H} = \sum_k A_k a_k^\dagger a_k + B b_k^\dagger b_k + F(a_k b_{-k} - a_k^\dagger b_{-k}^\dagger) \qquad (5.18)$$

Eq. (5.18) has the same form as (4.6). Hence it can be diagonalized with (4.8) - (4.10). The result is

$$\mathcal{H} = \sum_k \hbar\,\omega_{\alpha k} \alpha_k^\dagger \alpha_k + \hbar\omega_{\beta k} \beta_k^\dagger \beta_k \qquad (5.19)$$

$$\hbar\,\omega_{\substack{\alpha k\\ \beta k}} = \frac{A_k - B}{2} \pm \left[\left(\frac{A_k - B}{2}\right)^2 - F^2\right]^{1/2}$$

and

$$a_k = u_k \alpha_k + v_k \beta_{-k}^\dagger$$

$$b_{-k}^\dagger = v_k \alpha_k + u_k \beta_{-k}^\dagger \qquad (5.20)$$

$$A_k = \hbar(\omega_k + A\langle I^z\rangle),\ B = \hbar\omega_N = \hbar(\gamma_N H_0 + AS),\ F = \hbar A(S\langle I^z\rangle)^{1/2}$$

one can show that under the condition $\hbar\omega_N << k_B T$

$$\langle I^z \rangle \simeq \frac{\hbar\omega_N I(I+1)}{3K_B T}$$

In usual cases ω_k is in the microwave range ($2\pi \times 10^9 - 2\pi \times 10^{11}$ sec^{-1}, ω_N is an order of magnitude smaller and $F/\hbar$ is even smaller due to the factor $\langle I^z \rangle^{1/2}$. Therefore we can assume that $F/(A_k + B) << 1$ and the results (5.19) and (5.21) can be simplified to (use $\gamma_e H_n \equiv A\langle I^z \rangle$)

$$\omega_{\alpha k} = \hbar^{-1}\left(A_k - \frac{F^2}{A_k + B}\right) \simeq \omega_k + \gamma_e H_N - \frac{\omega_N \gamma_e H_N}{\omega_k}$$

$$\omega_{\beta k} = \hbar^{-1}\left(-B + \frac{F^2}{A_k + B}\right) \simeq -\omega_N \left(1 - \frac{\gamma_e H_N}{\omega_k}\right) \tag{5.22}$$

$$u_k \simeq 1 + \frac{1}{2}\frac{F^2}{(A_k + B)^2} \simeq 1$$

$$v_k \simeq -\frac{F}{(A_k + B)} \simeq -\left(\frac{\omega_N \gamma_e H_N}{\omega_k^2}\right)^{1/2} << 1 \tag{5.23}$$

Examination of (5.22) and (5.23) shows that the complete solution of the eigenmodes of the Hamiltonian (5.7) leads to two types of modes. The α_k mode ($\alpha_k \simeq a_k$) involves essentially the excitation of the electronic spins, whose frequency is slightly different than the pure electronic magnon frequency ω_k. The β_k mode ($\beta_k \simeq b_k$) represents collective excitations of the nuclear spins with frequency different than the usual NMR frequency ω_N. The behavior of these frequencies with wavenumber is shown in Fig. 15. The hyperfine field H_N is of the order of a few Oersted at low temperatures so that the fractional frequency pulling in ferromagnets is of the order of 0.1.

The Suhl-Nakamura interaction is also present in an antiferromagnetic spin arrangement, and so are nuclear magnons. We shall not go through the calculation of the frequencies in this case, but shall quote the results. In a two-sublattice antiferromagnet in the normal state, the nuclear spins have different resonance frequencies in the presence of an external field [23],

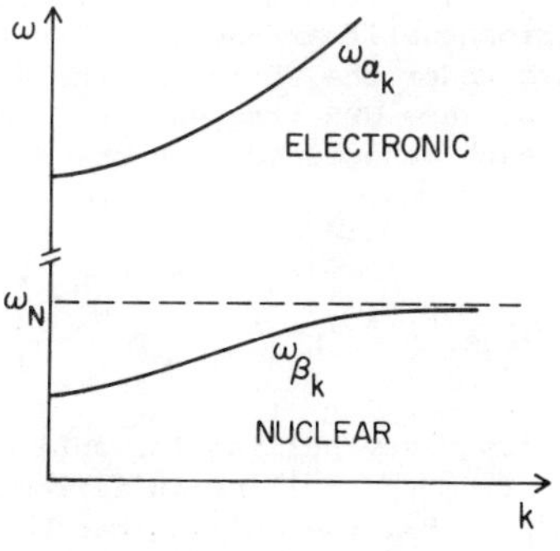

FIG.15. Electronic and nuclear magnon dispersion curves.

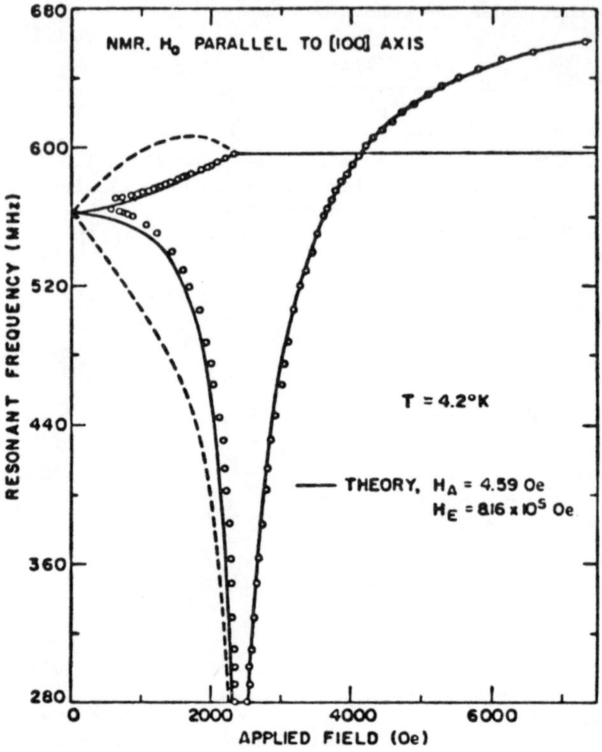

FIG.16. ^{55}Mn NMR frequencies in RbMNF$_3$. (After Ince and Morgenthaler [35]).

since they are oppositely oriented. Call $\omega_{N\uparrow,\downarrow}$ the two unpulled NMR frequencies. It can be shown that the two nuclear magnon frequencies are in this case

$$\omega_{\uparrow,\downarrow}(k) = \omega_{N\uparrow,\downarrow}\left(1 - \frac{2\gamma_e^2 H_E H_N}{\Omega_{\alpha k}\Omega_{\beta k}}\right)^{1/2} \tag{5.24}$$

where $\Omega_{\alpha k}$ and $\Omega_{\beta k}$ are the two electronic magnon frequencies (4.9).

In a flopped antiferromagnetic arrangement, in which the nuclear spins of the two sublattices are under the action of the same net field (external + hyperfine), they have the same NMR frequencies. In this case there are still two nuclear modes, each associated with one electronic magnon [30,34] and having frequencies

$$\omega_{\alpha,\beta}(k) = \omega_N\left(1 - \frac{2\gamma_e^2 H_E H_N}{\Omega^2_{\alpha k,\beta k}}\right)^{1/2} \tag{5.25}$$

Note that the fractional frequency pulling for antiferromagnetic magnons differs by a factor of H_E/Ω_k from the pulling in ferromagnets. Thus, in materials where at small k $\Omega_k \ll H_E$, the pulling can be large. This condition is satisfied in several low-anisotropy antiferromagnets, such as $RbMnF_3$ and $CsMnF_3$, whose electronic frequencies lie in the low microwave range. A

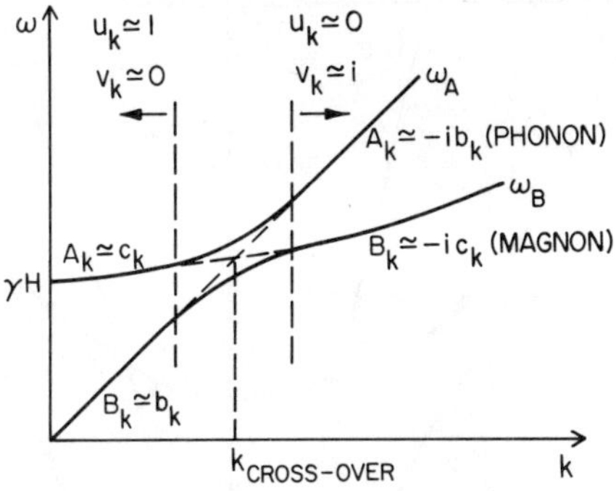

FIG.17. Magnetoelastic dispersion curves.

number of interesting experiments have been performed in these crystals, which allowed the measurements of several characteristics of nuclear magnon propagation. Fig. 16 shows, for example, measurements of the Mn^{55} NMR frequencies, i.e., the k=0 magnon frequency in $RbMnF_3$. Note that at fields close to the spin flop value (2400 Oe) the pulling is largest because the electronic magnon frequency is smallest.

B. Magnetoelastic Waves: The non-linear magnon-phonon interactions in (3.1) represent processes which can relax the magnetic excitations we have been studying. The linear term (3.6) however, represents processes in which a magnon is converted into a phonon with the same energy and momentum and vice-versa. Therefore it gives rise to well defined collective excitations with magnetic and lattice vibration admixtures [2,13,36,37]. In order to study these excitations we consider a magnon-phonon system described by the Hamiltonian

$$\mathcal{H} = \sum_k \hbar\omega_m(k)\, c_k^\dagger c_k + \hbar\omega_p(k)\, b_k^\dagger b_k + i\hbar\,\frac{\sigma_k}{2}\,(c_k^\dagger b_k - b_k^\dagger c_k) \tag{5.26}$$

where

$$\sigma_k = k\left(\frac{4b^2 S^3 N}{\hbar\,\omega_p(k)\rho V}\right)^{1/2} \tag{5.27}$$

In (5.26) the first term is the pure electronic magnon quadratic Hamiltonian, the second term is the pure phonon energy and the last one is the linear magnon-phonon interaction. This Hamiltonian can be diagonalized with the usual canonical transformations. The results are [38]

$$\omega_A(k) = \frac{1}{2}(\omega_p + \omega_m) + \omega_b$$

$$\omega_B(k) = \frac{1}{2}(\omega_p + \omega_m) - \omega_b$$

$$\omega_b = \left[\omega_d^2 + \frac{\sigma_k^2}{4}\right]^{1/2} \qquad \omega_d = \frac{1}{2}(\omega_p - \omega_m) \tag{5.28}$$

and

$$c_k = u_k A_k + v_k B_k$$

$$b_k = v_k A_k + u_k B_k \tag{5.29}$$

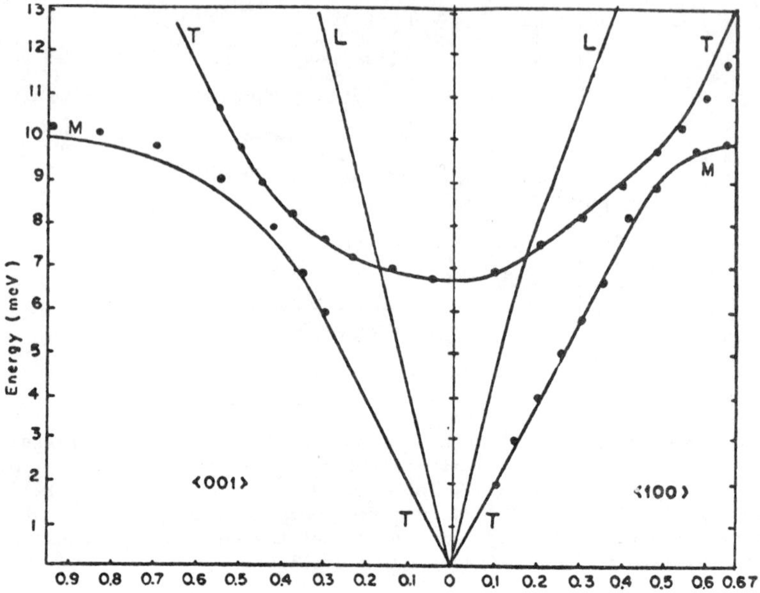

FIG.18. Measured dispersion relations for phonons and magnons in FeF_2 [43].

The dispersion curves for the modes A_k and B_k are shown in Fig. 17. They are essentially the pure magnon and phonon curves except in the region where they cross each other, where there is a splitting. In the cross-over point, defined by $\omega_m(k_c) = \omega_p(k_c)$, the difference between the two normal mode frequencies is $\omega_A(k_c) - \omega_B(k_c) = \sigma_k$. This frequency splitting is proportional to the magnetoelastic constant b and usually corresponds to a few percents of the cross-over frequency. Near the cross-over point the excitations are mixed, involving magnons and phonons.

The fact that the normal mode excitations of a magnetoelastic system contain either magnon or phonon contribution, depending on the values of the wavenumber or the frequency relative to the crossover values, allows one to convert magnons into phonons and vice-versa. Experiments of this type have been widely done in YIG [39-42]. This phenomenon has considerable potential for device application and has contributed to extensive investigations in this field.

Magnetoelastic waves can also exist in antiferromagnets as long as the phonon-magnon interaction is strong enough. This is the case of FeF_2, where the large spin-orbit coupling of the Fe^{2+} ions gives origin to a sizeable magnetostrictive coupling. The magnon and phonon dispersion curves have been measured directly in beautiful neutron scattering experiments [43]. The results, shown in Fig. 18, clearly demonstrate the magnetoelastic splitting of the dispersion curves. The values of the frequency splitting give direct information on the magnetoelastic constants b.

6. MAGNONS IN DISORDERED SYSTEMS

The crystals we have studied so far were considered magnetically perfect, i.e., formed by absolutely regular patterns of spins with a characteristic unit cell. Crystals with more than one magnetic specie distributed at random also present a great deal of interest. In case the concentration of one of the constituents is small, it is more properly referred to as a magnetic impurity. In this case the energy of the impurity spin excitation may lie outside the bands of the host spin waves, giving rise to localized magnons. In case the concentrations of the constituents are large, the excitations may contain admixtures of all the spins. For didactic purposes we consider the two situations separately.

A. The Impurity Problem: Localized Magnons: The general impurity problem in solids seems to have first been analyzed in depth by Lifshitz [44]. The most extensive work has been on the electron problem [45], in which one is mainly interested in the modifications of the electronic energy structure caused by the impurities. The phonon problem [46] has also been largely studied and more recently the magnetic problem is attracting attention [47]. A magnetic impurity in a strongly magnetic host is an ion with a spin S' different from the ones of the host and/or with an exchange interaction with the host spins characterized by a different constant J'. The simplest problem and the only one we shall treat is that of an impurity in a ferromagnetic insulator. We shall use the single-impurity approximation, which is valid for concentrations up to a few percents. The treatment presented here [48] is somewhat simpler than the usual [49]. Consider a ferromagnetic crystal described only by a nearest neighbor exchange Hamiltonian

$$\mathcal{H} = - \sum_{i,\delta} J_{i,i+\delta} \vec{S}_i \cdot \vec{S}_{i+\delta} \tag{6.1}$$

Assume that the spin at the lattice site i=0 is S'; its exchange constant with the nearest neighbors is J', whereas all the other spins are the same S, with exchange constant J. We may rewrite (6.1) as

$$\mathcal{H} = - J \sum_{i,\delta} \vec{S}_i \cdot \vec{S}_{i+\delta} + 2J \sum_{\delta} \vec{S}_0 \cdot \vec{S}_\delta - 2J' \sum_{\delta} \vec{S}'_0 \cdot \vec{S}_\delta \tag{6.2}$$

The first term now represents the Hamiltonian for a crystal with no impurities. If J,J' > 0, the ground state has all spins parallel. So we can introduce boson spin deviation operators in the usual manner. We have for the host and impurity spins,

$$\begin{aligned} S_i^+ &= \sqrt{2S}\, a_i & S'^+_0 &= \sqrt{2S'}\, a_0 \\ S_i^- &= \sqrt{2S}\, a_i^\dagger & S'^-_0 &= \sqrt{2S'}\, a_0^\dagger \\ S_i^z &= S - a_i^\dagger a_i & S_0^z &= S' - a_0^\dagger a_0 \end{aligned} \tag{6.3}$$

As in Sec. 1.B we can express (6.2) in terms of the new operators and retain only the quadratic part. The result can be written as

$$\mathcal{H} = \mathcal{H}_0 + V \tag{6.4a}$$

$$\mathcal{H}_0 = 2JS \sum_{i,\delta} (a_i^\dagger a_i - a_i^\dagger a_{i+\delta}) \tag{6.4b}$$

$$V = 2JSz\varepsilon a_0^\dagger a_0 + 2JS\rho \sum_\delta a_\delta^\dagger a_\delta - 2JS\gamma \sum_\delta (a_0^\dagger a_\delta + a_\delta^\dagger a_0) \tag{6.4c}$$

where

$$\varepsilon = (\frac{J'}{J} - 1) \qquad \rho = (\frac{J'S'}{JS} - 1) \qquad \gamma = \left(\frac{J'}{J}\sqrt{\frac{S'}{S}} - 1\right) \tag{6.5}$$

Notice that the term $\mathcal{H}_0$ is the same Hamiltonian for the pure crystal studied in Section 1.B, except that here we have considered the applied field zero. Ther term V has the same structure as $\mathcal{H}_0$. Therefore we can diagonalize the complete Hamiltonian with the more general canonical transformations

$$a_i = \sum_\lambda \Gamma_\lambda(i) a_\lambda \qquad a_i^\dagger = \sum_\lambda \Gamma_\lambda^*(i) a_\lambda^\dagger \tag{6.6}$$

where the eigenfunctions $\Gamma_\lambda(i)$ satisfy the completeness and orthonormality conditions

$$\sum_i \Gamma_\lambda(i)\Gamma_{\lambda'}^*(i) = \delta_{\lambda\lambda'} \qquad \sum_\lambda \Gamma_\lambda(i)\Gamma_\lambda^*(i') = \delta_{ii'}^\dagger \tag{6.7}$$

which are in agreement with the commutation relations for the new operators

$$[a_\lambda, a_{\lambda'}^\dagger] = \delta_{\lambda\lambda'} \qquad [a_\lambda, a_{\lambda'}] = [a_\lambda^\dagger, a_{\lambda'}^\dagger] = 0 \tag{6.8}$$

The transformations (6.6) and the conditions (6.7) allow the total Hamiltonian to be written in the form

$$\mathcal{H} = \sum_\lambda E_\lambda a_\lambda^\dagger a_\lambda \tag{6.9}$$

where the eigenfunctions $\Gamma_\lambda(i)$ and the energies E_λ are subjected to the eigenvalue equation

$$E_\lambda \Gamma_\lambda(i) = 2\sum_\delta J_{i,i+\delta}\left[S_{i+\delta}\Gamma_\lambda(i) - (S_i S_{i+\delta})^{1/2}\Gamma_\lambda(i+\delta)\right] \tag{6.10}$$

where $J_{i,i+\delta} = J$ for exchange between host atoms or J' for the exchange between a host atom and the impurity and S_i is the spin at site i, which is S for i ≠ 0 and S' for i = 0. This eigenvalue equation reduces to the one for a perfect ferromagnet when S = S' and J = J'. Its solution here, however, is more complicated because of the lack of translational symmetry. In order to solve (6.10) we rewrite it as

$$E_\lambda \Gamma_\lambda(i) = \sum_j (\mathcal{H}_{ij}^0 + V_{ij})\Gamma_\lambda(j) \tag{6.11}$$

where $\mathcal{H}_{ij}^0$ and V_{ij} are the elements of N x N matrices given by

$$\mathcal{H}_{ij}^0 = 2JS\,\delta_{ij} + 2JS\,\delta_{i-j,\delta} \tag{6.12a}$$

$$V_{ij} = 2JS\{z\varepsilon\delta_{ij}\delta_{i,0} + \rho\delta_{ij}\delta_{i\delta} - \gamma\delta_{i-j,\delta}(\delta_{i,0} + \delta_{j,0})\} \tag{6.12b}$$

Eq. (6.11) can be written in matrix form defining the N x 1 eigenfunction matrix Γ and the N x N matrices (6.12). It then becomes

$$E_\lambda \Gamma = (\mathcal{H}_0 + V)\Gamma \tag{6.13}$$

which can also be cast in the form

$$(I + G_0 V)\Gamma = 0 \tag{6.14}$$

where I is the N x N unit matrix and G_0 is the Green's function matrix for the pure crystal (so-called because its poles are the energies of the excitations of the pure system).

$$G_0 = (E_\lambda - \mathcal{H}_0)^{-1} \equiv \frac{1}{E_\lambda - \mathcal{H}_0} \tag{6.15}$$

Now we are led to define the Green's function for the impure crystals as

$$G = \frac{1}{E_\lambda - \mathcal{H}} = \frac{1}{E_\lambda - \mathcal{H}_0 - V} \tag{6.16}$$

which can be rewritten as

$$G = G_0 + G_0 V G \tag{6.17}$$

The advantage of using the Green's functions is now apparent. Eq. (6.17) has the form of a Dyson equation, which can be solved with methods similar to those used in quantum physics problems. V is the perturbation of the system and G_0 is the Green's function of the unperturbed system. The interpretation of (6.17) is that a spin excitation can propagate from site i to j either freely (by the propagator G_0) or by scattering at an intermediate site ℓ, one or more times. The energies and eigenfunctions of the unperturbed system are the same as the ones given in Sec. 1.B,

$$E_k = \hbar \omega_k = 2JSz(1-\gamma_k)$$
$$\Gamma_k^{(0)}(i) = \frac{1}{\sqrt{N}} e^{i\vec{k}\cdot\vec{r}_i} \tag{6.18}$$

Note that the Green's functions for the pure crystal obey the equation

$$(E_\lambda - \mathcal{H}_0) G_0 = I \tag{6.19a}$$

or

$$\sum_j (E_\lambda - \mathcal{H}_0)_{ij} G^0_{jn} = \delta_{i,n} \tag{6.19b}$$

With a Fourier transform this is easily solved to give

$$G^0_{ij} = \frac{1}{N} \sum_k \frac{e^{i\vec{k}\cdot(\vec{r}_i - \vec{r}_j)}}{E_\lambda - E_k} \tag{6.20}$$

The energies of the "perturbed" system are then given by (6.14), which leads to

$$\det(I - G_0 V) = 0 \tag{6.21}$$

Notice that V is a matrix with only (z+1) x (z+1) non-zero elements. Therefore having found G_0 the eigenvalue problem is greatly reduced with respect to the original one. Eq. (6.21) has 7 roots in the case of a simple cubic crystal and can be solved with the help of the point group symmetry of the impurity. Using group theory techniques one can find the unitary transformation which reduces (6.14) to the form of blocks, which characterize the linear combinations of the wavefunctions that are the eigensolutions of the problem. For a simple cubic structure (point group Oh) the unitary transformation gives [47] two s-like linear combinations, three p-like combinations and two d-like combinations. The determinant (6.21) can then be written as the product of three determinants

$$\det(I - G_0V) = D^2(s)D^3(p)D^2(d) = 0 \quad (6.22a)$$

where

$$D(s) = 1 + \varepsilon + \frac{\rho}{2JSz}E - \left[(\varepsilon - \rho)E + \frac{\rho}{2JSz}E^2\right]G^0_{00}(E)$$

$$D(p) = 1 - 2JS\rho\left[G^0_{00}(E) - G^0_{12}(E)\right]$$

$$D(d) = 1 - 2JS\rho\left[G^0_{00}(E) + G^0_{12}(E) - 2G^0_{13}(E)\right] \quad (6.22b)$$

Eq. (6.22) gives the normal mode energies E_ν of the complete system. In order to understand better the nature of the eigenmodes let us look at the behavior of the density of states of the excitations. First we recall that the density of states is given by

$$\rho(\omega) = \frac{1}{N}\sum_\lambda \delta(\omega - E_\lambda) \quad (6.23)$$

Note that for the spin wave band modes k, the density of states is proportional to the imaginary part of the Green's function in (6.20) for i = j and and $E_\lambda \to \omega + i\varepsilon(\varepsilon \to 0)$. (Which just gives the delta function in (6.23)). Now, if the energies associated with the impurity modes, given by (6.22), lie outside the magnon band, i.e., $E_\nu > E_k$, the Green's functions (6.20) are real for $E = E_\nu$. Therefore the roots of (6.22) can be real. The density of states of these modes, which are now associated with the imaginary part of (6.16), will correspond to delta-like density of states as shown in Fig. 19a. The wavefunctions associated with these states are non-zero only around the impurity ion and the spin excitations are localized. On the other hand, if E_ν lies inside the magnon band the Green's functions (6.20) are complex. Therefore, the roots will have necessarily real and imaginary parts and the density of states is not delta-like but shows the resonance behavior depicted in Fig. 19b. Therefore the modes show a finite lifetime, even in the absence of losses. These modes are called resonant, and the spin excitations associated with them are not localized. Localized magnons have been extensively studied experimentally by several techniques, such as NMR [50], light scattering [51], infrared absorption [52], neutron scattering [53], and recently by direct far-infrared laser excitation [54]. Most experiments have been performed on the fluoride antiferromagnets. The analysis presented here can be applied to a two-sublattice antiferromagnet without much additional complication [49].

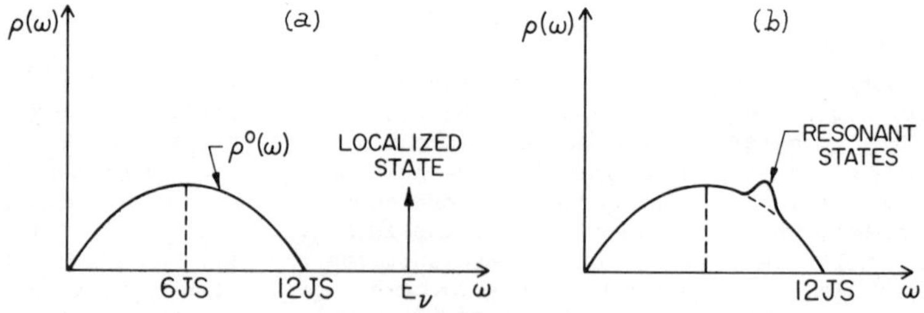

FIG.19. (a) Density of states of impure ferromagnet with localized magnons. (b) Density of states with resonant (virtual) modes.

B. Excitations in Disordered Systems: In randomly mixed magnetic crystals we expect that the spins of all the magnetic species interact with each other, resulting in spin wave excitations with spectra more complicated than the one studied in the previous section. Of course, since the exact specie configuration cannot be known, the problem here does not have an exact solution. Fortunately several of the measurable quantities are macroscopic; i.e., they involve sums over a large number of sites, so that averaging techniques may be used to calculate them in an approximate way. The most successful technique for dealing with the excitations in disordered magnetic systems is the so-called coherent potential approximation (CPA). This method has been used to study other excitations in disordered crystals, as recently reviewed by Elliott, et al. [55]. Here we will just outline the general CPA procedure and present some results obtained for spin wave excitations.

The starting point can be the Dyson equation (6.17) with the perturbation potential at each site now allowing for the possibility of different species being on neighboring sites. Since one wants to calculate macroscopic quantities, one can average the equation over all possible specie configurations. This averaging restores the translational invariance to the Green's

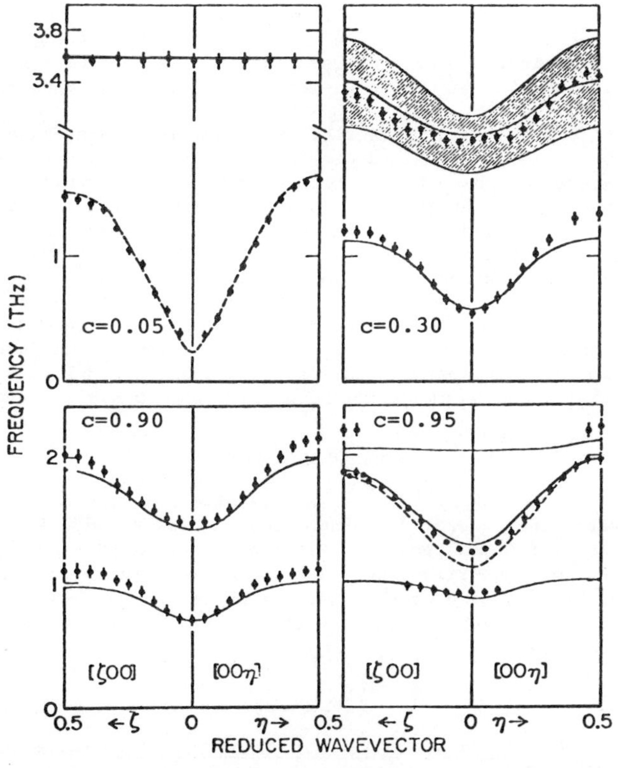

FIG.20. Dispersion relations of spin excitations in the disordered antiferromagnet $(Co)_c(Mn)_{1-c}F_2$. The solid lines are CPA calculations [57]. The dashed lines are the known dispersion curves for the pure materials. Dots are experimental points [60, 61]. The shaded region for c = 0.30 gives the width predicted by calculations. (After Cowley [60]).

function G(E) for the disordered system. One can now seek solutions to the Dyson equation of the form

$$G^c(E,k) = G_0(E,k) + G_0(E,k)\Sigma(E,k)G^c(E,k) \qquad (6.24)$$

where the superscript c indicates a configurational average and Σ is the perturbation as yet to be determined. Now assume that the medium can be characterized by an effective Green's function G^e, defined by $G^e=G_0+G_0\Sigma G^e$. The CPA condition is that in this random lattice there is no net scattering of the spin waves described by the propagator G^e. We can eliminate G_0 between this equation and (6.17) to obtain (a N x N matrix equation)

$$G = G^e + G^e(V - \Sigma)G \qquad (6.25)$$

Note that Σ behaves as an effective potential on each site. If we write $\Delta \equiv V-\Sigma$ and require that $G^c = G^e$, we obtain the condition for the scattering to be zero. This condition can be written in terms of a T-matrix for the site i

$$\langle T_i \rangle_c = \langle \Delta_i (1-\Delta_i G^e)^{-1} \rangle_c = 0 \qquad (6.26)$$

This configurational averaging is performed under suitable approximations which restrict the number of terms in the sum. With (6.24) - (6.26) the coherent potential and the Green's function can be determined self-consistently. The latter can then be used to calculate quantities of interest such as the frequencies of the spin excitation, the lineshapes, density of states, etc. [56-60].

In recent years a large number of experimental results have been obtained [55,60] on the excitations of mixed and diluted magnetic systems. Nearly all experiments have been done with the fluoride antiferromagnets, since they have simple lattice and magnetic structures, are relatively easy to prepare and can be made with a variety of spin configurations with the basic parameters of the spin Hamiltonian known. Fig. 20 illustrates some results obtained by neutron scattering on the mixed system $(Mn)_{1-c}(Co)_cF_2$ [60,61]. Note that for c = 0.05 the dispersion curves are essentially those for the host spin waves in MnF_2 (the same band shown in Fig. 12) and the dispersionless curve for the Co impurity mode. At larger Co concentration two mixed spin excitation dispersion curves are present. Finally, for c = 0.95 we see essentially the Mn impurity modes below the CoF_2 spin wave band.

ACKNOWLEDGEMENTS

I would like to acknowledge all my colleagues at the Departamento de Fisica, Universidade Federal de Pernambuco, in Recife, and, in particular, Cid B. de Araujo, E.A. Soares, I.P. Fittipaldi, L.C.M. Miranda, M.A.C.G. Moura and M.D. Coutinho Filho, who contributed largely to my understanding of a considerable fraction of the material presented in these notes. I am also grateful to Ms. Jennifer Fawcett and Ms. Sandy Paredes for patiently typing the manuscript.

REFERENCES

[1] Morrish, A.H., The Physical Principles of Magnetism (Wiley, New York, 1965). This book presents at an introductory level a large number of phenomena in magnetism.

[2] Keffer, F. "Spin Waves" in Handbuch der Physik, Vol. XVIII/B, Ed. Flugge, S., (Springer-Verlag, 1966). This is quite a complete review of the subject up to 1966.

[3] Bloch, F., Z. Physik 61 (1930), 206; 74 (1932), 295.
[4] Tyablikov, S.V., Methods in the Quantum Theory of Magnetism (Plenum, New York, 1967).
[5] Holstein, T. and Primakoff, H., Phys. Rev. 58 (1940) 1098.
[6] Sparks, M. Ferromagnetic Relaxation Theory (McGraw-Hill, New York, 1964).
[7] Dyson, F.J., Phys. Rev. 102 (1956) 1217.
[8] White, R.M., Quantum Theory of Magnetism (McGraw-Hill, New York, 1970)
[9] Low, G.G., Proc. Phys. Soc. 82 (1963) 992.
[10] Glauber, R.J., Phys. Rev. 131 (1963) 2766.
[11] Rezende, S.M. and Zagury, N., Phys. Letters 29A (1969) 47.
[12] Zagury, N. and Rezende, S.M., Phys. Rev. B4 (1971) 201.
[13] Akhiezer, A.I., Bar'yakhtar, V.G. and Peletminskii, S.V., Spin Waves (North-Holland, Amsterdam, 1968).
[14] Zubarev, D.N., Soviet Phys. Uspekhi 3 (1960) 320.
[15] Kittel, C., Quantum Theory of Solids (Wiley, New York, 1963).
[16] Rezende, S.M. and White, R.M., Phys. Rev. B (to be published).
[17] Haas, C.W. and Callen, H.B., in Magnetism I, ed. by Rado, G.T. and Suhl, H. (Academic Press, New York, 1963).
[18] Sparks, M., Phys. Rev. 160 (1967) 364.
[19] Le Craw, R.C. and Spencer, E.G., J. Phys. Soc. Japan, Suppl. 17 (1962) 401.
[20] See, for example, Ince, W.J., Phys. Rev. 184 (1969) 574 and Platzker, A. and Morgenthaler, F.R., Phys. Rev. Letters 26, (1971) 442.
[21] Johnson, F.M. and Nethercott, A.H., Jr., Phys. Rev. 114 (1959) 705.
[22] Kotthaus, J.P. and Jaccarino, V., 18th Ann. Conf. Mag. and Mag. Materials, AIP Conference. Proceedings 10 (1973) 57.
[23] Paquette, D., King, A.R. and Jaccarino, V., Phys. Rev. B11 (1975) 1493.
[24] Jaccarino, V."Nuclear Resonance in Antiferromagnets", in Magnetism II-A, ed. Rado, G.T.and Suhl, H.(Academic Press, New York, 1965).
[25] Ohlmann, R.C. and Tinkham, M., Phys. Rev. 123 (1961) 425.
[26] Fleury, P.A., Proc. 2nd Int'l. Conf. Light Scattering, (Ed., M. Balkanski, Flamarion, Paris, 1971), p. 151.
[27] Low, G.G., Proc. Symposium on Inel. Scattering of Neutrons in Solids and Liquids, 1964 (IAEA, Vienna 1965) and Turberfield, K.C., Okazaki, A. and Stevenson, R.W.H., Proc. Phys. Soc. 85 (1965) 743.
[28] Suhl, H., Phys. Rev. 129 (1958) 606.
[29] Nakamura, T., Prog. Theoret Phys. (Kyoto) 20 (1958) 542.
[30] de Gennes, P.G., Pincus, P.A., Hartmann-Boutron, F. and Winter, J.M., Phys. Rev. 139 (1963) 1105.
[31] Ruderman, M.A. and Kittel, C., Phys. Rev. 96 (1954) 99.
[32] White, R.M., Sparks, M., and Ortenburger, I., Phys. Rev. 139 (1965) A450.
[33] Ninio, F., and Keffer, F., Phys. Rev. 165 (1968) 735.
[34] Richards, P.M., Phys. Rev. 173 (1968) 581.
[35] Ince, W.J., and Morgenthaler, F.R., Phys. Letters 29A (1969) 106.
[36] Kittel, C., Phys. Rev. 110 (1958) 836.
[37] Schlomann, E., J. Appl. Phys. 31 (1960) 1647.
[38] Almeida, J.L.R. and Rezende, S.M., Phys. Stat. Solidi b73 (1976) 661.
[39] Schlomann, E., and Joseph, R.I., J. Appl. Phys. 35 (1964) 2382.
[40] Eshbach, J.R., J. Appl. Phys. 34 (1963) 1298.
[41] Rezende, S.M. and Morgenthaler, F.R., J. Appl. Phys. 40 (1969) 524.
[42] Auld, B.A., Magnetostatic and Magnetoelastic Wave Propagation in Solids, in Applied Solid State Science, Vol. 2, Ed. R. Wolfe (Academic Press, New York, 1971).
[43] Rainford, B.D., Houmann, J.G., and Guggenheim, H.J., Proc. Symposium on Inel. Scattering of Neutrons in Solids and Liquids (IAEA, Vienna, 1972) p. 655.

[44] Lifshitz, J.M., J. Exp. Theor, Phys. (JETP) 17 (1947) 1017; Ibid., 17 (1947) 1076; Ibid., 18 (1948) 293.

[45] Koster, G.F. and Slater, J.C., Phys. Rev. 96 (1954) 1208.

[46] Dawber, P.G. and Elliott, R.J., Proc. Phys. Soc.273 (1963) 222.

[47] Wolfram, T. and Callaway, F., Phys. Rev. 130 (1963) 2207.

[48] Fittipaldi, I.P., Rezende, S.M. and Miranda, L.C.M., Solid State Communications 13 (1973) 1797; in more detail in Fittipaldi, I.P. and Rezende, S.M., Phys. Rev. B12 (1975) 5081.

[49] For a review of the theory, see, for example: Hone, D., "Localized Magnons" in Localized Excitations in Solids, Wallis, R.F., Ed. (Plenum Press, New York, 1968).

[50] Butler, M. Jaccarino, V., Kaplan, N., and Guggenheim, H.J., Phys. Rev. B1 (1970) 3058.

[51] Oseroff, A.R. and Persham, P.S., Proc. 2nd Int. Conf. Light Scattering (Ed., Balkanski, M., Flammarion, Paris, 1971), p. 156.

[52] Enders, B., Richards, P.L., Tennant, W.E., and Catalano, E., 19th Ann. Conf. Magnetism and Magnetic Materials (1973), AIP Conf. Proc. 18 (1974) 179.

[53] Cowley, R.A. and Buyers, W.J.L., Rev. Mod. Phys. 44 (1972) 406.

[54] Durr, U. and Button, K.J., Sol. St. Commun. 16 (1975) 695.

[55] Elliott, R.J., Krumhansl, J.A. and Leath, P.L., Rev. Mod. Phys. 46 (1974) 465.

[56] Tahir-Kheli, R.A., Phys. Rev. B6 (1972) 2808.

[57] Buyers, W.J.L., Pepper, D.E., and Elliott, R.J., J. Phys. C: Solid St. Phys. 5 (1972) 2611.

[58] Harris, A.B., Leath, P.L., Nickel, B.G. and Elliott, R.J., J. Phys. C: Solid St. Phys. 7 (1974) 1693.

[59] Holcomb, W.K., J. Phys. C: Solid St. Phys. 7 (1974) 4299.

[60] Cowley, R.A., 21st Ann. Conf. on Magnetism and Magnetic Materials (1975), AIP Conf. Proc. (to appear).

[61] Buyers, W.J.L., Holden, T.M., Svensson, E.C., Cowley, R.A. and Stevenson, R.W.H., Phys. Rev. Lett. 27 (1972) 1442.

IAEA-SMR-20/56

PLASMONS IN CRYSTALLINE MEDIA, PARTICULARLY SEMICONDUCTORS

E. TOSATTI
GNSM-CNR, Istituto di Fisica,
Università di Roma,
Rome,
Italy

Abstract

PLASMONS IN CRYSTALLINE MEDIA, PARTICULARLY SEMICONDUCTORS.

The theory of plasmons in periodic crystals is discussed. Besides the phenomenological theory, both the dielectric matrix and the equation of motion approach are reviewed. Plasmon bands are discussed in two separate models, the single pole model and the nearly free plasmon model, derived from these two approaches. The resulting plasmon gaps, and the possibility of their experimental observation, especially in semiconductors, are discussed.

1. INTRODUCTION

Plasmons are usually described as charge density oscillations of a *homogeneous* electron gas [1]. A uniform positive background, the jellium, is supposed to exist in order to ensure charge neutrality. The plasmon energy values and dispersion with wavevector $\vec{q}$ thus obtained, compare reasonably well with the experimental results for simple metals and often also d-band metals and semiconducting and insulating systems [1, 2]. This justifies the initial neglect of the periodic distribution of ions in real crystals. It is clear, however, that a more detailed and complete account of all the plasmon properties in crystals cannot be obtained without renouncing the jellium approximation.

The modifications which specifically arise from the presence of a periodic lattice may roughly be divided into effects at $q \approx 0$ and effects at $q \approx G/2$, G being a reciprocal lattice vector. There is actually no fundamental difference between them in that both arise because of the substitution of a free electron parabola $E^0 = k^2/2m$ with a real band structure $E = E(k)$, and both consist of a shift of the plasmon energy, plus a broadening (i.e. a complex plasmon self-energy). However, the physical implications and the end result are distinct for the two cases.

In this paper we describe the general framework for a theory of plasmons in periodic media, considering both dielectric matrix and equation of motion approaches. We do not discuss at length the results available for near $q = 0$, which is the situation best investigated so far. Relevant for this case are Refs [1, 34, 35] and [36], and previous works cited therein. We apply the two approaches introduced earlier to study $q = G/2$, which has received virtually no attention in the past. In particular, we calculate the gap in the plasmon band structure near $q = G/2$. A discussion of the results and of their experimental implications, particularly for what concerns group IV semiconductors, concludes the paper.

2. MACROSCOPIC DESCRIPTION: ELECTROMAGNETIC NORMAL MODES

A plasmon is, first of all, an elementary excitation, that is an approximate eigenstate of the microscopic system hamiltonian. Thus its energy can, in principle, only be obtained as a result of a microscopic calculation. However, its nature is collective [1] rather than quasi-particle and this, in turn, implies macroscopically observable quantities connected with it. For a charge density oscillation, one such quantity is the electric field. In the presence of an undamped plasmon of momentum q and energy ω in a general medium a self-sustaining electric field is expected of the plane-wave like form $E_{q\omega}\exp[i\vec{q}\cdot\vec{r}-i\omega t]$. This means the plasmon is also an electromagnetic normal mode and as such is contained in the macroscopic Maxwell's equations. We start from [3]

$$i\vec{q}\wedge\vec{E}=i\frac{\omega}{c}\vec{B} \tag{1a}$$

$$i\vec{q}\wedge\vec{B}=-i\frac{\omega}{c}(\overleftrightarrow{\epsilon}\cdot\vec{E})+\frac{4\pi}{c}\vec{j}^{\,ext} \tag{1b}$$

in Fourier space, where $\vec{E}$, $\vec{B}$, $\vec{j}^{\,ext}$ and the dielectric tensor $\overleftrightarrow{\epsilon}$ are understood to be functions of $\vec{q}$ and ω. Combining these equations one eliminates $\vec{B}$ (the magnetic permeability tensor is included in the tensor $\overleftrightarrow{\epsilon}$, as was first done by Lindhard [4–6], yielding the normal equation connecting electric field and external current

$$F_{ij}E_j=i\frac{4\pi\omega}{c^2}\,j_i^{ext} \tag{2}$$

where i, j = x, y, z, summation over repeated indices is understood, and

$$F_{ij}=q^2\delta_{ij}-q_iq_j-(\omega/c)^2\epsilon_{ij} \tag{3}$$

may be called the "Fresnel tensor", characterizing the microscopic properties of the system. Electromagnetic normal modes are obtained by asking for $E\neq 0$ with $j^{ext}=0$. That is only possible in the linear problem (2) if

$$\det F_{ij}=0 \tag{4}$$

This is, or course, the usual wave equation, also called in optics the Fresnel equation. Plasmons must be contained among its solutions. To look at what they are, it is useful to introduce longitudinal and transverse projectors

$$\begin{aligned} P_{ij}^L &= q_iq_j/q^2 \\ P_{ij}^T &= \delta_{ij}-q_iq_j/q^2 \end{aligned} \tag{5}$$

In isotropic media one can define longitudinal and transverse scalar dielectric functions

$$\epsilon_{ij}(q,\omega)=\epsilon^T_{(q,\omega)}P_{ij}^T+\epsilon^L_{(q,\omega)}P_{ij}^L \tag{6}$$

which are related to the conventional ϵ and μ simply as

$$\begin{aligned} \epsilon^L &= \epsilon \\ \epsilon^T &= \epsilon+\frac{c^2q^2}{\omega^2}(1-\mu^{-1}) \end{aligned} \tag{7}$$

whence $\epsilon^L=\epsilon^T$ for $q\to 0$. The Fresnel operator is

$$F_{ij}=[q^2-(\omega/c)^2\epsilon^T]P_{ij}^T-(\omega/c)^2\epsilon^LP_{ij}^L \tag{8}$$

yielding the isotropic Fresnel equation

$$-(\omega/c)^2\epsilon^L[q^2-(\omega/c)^2\epsilon^T]^2=0 \tag{9}$$

Apart from $\omega=0$, which is not a physical solution unless the system is unstable (that is, out of thermodynamical equilibrium [4]), this equation factors into a twofold degenerate mode

$$q^2-(\omega/c)^2\,\epsilon^T(q,\omega)=0 \tag{10}$$

which is transverse since it implies

$$F_{ij}E_j=-(\omega/c)^2\epsilon^L P^L_{ij}E_j=0$$

and a nondegenerate mode

$$\epsilon^L(q,\omega)=0 \tag{11}$$

which is longitudinal. Only the latter involves a charge density oscillation, as implied by, for example,

$$i\vec{q}\cdot\vec{E}=4\pi\rho^{tot} \tag{12}$$

Thus expression (11) should be identified with the dispersion relation of undamped plasmons in isotropic media. In real media, where damping is present, this is replaced by

$$\mathrm{Im}\,[\epsilon^L(q,\omega)]^{-1}=\text{maximum} \tag{13}$$

Finally, the generalization to anisotropic media is easily obtained in the form

$$\mathrm{Im}\,[q_i\epsilon^L_{ij}(q,\omega)q_j]^{-1}=\text{maximum} \tag{14}$$

Here the field is not strictly longitudinal, unless $\vec{q}$ is in a principal direction, in which case Eq.(14) degenerates into a plasmon *and* a transverse mode. For a general direction, the solution of Eq.(14) is just the extraordinary ray of optics.

3. MICROSCOPIC DESCRIPTION: DIELECTRIC MATRIX

So far we have assumed macroscopically homogeneous media; this is equally correct for the truly homogeneous electron gas as for a liquid or a crystal, where the macroscopic electron distribution is not uniform. Differences between these cases arise, of course, in a microscopic description of the dielectric response. Let us look at the longitudinal response, by constructing the screened potential V(r) starting from a bare perturbation $V^0(r)$.

In a truly homogeneous medium

$$V(r)=\int d^3r'\epsilon^{-1}(r,r')V^0(r') \tag{15}$$

where ϵ^{-1} is the inverse dielectric function and the dependence of $r-r'$ alone is required by full translational invariance. By the convolution theorem, the Fourier transform of Eq.(15) is just a product

$$V_q = \epsilon_q^{-1} V_q^0 \tag{16}$$

This is no longer true for a general inhomogeneous medium where Eq.(15) becomes

$$V(r) = \int d^3r' \epsilon^{-1}(r, r') V^0(r') \tag{17}$$

yielding

$$V_q = \sum_{q'} \epsilon_{q,q'}^{-1} V_{q'}^0 \tag{18}$$

where

$$\epsilon_{\vec{q},\vec{q}'}^{-1} = \int d^3r\, d^3r'\, e^{i\vec{q}\cdot\vec{r} + i\vec{q}'\cdot\vec{r}'} \epsilon^{-1}(\vec{r}, \vec{r}') \tag{19}$$

Owing to inhomogeneity, the dielectric function is now replaced by a dielectric *matrix* (DM) depending on two wavevectors q and q′. For a fully inhomogeneous case the matrix is continuous, i.e. all values q and q′ are coupled by a finite response. On the other hand, in a fully homogeneous situation, only the diagonal elements are non-zero $\epsilon_{qq}^{-1} = \epsilon_q^{-1}$. We shall be interested in crystals, which are intermediate between the two cases, being invariant only under finite translations of multiples of lattice vectors. It is useful in this case to split the full wavevector in the form $\vec{q} + \vec{G}$, where $\vec{q}$ is now restricted to the first Brillouin zone and $\vec{G}$ is a reciprocal lattice vector. One easily finds that the screened potential in this case is

$$V_{\vec{q}+\vec{G}} = \sum_{\vec{G}'} \epsilon_{\vec{q}+\vec{G},\vec{q}+\vec{G}'}^{-1} V_{\vec{q}+\vec{G}'}^0 \tag{20}$$

The dielectric matrix of a crystal is thus infinite but discrete. Its rows and columns are labelled by reciprocal lattice vectors.

In principle, the dielectric response is calculable once the exact eigenstates and energy eigenvalues of the crystal are known [1]. In practice however, a reasonable knowledge of the one-electron bands and wavefunctions are about the best one can obtain (with only a few exceptions, like the electron gas). With this input, it is still possible to calculate the DM to an approximation which is often called the SCF (self-consistent field) or RPA (random-phase approximation) and is given by [7]

$$\epsilon_{q+G,q+G',\omega} = \delta_{GG'} + \frac{4\pi e^2}{(q+G)^2} \sum_{k\ell\ell'} \frac{f_{k\ell} - f_{k+q\ell'}}{E_{k+q\ell'} - E_{k\ell} - \omega + i0^+}$$

$$\times \langle k\ell | e^{-i(\vec{q}+\vec{G})\cdot\vec{r}} | k+q\ell' \rangle \langle k+q\ell' | e^{i(\vec{q}+\vec{G}')\cdot\vec{r}} | k\ell \rangle \tag{21}$$

Here $E_{k\ell}$ and $|k\ell\rangle$ denote one electron band energies and Bloch functions and $f_{k\ell}$ is their Fermi occupation.

This DM is in general complex. In a filled band situation, however, such as in semiconductors and insulators, all elements are real for ω real and smaller than the energy gap since then the energy denominator is never vanishing.

Many properties of this DM can be found in the original articles by Adler and Wiser [7] and in other more recent papers [8, 9]. Among them, we notice an important sum rule which concerns the $\omega \to \infty$ behaviour and is perfectly general

$$\epsilon_{\vec{q}+\vec{G},\vec{q}+\vec{G}',\,\omega\to\infty} = \delta_{GG'} - \frac{\omega_p^2}{\omega^2}\frac{(\vec{q}+\vec{G})\cdot(\vec{q}+\vec{G}')}{|\vec{q}+\vec{G}||\vec{q}+\vec{G}'|}\frac{\rho_{G-G'}}{\rho_0} \tag{22}$$

where $\omega_p^2 = 4\pi ne^2/m$ is the free electron plasma frequency and ρ_G denotes Fourier components of the crystal electron charge density.

The next point is how to make connection with the macroscopic description of the previous section. The *macroscopic* dielectric function is a scalar (built from this *microscopic* matrix) which describes the screening of very long wavelength perturbations only. From Eq.(20) one sees that [7]

$$\epsilon^{\text{macro.}}_{q,\omega} = 1/\epsilon^{-1}_{q,q,\omega} \tag{23}$$

Finally, we ask what is the plasmon dispersion relation for a crystal. By the arguments of section 2, it must be just $\epsilon^{\text{macro.}}_{q,\omega} = 0$, which, applied to Eq.(23) yields

$$\det(\epsilon_{q+G,q+G',\omega}) = 0 \tag{24}$$

We shall consider an application of this formula to a model plasmon band calculation in section 5.

4. ALTERNATIVE MICROSCOPIC APPROACH: EQUATION OF MOTION

A related, but practically distinct, approach to the study of plasmons in crystals is that of looking at the equation of motion for the electron density. This is supposed to be made up of a static part ρ_0 plus a small time-varying plasmon part ρ_1

$$\rho(r,t) = \rho_0(r) + \rho_1(r,t) \tag{25}$$

March and Tosi [10] have derived an equation of motion for ρ_1 as a straightforward generalization of the well-known homogeneous case. Starting from the equation of motion for the single-particle density matrix and proceeding as in Singwi et al. [11] they obtain

$$\ddot{\rho}_1(r,t) + \omega_p^2(r)\rho_1(r,t) = m^{-1}\nabla\rho_0(r)\cdot\nabla V_1(r,t) + m^{-1}\rho_1(r,t)\cdot\nabla^2 V_i(r)$$

$$+ m^{-1}\nabla\rho_1(r,t)\cdot\nabla V_i(r) - m^{-2}\int dx'dt' D(r,x',t-t')\cdot V_1(x',t') \tag{26}$$

which is valid within RPA and so long as $|e_1| \ll \langle e_0 \rangle$.

In the formula $\omega_p^2(r) = 4\pi e^2\rho_0(r)/m$, V_1 is the time-dependent Hartree potential set up by ρ_1,

$$V_1(\vec{r},t) = e^2\int d^3r'\rho_1(\vec{r}',t)/|\vec{r}-\vec{r}'| \tag{27}$$

m is the free electron mass, $V_1(r)$ is the periodic potential set up by $\rho_0(r)$, and D is given by

$$D(r, x', t-t') = i\hbar m^{-1} \sum_{\alpha\beta} \nabla_r^\alpha \nabla_r^\beta \nabla_{r'}^\alpha \nabla_{r'}^\beta \sum_{kk'\Omega} \Psi_k^* \left(r + \frac{r'}{2}\right) \Psi_{k'} \left(r - \frac{r'}{2}\right)$$

$$\times \Psi_{k'}^*(x') \Psi_k(x') e^{-i\Omega(t-t')} [f_{k'} - f_k](E_{k'} - E_k - \Omega - i\eta)^{-1} \qquad (28)$$

where r' is to be set equal to zero after differentiation, α and β denote Cartesian components and the index k subsumes both electron momentum and band index. In the limiting case of a homogeneous electron gas, the first three terms on the right-hand side of Eq.(26) vanish, while it can be shown [10] that the fourth is a kinetic piece that remains and gives for $q \to 0$ the ordinary plasmon dispersion, i.e. in Fourier space

$$-\omega^2 \rho_1 + \omega_p^2 \rho_1 = -\frac{3}{5} q^2 v_F^2 \rho_1 \qquad (29)$$

Although in principle completely equivalent to the dielectric formulation, the equation of motion approach emphasizes one interesting aspect of the plasmon problem. This is that the electron density ρ_1, a function of position and time, behaves like a sort of 'plasmon wave function', in that it has to be the eigenfunction of a linear homogeneous equation like Eq.(26) whose eigenvalue is the square plasma frequency,

$$(K - \omega^2)\rho_1(q, \omega) = 0 \qquad (30)$$

The 'inverse plasmon resolvent' $K - \omega^2$, or the 'plasmon hamiltonian operator' K, contains all the crystal symmetry and periodicity. Thus the eigendensities $\rho_1(q, \omega)$ should also be classified according to the irreducible representations of the crystal symmetry at the point q of the Brillouin zone. 'Plasmon band structure' calculations can be thought up as diagonalization schemes for the hamiltonian K, following closely the methods of one-electron band theory. A cellular method has already been proposed in the original work of March and Tosi [10] and a 'nearly free plasmon' perturbative scheme, discussed by Girlanda, Parrinello and Tosatti [12], is presented in detail in section 6. Other promising possibilities would perhaps be a tight binding of localized plasmons and also variational approaches, but no work is yet available in this direction.

The important fact remains that one can handle plasmon eigencharges as though they were wavefunctions, while, of course, the true plasmon eigenfunction [1] depends upon all the electron co-ordinates and appears extremely cumbersome to handle.

5. PLASMON BANDS IN THE SINGLE OSCILLATOR MODEL

In this section we discuss a model calculation of plasmon energies of a semiconductor in a framework that borrows ideas from the so-called Penn model [13], which so far has been very useful in describing the diagonal dielectric function. What we do is construct a dielectric matrix by inserting into Eq.(21) a very crude band structure consisting of two bands separated by a gap, and by fixing the matrix element to the sum rule (22). The plasmon eigenvalues can then be obtained from Eq.(24) which, particularly near a Brillouin zone border, reduces to a 2 × 2 problem. Though the model is obviously oversimplified in some respects, it serves very well the purpose of showing how plasmons form bands in crystals as suggested in principle by various authors [10, 12, 18].

We first briefly introduce the original Penn model and its result for the diagonal dielectric function of a semiconductor. Following the work of Johnson [9] we shall then extend this approximation to the off-diagonal elements of the DM. The formation of plasmon bands can then be discussed by looking at the zeros of the DM determinant as shown in section 3.

The Penn model is a two-band model of a semiconductor

$\ell = +$, valence

$-$, conduction

$$E_k^{\pm} = \tfrac{1}{2}[(E_k^0 + E_{k'}^0) \pm \sqrt{(E_k^0 - E_{k'}^0)^2 + E_g^2}] \tag{31}$$

$$|k \pm\rangle = (1 + |\alpha_k^{\pm}|^2)^{-1/2}(e^{i\vec{k}\cdot\vec{r}} + \alpha_k^{\pm} e^{i\vec{k}'\cdot\vec{r}}) \tag{32}$$

where

$$E_k^0 = k^2/2m$$

$$\alpha_k^{\pm} = \frac{E_g/2}{E_k^{\pm} - E_{k'}^0} \tag{33}$$

This is easily recognized as the one-dimensional 'nearly free electron' (NFE) model for electrons in a periodic potential of Fourier component $E_g/2$ [14]. In the present case, a 3-dimensional Fermi momentum is assumed for the unperturbed electrons and the gap E_g is taken to open up just at this wavevector, so as to create one completely filled (+) and one completely empty (−) band. However, in a real 3-dimensional crystal, gaps form on Bragg planes, and they cannot fit as exactly the spherical Fermi surface, as in one dimension. The Penn model is a spherical extension of the one-dimensional case and the gap is supposed to fall always on the Fermi sphere. This is achieved by coupling each point k on the sphere with its opposite

$$\vec{k}' = \vec{k} - 2k_F\vec{k}/|\vec{k}| \tag{34}$$

Thus in this model there is, so to speak, a continuous infinity of reciprocal lattice vectors, but they all have the same modulus which is $2k_F$. Penn showed that although this model is quite unrealistic for $\omega \sim E_g$ where the approximations made are too crude, it could nonetheless provide quite a sensible result for $\omega \ll E_g$, and in particular he calculated $\epsilon_{q,q,0}$ to be approximately given by

$$\epsilon_{q,q,0} = 1 + \omega_p^2/\Omega_{qq}^2 \tag{35}$$

where

$$\Omega_{qq} \approx E_g\left\{1 + \frac{E_F}{E_g}\cdot\frac{q^2}{k_F^2}\right\}, \qquad E_F = k_F^2/2m \tag{36}$$

The result (35) has a simple pole structure at the effective 'Penn gap' Ω_{qq} which concentrates all the oscillator strength of what is in reality a branch cut extending from the minimum gap upwards. This is of course very reminiscent of Lorentz's oscillator model with the frequency of the oscillator equal to the Penn gap. The new interesting feature is that the pole energy depends on wavevector, which provides the 'spatial dispersion' (i.e. the q-dependence) of ϵ that one was looking for.

At this point, we notice that there exists, besides $\omega \sim 0$, another frequency regime where the single pole approximation is expected to work well. This is the high-frequency regime, where

$\omega \gg E_g$. At very high frequencies the valence oscillator strength is practically exhausted and dispersion is again well described by a single oscillator [15]. Thus we may write

$$\epsilon_{q,q,\omega} \cong 1 + \frac{\omega_p^2}{\Omega_{qq}^2 - \omega^2} \tag{37}$$

Neglecting off-diagonal DM elements, this would give a plasma frequency of the form

$$\omega^2(q) = \omega_p^2 + \Omega_{qq}^2 \tag{38}$$

which is a well-known result, in the absence of local field corrections [15–17]. The crucial ingredients of Eq.(37) are that the pole strength ω_p^2 is fixed by the f-sum rule (22) requiring an asymptotic behaviour like $1 - \omega_p^2/\omega^2$, and that the oscillator frequency is fixed by the band structure as an effective gap, which in turn is given by the static limit (35) and this is just the Penn gap.

We now extend this model to obtain the off-diagonal DM elements. First, we notice that a single-pole approximation is equally justified for them because their imaginary part, coming from the vanishing of energy denominator in Eq.(21), will be non-zero in exactly the same spectral range as for the diagonal elements. Only the matrix elements are different. Second, if a single-pole approximation is assumed, then the pole strength is again uniquely fixed by the generalized f-sum rule (22). We obtain for large ω and $G \neq G'$

$$\epsilon_{\vec{q}+\vec{G},\vec{q}+\vec{G}',\omega} \cong \frac{(\vec{q}+\vec{G})\cdot(\vec{q}+\vec{G}')}{|\vec{q}+\vec{G}||\vec{q}+\vec{G}'|} \left(\frac{\rho_{G-G'}}{\rho_0}\right) \frac{\omega_p^2}{\Omega^2_{\vec{q}+\vec{G},\vec{q}+\vec{G}'} - \omega^2} \tag{39}$$

For the effective off-diagonal gap $\Omega_{q+G,q+G'}$ we may take the approximation suggested by Johnson [9] who studied the static case,

$$\Omega_{\vec{q}+\vec{G},\vec{q}+\vec{G}'} \approx E_g \left\{1 + \frac{E_F}{E_g}\left[\frac{\vec{q}+\frac{1}{2}(\vec{G}+\vec{G}')}{k_F}\right]^2\right\} \tag{40}$$

We now have in Eqs (37) and (39) the full DM in the single-pole approximation. We can apply it to plasmons for all those cases where the plasma frequency is expected to lie well above the effective gaps. This is certainly true of the metals, but also of several semiconductors and moderate gap insulators. Roughly speaking, the criterion of applicability is that the static dielectric constant $\epsilon_{0,0}$ should be much larger than 2, because we know that this will imply through Penn's formula (35) that

$$\omega_p^2/\Omega_{qq}^2 \gg 1 \tag{41}$$

Looking at the valence semiconductors, for instance, we see that this is well fulfilled for α-Sn ($\epsilon_0 = 24$), Ge ($\epsilon_0 = 16$), Si ($\epsilon_0 = 12$), and even diamond ($\epsilon_0 = 5.5$).

The plasmon bands can now be obtained from

$$\det(\epsilon_{q+G,q+G',\omega}) = 0 \tag{24'}$$

which, in principle, involves an infinite matrix. In practice, the ω^2 in the denominator of Eqs (37) and (39) cuts off rather rapidly, so that even small size matrices are already well converged. For illustration, we show the result obtained by restricting to just 2 × 2. This is of course a rather inaccurate evaluation, but it has all the essential features in it, and it is best suited for discussion. We now involve only 0 and G and call q + G = q′,

$$\begin{pmatrix} 1+\dfrac{\omega_p^2}{\Omega_{qq}^2-\omega^2} & \dfrac{\omega_p^2(\rho_G/\rho_0)(\vec{q}\cdot\vec{q}\,')/qq'}{\Omega_{qq'}^2-\omega^2} \\ \dfrac{\omega_p^2(\rho_G/\rho_0)(\vec{q}\,'\cdot\vec{q})/qq'}{\Omega_{qq'}^2-\omega^2} & 1+\dfrac{\omega_p^2}{\Omega_{q'q'}^2-\omega^2} \end{pmatrix} = 0 \qquad (42)$$

Since in general $\Omega_{qq'} \neq \Omega_{qq}$, this equation is of fourth degree. It can be simplified further by noticing that at high ω

$$\frac{(\omega^2-\Omega_{qq}^2)(\omega^2-\Omega_{q'q'}^2)}{(\omega^2-\Omega_{qq'}^2)(\omega^2-\Omega_{q'q}^2)} \approx 1$$

with the end result

$$\omega_\pm^2 = \omega_p^2 + \tfrac{1}{2}(\Omega_{qq}^2+\Omega_{q'q'}^2) \pm [\tfrac{1}{4}(\Omega_{qq}^2-\Omega_{q'q'}^2)^2 + \omega_p^4(\rho_G/\rho_0)^2]^{1/2} \qquad (43)$$

The two plasmon bands obtained correspond, as sketched in Fig.1, to the previously unperturbed plasmon branch $\omega_p^2 + \Omega_{qq}^2$ folded back at the Brillouin zone border q = G/2. The plasmon gap at this point is in this approximation:

$$2\Delta = \omega^+(G/2) - \omega^-(G/2) = \frac{\omega_p^2(\rho_G/\rho_0)}{\sqrt{\omega_p^2+\Omega_{G/2,G/2}^2}} \qquad (44)$$

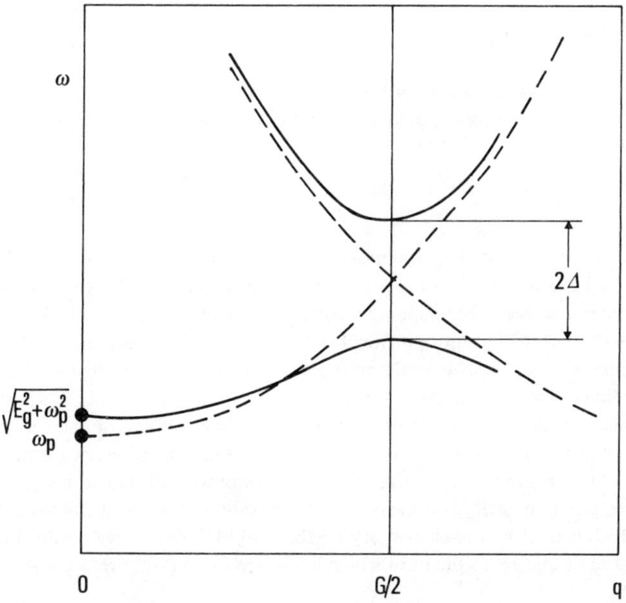

FIG.1. Sketch illustrating the formation of two plasmon bands (solid lines) due to Bragg diffraction at G/2 of an unperturbed plasmon at G/2 with one at −G/2 (dashed lines). The real plasmon gap is also indicated.

This is just the result obtained by Pandey et al. [18], who neglect Penn gaps altogether. With the present scheme, the accuracy can be increased at will, by extending Eq.(42) to include successive shells of reciprocal lattice vectors, and solving the resulting algebraic equation without further approximations. The only input needed for this simple plasmon band structure calculation is the crystal charge density.

It should be noted, however, that so long as one retains *real* pole energies $\Omega_{qq'}$ plasmon damping is completely neglected, which is in general quite unrealistic, except for the lowest plasmon band. We shall return to this problem in the discussion of section 7.

6. THE NEARLY FREE PLASMON MODEL

Here we study the behaviour of plasmon bands in a different model crystal from that of section 5. We now consider an electron gas plus a weak superposed periodic potential

$$v(\vec{r}) = \sum_{\vec{G}} v_G e^{i\vec{G}\cdot\vec{r}} \tag{45}$$

In the free gas ($v_G = 0$) plasmons propagate freely with equation of motion

$$(K_0(q, \omega) - \omega^2)\rho_1(q, \omega) = 0 \tag{46}$$

The free "hamiltonian" K_0, (which is obtained, for example, from Eqs (26) and (28) by setting $\nabla\rho_0 = \nabla V_i = 0$ and assuming free electrons) is diagonalized by plane wave eigencharges, which we denote as $|q\rangle$. Its eigenvalues may be written quite generally as functions of the free electron (Lindhard) dielectric function $\epsilon_{q,\omega}$ in the form

$$\langle q|K_0(q, \omega)|q\rangle = \omega^2(1 - \epsilon_{q\omega}) \tag{47}$$

which is obtained by direct comparison of Eqs (46) and (11).

The effect of the weak periodic potential v_G is to add a perturbation to the plasmon hamiltonian

$$K(q, \omega) = K_0(q, \omega) + u(g, \omega) \tag{48}$$

The operator u may be obtained by subtraction from the equation of motion (26) and can, in general, be expressed as a power series of v_G by direct expansion of ρ_0, V_i, E_k and ψ_k. The effect of the perturbation u is to cause Umklapp scattering of plasmons, i.e. gain or loss of any reciprocal lattice vector G, with a probability proportional to $\langle q|u|q \pm G\rangle$. The result on the plasmon energy spectrum is of course the appearance of plasmon bands, originating from the initial single plasmon branch. The banding effect is strongest at the Brillouin zone border, $q = \pm G/2$, where Bragg scattering takes place, with consequent formation of standing waves and of an energy gap. For a weak potential, Bragg scattering is also weak, and the plasmon energy gap is accurately described, in the expansion of u in powers of v, by the lowest term only which is linear in v_G. This may be called a nearly free plasmon case, for it bears strong analogies with Bragg diffraction of nearly free electrons, which is discussed in elementary textbooks [14]. Below, we outline how the plasmon gap can be evaluated in this approximation which lends itself to a rather transparent physical discussion [19].

The nearly free plasmon gap results from degenerate perturbation theory coupling of plane wave eigencharges at $-G/2$ and $G/2$. A 2×2 secular problem can formally be set up

$$\begin{pmatrix} \langle G/2|K_0|G/2\rangle - \omega_2 & \langle -G/2|u|G/2\rangle \\ \langle G/2|u|-G/2\rangle & \langle -G/2|K_0|-G/2\rangle - \omega^2 \end{pmatrix} = 0 \tag{49}$$

A word of justification concerning straightforward application of simple perturbation theory is needed here. Its derivation in fact is usually based on a hermitian hamiltonian which ensures the orthonormality of eigenstates. However, the present hamiltonian $K_0 + u$ is the general non-hermitian corresponding to complex eigenvalues. This is necessary in order to account for the fact that the free plasmon is already *damped* in certain wavevector regimes and that further Umklapp damping is introduced by u. Assuming, as we do, a single plasmon pole, damping may be accounted for with a complex eigenvalue which implies non-hermiticity of K_0 and u. Nevertheless, it is easily verified that at $q = G/2$ the non-hermiticity is of the simple type $K_{ij}(\omega) = K_{ji}^*(-\omega^*)$. In this case, the orthonormality of the spectrum, and thus applicability of ordinary perturbation theory, is recovered [20] by defining scalar products as $(\psi_i(-\omega^*), \psi_j(\omega))$.

The second point to stress about Eq.(49), is that this is not just a second order equation for ω^2, as it is in the nearly free electron problem. In fact, here K_0 and u themselves depend on ω, and the solutions have to be determined selfconsistently. This can be done by making use of Eq.(47) and by recalling that the unperturbed plasmon eigenvalue $\omega_0^2(G/2)$ is defined as the zero of $\epsilon_{G/2,\omega}$. One obtains

$$\omega - \omega_0 = \pm \left[\omega_0^2 \frac{\partial \epsilon_{G/2,\omega}}{\partial \omega} \bigg|_{\omega_0} \right]^{-1} \langle -G/2|u(\omega_0|G/2\rangle \tag{50}$$

with all quantities complex in general. The complex plasmon gap which we call 2Δ is just twice the right-hand side. When u is obtained, as explained above, by lowest order expansion of Eqs (26) and (28), and matrix elements are taken, the result is

$$\langle -G/2|u(\omega_0)|G/2\rangle = [4\epsilon_{G,0} - 3] \frac{G^2 v_G}{4m} \left\{ 1 - \frac{G^2}{16m} \left(\frac{1}{G^2/8m - \omega_0} + \frac{1}{G^2/8m + \omega_0} \right) \right\} \tag{51}$$

with no further approximations, other than RPA. Thus, the plasmon eigenvalues at G/2 are $\mathrm{Re}\,\omega_0 \pm \Delta_R - i(\mathrm{Im}\,\omega_0 \pm \Delta_I)$, with

$$\Delta_R - i\Delta_I = \frac{G^2 v_G (4\epsilon_{G,0} - 3)}{4m\omega_0^2 [\partial \epsilon_{G/2,\omega}/\partial\omega]_{\omega_0}} \left\{ 1 - \frac{G^2}{16m} \left(\frac{1}{G^2/8m - \omega_0} + \frac{1}{G^2/8m + \omega_0} \right) \right\} \tag{52}$$

The corresponding eigendensities at G/2 are of course

$$\rho_1^{\pm} = 2^{-1/2} [|-G/2\rangle \pm |G/2\rangle]$$

and have a maximum *between* the atoms (+) or *on* the atoms (−) respectively. The gap 2Δ may correspondingly be viewed as due to the different static densities in these two regions.

7. DISCUSSION AND RELEVANCE TO THE EXPERIMENTAL SITUATION

We have qualitatively described the formation of plasmon bands in crystals by means of two different models: the single pole model of section 5 and the nearly free plasmon model of section 6. The single pole model is a high frequency approximation which should be good in all cases where interband transitions – centered about the "Penn gap" E_g – occur well below the plasma frequency, i.e. $E_g/\omega_p \ll 1$. No approximation is involved concerning the strength of the periodic potential, which enters exactly through the sum rule (22), to determine the periodic charge density. The whole plasmon band can be calculated, to any chosen accuracy of convergence, by properly increasing the size of the DM whose determinant is required to vanish. However, no allowance is made for plasmon damping, and the method as it stands is thus applicable only so long as a well-defined plasmon pole, or a narrow plasmon resonance, can be identified in the system. On the other hand, the nearly free plasmon model contains no approximations so far as frequency is concerned, but applies only to crystals whose plasmon properties can be assimilated to those of a weakly perturbed electron gas. A closer inspection of the result of Eq.(52) for the plasmon gap leads to the conclusion that the parameter which must be perturbatively much smaller than unity is $v_G E^0_{G/2}/\omega_p^2$. This, somewhat fortunately, turns out to be a rather small quantity, even for crystals which are in fact a long way away from a nearly free electron gas, at least so far as their low energy properties are concerned. Plasmon damping, which is a very important feature, is included. This is done, however, only within RPA, which is rather crude in this respect, and has no damping for $q < q_c$, the Landau damping critical wavevector. Unlike the single pole model, the validity of the nearly free plasmon model is also restricted to a small neighbourhood of $q = G/2$, where corrections of second and higher orders in v_G may be neglected since they do not contribute to the gap. Away from this region, calculations of plasmon bands require going to second order at least and the method does not appear practicable any longer.

The two models thus have some complementarity and a comparison of their results may be of interest. We note that in principle both should apply not only to simple metals, but also to at least some d-band metals, and valence semiconductors, which largely fulfil the limitations of both.

But before we go on to a quantitative evaluation of the plasmon gaps, it is necessary to somewhat clarify the experimental situation. Plasmon energies and their dispersion near $q = 0$ have been investigated in a large class of materials [1, 2]. Much less work, essentially restricted to metals [21–30], is available on the spectrum near G/2, where pairs of plasmon bands should be better observable. Neither electron energy loss [21–26] nor inelastic X-ray scattering [27–30] has shown any evidence for plasmon bands, though often a very damped plasmon, or its remnants, persists at larger momenta than G/2. This has provided the basis for the belief that plasmon gaps, if they exist, are exceedingly small and generally negligible. The numerical evaluations we shall give below, based on the two models discussed, show that this should very often, *but not always,* be the truth. It may just be necessary to apply more effort to somewhat less natural candidates than the simple metals, such as the semiconductors. The principal obstacle to experimental observation, just as it is to theoretical calculation, is plasmon damping. Its presence does not only cause a broadening of the plasmon peak, which hinders the possible resolution of a doublet structure near G/2; it also fatally quenches Bragg diffraction as soon as the plasmon mean free path is comparable and shorter than the lattice parameter. We have pointed out [12] that it may be necessary to introduce some kind of q-modulation technique to resolve the doublet.

Table I reports the plasmon gaps as calculated by Girlanda, Parrinello and Tosatti [12] on the nearly free plasmon model Eq.(52) for a choice of materials, and those obtained from the approximate result of the single pole model Eq.(44) for Ge and α-Sn. In this latter model, we have not attempted to calculate for the other materials, either because the known damping is much too large for it to be applicable, or simply because we do not know ρ_G. The basic ingredient of the nearly free plasmon calculation is a carefully chosen local pseudopotential form factor v_G, which is taken from Ref.[31], as discussed in Ref.[12]. The results show exceedingly small gaps in all the simple metals, as a consequence of one or more of the following three circumstances: (a) the Brillouin zone border G/2 falls beyond the Landau damping onset wavevector q_c, and the plasmon is washed out; (b) v_G is "accidentally" zero or very small for the first G. (This is of course not really an accident in a simple metal!); (c) non-locality reduces the *repulsive* part of v(g) very much, while it enhances the *attractive* part [31]. For group I and II metals v_G is weakly

TABLE I. COMPLEX ENERGY GAPS 2Δ OF PLASMONS AT $q = G/2$, IN THE SINGLE POLE MODEL (SPM) AND IN THE NEARLY FREE PLASMON MODEL (NFP) (AFTER REF.[12])

	RPA $\omega_0(0)$ (eV)	G $(2\pi/a)$	G/2 (Å^{-1})	q_c (Å^{-1})	RPA $\omega_0(G/2)$ (eV)	NFP $2\Delta_R$ (eV)	NFP $2\Delta_I$ (eV)	SPM $2\Delta_R$ (eV)
Li (bcc)	7.9	110	1.25	0.96	16.6 − i0.37	0.4	0.1	
Cs (bcc)	3.5	110	0.74	0.56	5.6 − i0.06	0.01	0.007	
Be (hcp)	18.4	100	1.59	1.38	33.0 − i3.2	0.06	−0.3	
Mg (hcp)	10.9	100	1.14	1.12	16.6 − i0.28	0	0	
Ca (hcp)	7.8	100	0.91	0.96	10.6	−0.03	0	
Al (fcc)	15.7	111 200	1.34 1.55	1.28	24.7 − i0.07 30 − i4.4	0.01 0.17	0.005 −0.21	
β-Sn (tetr)	14.4	200 101	1.08 1.14	1.25	18.9 19.4	0.9 0.3	0 0	
Pb (fcc)	13.4	111	1.10	1.20	18.0	0.6	0	
Ge (diam)	15.6	111	0.96	1.28	19.0	−2.1	0	−2.8
α-Sn (diam)	12.7	111	0.84	1.18	15.0	−2.0	0	−2.7

repulsive while for group III it is essentially zero. On the other hand, for group IV semiconductors and metals, $G/2 < q_c$ and at the same time v_G is attractive, which implies an enhancement from non-locality. This is in essence the reason why the calculated nearly free plasmon gap in the latter case is up to an order of magnitude larger. This result is essentially confirmed by the single pole model estimate, which is based on pseudopotential charge densities [32], and gives very similar gaps, between 2 and 3 eV in α-Sn and Ge. Wider gap insulators have not been considered because of their larger plasmon damping. Anisotropic crystals, such as layer or chain structures, and also charge density wave superlattices, such as those of $1T\text{-}TaS_2$ [33], can also be considered once their relevant ρ_G is known. For the time being the conclusion is that valence semiconductors are perhaps worth a more detailed study near G/2. It should be borne in mind, however, that neither of the models discussed incorporates the amount of plasmon damping which is very often already present at small q. It seems very difficult at this stage to think of a manageable approximation which includes both Umklapp and non-RPA damping in a satisfactory way, although attempts have been made [34–37]. This is still a direction in which more theoretical work is required in the future.

ACKNOWLEDGEMENTS

The author is grateful to K. Girlanda, M. Parrinello and M.P. Tosi for numerous discussions, and to Luisa Sossi for her generous help.

REFERENCES

[1] PINES, D., Elementary Excitations in Solids, Benjamin, New York (1963) Ch.4.
[2] VON FESTENBERG, C., DANIELS, J., RAETHER, H., ZEPPENFELD, K., Springer Tracts in Modern Physics, vol.54 (1970).

[3] See also BASSANI, F., IAEA-SMR-20/28, these Proceedings, Vol.II. The present author follows the notations of Refs [4–6].
[4] TOSATTI, E., Nuovo Cimento **63B** (1969) 54; TOSATTI, E., in Proceedings of the International School "Enrico Fermi" of Varenna, Course 52 (BURSTEIN, E., Ed.) Academic Press (1971).
[5] LINDHARD, J., Kgl. Danske Videnskab. Selskab., Mat-Fys. Medd. **28** (1954) 8.
[6] RUKHADZE, A.A., SILIN, V.P., Sov. Phys. Uspekhi **4** (1961) 459.
[7] ADLER, S.L., Phys. Rev. **126** (1962) 413; WISER, N., Phys. Rev. **129** (1963) 62.
[8] RAUH, A., Z. Phys. **251** (1972) 195.
[9] JOHNSON, D.L., Phys. Rev. **B9** (1974) 4475.
[10] MARCH, N.H., TOSI, M.P., Proc. Roy. Soc. (London) **A330** (1972) 373.
[11] SINGWI, K.S., SJÖLANDER, A., TOSI, M.P., LAND, R.H., Phys. Rev. **B1** (1970) 1044.
[12] GIRLANDA, R., PARRINELLO, M., TOSATTI, E., Phys. Rev. Lett. **36** (1976) 1386.
[13] PENN, D.R., Phys. Rev. **128** (1962) 2093.
[14] See, for example, ZIMAN, J., Principles of the Theory of Solids, Cambridge University Press (1965) Ch.3.
[15] FANO, U., Phys. Rev. **118** (1960) 451.
[16] HORIE, C., Prog. Theor. Phys. **21** (1959) 113.
[17] GIAQUINTA, P.V., PARRINELLO, M., TOSATTI, E., TOSI, M.P., J. Phys. **C9** 11 (1976) 2031.
[18] PANDEY, K.C., PLATZMAN, P.M., EISENBERGER, P., FOO, E.-Ni, Phys. Rev. **B9** (1974) 5046.
[19] In this section we largely follow Ref.[12].
[20] See, for example, DES CLOIZEAUX, J., Phys. Rev. **A135** (1964) 685.
[21] ZACHARIAS, P., Z. Phys. **256** (1972) 92.
[22] KLOOS, T., Z. Phys. **265** (1973) 225.
[23] ZACHARIAS, P., J. Phys. **C7** (1974) L26.
[24] HÖHBERGER, H.J., OTTO, A., PETRI, E., Solid State Commun. **16** (1975) 175.
[25] ZACHARIAS, P., J. Phys. **F5** (1975) 645.
[26] GIBBONS, P.C., SCHNATTERLY, S.E., RITSKO, J.J., FIELDS, J.R., Phys. Rev. **B13** 6 (1976) 2451.
[27] MILIOTIS, D.M., Phys. Rev. **B3** (1971) 701.
[28] EISENBERGER, P., PLATZMAN, P.M., PANDEY, K.C., Phys. Rev. Lett. **31** (1973) 311.
[29] PLATZMAN, P.M., EISENBERGER, P., Phys. Rev. Lett. **32** (1974) 152.
[30] EISENBERGER, P., PLATZMAN, P.M., SCHMIDT, P., Phys. Rev. Lett. **33** (1975) 18.
[31] BERTONI, C.M., BORTOLANI, V., CALANDRA, C., NIZZOLI, F., J. Phys. **F4** (1974) 19.
[32] BALDERESCHI, A., private communication.
[33] WILSON, J.A., DI SALVO, F.J., MAHAJAN, S., Adv. Phys. **24** (1975) 117.
[34] DUBOIS, D.F., KIVELSON, M.G., Phys. Rev. **186** (1969) 409.
[35] PAASCH, G., Phys. Status Solidi **38** (1970) K123.
[36] HASEGAWA, M., J. Phys. Soc. Japan **31** (1971) 649.
[37] HAQUE, M.S., KLIEWER, K.L., Phys. Rev. **B7** (1973) 2416.

X-RAY EMISSION AND ABSORPTION

D.C. LANGRETH
Nordita,
Copenhagen,
Denmark
and
Rutgers University*,
New Brunswick,
New Jersey,
United States of America

Abstract

X-RAY EMISSION AND ABSORPTION.

Some of the many-body effects which occur in X-ray experiments are discussed in an elementary way. Particular attention is paid to X-ray photoemission.

1. INTRODUCTION

Of all the vast number of topics that could be included under the heading "X-ray Emission and Absorption", I have chosen to include in these published lecture notes only certain theoretical aspects of a certain type of x-ray absorption experiment, namely that in which the absorption is monitored by measuring the momentum distribution of the escaping electrons. This goes by a variety of names and initials, but we will call it here x-ray photoemission. What I will do is to give my own personal view of some narrow, mostly many-body, aspects of this one experiment. I do not promise that my treatment will be unbiased and representative. I hope to compensate by making it instructive. It is written on an elementary level so that all can understand.

This is not a review, and no attempt is made to be complete. There exist two excellent reviews of the many-body aspects of this problem [1, 2] and the reader is referred to them. The reader is also referred to articles by Langreth [3, 4], Nozières and de Dominicis [5], and Minnhagen [6]. These have a high "impedence match" with the present lecture notes.

2. BASIC FORMULAS

The theoretical formulas for x-ray absorption and emission are most easily derived by applying lowest order perturbation theory to the quantized photon field as done in many standard textbooks. One finds then, for absorption of photons, that the

---

* Permanent address.

rate at which the system changes from its ground state "0" to an exacted state "n" is

$$R_{no} = \left(\frac{2\pi e}{m}\right)^2 \frac{1}{\omega} \; |\langle n|\vec{j}_k^{\,+} \cdot \hat{u}|0\rangle|^2 \delta(\omega-\omega_{no}) \qquad (2.1a)$$

where ω_{no} is the energy difference between the states

$$\omega_{no} = \epsilon_n - \epsilon_o \qquad (2.1b)$$

We always use units such that $\hbar = 1$ and hence speak interchangeably about frequencies and energies, momenta and wave vectors, and so on. In (2.1) the quantities ω and $\hat{u}$ are respectively the frequency and (unit) polarization vector of the photon, and $\vec{j}$ is the Fourier transform of the number current density for the electrons in the system

$$\left[\vec{j}_k = \int d^3 x\, e^{-i\vec{k}\cdot\vec{x}}\, \vec{j}(\vec{x})\right] .$$

In this chapter we will derive expressions for the theory of two types of x-ray absorption experiments. The first of these measures the total absorption or attenuation rate of photons and is described by the imaginary part ϵ_2 of the transverse dielectric function. For the purposes of these notes we will call such an experiment x-ray attenuation (without meaning to imply any particular experimental arrangement), in order to distinguish it from the second type of x-ray absorption where the momentum distribution of ejected electrons is measured instead. This latter type of x-ray absorption experiment we will term x-ray photoemission, which is just one of several common names for this in the jargon of the specialists.

For x-ray attenuation we need the total absorption rate $1/\tau$, given that there was originally a photon, which is just

$$1/\tau = \sum_n R_{no} \qquad (2.2)$$

Then, according to usual argument, of electromagnetic theory, we have that the real part of the conductivity $\sigma_1 = (4\pi\tau)^{-1}$, which implies that the imaginary part of the conductivity is

$$\epsilon_2(k,\omega) = (\omega\tau)^{-1} = \frac{4\pi^2 e^2}{m^2\omega^2} \sum_n \left|\left(j_k^{\,+}\right)_{no}\right|^2 \delta(\omega-\omega_{no}) \qquad (2.3)$$

for $\omega = 0$. The real part $\epsilon_1(k,\omega)$ of ϵ may now be obtained from the Kramers-Kronig relation

$$\epsilon_1(k,\omega) = 1 + \int_0^\infty \frac{d\omega'}{\pi} \frac{2\omega'}{\omega'^2-\omega^2} \epsilon_2(k,\omega) \tag{2.4}$$

At high frequencies (2.4) has the limiting form $\epsilon_1 \to 1 - \omega_p^2|\omega^2$, where ω_p is the electronic plasma frequency. Since $\omega >> \omega_p$ at x-ray frequencies, we will take $\epsilon_1 \approx 1$ and regard x-ray attenuation as the theory of ϵ_2 . Eq. (2.3) may also be written in the fluctuation-dissipation theorem form by writing (2.3) in terms of the Heisenberg j operators and Fourier transforming

$$\epsilon_2(k,\omega) = \frac{4\pi^2 e^2}{m^2\omega^2} \int_{-\infty}^{\infty} \frac{dt}{2\pi} <0|j_k(0) j_k^+(t)|0> \tag{2.5}$$

With $j_k(t) = e^{iHt} j_k e^{-iHt}$, where H is the Hamiltonian of the system without photon field. Formulas like (2.5) are sometimes also knowns as Kubo formulas.

X-ray photoemission can be described theoretically in roughly the same way. What is measured in principle is a steady state average of the momentum decomposition of the asymptotic current outside the sample. If we let $\vec{\nu}_n$ be the value of this latter quantity in the n state of the system, then $<\vec{\nu}>$, the quantity we derive, is given by

$$\begin{aligned} <\vec{\nu}> &= \sum_n \vec{\nu}_n R_{no} \\ &= \left(\frac{2\pi e}{m}\right)^2 \frac{1}{\omega} \sum_n \vec{\nu}_n |(j_k^+)_{no}|^2 \delta(\omega-\omega_{no}) \quad . \end{aligned} \tag{2.6}$$

This is very similar to the expression for ϵ_2 , aside from the $\vec{\nu}_n$ which samples the final state, asymptotic current. It can also be expressed as a 3 current correlation function for which the reader should see the references for details.

The theory of x-ray emission is in principle much more complicated. First one creates holes, say by electron bombardment; secondly, one measures the x-ray spectrum as the hole decays. The simplest theoretical assumption is that of complete relaxation around the hole before emission. This means that the excitations created in the excitation process are assumed to have moved away from the hole and that nearby electrons have had a chance to relax to their "ground-state" in the presence of the hole. Said another way, this means that the lifetime of the hole is long enough so that when it finally decays, there is no memory of the excitation process.

Then one can take as the "initial state" the one with the hole present, and do lowest order perturbation theory in the photon field for the decay of this hole with the emission of a photon. As for the absorption, one finds that the emission rate per initial hole is

$$\frac{1}{\tau_{\ell m}} = \left(\frac{2\pi e}{m}\right)^2 \frac{1}{\omega} \sum_{\tilde{n}} \left|(j_k)_{\tilde{n}\tilde{o}}\right|^2 \delta(\omega+\omega_{\tilde{n}\tilde{o}}) \tag{2.7a}$$

where

$$\omega_{\tilde{n}\tilde{o}} = \epsilon_{\tilde{n}} - \epsilon_{\tilde{o}} \tag{2.7b}$$

as in Eq. (2.1). The essential difference between this and (2.3) is that here $\tilde{o}$ represents the "ground" state in the presence of the hole, while $\tilde{n}$ represents an excited state without it. Note that $\epsilon_{\tilde{o}} > \epsilon_{\tilde{n}}$ so that $\omega_{\tilde{n}\tilde{o}}$ is negative. Deviations from this <u>complete relaxation</u> approximation have been considered in the literature, but will not be considered here.

3. CORRELATION FUNCTIONS AND GREEN'S FUNCTIONS

Basically the theoretical question which the x-ray experiments shed light on is how much energy is required to move an electron from one class of states to another. It is convenient however to first consider the simple question of how much energy is required to remove an electron from the system entirely. This latter quantity we will denote by $G_i^<(\omega)/2\pi$, which is defined to be the probability that an energy ω is required to remove an electron from the single particle state i, and can be thought of as the "density" of single particle states, a correspondence which becomes exact when interparticle interactions are turned off. Mathematically it is defined as

$$G_i^<(\omega) = \int dt e^{i\omega t} \left[\langle 0|C_i^+(0)\ C_i(t)|0\rangle\right] \tag{3.1}$$

where $C_i(t)$ is the Heisenberg destruction operator for the state i. As we will see later, this quantity (3.1) is almost directly measured in an idealized x-ray photoemission experiment. Similarly we define $G_i^>(\omega)$, which is similarly related to the probability energy spectrum for adding a particle to the state i:

$$G_i^>(\omega) = \int dt e^{i\omega t} \left[\langle 0|C_i(t)\ C_i^+(0)|0\rangle\right] \tag{3.2}$$

In the absence of interaction, $G^{\gtrless}$ of course are both proportional to delta functions

$$G^{\gtrless}(\omega)/2\pi \propto \delta(\omega - \epsilon_i) \tag{3.3}$$

where the coefficient of the delta function of $G^{<}$ ($G^{>}$) in (3.3) is the average number of particles (holes) in that state. In real systems the structure of $G^{\gtrless}$ becomes complex, and it is this complex structure which we wish to interpret in the various experiments under consideration here. It is also sometimes useful to talk about the Fourier transforms $G_i^{<}(t)$ and $G_i^{>}(t)$; these are just the quantities in square brackets in (3.1) and (3.2), respectively.

At this point, let us make contact with diagrammatic many-body perturbation theory. A knowledge of the latter will in no way be assumed for the understanding of these notes, but hopefully such a discussion will form a useful bridge to some of the theoretical literature which uses such techniques. The basic quantity in this theory is the single particle Green's function, defined by

$$\begin{aligned} G_{ii}(\omega) &= -i\, G^{>}(t) \qquad \text{for } t > 0 \\ &= +i\, G^{<}(t) \qquad \text{for } t < 0 \\ &\equiv -i \langle T\, C_i(t)\, C_i^{+}(0)\rangle \end{aligned} \tag{3.4}$$

where the meaning of the symbol T is defined by the last identity. We may also consider off-diagonal Green's functions

$$G_{ij}(\omega) = -i \langle T\, C_i(t)\, C_j^{+}(0)\rangle \quad .$$

In a similar way one can define 2 or more particle Green's functions, for example

$$\begin{aligned} &G^{(2)}_{ijk\ell}(t_1, t_2, t_3, t_4) \\ &= (-i)^2 \langle T\, C_i(t_1)\, C_j(t_1)\, C_k^{+}(t_1)\, C_\ell^{+}(t_4)\rangle \end{aligned} \tag{3.5}$$

We now express ϵ_2 in these terms. Using (2.5) plus the fact that j_k is of the form

$$j_k = \sum_{\substack{pp' \\ bb'}} (j_k)_{\substack{pp' \\ bb'}} C^{+}_{pb}\, C_{p'b'} \tag{3.6}$$

one finds that

$$\epsilon_2 \propto \sum (j_k)^{bb'}_{pp'} \quad (j_k)^{bb'}_{\bar{p}\bar{p}'} \int dt e^{-i\omega t}$$

$$\times \langle 0 | \, C^+_{pb} \, C_{p'b'} \quad C^+_{\bar{p}'\bar{b}'}(t) \, C_{\bar{p}\bar{b}} \, (t) | 0 \rangle \tag{3.7}$$

where p refers to a Bloch-state wave vector and b is a band index. Note that the quantity on the second line of (3.7) is related to $G^{(2)}$ of Eq. (3.5). Consider first the case of non-interacting particles; then the second line of (3.7) factors into $G^>{}_{p+k}(t) \, G^<{}_p(t)$, where $G^>$ for a lower one (possibly a sum over bands) will be required. We then have

$$\epsilon_2(k, \omega) = \sum_{pbb'} \quad f \int_{-\infty}^{\infty} \frac{d\bar{\omega}}{2\pi} \, G^>_{b',p+k} \, (\omega+\bar{\omega}) \; G^<_{bp} \, (\bar{\omega}) \tag{3.8}$$

where

$$f = \frac{2\pi e^2}{m^2 \omega^2} \quad \left| (j_k)_{p+k,p} \right|^2 \quad . \tag{3.9}$$

For free particles (3.8) has the usual joint density of states interpretation, that is to say the photon energy is the energy required to remove a particle from the state pb $(G^<)$, less the energy gained by adding the particle to state b', $p+k$ $(G^>)$.

The derivation is only exact for the case of free particles, in which case there are simpler ways to obtain the result; what is more important, however, is that for real systems for which $G^{\gtrless}$ are no longer delta functions, (3.8) defines the first term in a well-defined perturbation series. In ϵ_2 it turns out that the higher terms in the expansion are often very important, so that we need more than $G^<$ and $G^>$. This will not be dealt with in these notes which consider only x-ray photoemission. For the latter, however, an analogous expression exists. Here it turns out that the approximation analogous to (3.8), is often a very good starting point. It thus behooves us to consider the theory for $G^>$ and $G^<$, even though they provide the answer to simpler questions than just what the experiments measure. We concentrate on $G^<$, since in an x-ray experiment this is where most of the structure lies.

4. THE CORE-HOLE PROBLEM

We consider a general class of model Hamiltonian which represent a deep atomic electron, in a state say 50 eV below the Fermi level of a metal, but which when removed leaves a hole which interacts with all the "other electrons". The treatment follows Langreth [3] quite closely for a while. The model Hamiltonian is

$$H = \epsilon\, C^+C + CC^+V + H_e \tag{4.1}$$

where C^+ creates a deep electron, so that the number of deep electrons $n_d = C^+C$, while CC^+ is the number of deep holes. The quantity V represents the interaction between the deep hole and the "other electrons" while H_e is the Hamiltonian for the "other electrons" alone, and in general contains the interaction between them. Eq. (4.1) says essentially that when there exists a hole in the deep state, this shakes up the other electrons. The above H contains almost all the physics we would ever want to take into account, except for the lifetime of the hole.

Noting that in this case

$$G^<(\omega) = \int \frac{dt}{2\pi}\, e^{i\omega t} \langle 0|C^+(0)\; C(t)|0\rangle \tag{4.2}$$

and that $[C^+C, H] = 0$, we can define two separate Hamiltonians

$$H^{(0)} \equiv H|_{n_a = 0} = H_e + V \tag{4.3a}$$

and

$$H^{(1)} \equiv H|_{n_d = 1} = H_e + \epsilon \quad , \tag{4.3b}$$

and thus (4.2) becomes

$$G^<(\omega) = \int \frac{dt}{2\pi}\, e^{i\omega t} \langle 0|\; e^{i(H^{(0)}-\epsilon-E_o^e)t}|0\rangle \tag{4.4}$$

where

$$H_e|0\rangle = E_o^e|0\rangle \quad .$$

In general this is as far as we can go. But stop to consider the type of excitation that is likely to be produced: a) electron-hole pairs (either free electron-like or bound) b) collective excitations or plasmons, c) phonons. Note that these are all bosons or boson-like. The basic approximation which has been apparently rather successful but which has not been checked except for certain cases [7], is to neglect the interaction between the bosons. So we specialize (4.1) by an independent boson-model

$$H = \epsilon \, c^+c + cc^+ \sum_q g_q \, (b_q + b_q^+) + \sum_a \omega_q \, b_q^+ \, b_q \, . \qquad (4.5)$$

Note that in the notation of (4.3) this means that

$$H_e = \sum_q \omega_q \, b_q^+ \, b_q \quad \text{and} \quad V = \sum_q g_q \, (b_q + b_q^+) \quad .$$

Here b_q^+ is a boson creation operator, that is

$$[b_q \, , \, b_{q'}^+] = \delta_{qq'} \ ;$$

ω_q in the boson frequency, which for plasmons would be ω_p as $q \to 0$; and g_q is the coupling constant which for plasmons is

$$(4\pi e^2 \, \omega_p/2q^2)^{\frac{1}{2}}$$

in the same limit. The electron hole case is similar and will be discussed later.

Within this model the solution is trivial. One substitutes the pieces of (4.5) into (4.4) and notes that $|0>$ is now the zero boson state. The algebra is straight-forward and is contained in ref. [3]; the result is that

$$G^<(\omega) = \int_{-\infty}^{\infty} dt \; e^{i(\omega - \epsilon - \Delta\epsilon)t} \; e^{B(t)} \qquad (4.6)$$

where the so called "satellite generator" is given by

$$B(t) = - \sum_q (g_q{}^2/\omega_{q^2})(1 - e^{i\omega_q t}) \qquad (4.7)$$

and the "relaxation shift" is given by

$$\Delta\epsilon = \sum_q g_q{}^2/\omega_q \quad . \tag{4.8}$$

One should note that the exact correlation function $G^<(t)$ is given by

$$G^<(t) = G_o^<(t) \exp C(t) \tag{4.9a}$$

where $G_o^<(t)$ is the bare propagator in the absence of interactions and

$$C(t) = B(t) - i\Delta\epsilon t \ . \tag{4.9b}$$

To see the nature of the spectrum let us assume a single boson frequency $\omega_q = \omega_o$, and let

$$a = \sum_q g_q{}^2/\omega_o{}^2 \tag{4.10}$$

Then by expanding (4.6) in powers of $\exp(i\omega_o t)$ and carrying out the Fourier transform, one finds

$$G^<(\omega) = 2\pi \sum_n P_n \, \delta(\omega - \epsilon - a\omega_o + n\omega_o) \tag{4.11}$$

where

$$P_n = e^{-a} a^n/n! \tag{4.12}$$

The factor P_n is clearly a Poisson probability distribution and reflects the fact that the bosons are independent; one should note now that we are talking about boson with respect to the displaced coordinates in the presence of V , and these displaced bosons, like the original bare ones, are uncorrelated with each other because the hole cannot recoil or communicate information, - and hence the Poisson distribution of the satellites, each of which represents an additional displaced boson in addition to the ground state of $H^{(0)} = H_e + V$. The so-called relaxation shift [$= a\omega_p$] of the zero boson line to the right occurs also because of the presence of the hole potential V .

Although (4.11) has the general features of a more realistic model - that is a relaxation shift to the right and a series of boson satellites to the left - generally dispersion and lifetime effects cause the satellites to be overlapping, and of course in the case of electron-hole pairs in a metal they are continuous. For plasmons, where one still expects to see some discrete features, the reader will find plots of more realistic calculations in references [3, 8, 6]. The electron-hole case is discussed in more detail later.

It is tempting, since we know that the deep hole just sits there and produces a potential, to regard P_n as the probability that a time dependent potential

$$V(t) = \Theta(t) \sum_q g_q (a_a + a_q)\ e^{-\eta t} \qquad (4.13)$$

produces n bosons as $t \to \infty$, where η is an infinitesimal adiabatic switching off parameter. This does indeed give the right answer in this case, and is a popular argument to use in more complicated cases. One must be most careful, however, because generally the argument is not correct, and for example gives qualitatively the wrong answer when the deep hole is allowed to decay, even when the time dependent potential V is modified to properly account for this decay.

5. RELAXATION AND SATELLITES

In the previous model, the spectrum starts at - that is, the quasiparticle line is at -

$$\epsilon + \sum_q g_q{}^2/\omega_q \quad . \qquad (5.1)$$

This is the ground state energy of $H^{(0)}$. The hole is produced and the other electrons relax around it, thus changing its energy. The second term in (5.1) is known as the relaxation shift.

The question that almost always comes up is how this can be, for after all the transition is assumed to be fast, and indeed $G^<(\omega)$ gives, for example, the energy distribution of an x-ray photoelectron in the sudden approximation (with respect to the photon's energy). The question is then how the electrons have time to relax. The answer is that by the uncertainty principle, when we are looking at the exact position of a feature in energy, this implies that we must be looking at long time behavior. The threshold is always at the relaxed position, even if the transition is fast. This may be seen by introducing the eigenstates of the zero deep-electron Hamiltonian $H^{(0)}$ [see (4.3a)]:

$$H^{(0)} |\phi_n> = E_n^{(0)} |\phi_n> \quad . \tag{5.2}$$

Then introducing these as a complete set in (4.4) and performing the indicated Fourier transform gives

$$G^{<}(\omega) = \sum_n |<0| \phi_n>|^2 \, 2\pi \, \delta(\omega - \epsilon - E_o^e + E_n^{(0)}) \quad . \tag{5.3}$$

If $n = 0$ denotes the ground state, then the threshold clearly occurs at ω_r , where from now on the subscript "r" means "relaxed". The "relaxed" threshold ω_r is given by

$$\omega_r = E_o^{(1)} - E_o^{(0)} \tag{5.4}$$

where

$$E_o^{(1)} = \epsilon + E_o^e$$

is the ground state of the one electron Hamiltonian. Thus the energy threshold energy required to remove is the difference between the two ground state energies before and after, which means that the final state electrons have had time to relax even though the transition was fast. Eq. (5.4) should be contrasted with the frozen energy: if the electrons had not had time to adjust, then the energy difference corresponding to Eq. (5.4) would have been

$$\omega_{frozen} = E_o^{(1)} - E_{frozen}^{(0)} \tag{5.5}$$

where $E_{frozen}^{(0)}$ is the expectation value of the "new" Hamiltonian with the "old" wave functions, i.e.

$$E_{frozen}^{(0)} = <0| H^{(0)} |0> = E_o^e + <0|V|0> \tag{5.6}$$

The fact that the transition is fast is on the other hand reflected in what you see if you look at the spectrum with poor energy resolution. For example, if one views the spectrum so myopically that one can only determine its first moment, then because the transition was in fact fast, this first moment (or position of the spectral "line" as a whole) should be at the "frozen" energy. Thus we expect a sum rule

$$\int_{-\infty}^{\infty} \frac{d\omega}{2\pi}\, \omega\, G^{<}(\omega) = \omega_{\text{frozen}} \tag{5.7}$$

This is indeed a rigorous result for all Hamiltonians of the form (4.1), as may be proved [9] by noting that

$$\int_{-\infty}^{\infty} \frac{d\omega}{2\pi}\, \omega\, G^{<}(\omega) = i \frac{d}{dt} G^{<}(t) \Big|_{t=0} \tag{5.8}$$

and then using (4.4) for $G^{<}(t)$. Note that the factor 2π results only from our choice of normalization. $G^{<}/2\pi$ is a probability, and

$$\int_{-\infty}^{\infty} \frac{d\omega}{2\pi}\, G^{<}(\omega) = 1 \tag{5.9}$$

is the usual sum rule for the spectral weight function, which is of course identical to $G^{<}$ in this core-hole case.

The sum rule (5.7) means that there is an intimate connection between the presence of satellites and a relaxation shift. When one is present so is the other. Note that in this discussion we use the word satellites in a generalized sense, as in some cases these satellites might be merged into a continuous tailing off at large distances from threshold.

6. GENERALIZATION OF THE BOSON MODEL

One would of course like to generalize the boson model of section 4 to real metals. One method of obtaining this generalization which I gave many years ago [3] will be discussed here. Comparison with Eq. (4.7) and (4.9) allows us to write the exponent $C(t)$ associated with $G^{<}$ as

$$C(t) = -\sum_{q} \int \frac{d\omega}{2\pi} \left[g_q^{\,2}\; 2\pi\, \delta(\omega-\omega_q) \right] \left[(1+i\omega t - e^{i\omega t})/\omega^2 \right] \tag{6.1}$$

We have written (6.1) in a suggestive form. The δ function in the first set of square brackets is essentially the density of boson states of energy ω and momentum q , and is non-zero along a line in ω vs. q space. In a real material the analogue of our free bosons are the density fluctuation excitations. Even for a simple metal the locus of these in the ω, q plane is not a line, although the plasmon part approximates this for small q. The electron-hole pair state density on the other hand is non-vanishing (in the simplest approximation) whenever ω and q are in the region between the parabolic segments $\omega = qv_f + q^2/2m$ and $\omega = -qv_f + q^2/2m$. The effective state density for these boson-like states is the so-called dynamic form factor $S(q, \omega)$, that is the Fourier transform of the density-density correlation function

$$S(q,\omega) = \int_{-\infty}^{\infty} \frac{dt}{2\pi} \langle 0|\, n_q(t)\, n_q^+(0)\, |0\rangle \qquad (6.2)$$

where

$$n_q(t) = \int d^3x\, n(\vec{x},t)\, e^{-i\vec{q}\cdot\vec{x}}$$

and where $n(\vec{x},t)$ is the Heisenberg electron number-density operator. This is the analogue of the δ function in (6.1). The analogue of g_q is some sort of pseudopotential v_q with which the change when the hole is produced interacts with the other electrons. We make these replacements heuristically here, and write

$$C(t) = -\sum_q v_q^2 \int d\omega\, S(q,\omega)(1 + i\omega t - e^{i\omega t})/\omega^2 . \qquad (6.3)$$

In ref. [3], Eq. (6.3) has been derived by summing a certain subclass of terms in perturbation theory. To some degree this approximation has been justified after the fact by Minnhagen's numerical calculations [6].

One notes that $S(q,\omega)$ has a plasmon piece, which at small q becomes

$$S_{plasmon} \rightarrow \omega_p(q^2/8\pi e^2)\, \delta(\omega-\omega_p) \qquad (6.4)$$

so that in this case the boson model is recovered. But there is also a contribution from the particle-hole pairs, which is most easily seen if we write $S(q,\omega)$ in terms of the dielectric function $\epsilon(q,\omega) = \epsilon_1 + i\epsilon_2$,

$$S(q,\omega) = \frac{q^2}{4\pi^2 e^2} \frac{\epsilon_2(q,\omega)}{|\epsilon(q,\omega)|^2} \Theta(\omega) \qquad (6.5)$$

Using the Lindhard value for ϵ_2

$$\epsilon_2(q,\omega) = \frac{1}{2}\pi \frac{\omega}{qv_f} \frac{k_{ft}^2}{q^2} \quad \text{for} \quad \omega < qv_f - q^2/2m \tag{6.6}$$

we are able to evaluate a significant contribution to C. Before doing this we discuss in the next section the zero quasiparticle line to which (6.6) makes a special contribution.

7. THE ANDERSON THEOREM AND THE QUASIPARTICLE LINE

Return to Eq. (5.3) for $G^<$ and let $n = 0$ refer to the ground state of $H^{(0)}$. Then the $n = 0$ contribution to the sum is

$$G^<(\omega) = |\langle 0|\phi_o\rangle|^2 \; 2\pi\delta(\omega - \omega_{\text{relaxed}}) + \text{other terms} . \tag{7.1}$$

This is in the context of Fermi liquid theory known as the "quasiparticle line" and in the context of phonon interaction as the "zero-phonon line". Its weight is just $|\langle 0|\phi_o\rangle|^2$, that is the square of the projection of the ground states of the "other" electrons with and without the potential due to the deep electron in question. Usually this is a finite constant, but in certain cases, namely metals, this overlap is zero [10]. In such cases we expect the spectrum of $G^<$ near threshold to be less singular than the δ function implied by (7.1) and such is indeed the case.

We can obtain a simple expression for this overlap by examining (6.3). Consider the factor in the parenthesis $(1 + i\omega t - e^{i\omega t})$. The middle $(i\omega t)$ term gives the relaxation shift, while the final term $(e^{i\omega t})$ gives the contribution of one or more density fluctuation excitations, just as in the reasoning that precedes (4.11). The first term (1), however, gives a delta function at threshold (i.e. the quasiparticle line). Therefore its weight is

$$|\langle 0|\phi_o\rangle|^2 = \exp\left[-\sum_q v_q^2 \int d\omega \; S(q,\omega)/\omega^2\right] \tag{7.2}$$

Using (6.5) and (6.6) it is easy to see that the exponent diverges at small frequency as

$$|\langle 0|\phi_o\rangle|^2 \propto \exp\left[-\alpha \int_0 \frac{d\omega}{\omega}\right] \tag{7.3}$$

where

$$\alpha = \sum_{q<q_f} \left| v_q/\epsilon(q,0) \right|^2 \rho(\epsilon_f)/qv_f \tag{7.4}$$

where $\rho(\epsilon_f)$ is the density of states at the Fermi level. It is straightforward to show that (7.4) is the Born approximation to

$$\alpha = \sum_{\ell m\sigma} (\delta \ell m\sigma/\pi)^2 \tag{7.5}$$

where δ is the phase shift produced in the ℓ , m , σ'th partial wave by the statically screened potential $v_q/\epsilon(q,0)$. It is clear then (because the integral in (7.3) diverges at lower limit that

$$|\langle 0|\phi_o\rangle|^2 = 0$$

for our infinite system. For a finite system, on the other hand, there is a finite minimum excitation energy for particle-hole pairs

$$\sim v_f \, q_{min} \sim v_f/L$$

where L is a typical lineal dimension of the sample $L \sim V^{1/3}$. Then we see that

$$|\langle 0|\phi_o\rangle|^2 \propto e^{-\alpha \, \ell n \, V^{1/3}} = V^{-\alpha/3} \tag{7.6}$$

which is Anderson's [10] well-known orthogonality theorem.

8. THE THRESHOLD SINGULARITY

Armed with the knowledge that there is a singularity at threshold because of (7.6), it is easy to evaluate the form of $G^<$ near threshold, because it is controlled by the long time behavior of $C(t)$ or better yet $B(t)$, which is as before obtained from $C(t)$ [Eq. (6.3)] by dropping the $i\omega t$ term. This limit for large t becomes

$$\begin{aligned} B(t) &\longrightarrow -\alpha \int_0^t (1 - e^{i\omega t}) \frac{dt}{t} \\ &\longrightarrow -\alpha \ln(-it) \end{aligned} \tag{8.1}$$

where we have used (6.3) and (7.4). We have also neglected real constant terms in (8.1) and in particular the energy which multiplies t inside the logarithm in (8.1) to make it dimensionless. We will call this number D; the actual calculation of it is difficult, and has first effectively been carried out by Minnhagen [6]. This value is difficult to read off his graphs, but the scale for it is set by the plasma frequency ω_p.

Restoring this effective cutoff D and performing the Fourier transform [Eq. (4.6)] gives for $\omega < \omega_{threshold}$

$$G^<(\omega) = \alpha D^{-\alpha} (\omega_{threshold} - \omega)^{\alpha - 1} \tag{8.2}$$

and zero for $\omega > \omega_{threshold}$. This is valid in the asymptotic limit near threshold, and was first derived in a different manner by Nozières and de Dominicis [5].

9. HOLE-LIFETIME EFFECTS

Before comparing (8.2) with experiment we must account for certain decay processes, which occur in practice, but which are not contained in the model Hamiltonian (4.1). The most important of these for the class of materials we are most interested in here is the Auger decay term, which allows a deep hole once created to decay into another hole (say in the conduction band) plus an electron-hole pair via the Coulomb matrix-element. This process when acting alone would give roughly the standard Lorentz shape

$$G^<(\omega) = \frac{2\Gamma}{(\omega - \omega_{th})^2 + \Gamma^2} \tag{9.1}$$

where

$$2\Gamma = 2\pi |M.E.|^2 \rho \tag{9.2}$$

where ρ is here the density of 2 electron-one hole final states. The question is what happens when one has both the coupling to the decay product in a deep-hole nonconserving way, and to the boson-like excitations of the other electrons (and phonons) in a deep-hole conserving way.

To simplify the discussion, consider the single frequency independent boson model, whose spectral function $G^<$ would in the absence of lifetime effects be given by Eq. (4.11). The combined effect of the lifetime effect (9.2) and the multiple boson excitation is usually handled by the ansatz of Doniach on Šunjić [11], which is that the correct result should be given by the convolution of (9.1) and (4.11), that is

$$G^<_{\Gamma\neq 0}(\omega) = \int \frac{d\bar{\omega}}{2\pi} \frac{2\Gamma}{(\omega-\bar{\omega})^2+\Gamma^2} G^<_{\Gamma=0}(\bar{\omega}) \tag{9.3}$$

which for the single frequency boson model becomes

$$G^<(\omega) = \sum_n P_n \frac{2\Gamma}{(\omega-\omega_{th}+n\omega_o)^2+\Gamma^2} \tag{9.4}$$

where $\omega_{th} = \omega_{threshold}$. This convolution procedure (or alternatively simple multiplication in time-space) of course neglects any possible interference effects between the two processes, and several recent preprints have attempted to treat these. However, it can be shown that the convolution procedure (9.3) is essentially exact in this case. The key to proving this is first to show that the rate of change of Γ with frequency is of the order of $\Gamma/(|\omega_{th}| - \epsilon_f)$ which is negligible, and then to show that the only corrections to (9.3) are of order $\partial\Gamma/\partial\omega$. The details of the proof will be published elsewhere [12], but the hints given here should be sufficient to allow the interested reader to reconstruct the proof. The thing to beware of is that P_n is not the probability that n bosons are emitted at time ∞ by a potential of the type (4.13) (with say η replaced by Γ). The latter number is much smaller than the actual coefficient P_n in (9.4), if Γ is comparable or larger than ω_o .

Applying the convolution argument, which we now know to be reasonable, to the lineshape (9.2), gives what is known as the Doniach-Šunjić formula for the x-ray photoemission lineshape. Before actually comparing with experiment one must also convolute (9.2) with a Gaussian shape representing the phonons. The most recent calculation of this effect is due to Hedin [13]. In this case there is no question of the validity of merely convoluting the results for the phonon shake-up spectrum with (8.2), at least to the extent that the interaction between the phonons and the "other electrons" is neglected.

TABLE I. COMPARISON OF EXPERIMENTAL AND THEORETICAL VALUES FOR THE EXPONENT α. All numbers rounded to two significant figures.

	Expt.[a]	Expt.[b]	Theory[c]	Theory[d]
Na	.19		.19	.21
Li	.25	.18	.16	.24
Mg	.13	.13		.13
Al	.12	.16	.13	.11

a. P. CITRIN, G. WERTHEIM, Y. BAER, Phys. Rev. Lett. 35 (1975) 885 and to appear

b. L. LEY, S.P. KOWALEZYK, J.G. JENKIN, D.A. SHIRLEY, Phys. Rev. B11 (1975) 600

c. C.O. ALMBLADH, U. von BARTH, Phys. Rev. B (to appear)

d. P. MINNHAGEN, Phys. Lett. 56A (1976) 327

By comparing the doubly convoluted theoretical formula with the experimental data convoluted with the known experimental resolution function, one can extract the exponent α. A comparison of the results of experiments of two groups with two different theoretical calculations is shown in Table I. One notes that the agreement is generally fairly good, especially considering that the experimental quantity actually being determined (hopefully), is $1-\alpha$ where α is quite a bit smaller than unity.

10. EFFECTS AWAY FROM THRESHOLD

Note that any of these fits may well depend on what happens away from threshold, and not just on the asymptotic forms (8.2), especially once the lifetime effects are folded in. Some of the effects which may come in are as follows: 1) The matrix elements and particle-hole state densities depend on energy in a way that differs from the linear implied by the ω in Eq. (6.6); 2) Dynamic screening may be important, that is it takes a finite time for the electrons to respond to produce the screened potential caused by the sudden appearance of the deep hole. 3) There may be deviations from the sudden approximation (which we assume by calculating just $G^<$), because the deep hole potential develops gradually, not suddenly, since the escaping electron moves away with a finite, not infinite, speed. 4) There may be other many-body effects, including the so-called intrinsic effects, that is the energy loss of the escaping photoelectron, as well as various quantum mechanical interference effects which have been discussed in detail elsewhere [4, 7, 14, 15].

The dynamic screening and matrix element-state density effect has been considered recently by Minnhagen [6], who numerically evaluated the following expression first written down by Langreth [3]:

$$B(t) = -\sum_{q<2q_f} \int^{qv_f-q^2/2m} d\omega \; |v_q/\epsilon(q,\omega)|^2$$

$$\times\ (\omega/qv_f)\rho(\epsilon_f)\ (i - e^{i\omega t})/\omega^2 \qquad (10.1)$$

$$+\ \text{other terms}$$

This can be derived by inspection of Eq. (6.3) ff . The "other terms" in (10.1) arise from the deviations of ϵ_2 from its small q and ω form, as well as from deviations from the RPA. Of course if v_q cannot be treated as small there will be deviations on this account as well. In the limit $t \to \infty$, which is sufficient to obtain the threshold exponent (but not the amplitude), it is exact to put $\omega = 0$ in the denominator of (10.1), and this says that static screening is sufficient to obtain the exponent, an approximation made in all other calculations of α I know of. But away from threshold one cannot do this.

Minnhagen [6] has recently evaluated $G^<$ for the case of finite ω. By using several different pseudopotentials he finds, for several different materials, that the effects of finite ω and the q variation of the matrix elements, etc., when taken together, leave a form for $G^<(\omega)$ which is very similar to what would have been obtained if the power law were continued away from threshold. If the dynamic screening effect were neglected, and ω put equal to zero in $\epsilon(q,\omega)$ in the denominator of (9.1), then there would be a significant deviation from the asymptotic form.

The assumption that the photo-current $\propto G^<$ is known as the sudden approximation;by neglecting the photoelectron's interactions with the other electrons, it is effectively put instantaneously into its final state. In an atom, it is valid at high photon energies. In an infinite solid it is never strictly valid because of "extrinsic" effects which I discuss later - but these can be calculated separately. There may still be in addition "intrinsic" deviations from the sudden approximation if v, the velocity of the escaping electron is not high enough. This effect has been calculated for plasmon satellites [7], but not for the threshold singularity. But in a recent calculation Gadzuk and Šunjić [16] simulate the effect with a parameterized time dependent potential for the hole

$$V(r, t) = 0 \qquad \text{for} \quad t < 0$$
$$= V_h(r)\ (i - e^{-\eta t}) \qquad \text{for} \quad t > 0 , \tag{10.2}$$

where η is a parameter measuring the speed of turning on, and $V_h(r)$ is the hole potential, which in our previous notation would be $\Sigma_q (v_q/\epsilon(q, 0) e^{iq\cdot r}$ if one is satisfied with a Born-approximation pseudopotential. In (10.2) the sudden approximation corresponds to $\eta = \infty$. The authors do indeed find deviations from the Doniach-Šunjić formula, but for reasonable values of η, for x-ray photoemission, these deviations are small.

In concluding this section, we give a summary of some of the factors affecting the asymmetry of the x-ray photoemission line.

1) The basic asymmetry mechanism - that is the emission of soft electron-hole pairs. This makes the line asymmetric with tailing on the low energy side.

2) Dependence of matrix elements and density of e - h states on energy - that is the deviation from linear increase. Dependence must eventually fall off from this, which will make lineshape less asymmetric. This has been verified by Mahan [17].

3) Dynamic Screening. This will make lineshape more asymmetric because higher ω components of the potential will be less perfectly screened than for static screening, and so the basic asymmetry mechanism (1) is reduced in strength. Minnhagen's recent calculation indicates that the cancellation between 2) and 3) may be quite good in some cases.

4) Deviation from the sudden approximation (excluding "extrinsic" effects). This reduces the energy available for higher energy shake-up structure and thus makes line more symmetric. The magnitude of the effect depends on the electron's velocity and is probably not too large at x-ray energies.

5) Extrinsic Effects. Multiple excitation of soft excitations by the escaping electron. This will make line more asymmetric, but since it increases as a power of frequency it is not important until one gets a bit away from threshold; it can perhaps be experimentally distinguished from the other effects.

6) Phonons. Because of their Gaussian distribution they give a large deviation from the Doniach-Šunjič formula. They could also give additional asymmetry at low temperatures.

7) Deviation from the convolution approximation for the various width mechanisms. According to our previous arguments this effect is probably small.

8) Other many-body effects and deviations from basic model Hamiltonian (4.1). Although we think we have isolated the important mechanisms, one never knows for sure what is lurking outside.

11. EFFECTS FAR FROM THRESHOLD - PLASMONS

As one moves away from threshold then the plasmon contributions, of which Eq. (6.4) is the small q limit, begin to contribute to $G^<$. Essentially we expect satellites at the plasmon frequency [3], as has recently been verified by Minnhagen [6] and the reader is refered to figure (3) of the latter papers. Here the plasmons are acting like free bosons of the simple model in addition to the continuum of electron hole pairs.

Unfortunately in this energy range, the energy loss of the escaping photoelectron (often termed an "extrinsic" effect) becomes important. Its effect has been derived in various ways. We present here a rate equation method which is particularly physical. The rate equation, which we write down physically, can be derived rigorously [7, 14, 18], including corrections.

Let r_p be the rate at which state p high in the conduction band is populated due to the x-ray absorption process

$$r_{\vec{p}} = 2\pi \, |M|^2 \, \delta(\Omega - \epsilon_p) \tag{11.1}$$

where M is a dipole matrix element, $\epsilon_p = p^2/2m$ is the electron's energy, and Ω is the difference between the phonon energy plus initial hole energy and the "intrinsic" loss processes included in $G^<$ as discussed earlier. Let $R_{\vec{p}'\leftarrow\vec{p}}$ be the rate at which state $\vec{p}$ (given that it is populated) is scattered into state $\vec{p}'$ due to electron-electron interactions, plasmons, and all other processes. Within the Born approximation which is valid for these assumed high-energy electrons ($\epsilon_p \sim 1.5$ keV), one can write

$$R_{\vec{p}-\vec{q}\leftarrow\vec{p}} = \int d\bar{\omega} \left(\frac{4\pi e^2}{q^2}\right) S(q, \bar{\omega}) \; 2\pi \, \delta(\omega + \epsilon_{\vec{p}-\vec{q}} - \epsilon_{\vec{p}}) \; . \tag{11.2}$$

Using (11.1) and (11.2) we can write down a rate equation for $n_{\vec{p}}$, the population of state $\vec{p}$:

$$\gamma_{\vec{p}} n_{\vec{p}} = r_{\vec{p}} + \sum_q R_{\vec{p}\leftarrow\vec{p}-\vec{q}} \; n_{\vec{p}-\vec{q}} \tag{11.3}$$

where γ_p is the lifetime of state $\vec{p}$

$$\gamma_p = \sum_q R_{\vec{p}-\vec{q} \leftarrow \vec{p}} \quad . \tag{11.4}$$

To make some progress in solving (11.3), we assume that p is sufficiently large that the momentum transfer q is small in comparison with p $(q << p)$. Then for practical purposes the photoelectron continues in a straight line, and the only active variables in (11.3) are the energies $\nu = \epsilon_p$; $\nu' = \epsilon_{p-q}$, so that we have

$$\gamma(\nu)\ n(\nu) = \bar{r}\ \delta(\nu - \Omega) + \int d\nu'\ R(\nu \leftarrow \nu')\ n(\nu') \tag{11.5}$$

where

$$\bar{r} = 2\pi\ |M|^2\ \rho(\nu)$$

and

$$R(\nu \leftarrow \nu') = \int \frac{d\Omega_{\vec{p},\vec{p}-\vec{q}}}{4\pi}\ R_{\vec{p},\ \vec{p}-\vec{q}}\ \rho(\nu) \tag{11.6}$$

where $\rho(\nu)$ is the density of states and the integral over $\Omega_{\vec{p},\ \vec{p}-\vec{q}}$ means a solid angle average. In other words we are neglecting any effects from the fact the scattering changes direction of the particle.

To my knowledge, even the simplified equation (11.5) has never been solved for the energy spectrum. To get an idea of how the solution goes however, let us discuss the energy dependence of $R(\nu \leftarrow \nu')$. First it is a strong function of $\nu' - \nu$ but only a weak function of $\nu' + \nu$ (because $(\nu' + \nu)/2 >> \omega_p$) , and we therefore neglect its dependence on this latter variable. As a function of $\nu' - \nu$ it has a smeared out slowly varying part which extends over a range of a number of plasmon frequencies and at high $\nu + \nu'$ has around 1/3 of the weight. This part arises from particle-hole excitations. Superposed on this is a sharp rapidly varying peak beginning sharply at ω_p and then falling off fairly rapidly (within one plasma frequency) thereafter, and which contains roughly 2/3 of the weight. This part arises mainly from plasmon excitation. Therefore let us make a crude approximation which exhibits these features:

$$R \simeq \gamma^{(1)}\ \delta(\nu' - \nu - \omega_p) + R^{\text{pair}}\ (\nu - \nu') \tag{11.7}$$

where R^{pair} is the slowly varying particle-hole part and the δ function is to simulate the "plasmon" part. Since the plasmons merge continuously into the particle hole continuum at high frequency, there is no rigorous way to make this separation, but we can do it qualitatively. The rate $\gamma^{(1)}$ is then the total rate for scattering due to "plasmon" production and is about 2/3 of the total area under R , i.e. $\gamma^{(1)} \sim (2/3)\gamma$. Let this ratio be α ($\equiv \gamma^{(1)}/\gamma$) .

With the ansatz (11.7), the integral equation (11.5) is trivially solved for the rapidly varying parts - that is those still proportional to δ functions. We find

$$n(\epsilon) = (\bar{r} / \gamma) \sum_{m=0}^{\infty} \alpha^m \, \delta(\epsilon - \Omega - m\omega_p) \qquad (11.8)$$

$$+ \text{ slowly varying terms}$$

This says that the plasmon satellites produced by this extrinsic energy loss effect fall off in constant ratio and are fairly strong even when the electron's velocity is high. The result (11.8) was first given in ref. [19] for the inverse [admittance rather than escape] and the derivation for admittance and escape presented in refs. [14, 7, 18]. It has effectively been derived in another way in ref. [20]. The recent statement by Pardee et al.[21] that the Mahan theory is fundamentally different from the Chang-Langreth theory is incorrect.

To obtain a prediction for the actually observed spectrum, one must convolute the result for $G^<$ (say from Minnhagen's calculation [6]) with the so-far non-existent numerical solution of (11.5) [to which (11.8) is a crude approximation]. So far the comparisons with experiment [4, 21] have only been on the level of (11.8), combined with a Poisson distribution [see Eq. (4.12)] for the plasmon satellites that occur in $G^<$. Because of the sensitivity of such comparisons to the assumptions made about the background and the smooth part of R , the conclusions must be regarded as tentative.

Finally it should be noted that interference terms between $G^<$ and the energy loss of the escaping photoelectron are not completely negligible, because in practice the electron's velocity is actually often not high enough. These were first considered on a level of approximation consistent with (10.8) by Chang and Langreth [14, 7] and later by Sunjić [15]. If one lets $\frac{1}{2} mv^2 = 1.5$ keV in Eq. (29) of [14] or Eq. (53) of [7], one gets a number of 0.15 which, for example is to be subtracted from the strength of the first satellite and added to the strength of the quasiparticle peak. Although this may seem rather small in comparison with 0.7 for the strength of the first extrinsic satellite combined with the 0.3 - 1.0 strength for the first $G^<$ satellite as suggested by earlier estimates [22, 4], the detailed recent calculation by Minnhagen [6] shows for Aℓ a $G^<$ satellite of weight only around 0.2 , of which the subtraction 0.15 is a significant fraction. The surface effects are of the same order of magnitude [7, 15].

REFERENCES

[1] HEDIN, L., X-ray Spectroscopy, (AZÁROFF, L.V., Ed.), McGraw Hill, New York (1974)
[2] MAHAN, G.D., Solid State Physics (EHRENREICH, H., SEITZ, F., TURNBULL, D., Eds.) Academic Press, New York and London, 29 (1974) 75
[3] LANGRETH, D.C., Phys. Rev. B1 (1970) 471
[4] LANGRETH, D.C., Collective Properties of Physical Systems (Proc. 24th Nobel Symposium) (LUNDQVIST, B., LUNDQVIST, S., Eds.) Academic Press, New York and London, 24 (1973) 210
[5] NOZIÈRES, P., de DOMINICIS, C.J., Phys. Rev. 178 (1969) 1097
[6] MINNHAGEN, P., Phys. Lett. 56A (1976) 327
[7] CHANG, J.J., LANGRETH, D.C., Phys. Rev. B 8 (1973) 4638
[8] HEDIN, L., LUNDQVIST, B., LUNDQVIST, S., Electron Density of States, (BENNETT, L.H., Ed.), National Bureau of Standards Spec. Publ. 323, Washington (1971) 223
[9] That there must be such a sum-rule was noted by LUNDQVIST, B.I., Phys. Kondens Materie 9 (1969) 236. It was proved for the general type of Hamiltonian of Eq. (4.1) by LANGRETH, D.C., ref. [3] above. It was proved in the Hartree-Fock approximation by MANNE, R., ÅBERG, T., Chem. Phys. Lett. 7 (1970) 282
[10] ANDERSON, P.W., Phys. Rev. Lett. 18 (1967) 1049
[11] DONIACH, S., ŠUNJIĆ, M., J. Phys. C 3 (1970) 285
[12] LANGRETH, D.C. (to be published)
[13] HEDIN, L. (to be published)
[14] CHANG, J.J., LANGRETH, D.C., Phys. Rev. B 5 (1972) 3512
[15] ŠUNJIĆ, M., ŠOKČEVIĆ, D., Solid State Comm. 18 (1976) 373
[16] GADZUK, J.W., ŠUNJIĆ, M., J. Phys. Rev. B 12 (1975) 524
[17] MAHAN, G.D., Phys. Rev. B11 (1975) 4814
[18] LANGRETH, D.C., Linear and Non-Linear Electronic Transport in Solids (Proc. Antwerp Advanced Study Institute, Antwerp, 1975) (to be published)
[19] LANGRETH, D.C., Phys. Rev. Lett. 26 (1971) 1229
[20] MAHAN, G.D., Phys. Stat. Sol. (b) 55 (1973) 703
[21] PARDEE, W.J., et al., Phys. Rev. B 11 (1975) 3614
[22] See LUNDQVIST, B.I., ref. [9] above.

PRINCIPLES AND APPLICATIONS OF ELECTRON SPECTROSCOPY

H. SIEGBAHN
Institute of Physics,
University of Uppsala,
Uppsala,
Sweden

Abstract

PRINCIPLES AND APPLICATIONS OF ELECTRON SPECTROSCOPY.

A brief historical background to electron spectroscopy is given, describing the general aspects of experiment and theory. The photoelectric process is treated with particular emphasis on core electron spectroscopy, and the characteristics of the shake-up, shake-off spectrum are described. Expressions for the chemical effects on the core electron spectra are derived based on potential model approximations. On the simplest level these models neglect electronic relaxation in the ionization process. To include relaxation, a transition potential model is introduced and an expression for the relaxation energy derived. The intensities of the electron lines from gases and solids are discussed, starting from one-electron expressions for the photoelectric differential cross-sections. Results on angular distributions are chosen as examples of recent experimental development. The Auger and autoionization electron spectra are described with respect to final vacancies either in the valence electron shell or in the core electron region. In the former case, an expression is presented to treat Auger electron spectra from simple molecules such as H_2O. For core electron Auger spectra, shifts in the Auger lines are treated on the basis of transition potential models and a relation is derived which connects the Auger shifts with binding energy shifts. Finally, the shapes and widths of electron lines are discussed. It is shown that important physical and chemical information on the substance can be gained from a consideration of these features in the electron spectra.

1. INTRODUCTION

The purpose of this review is to present the basic aspects of ESCA and its applications on the electronic structure of matter. Since the field has undergone and is still undergoing substantial developments in large areas of physics and chemistry, some topics that may have deserved more space have been treated in limited detail. For the reader interested in particular aspects, the list of references may serve to supply the necessary information on recent publications in each area. The specific examples chosen are from studies of all states of aggregation of matter to illustrate the various theoretical and experimental concepts involved in ESCA.

The development of ESCA has gone through several stages, starting in the nineteen forties and fifties with the improved experimental possibilities for electron analysis, developed chiefly for work in nuclear physics [1]. Experiments with the photoeffect had, of course, been in progress earlier than this, but the realization that results from such experiments could be used to obtain chemical information did not begin to evolve until this period. When the first experiments in this direction were performed at the University of Uppsala [2, 3] it was found that an essentially symmetric photoelectron line of small width could be resolved from the high-energy edge of the electron energy loss continuum. Later on, the energy loss continuum could be further suppressed by the use of soft X-ray sources such as Mg ($h\nu$ = 1253.6 eV) and Al ($h\nu$ = 1486.6 eV) for excitation of the photoelectron spectra. The use of these sources was also advantageous from the point of view of a smaller inherent line width and thus a more narrow photoelectron line. Having thus achieved an electron *line* spectrum instead of the broad energy distributions previously observed in photoelectron experiments, systematic measurements of the energy levels of a large number of the elements of the periodic system were performed [4]. The attractive feature of photoelectron spectroscopy, that of being able to measure *directly* the energy levels of a system rather than having to use the *indirect* methods of photon spectroscopy, was thus established. In several cases substantial corrections were found for values of binding energies previously obtained by X-ray spectroscopy [5]. In the course of these measurements an observation was made which forms the basis of one of the most important applications of ESCA: that core electron lines from an element shift upon change in the chemical state. The first chemical shift of a core electron line was observed in a study of copper and the copper oxides [6–8]. Subsequently chemical shifts were and are being investigated and related to chemical structure for a large number of elements. The main applications have been for the lighter elements where, during recent years, a total of more than one thousand compounds have been studied for carbon, nitrogen and oxygen [4, 9, 10]. This more extended mapping of the energy levels of compounds has taken place mainly since the beginning of the sixties. In parallel with this work using X-ray excitation, the use of alternative modes of excitation, chiefly resonance radiation from helium ($h\nu_I$ = 21.2 eV and $h\nu_{II}$ = 40.8 eV) and neon ($h\nu$ = 16.8 eV), was developed to study electron energy levels in free atoms and molecules [4, 11, 12] and the band structure of solids [13, 14]. Due to the energy limitation, these latter radiations can only be used to study the outer valence levels of the system. This is not the case with X-ray excitation, but the resolution will in general be much higher with u.v. radiation, because of lower electron kinetic energy and lower inherent width of the exciting radiation. Thus, for molecules, the vibrational progressions observed in the u.v. excited electron spectra are a chief source of chemical information gained from this mode of excitation. As will be discussed further, the two modes of excitation complement each other and the combined use of the two gives increased possibilities of assigning and interpreting spectra from gases and solids.

Since electron spectroscopy in general implies excitation of a specimen by some means (u.v., X-ray, electrons) and the subsequent analysis of the ejected electrons, the electrons originating from the de-excitation processes (Auger, auto-ionization) will be a natural ingredient in electron spectra. These aspects of the electron spectra have been subjected to numerous applications of ESCA. The first of these involved the detailed study of the diagram lines in KLL Auger spectra for several elements [15, 16]. The energies and intensities of the lines were accurately measured and compared with current theories and important conclusions could be drawn as to their validity. The number of such studies has been rapidly increasing, as have studies of auto-ionization spectra [4, 9, 17–24].

The theoretical understanding of the processes involved has essentially gone hand in hand with experimental development. The collection of data on chemical shifts, for instance, has led to intensified theoretical research on successively higher levels of approximation for a detailed account of their origin. It may be said that the basic phenomena in ESCA (such as chemical shifts, intensities, linewidths) are already rather well understood. It is the fact that they may be described in fairly simple and easy-to-handle terms that constitutes the power of electron spectroscopy.

Electron spectroscopy is now applicable, by means of a large number of different radiations, to gases, liquids and solids. Applications range from biological (metal-containing enzymes, detection of trace elements in environmental research, etc.) to chemical (structure determination, correlations with chemical reaction parameters, investigations of catalysts, etc.) and physical (photoionization process, energy levels and charge distributions in atoms, molecules and solids, vibrational excitations, etc.). A number of books that summarize the status of the field from different points of view have recently been published, including conference proceedings [25–27] and ordinary text-books [4, 9, 12, 28–30].

2. PHOTOELECTRIC PROCESS

2.1. Shake-up and shake-off transitions

When an electronic system is subjected to radiation whose energy exceeds the first ionization potential of the system, photoionization may occur through the photoelectric effect. This phenomenon was explained by Einstein in 1905 by means of quantum theory, and the famous result of his paper [31] is summarized in the photoelectric law which, for electron spectroscopy purposes, may be stated as:

$$E_{kin} = h\nu - E_B - \phi_s \tag{1}$$

where $h\nu$ is the incident photon energy, E_{kin} the electron kinetic energy, E_B the binding energy with respect to the Fermi level of the system and ϕ_s is the work function. The task of electron spectroscopy, based on Eq.(1), is to answer the following questions: How many electrons are ejected within a certain interval of energy and angle of emission? What information on the nature of the process and the electronic structure of the substance can be drawn from this knowledge?

I shall start with the photoelectron distribution in energy. As is well known, the electronic properties of a system split up in a more or less well-defined way into those of the core electron region and those of the valence electron region. The core states are characterized by well localized wave functions close to the nuclei of the system. The valence wave functions, on the other hand, which are responsible for the bonding, are spread out in space and have in general large amplitudes between atomic sites. It is obvious that the hole state created by photoionization of a core electron will be of a different character than if a valence electron is ejected. The core ionization is characterized by a strong perturbation of the remaining system.

I shall assume that both initial and final states of the photoionization may be given the form of antisymmetrized products of a one-particle state and an (N-1)-particle state (A = antisymmetrizer):

$$\Psi_{in} = A\varphi_i \, \Psi_i^G(N\text{-}1) \tag{2a}$$

$$\Psi_{fi}^j = A\varphi_k \, \Psi_i^{+j}(N\text{-}1) \tag{2b}$$

where φ_i is the one-particle core state, which is removed from the ground state. $\Psi_i^G(N\text{-}1)$, formed from the ground-state function, has the character of a frozen state which has not reacted to the removal of the core state φ_i. φ_k in (2b) represents the outgoing electron and $\Psi_i^{+j}(N\text{-}1)$ the remaining ion system with core hole i, which in general may also be excited among the outer electrons (designated by index j). $\Psi_i^{+j}(N\text{-}1)$ satisfies the Schrödinger equation of the ion:

$$H(N\text{-}1)\, \Psi_i^{+j}(N\text{-}1) = E_i^j \, \Psi_i^{+j}(N\text{-}1) \tag{3}$$

The probability for photoionization from the ground state to the ij state is given by the dipole matrix element:

$$P_{in \to fi} \propto |\langle \Psi_{in} | \vec{r} | \Psi_{fi}^j \rangle|^2 = |\langle A\varphi_i \, \Psi_i^G(N\text{-}1) | \vec{r} | A\varphi_k \, \Psi_i^{+j}(N\text{-}1)\rangle|^2 \tag{4}$$

The function $\Psi_i^G(N\text{-}1)$ may be expanded in the ion set:

$$\Psi_i^G(N\text{-}1) = \sum_{j'=0} \langle \Psi_i^{+j'} | \Psi_i^G \rangle \, \Psi_i^{+j'}(N\text{-}1) \tag{5}$$

With the assumptions (2) as well as the assumption that the one-electron matrix element $\langle \varphi_i | \vec{r} | \varphi_k \rangle$ does not change appreciably with photoelectron energy over the range of possible final states, (4) reduces to:

$$P_{in \to fi} \propto |\langle \Psi_i^{+j} | \Psi_i^G \rangle|^2 \tag{6}$$

An interesting sum rule is obtained [32, 33] if we consider the matrix element:

$$\langle \Psi_i^G | H(N\text{-}1) | \Psi_i^G \rangle$$

which, according to (3) and (5), is equal to:

$$\langle \Psi_i^G | H(N\text{-}1) | \Psi_i^G \rangle$$

$$= \sum_{j=0} \langle \Psi_i^G | \Psi_i^{+j} \rangle \langle \Psi_i^G | H(N\text{-}1) | \Psi_i^{+j} \rangle = \sum_{j=0} E_i^j |\langle \Psi_i^G | \Psi_i^{+j} \rangle|^2 \tag{7}$$

The same matrix element may also be written:

$$\langle \Psi_i^G | H(N\text{-}1) | \Psi_i^G \rangle = E_0 - \epsilon_i^{fr} \tag{8}$$

where E_0 designates the ground-state total energy and $-\epsilon_i^{fr}$ the total energy increase due to the removal of the core state i. $-\epsilon_i^{fr}$ thus represents a measure of the binding energy in a frozen approximation. Eq. (7) then yields:

$$-\epsilon_i^{fr} = \sum_{j=0} |\langle \Psi_i^G | \Psi_i^{+j} \rangle|^2 E_B^j \tag{9}$$

or with respect to the hole ground state (j = 0):

$$-\epsilon_i^{fr} - E_B^0 = -E_B^0 + \sum_{j=0} |\langle \Psi_i^G | \Psi_i^{+j} \rangle|^2 E_B^j = \sum_{j=1} |\langle \Psi_i^G | \Psi_i^{+j} \rangle|^2 (E_B^j - E_B^0) \tag{10}$$

where E_B^j is the electron binding energy of state j. Equations (9) and (10) then state that if one measures the entire spectral weight function of a level and takes the first moment of this, the result obtained will be $-\epsilon_i^{fr}$. In particular, if Ψ_i^G is adequately described by a Hartree-Fock one-determinantal function, $-\epsilon_i^{fr}$ is equal to the orbital energy of state i. Thus, within the assumptions above and the Hartree-Fock approximation, orbital energies could in principle be experimentally obtained in this way. This would be an interesting possibility, since shifts in orbital energies between different chemical species may be more readily compared with other types of measurement than shifts in the main line energies. Unfortunately, this is not feasible with any higher accuracy for another reason: that Eqs (9) and (10) imply summation over discrete states, 'shake-up states' *as well as* integration over continuum, 'shake-off states'. This entire spectrum effectively ranges at least 50–100 eV below the leading photoelectron line, the shake-off continuum being a rather flat structure. If this continuum is not properly taken care of in calculation of the first moment this shifts the orbital energy values considerably. Since it is experimentally extremely difficult to measure the continuum with the desired accuracy, this severely limits the use of Eqs (9) and (10) for practical quantitative work. They are useful, however, from a qualitative point of view, since they give an indication of the magnitude of the perturbation of the system caused by the ionization. A measure of this perturbation is given by the quantity $(-\epsilon_i - E_B^0)$, which is usually termed the relaxation energy. It is the lowering of the binding energy resulting from the reorganization of the

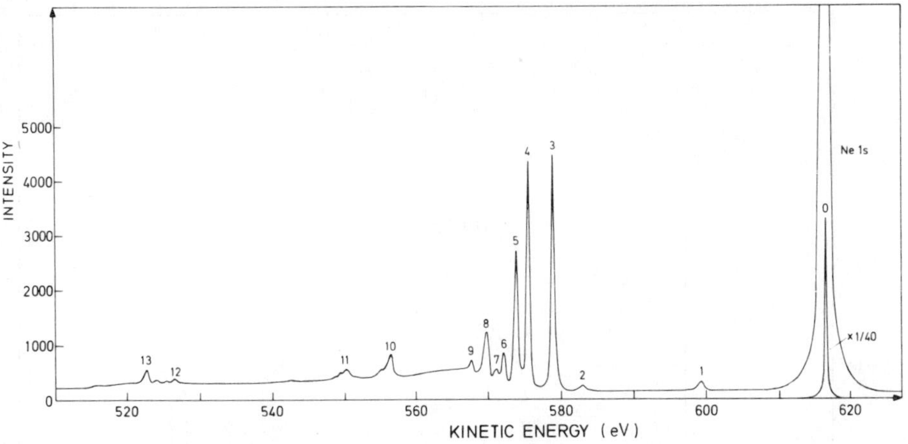

FIG.1. Shake-up, shake-off spectrum of neon (Ne 1s).

remaining electrons. From Eq. (10) it appears that the magnitude of the relaxation energy is related to the magnitude of the shake-up, shake-off spectrum. Thus, in core electron spectra one usually encounters substantial shake-up, shake-off spectra.

2.2. Shake-up spectra of Ne and H_2O

A typical shake-up spectrum of a core level is shown in Fig. 1 [34], which shows the Ne 1s line and low kinetic energy region excited by monochromatized Al Kα radiation (hν = 1486.6 eV). The lines 3–9 correspond to transitions to final states of the type $(1s)^1(2s)^2(2p)^5(np)^1(^2S)$, n = 3, 4, 5, 6. Due to the form of the excitation probability, Eq.(6), the final state in the shake-up transition must have the same symmetry as the frozen state Ψ^G (since the operator in this matrix element is equal to the unity operator, the transitions are referred to as monopole). For each final state configuration of the above type, two different couplings of the electrons are possible, corresponding to a singlet and a triplet of the valence shell. These states, in turn, interact to form one "Ψ_{1s}^{+np} (upper)" state and one "Ψ_{1s}^{+np} (lower)" state. Thus, peaks 3 and 4 correspond to a 3p-excitation, lower and upper, respectively, 5 and 8 to a 4p-excitation, and 6 and 9 to a 5p-excitation. The energies of these peaks have been found to be very well reproduced when the wave functions are constructed in the above fashion in MCHF calculations [9, 34]. The intensities of the peaks are, however, sensitive to the quality of the wave functions [35–38]. Since the matrix element (6) is in general small, only minor admixtures of additional configurations may lead to substantial effects, depending on the orbital basis used. Procedures have been found, however, involving configuration interaction in both initial and final states [35–38], that lead to good agreement with experiment. These procedures essentially retain the description of the mechanism of the transition in terms of orbital excitations, which is desirable from the point of view of comparison with other experimental data, such as electron impact and u.v. absorption spectra. Another interesting feature is seen in the neon spectrum, namely the low-intensity line No. 2. This line cannot, on energetic grounds, be ascribed to a monopole shake-up, but is instead interpreted as due to a final-state $1s\,2s^2\,2p^5\,3s\,(^2P)$-configuration. The existence of this line is an indication of a break-down of the assumption (2b). The final state configuration ^2P implies that the continuum electron must have

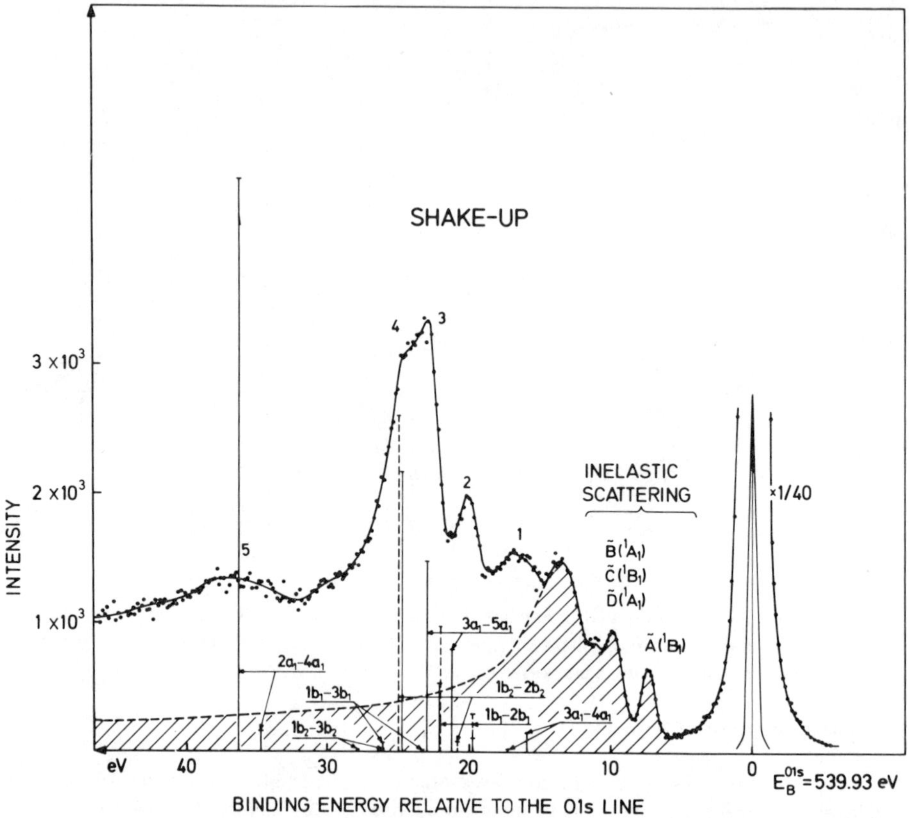

FIG.2. Shake-up, shake-off spectrum of water (O 1s).

s symmetry. The monopole shake-up configurations, on the other hand, lead to a p-wave continuum electron. The intensity of the line No. 2 is thus provided for by interchannel coupling in the continuum.

As an example of a shake-up spectrum from a molecule, Fig. 2 shows the O 1s spectrum of H_2O [36, 39]. The first three peaks below the main line are interpreted as due to inelastic scattering of the outgoing electron against neutral molecules and an estimate of these contributions has been subtracted in Fig. 2. Results of theoretical calculations of the energies and intensities [36] are indicated by the vertical bars. The dashed and solid bars correspond to results obtained using different MO-bases in the CI. It appears for some of the transitions that this choice of basis may lead to rather large effects in the calculated intensities, whereas the energies are very little affected. The calculations give an overall agreement with the observed spectrum which satisfactorily accounts for the main structure. In particular we note that the trend observed for neon does not seem to hold in this case, namely that the higher the transition energy of a given symmetry, the lower the intensity.

2.3. Characteristics of the main line

Since the shake-up intensities and energies are functions of the spectrum of unoccupied levels of the ion, spectra from a specific element will exhibit a strong dependence on the chemical bonding

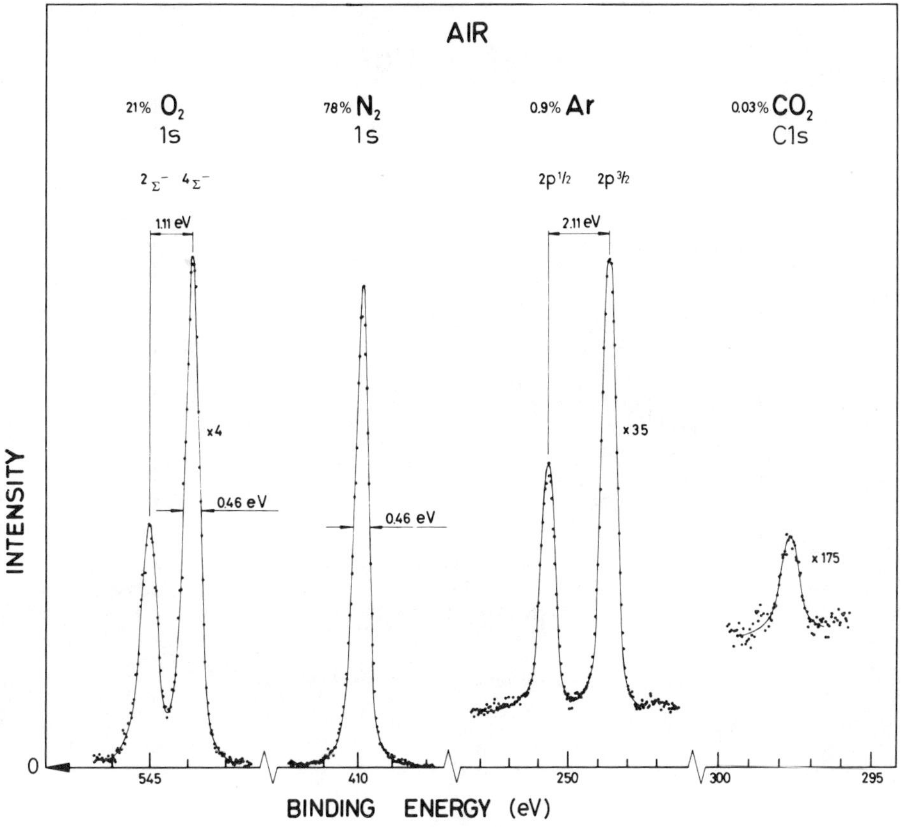

FIG.3. ESCA spectrum of air.

in which the atom is involved. The main source of information in core electron spectra is the leading photoelectron peak, however, and our analysis will be concentrated on this dominating feature. To begin with, the core line spectrum of an element is a unique property, a 'fingerprint' of the element, which can be used for qualitative and quantitative chemical analysis. As an example of this analytical applicability, Fig. 3 shows the ESCA spectrum of air. Apart from the oxygen and nitrogen lines, the 2p-lines of argon appear with good intensity (abundance 0.8%) as well as the carbon signal from CO_2 (abundance 0.03%). The amounts of material needed to detect an ESCA signal are actually very low, which has important implications in the study of solid samples. Since the electrons that contribute to the photoelectron line are those that have not undergone any inelastic collisions, they are effectively created within a distance from the sample surface equal to the electron mean free path. This so-called escape depth is generally within 10 to 50 Å for the electron kinetic energies relevant in electron spectroscopy. This implies that ESCA is a surface technique with high sensitivity to the detailed conditions of the outermost atomic layers. In particular, chemisorption processes and surface reaction dynamics, like catalysis etc., may be studied.

The highest surface sensitivity reported so far, by Brundle et al. [40], is 2×10^{-3} atomic layers. Techniques that make full use of this surface sensitivity for analytical purposes have been developed [41, 42]. The general idea is to deposit the substance to be detected, which is present in a solution

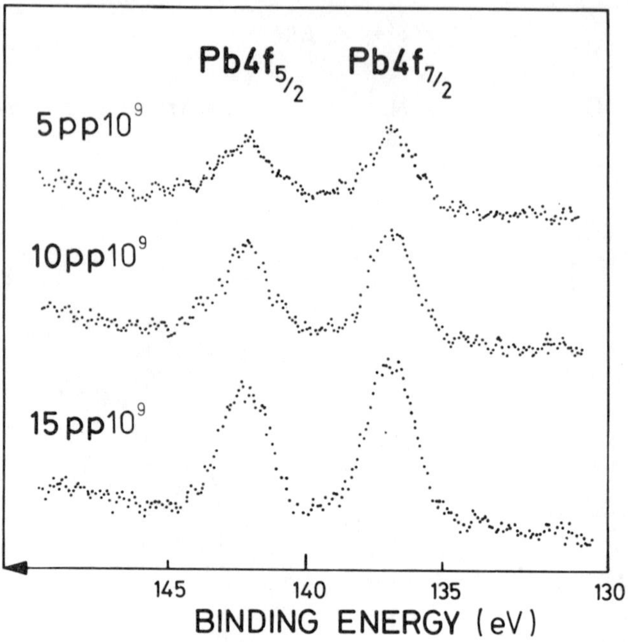

FIG.4. ESCA spectra from three solutions of lead ions.

at a low concentration, onto a backing and subsequently to measure its electron spectrum. Hercules et al. [41] have employed a fibreglass as backing, having organic functional groups on its surface, which bind in a more or less non-selective fashion the ions in the solution. In this way it was demonstrated that it is possible to reach detection limits below 10 ppm for several elements. Figure 4 shows a result from our own laboratory with the Hercules technique, which shows the $Pb4f_{5/2,7/2}$ lines obtained from solutions with different concentrations of lead ions, the smallest being as low as 5 $pp10^9$. The actual amount of lead needed to obtain these signals was 10^{-8} g. We note that the intensities of the lines increase in proportion to the amounts of lead ions. For this case such a proportionality would be expected to hold rigorously, since one is only concerned with one chemical state of the lead atom. It must be borne in mind, however, that in general it is rather the entire core electron spectrum, i.e. including shake-up and shake-off, that has this simple proportionality and not the leading core-hole ground-state peak. So, if one measures the intensity of this peak for an element in *different* chemical forms, deviations from such a simple relationship occur, since the shake-up spectrum is chemically dependent, as mentioned. It is found, however, that the relaxation energy for a core hole in different chemical species is largely unchanged (a point to which I shall return), which according to Eq.(10) implies that the total intensity of the shake-up, shake-off spectrum is restricted in its variation. This in turn implies that, within reasonable error limits, the intensities of the leading electron line follow the stoichiometric relationships. With this restriction in mind, the core electron spectrum may be used for quantitative analysis as well as the qualitative analysis described above.

3. CHEMICAL SHIFTS

3.1. Ground-state potential models

To deal with the chemical influences on core electron line energies, we shall base our argument on a one-electron picture and write the binding energy as:

$$E_B = -\epsilon - E_{Relax} + \Delta E_{Correlation} + \Delta E_{Relativistic} \tag{11}$$

where ϵ is the Hartree-Fock orbital energy, E_{Relax} is the relaxation energy and $\Delta E_{Correlation}$ and $\Delta E_{Relativistic}$ are the differences in correlation and relativistic corrections, respectively, between initial and final state of the ionization process. For core electron ionization, the two latter quantities are at least an order of magnitude smaller than the other two, so in this context of *differences* in binding energies, one can safely neglect them. (For O 1s ionization in H_2O, $-\epsilon = 559.5$ eV, $E_{Relax} = 20.4$ eV, $\Delta E_{Correlation} = 0.5$ eV and $\Delta E_{Relativistic} = 0.4$ eV.) This implies that the ΔSCF level of accuracy is sufficient for the calculation of chemical shifts:

$$\Delta E_B \approx \Delta E_B(SCF) = -\Delta\epsilon - \Delta E_{Relax} \tag{12}$$

Let us consider first the shift in orbital energy, $\Delta\epsilon$. If we assume the wave function to consist of doubly occupied spatial orbitals, the orbital energy of a core electron i situated on atom A in a molecule can be expressed as:

$$\epsilon_i = \langle i|-\tfrac{1}{2}\nabla_1^2|i\rangle + \sum_B \left\langle i\left|\frac{-Z_B}{r_{B1}}\right|i\right\rangle + \sum_j (2J_{ij} - K_{ij}) \tag{13}$$

where J_{ij} and K_{ij} are direct and exchange integrals, respectively. To analyse the different contributions, we shall make a division of this expression in the following fashion:

$$\epsilon_i = \left\langle i\left|-\tfrac{1}{2}\nabla_1^2 - \frac{Z_A}{r_{A1}}\right|i\right\rangle + \sum_{\substack{j = \text{core orb.}\\ \text{on A}}} (2J_{ij} - K_{ij}) - \sum_{\substack{j \neq \text{core orb.}\\ \text{on A}}} K_{ij} + \sum_{j = \text{val. orb.}} 2J_{ij} + \sum_{\substack{j = \text{core orb.}\\ \text{on B}}} 2J_{ij} + \sum_{B \neq A} \left\langle i\left|-\frac{Z_B}{r_{B1}}\right|i\right\rangle \tag{14}$$

It has been shown in calculations [43], that the first term, i.e. the kinetic energy plus the Coulomb attraction energy of the core electron, is constant to a high accuracy and can be considered atomic quantities. The same is true for the next term. Taking the shift in orbital energy implies then that these terms vanish. The exchange integrals contained in the next term, the exchange between two core electrons on different atomic centres and the exchange between a core and valence electron are small quantities and may be considered negligible in the present context. One is thus left with the last three terms, which give any appreciable contribution to the chemical shift of the orbital energy. The last two may be grouped together in a simple way, since the interaction of a core orbital on one atom with those of another is very accurately given by the interaction of two point charges, due to the localized nature of core orbitals. This gives:

$$\Delta\epsilon_i = \Delta\left(\sum_{\substack{j = \text{val.} \\ \text{orb.}}} 2J_{ij} - \sum_{B \neq A} \frac{Z_B^*}{R_{AB}} \right) \tag{15}$$

where Z_B^* represents the nuclear charge of atom B minus the number of core electrons. If we now introduce an LCAO-expansion for the molecular orbitals:

$$\varphi_j = \sum_{k, B} C_k^j(B)\, \phi_k(B) = \sum_{\ell, C} C_\ell^j(C)\, \phi_\ell(C) \tag{16}$$

we get

$$\Delta\epsilon_i = \Delta\left(\sum_{\substack{j = \text{val.} \\ \text{orb.}}} \sum_{\substack{k, \ell \\ B, C}} 2C_k^j(B)\, C_\ell^j(C) \left\langle \phi_k(B)\, i \left| \frac{1}{r_{12}} \right| \phi_\ell(C)\, i \right\rangle + \sum_{B \neq A} \frac{Z_B^*}{R_{AB}} \right) \tag{17}$$

The integrals in the first sum, where $k \neq \ell$ or $B \neq C$, are going to be small and are neglected in a common approximate semi-empirical MO-scheme, CNDO (complete neglect of differential overlap). We can then split the first sum into two parts:

$$\Delta\epsilon_i = \Delta\left(\sum_{\substack{j = \text{val.} \\ \text{orb.}}} \sum_{k} 2(C_k^j(A))^2 \left\langle \phi_k(A)\, i \left| \frac{1}{r_{12}} \right| \phi_k(A)\, i \right\rangle + \right.$$

$$\left. + \sum_{B \neq A} \sum_{\substack{j = \text{val.} \\ \text{orb.}}} \sum_{\ell} 2(C_\ell^j(B))^2 \left\langle \phi_\ell(B)\, i \left| \frac{1}{r_{12}} \right| \phi_\ell(B)\, i \right\rangle - \sum_{B \neq A} \frac{Z_B^*}{R_{AB}} \right) \tag{18}$$

The integrals in the second sum can be approximated by:

$$\left\langle \phi_\ell(B)\, i \left| \frac{1}{r_{12}} \right| \phi_\ell(B)\, i \right\rangle = \frac{1}{R_{AB}} \tag{19}$$

which is simple but quite accurate for our purposes. The integrals in the first sum represent the interaction between core and valence electrons on the same atom. If these are given an average value, k_A, they can be moved out of the summation sign. We thus obtain:

$$\Delta\epsilon_i = \Delta\left(k_A \sum_{j, k} (C_k^j(A))^2 + \sum_{B \neq A} \frac{1}{R_{AB}} \sum_{j, \ell} (C_\ell^j(B))^2 - \sum_{B \neq A} \frac{Z_B^*}{R_{AB}} \right) \tag{20}$$

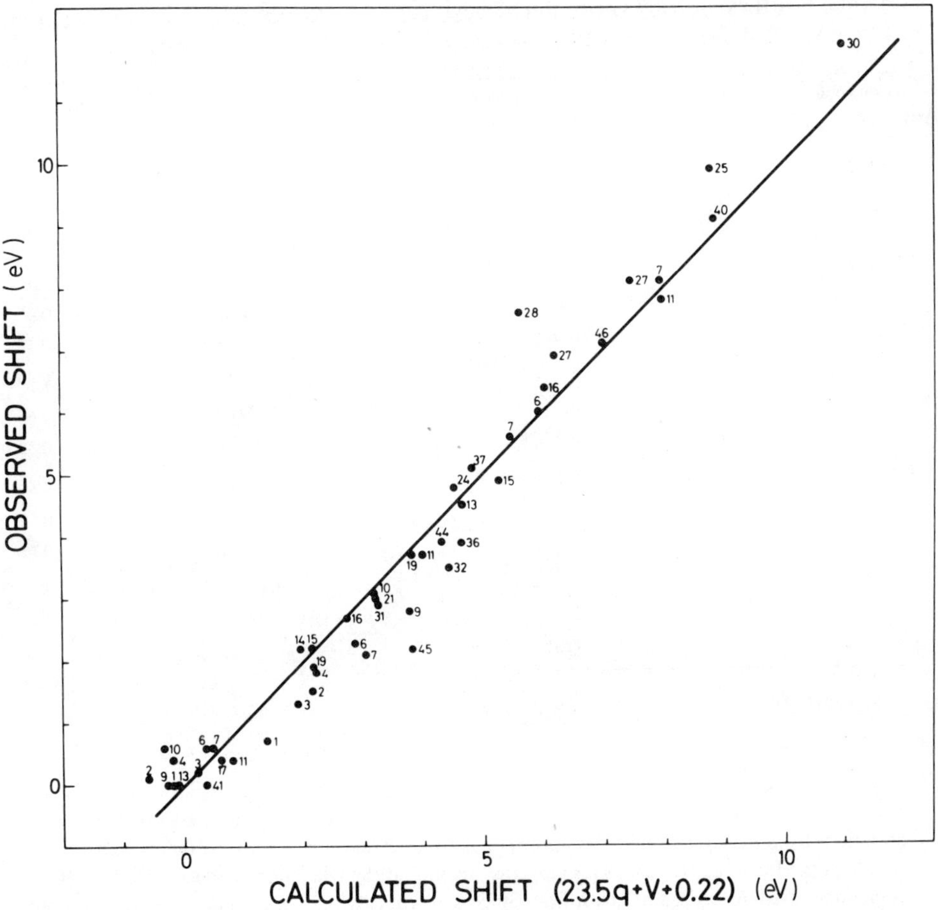

FIG.5. Observed chemical shifts of C1s line versus calculated shifts according to ground-state potential model [113].

The sums of MO-coefficients are the total electron populations on the respective atoms and, remembering that the charge q on an atom is given by:

$$q = Z - P$$

where P is the electron population, we obtain finally:

$$\Delta\epsilon_i = -\Delta\left(k_A q_A + \sum_{B \neq A} \frac{q_B}{R_{AB}}\right) \tag{21}$$

Although several approximations were made to obtain this simple expression it has been found surprisingly accurate. If we forget for the moment that it describes shifts in orbital energy only and try to correlate it with chemical shifts of binding energies, i.e. we neglect the shift in relaxation

TABLE I. CHARGES OBTAINED FROM OBSERVED 1s BINDING ENERGY SHIFTS AND FROM SEMIEMPIRICAL CALCULATIONS[a]

Compound	Atom	$\Delta E = \Delta E_i$ (Obs) – ΔE_i (Calc) (eV)	q_i	q_i (CNDO II)	
$C_2(CN)_4$	C_1	0.0	0.04(1)	0.05	
	C_2	0.0	0.17(1)	0.11	
	N	0.0	–0.19(1)	–0.14	
$C_2O(CN)_4$	C_1	0.0	0.25(1)	0.16	0.18
	C_2	0.0	0.12(1)	0.09	0.09
	N	0.0	–0.17(1)	–0.12	–0.13
	O	0.0	–0.33(1)	–0.18	–0.20
CO_2	C_1	0.1	0.68(3)	0.54	0.66
	O	0.2	–0.34(2)	–0.27	–0.33
C_3O_2	C_1	–0.2	–0.33(3)	–0.56	–0.68
	C_2	–0.3	0.45(6)	0.56	0.59
	O	–0.4	–0.29(3)	–0.29	–0.25
N_2O	N_1	–0.8	–0.07(7)	–0.18	
	N_2	–0.5	0.16(7)	0.52	
	O	–1.0	–0.10(6)	–0.34	

[a] From Ref. [50].

energy, ΔE_{Relax}, the agreement between estimates from Eq. (21) and experiment is strikingly good. The compounds investigated in Fig. 5 are a series of carbon compounds and we see that the maximum deviations from the straight line that corresponds to exact agreement between theory and experiment are not larger than, say, 1.5 eV. This indicates, then, that the relaxation energies for different compounds are fairly constant quantities. Since one generally refers the shifts to a specific compound, Eq. (21) may be written:

$$-\Delta\epsilon_i = k_A q_A + \sum_{B \neq A} \frac{q_B}{R_{AB}} + \ell_A \tag{22}$$

There are different ways that the relation (22) can be and have been used in numerous applications. First, the measured experimental shifts, geometries and charges obtained in semiempirical calculations (the CNDO/2 method is almost exclusively used in this context) may be inserted for a selection of compounds and the quantities k_A and ℓ_A may thus be obtained from (22) by least-squares fitting procedures. This has been done for B, C, N, O, F of the first row and Si, P, S, Cl of the second row, and a number of other elements in the periodic system, using other charge definitions [9, 44–49]. With access to these values one is in a position to invert the charge potential model (22) in two ways. Either one inserts geometries and charges and calculates theoretical shifts or one inserts measured shifts and geometries and calculates the charge distribution.

Some examples of these applications are shown in Table I. It appears that the agreement between the empirically obtained charges and those from CNDO/2 in general is quite good. This offers a possibility then, in the case of large molecular systems where theoretical calculations would be expensive, to calculate charge distributions based on ESCA shifts. The cost of doing this would be low, since the mathematical manipulations are simple. The aspect of calculating shifts from charges and geometries implies the feasibility of using ESCA as a structural tool. By combination of the calculated shifts, using different alternative geometries and information concerning line shapes and widths, the expected ESCA spectra can be deduced and used to distinguish between the various structure alternatives through comparison with the ESCA spectrum actually observed. As an example of this we may take a study made by Clark [51] on the resulting species of hydration of perfluoroindene. An investigation by NMR-spectroscopy showed the monohydroproduct to be a mixture of two isomers (see Fig. 6) in the ratio 4 : 1. However, it was not possible to decide which isomer was the dominating one in the mixture. By a simple application of the scheme outlined above this turned out to be feasible by means of ESCA. Figures 6 and 7 show the computer simulations of the ESCA spectra made by Clark; in Fig. 6 the spectra of the separate species and in Fig. 7 the spectra expected from the possible mixtures. The spectrum experimentally obtained (Fig. 7(c)) clearly shows that the mixture of isomers actually present is that of the lowest of the two theoretical alternatives, i.e. Fig. 7(b).

Another quite useful way of employing the potential model for practical purposes is to introduce the concept of group shifts. If we choose the reference compound in an appropriate manner, we may put $\ell_A = 0$ in Eq. (22). Combination of Eq. (22) with the electro-neutrality condition:

$$-\sum_{B \neq A} q_B = q_A \tag{23}$$

leads to the following expression for the shift:

$$-\Delta\epsilon = \sum_{B \neq A} \left(\frac{1}{R_{AB}} - k_A \right) q_B \tag{24}$$

The summation over the atoms B can be subdivided into partial sums over each of the groups G that are attached to atom A.

$$-\Delta\epsilon = \sum_{G} \left[\sum_{B \in G} \left(\frac{1}{R_{AB}} - k_A \right) q_B \right] = -\sum_{G} \Delta\epsilon_G \tag{25}$$

The general idea behind this version of the potential model is that one should be able to assign a group shift to a certain group that attaches to a specific element such that the addition of all these group shifts should give the total shift on the atom concerned. This assumes then that the bond geometry and the charge distribution of the group are always the same whenever the group is bonded to the element, irrespective of what other groups are bonded to it. This, of course, is an approximation which may be more or less serious, but if it is possible to introduce these figures to characterize the contribution of a group to the chemical shifts, this constitutes a simplification which is very useful. The example shown in Fig. 8 of a group shift analysis of phosphorus and arsenic compounds demonstrates the feasibility of this idea; the deviations from perfect agreement are rather small and quite acceptable considering the approximations made [114]. Here the group shifts have been treated as

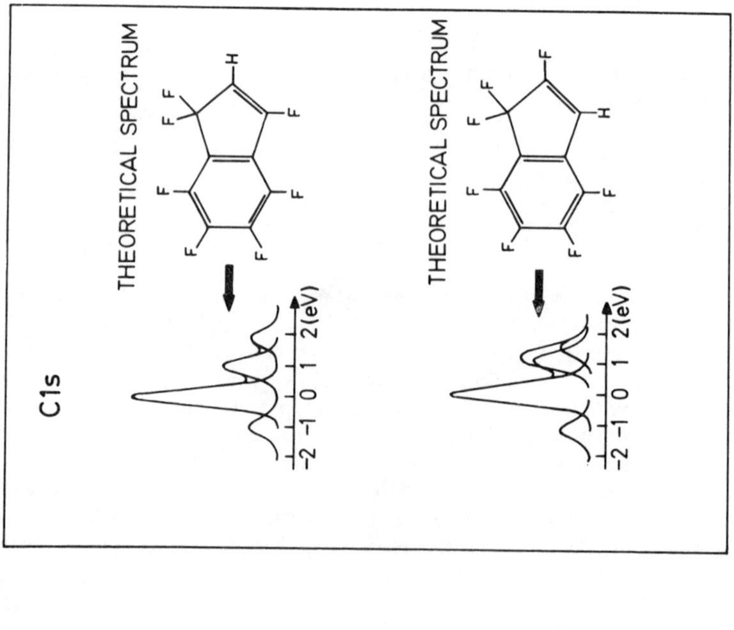

FIG.6. Simulated spectra of two possible isomers of monohydro product of perfluoroindene.

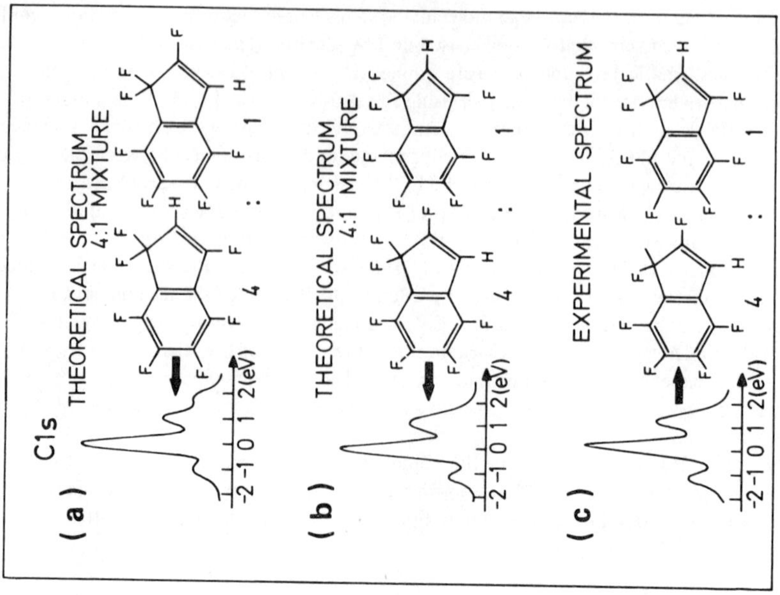

FIG.7. (a) and (b) Simulated spectra of mixtures of two isomers, and (c) experimental spectrum of monohydro product of perfluoroindene [51].

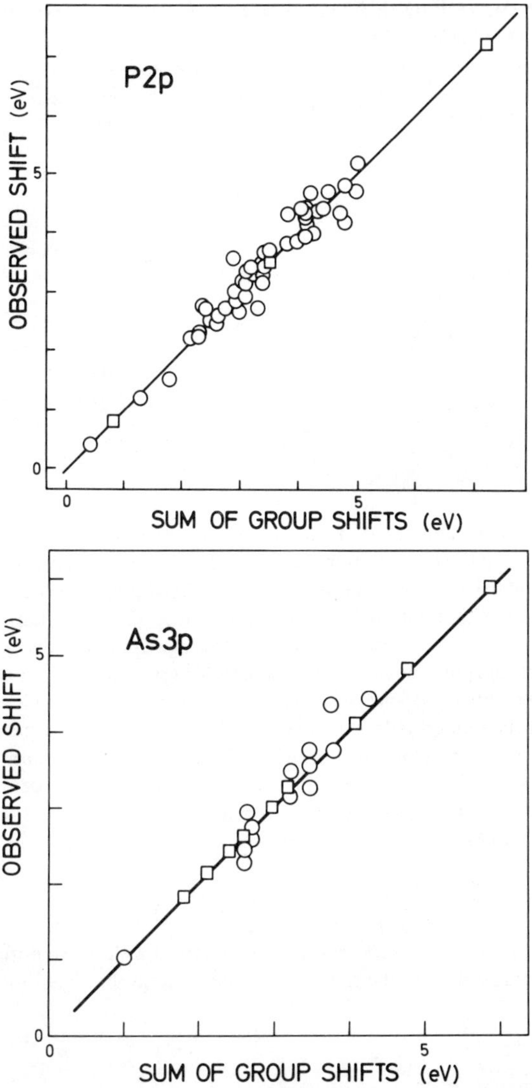

FIG.8. Correlations of sum of group shifts with observed shifts for phosphorus and arsenic compounds [114].

unknowns and least-squares fitted to the experimental data. The advantage of this concept is that, with access to a table of group shifts, it is a matter of "back of the envelope" work to calculate shifts for even the bulkiest of compounds such as $C_6H_{11}NH_3OSP(OCH_2C(CH_3)_2CH_2O)_2$ which is one of the compounds in the study of Fig. 8. It is thus a simple method for a practical working chemist.

3.2. Transition potential models

In the previous subsection the shifts have been handled on a frozen level of approximation, i.e. the shifts in relaxation energy, E_{Relax}, have been neglected. It may be concluded from the examples given that this is a rather good approximation. However, this is likely to hold true only

TABLE II. COMPARISON OF IONIZATION ENERGIES COMPUTED BY DIFFERENT METHODS FOR NEON AND WATER[a]

Hole state	Koopmans' theorem	ΔSCF	Transition operator	Experiment
Ne 1s	891.7	868.6	868.6	870.4
Ne 2s	52.5	49.3	49.3	48.4
Ne 2p	23.1	19.8	19.9	21.6
H_2O $1a_1$	559.5	539.1	539.1	540.2
H_2O $2a_1$	37.1	34.5	34.4	32.2
H_2O $3a_1$	14.9	13.3	13.2	14.7
H_2O $1b_2$	19.1	17.9	17.8	18.6
H_2O $1b_1$	13.2	11.2	11.0	12.6

[a] From Refs [9, 34, 53, 54].

when the compounds studied are similar in nature, such that one can expect the electronic response, 'conductivity', to the ionization to be fairly similar. Let us analyse this question further to arrive at a model which incorporates effects of relaxation in the calculated shifts. To do this, we make use of a formalism developed to handle optical excitations as well as ionization [52, 53]. On the SCF level, the usual approach is to make optimizations of initial and final states with *separate* sets of spin orbitals. If instead of doing this, one optimizes the mean value of the total energies of the initial and final states with a *common* set of spin orbitals, a set of one-electron equations for the spin orbitals is obtained, as in the case of a single-state optimization, but with a different Fock operator. This operator, which is referred to as the transition operator, is constructed such that each of the spin orbitals involved in the transition is occupied by half an electron. When the one-electron equations are solved, a set of orbital energies is obtained in the usual way and it is easy to show that:

$$\langle H\rangle_{\text{final}} - \langle H\rangle_{\text{initial}} = \epsilon_{\text{fi}}^{\text{T}} - \epsilon_{\text{in}}^{\text{T}} \tag{26}$$

where the expectation values on the left-hand side are obtained with the transition orbitals and the quantities on the right-hand side are the transition operator eigenvalues. For ionization this becomes

$$\langle H\rangle_{\text{final}} - \langle H\rangle_{\text{initial}} = -\epsilon_{\text{i}}^{\text{T}} \tag{27}$$

where i is the spin orbital involved. Now, since a restriction has been made on the spin orbitals, it is clear that the total energies of both the initial and final states will be higher than if optimizations were made on each of them separately. However, it turns out that both are high by almost exactly the same energy [52, 53], so that there is a cancellation of errors in the relations (26) and (27). Due to this cancellation, the transition operator eigenvalues simulate ΔSCF results to a very high accuracy. Results in Table II on neon and the water molecule demonstrate this fact quite clearly. Apart from the advantage that this model actually only needs one calculation per ionization energy, it has the appealing feature of retaining the one-electron picture of ionization at ΔSCF level of accuracy. This leads to simple formulas for the transition elements of dipole transitions and Auger transitions, including relaxation corrections involving only the orbitals participating in the transition – which would not be the case if the ordinary ΔSCF method was used for this purpose. It should also be noted that one does not in this connection use the term 'transition state', since the appearance of the transition operator derives not from the existence of such an artificial state, but from the optimization procedure.

In exactly the same way as above for the ground-state orbital energies, one can derive a potential model for the transition orbital energies [55, 56]:

$$-\epsilon_A^T = k_A^T q_A^T + \sum_{B \neq A} \frac{q_B^T}{R_{AB}} + \ell_A^T = k_A^T q_A^T + V_A^T + \ell_A^T \tag{28}$$

(Here the reference constant ℓ_A^T is chosen on an absolute energy scale.) The charges in this transition potential model (TPM) are then calculated using a quasi-atom, which has a core electron population half-way between the initial and final states. On the level of semiempirical calculations (CNDO/2) this implies increasing the nuclear charge by half an electronic charge. We rewrite (22) to distinguish it from (28):

$$-\epsilon_A^G = k_A^G q_A^G + \sum_{B \neq A} \frac{q_B^G}{R_{AB}} + \ell_A^G = k_A^G q_A^G + V_A^G + \ell_A^G \tag{29}$$

Under the assumption that (28) describes the full binding energy including relaxation effects, the relaxation energy is obtained as the difference between (29) and (28):

$$E_{Relax} = -\epsilon_A^G - (-\epsilon_A^T) = (k_A^G q_A^G - k_A^T q_A^T) - (V_A^T - V_A^G) + (\ell_A^G - \ell_A^T) \tag{30}$$

This can be written in terms of populations, P_A^G and P_A^T:

$$E_{Relax} = k_A^T P_A^T - k_A^G P_A^G - (V_A^T - V_A^G) + [(\ell_A^G - \ell_A^T) + (k_A^G - k_A^T) Z_A - k_A^T \cdot 0.5] \tag{31}$$

which reduces to:

$$E_{Relax} = (k_A^T - k_A^G) P_A^G + k_A^T \cdot \Delta P - (V_A^T - V_A^G) + \ell'_A \tag{32}$$

if one introduces the quantity:

$$\Delta P = P_A^T - P_A^G \tag{33}$$

and the constant:

$$\ell'_A = [(\ell_A^G - \ell_A^T) + (k_A^G - k_A^T) Z_A - k_A^T \cdot 0.5]$$

Equation (32) constitutes a very useful formula for discussing the physics in the electronic relaxation accompanying ionization. The first term represents, then, the energy gained by contraction of the valence charge density on the atom where the ionization takes place. The second and third terms represent the relaxation energy due to the reorganization of charge in the molecule as a whole. Specifically, the second term is the energy gained by the increase of charge, ΔP, on the ionized atom. This energy is proportional to the one-centre electron-electron repulsion integral k_A^T. Since this charge is taken from the rest of the molecule, the third term reflects from where the charge ΔP is effectively taken and transferred to the ionized atom. One may then write:

$$E_{Relax} = E_{Contr.} + E_{Flow} \tag{34}$$

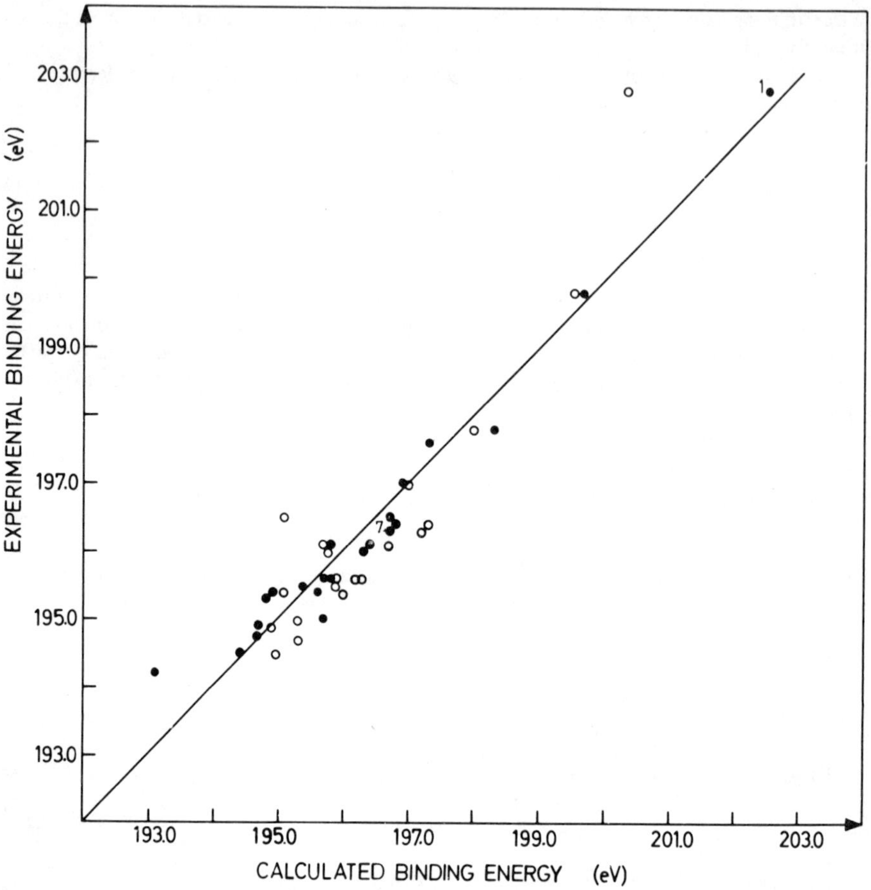

FIG.9. Observed B 1s binding energies versus calculated binding energies according to ground-state potential model (open circles) and transition potential model (full circles). Compound 1 refers to BF_3 and compound 7 to B_2H_6.

where the 'contraction part' is equal to:

$$E_{Contr.} = (k_A^T - k_A^G) \cdot P_A^G + \ell'_A \tag{35}$$

and the "flow part":

$$E_{Flow} = k_A^T \cdot \Delta P - (V_A^T - V_A^G) \tag{36}$$

To illustrate the importance of the relaxation energy and the variation of its two contributions (35) and (36), we choose as an example a series of boron compounds. Figure 9 shows the correlation between theory and experiment for these compounds. In Table III are to be found the relaxation energies as well as the relevant quantities to calculate them according to Eq. (32). The filled circles designate the binding energies calculated by the transition potential model (TPM) and

TABLE III. GROUND-STATE POPULATIONS AND POTENTIALS (eV), THEIR RESPECTIVE DIFFERENCES WITH TRANSITION OPERATOR AND RELAXATION ENERGIES (eV) FOR BORON 1s IONIZATION[a]

A			P_A^G	ΔP_A	V_A^G	ΔV_A	E_{Relax}
1.	BF_3		2.23	0.46	-8.52	5.04	8.5
2.	BCl_3		2.54	0.48	-3.86	4.06	10.6
3.	$B(CH_2CH_3)_3$		2.84	0.43	-2.10	3.03	11.3
4.	$ClB(CH_2CH_3)_2$		2.74	0.42	-2.75	2.98	10.9
5.	$B(OCH_3)_3$		2.38	0.49	-8.12	3.94	10.5
6.	$B(OCH_2CH_3)_3$		2.42	0.49	-8.11	3.80	10.8
7.	B_2H_6		3.12	0.29	2.03	3.02	9.1
8.	$B_2H_5N(CH_3)_2$		2.97	0.42	-0.89	3.65	10.7
9.	1, 5-$C_2B_3H_5$		2.98	0.35	0.29	2.75	10.3
10.	1, 6-$C_2B_4H_6$		3.02	0.39	0.29	2.90	11.0
11(a)	2, 4-$C_2B_5H_7$	(1, 7)	2.89	0.41	-0.45	3.00	11.0
(b)		(3)	3.07	0.35	1.98	2.33	10.9
(c)		(5, 6)	3.07	0.36	1.00	2.44	11.0
12(a)	1, 2-$C_2B_{10}H_{12}$	(3, 6)	2.96	0.39	1.36	2.47	11.3
(b)		(4, 5, 7, 11)	2.97	0.39	0.60	2.44	11.3
(c)		(8, 10)	2.99	0.37	-0.06	3.09	10.3
(d)		(9, 12)	2.98	0.39	-0.53	2.42	11.4
13(a)	1, 7-$C_2B_{10}H_{12}$	(2, 3)	2.96	0.39	1.42	2.50	11.3
(b)		(4, 6, 8, 11)	2.97	0.39	0.62	2.48	11.3
(c)		(5, 12)	2.96	0.39	0.15	2.52	11.2
(d)		(9, 10)	2.99	0.39	-0.07	2.41	11.4
14(a)	B_5H_9	(1)	3.15	0.36	0.67	2.76	10.8
(b)		(2, 3, 4, 5)	2.99	0.38	0.38	2.96	10.7

[a] From Ref. [56].

the open circles those calculated using the ground-state potential model (GPM). It is seen that the relaxation-corrected TPM binding energies are in better agreement with experiment than GPM ones.

One important thing to note in this context is that the one-centre electron repulsion integrals, k, chosen in Eqs (28) and (29) are not empirically optimized, as was done above in the case when only the GPM was used to calculate theoretical binding energies. It is clear that in doing the optimization of k (and ℓ) versus experiment, those parts of the relaxation energy linearly related to q are automatically incorporated in the calculations, although GPM is not strictly a model which includes relaxation. In particular, one sees from (35) that the entire contraction part is taken care of in such a procedure. It is the purpose for which one uses a potential model that decides whether or not this is a satisfactory procedure. If it is used to simulate spectra from ground-state charge densities or to calculate charge densities from spectra, the GPM with empirical k and ℓ may for most purposes be quite satisfactory. If, however, one is interested in the detailed contributions of ground-state properties and relaxation to the observed shifts, TPM and GPM with theoretical k-values (ℓ still needs to be fitted) are to be preferred. In the ordinary CNDO/2 programme with STO minimal basis, the k-values are obtained as $k = \xi/2$ (for first-row elements) and $k = \xi/3$ (for second-row elements), where ξ is the STO exponent.

In Figure 9 it is apparent that for particular compounds the improvement with respect to GPM is especially large, namely compounds Nos. 1 and 7, BF_3 and B_2H_6. These constitute two examples of different relaxation behaviour, which can be traced back to the electronic structure. The relaxation energies calculated from (31) for these compounds are found to be low with respect to the other compounds, 8.5 eV and 9.1 eV, respectively, against ~ 11.0 eV on the average for the other compounds. For BF_3 it is seen that the charge compensation, ΔP, is close to complete (0.46) and that the main part of the difference in relaxation energy with respect to the other molecules comes from the large value of ΔV. A comparison with the simplest of the other molecules of this type, BCl_3, explains this figure as being due to the magnitude of the B-F bond length. The difference in ΔV of 1 eV obtained between these two molecules is accounted for by the difference in bond length (1.30 Å for BF_3 and 1.73 Å for BCl_3). For B_2H_6, however, the low relaxation energy has another origin. In this case, the low charge compensation of 0.29 electrons is apparently the dominating factor in deciding the relaxation energy. This is a reflection of the typical chemistry of boron compounds, whose hydrides are characterized in their chemical behaviour by their electron deficiency. In the context of ionization it results in a low 'conductivity' of the B-H-B bonds. It is interesting to note that if one replaces one of the bridging hydrogens in B_2H_6 with a dimethylamino group (compound No. 8), the possibilities of charge flow increase in the system, by virtue of the B-N-B π-bonds, and the ΔP-value as well as the relaxation energy increase substantially.

Another interesting example of how the valence electron structure decides the relaxation energy is given by a comparison between methane and benzene. In the ground state, the carbon atoms in benzene are slightly more positive than in methane and the GPM indicates that the binding energy in benzene should be higher than that in methane by ~ 1 eV. Experimentally it is found to be lower by 0.5 eV. This can be explained if one carries out a calculation of the relaxation energies according to the foregoing. It appears that the charge compensation is close to complete in both cases, but that the value of ΔV for benzene is ~ 2 eV lower than for methane, which clearly indicates the importance of the π-system in benzene in carrying charge over extended distances. This difference in effective distance from where the compensation charge is taken accounts for the high relaxation energy in benzene as compared with methane and thus the shift between them.

It might be argued in this context that the analysis of the relaxation energy, not being a directly observable quantity, is more a game with words than a description of physical reality. However, this analysis is necessary from the point of view of the model chosen to explain the behaviour of the shifts. In other words, it is important to know to what extent we are justified in correlating the observed shifts to ground-state properties *only*. This, in turn, has implications on the relation between core electron spectroscopy and other types of measurement, such as NMR, Mössbauer spectroscopy, etc., as well as correlations with chemical reactivity data.

4. INTENSITIES OF ELECTRON LINES

4.1. Photoelectric cross-sections and angular distributions from atoms and molecules

The probability per unit time for ejection of a photoelectron into a continuum state is, according to time-dependent, first-order perturbation theory, given by:

$$P \propto |\langle\Psi_{in}|H_{int}|\Psi_{fi}\rangle|^2 \ \rho(E_{fi}) \qquad (37)$$

where we take the initial state as a closed shell HF-determinant and the final state contains a continuum orbital to describe the ejected electron. $\rho(E_{fi})$ is the density of states at the final electron energy. The interaction Hamiltonian, H_{int}, has the general form:

$$H_{int} = -\sum_i \left[\frac{e}{mc} (\vec{p}_i \cdot \vec{A}_i) - \frac{e^2}{2mc^2} |\vec{A}_i|^2 \right] \tag{38}$$

where $\vec{p}_i$ and $\vec{A}_i$ are the linear momentum and vector potential operators of electron i, respectively. Assuming the dipole approximation which amounts to neglect of photon momentum and thus a constant vector potential, $\vec{A}_0$, over the system volume, (37) reduces to

$$P \propto A_0^2 \left| \vec{u} \cdot \left\langle \Psi_{in} \left| \sum_i \vec{p}_i \right| \Psi_{fi} \right\rangle \right|^2 \rho(E_{fi}) \tag{39}$$

where $\vec{u}$ is the polarization vector of the incoming radiation. Since the incident number of photons is proportional to A_0^2, the differential cross-section is proportional to:

$$\frac{d\sigma}{d\Omega} \propto \frac{1}{\omega} \left| \vec{u} \cdot \left\langle \Psi_{in} \left| \sum_i \vec{p}_i \right| \Psi_{fi} \right\rangle \right|^2 \rho(E_{fi}) \tag{40}$$

To indicate the general form of the cross-section, one neglects relaxation in the final state and assumes an orthogonalized plane wave for the ejected electron of the form:

$$|k\rangle = N \left[|PW\rangle - \sum_j \langle j|PW\rangle |j\rangle \right] \tag{41}$$

By Slater's rules for the matrix element Eq. (40) then reduces to:

$$\frac{d\sigma}{d\Omega} \propto \frac{\rho(E_{fi})}{\omega} \left| \vec{u} \cdot \left[\langle \ell | \vec{p} | PW \rangle - \sum_j \langle \ell | \vec{p} | j \rangle \langle j | PW \rangle \right] \right|^2 \tag{42}$$

The plane wave is an eigenfunction to the momentum operator and one therefore obtains:

$$\frac{d\sigma}{d\Omega} \propto \frac{\rho(E_{fi})}{\omega} \left| \vec{u} \cdot \left[\hbar\vec{k} \langle \ell | PW \rangle - \sum_j \langle \ell | \vec{p} | j \rangle \langle j | PW \rangle \right] \right|^2$$
$$= \frac{\rho(E_{fi})}{\omega} \hbar \left| k \cos\varphi_{ku} \langle \ell | PW \rangle - \vec{u} \cdot \sum_j \langle \ell | \vec{p} | j \rangle \langle j | PW \rangle \right|^2 \tag{43}$$

This is the differential cross-section for a fixed position of the system with respect to the laboratory reference frame. To get the actual expression for a system with random orientation such as an atom or a molecule, Eq. (43) must be averaged over all possible orientations (the angle φ_{ku} kept fixed). The resulting general expression is of the form:

$$\frac{d\sigma}{d\Omega} = A + B \cdot \cos^2 \varphi_{ku} \tag{44}$$

which is usually written, employing a so-called asymmetry parameter, $\beta(E_{fi})$:

$$\frac{d\sigma}{d\Omega} = \frac{\sigma}{4\pi} \left(1 + \beta(E_{fi})\, P_2\, (\cos \varphi_{ku})\right) \tag{45}$$

where $P_2(x) = \frac{1}{2}(3x^2 - 1)$. Equation (45) is valid for linearly polarized light. Although the outlined derivation above of Eq. (44) employs particular assumptions such as the use of OPW's for the final state, it can be shown that it may be obtained under quite general conditions [57]. If non-polarized light is used, which is usually the case, the light can be considered as the incoherent superposition of two polarized beams with one polarization direction lying in the plane of the photon *and* outgoing electron directions and the other perpendicular to this plane. One then obtains for non-polarized light:

$$\frac{d\sigma}{d\Omega} = \frac{\sigma}{4\pi} \left[1 - \tfrac{1}{2}\beta(E_{fi})\, P_2\, (\cos \theta)\right] \tag{46}$$

where θ is the angle between the incident photon beam and the photoelectron directions. We note in the expressions (45) and (46) that there exist particular angles φ and θ, "magic angles", where the measurement of the differential cross-section directly yields the total cross-section divided by 4π irrespective of the value of the energy-dependent asymmetry parameter (φ_0 and $\theta_0 = 54.74°$). Also, since we must have a positive definite cross-section for all angles, it is seen from (45) that the allowed range of variation of β is between -1 and $+2$.

Extensive studies have been performed during the past few years, both theoretical and experiment on the study of cross-sections and angular distributions [58–66]. For atoms, calculations have been carried to a higher level of accuracy because of the relative ease of obtaining accurate wave functions for the bound as well as the continuum electrons. Thus, one is not restricted to plane wave functions but the radial Schrödinger equation for the continuum may be solved by standard numerical integration procedures [61–63]. The most thorough studies have been made on the noble gases, and satisfactory agreement has so far been found between theory and experiment [59, 62], although multielectron effects are important for a full understanding of the observed trends [59].

An extensive compilation of existing data on subshell cross-sections for a large number of elements using Al Kα-radiation ($h\nu = 1487$ eV) has been made by Leckey [67]. Figure 10 is reproduced from this paper and shows the subshell cross-sections as well as the total cross-sections for the elements. The crossing of the 2s and $2p_{3/2}$ cross-sections which is seen to occur between Z = 10 and 20 is apparent in a study made at our laboratory (Fig. 11), showing also the variation of the ratio (2s: $(2p_{1/2} + 2p_{3/2})$) using Cr Kα radiation ($h\nu = 5415$ eV) and Cu Kα ($h\nu = 8048$ eV) [4]. Figure 11 shows that the subshell cross-sections from an element are quite strong functions of energy, which is true also for the asymmetry parameter β. Figure 12 shows the calculated subshell cross-sections for carbon as a function of photon energy [68], which may serve to illustrate the general trends. We see that the cross-sections decrease with increasing photon energy well above threshold. Close to threshold, oscillations may occur both in cross-section (not present in this case) and β, depending on the radial nodes of the initial wave function and its overlap with the outgoing continuum wave function [62].

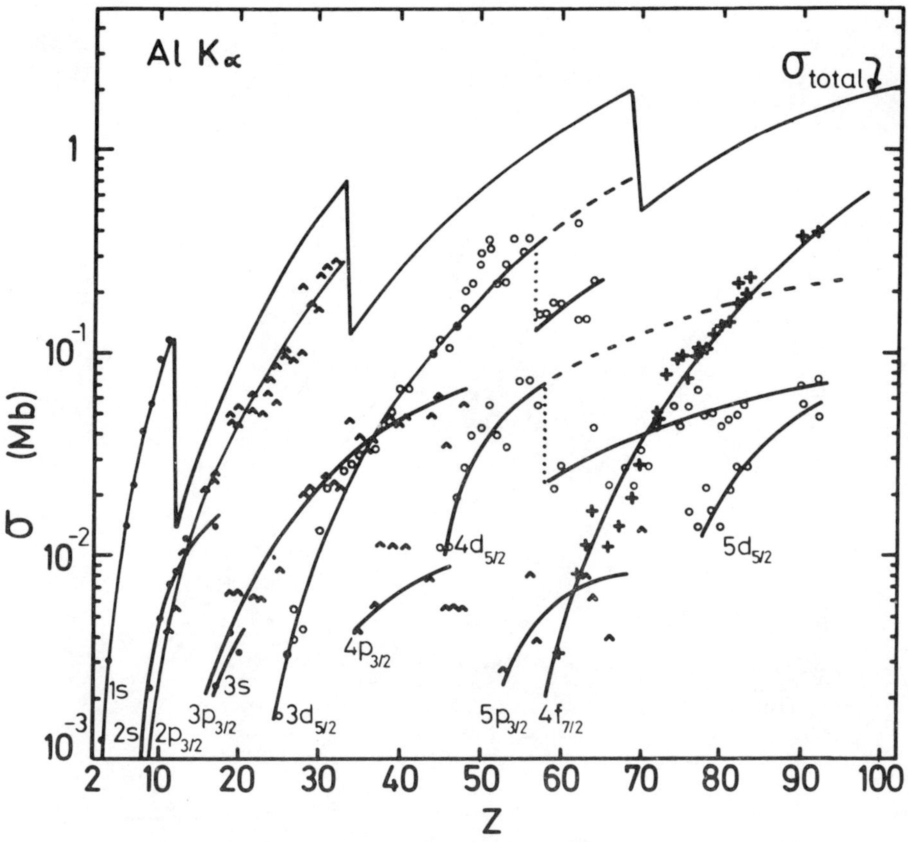

FIG.10. Compilation of existing data on subshell cross-sections for the elements using Al Kα-radiation (from Ref. [67]*).*

In general, the measurement of cross-sections and angular distributions requires good stability of excitation sources and detection facilities and a flexible management for variation of angular conditions in the experiments. The principle which has been almost exclusively employed in the measurement of β-values is the use of non-polarized radiation, whose direction of incidence on the sample is varied with respect to the outgoing electron direction. The analysis is then based on Eq. (46). A powerful alternative which has several attractive features has recently been developed. It is based on the use of polarized radiation in the u.v. region [69, 70]. Linear polarization of the He (I) radiation is achieved by means of successive reflections against gold mirrors. The geometry of the experiment is such as to keep the angle between the incoming photon direction and the outgoing electron direction constant (= 90°). The plane of linear polarization is then rotated with respect to the plane defined by the photon and electron directions. This geometry has the advantage that the excitation volume is kept constant in the angular variation. This cannot be achieved if the angle between photon and electron directions is varied and, if that mode of operation is chosen, the change in excitation volume has to be corrected for. Moreover, the use of polarized radiation allows measurements to be made over the entire angular interval from 0 to π, which leads

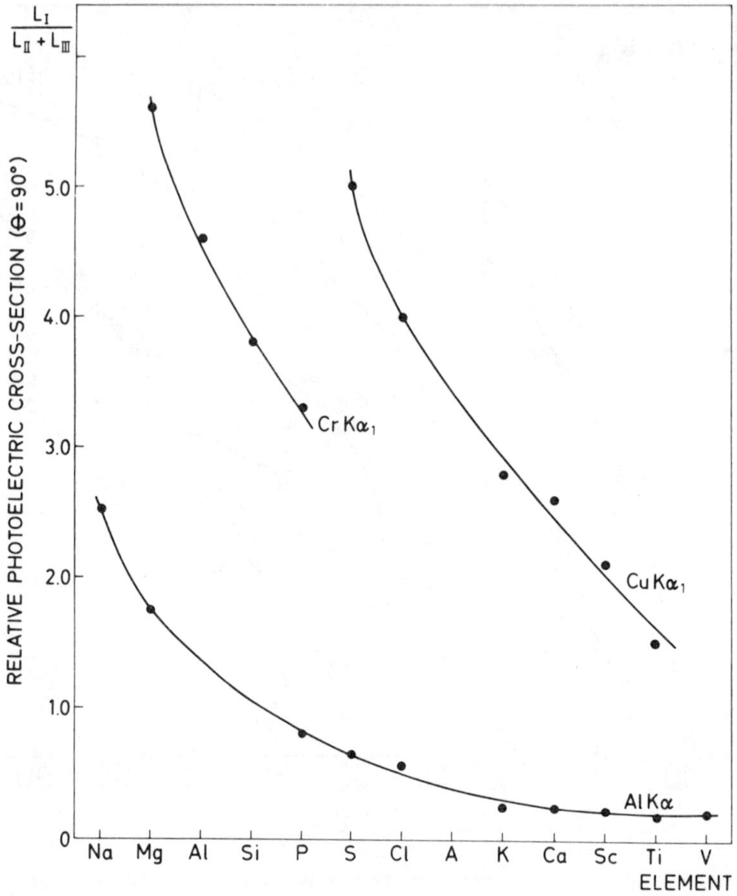

FIG.11. Relative photoelectron differential cross-section $L_I/(L_{II}+L_{III})$ for the elements Na to V.

to high precision in the determinations of β. With non-polarized radiation, angular variations are restricted from a practical point of view. Figure 13 shows results obtained at this laboratory [69] for the 4p lines in krypton with two different angles between the polarization and electron directions. It is seen that substantial changes in intensity occur. The β-values of the two spin-orbit components are also found to be somewhat different: 1.23 and 1.30. Figure 14 shows the entire angular variation. The experimental points fit well to the theoretically obtained curve indicated by the solid line.

It is expected that the experimental possibilities for studying the various aspects of the cross-sections as well as angular distributions will improve substantially with the advent of synchrotron radiation, which is not only highly polarized but is also variable in energy over a large interval, and well defined in direction. This will allow measurements of the above type to be made not only at one photon energy but over large continuous intervals of energy. Experiments at one photon energy are, however, often sufficient. One important such application in this context is the use of β-values for assignment of symmetry to valence electron levels in free molecules.

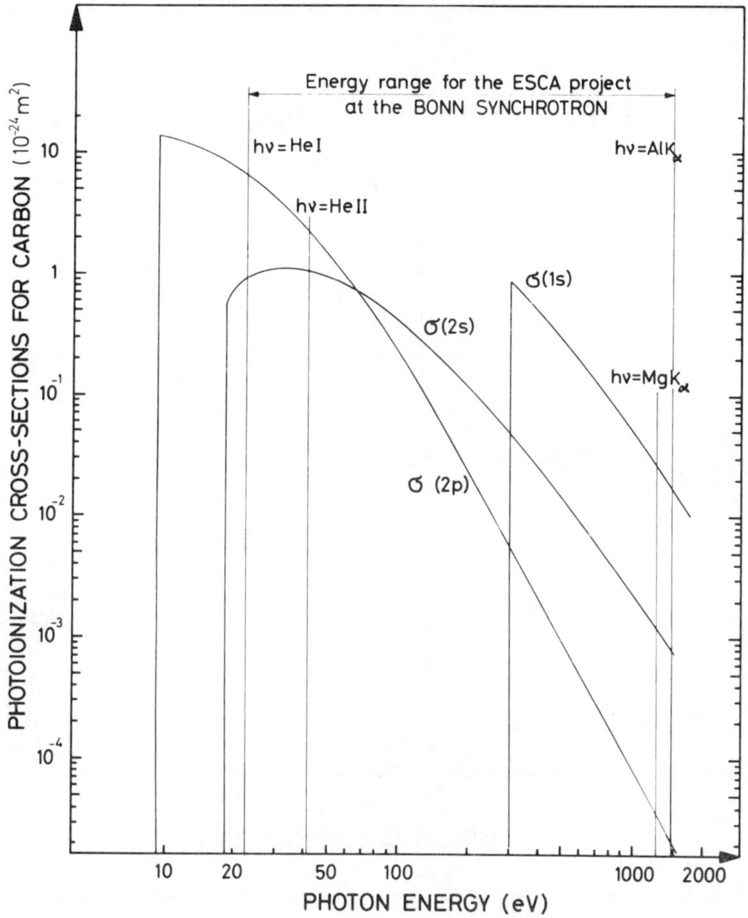

FIG.12. Theoretical cross-sections for the subshells in carbon versus photon energy. Also indicated is the photon energy range available at the Bonn synchrotron for the ESCA project [39].

In general, for molecular photoionization of the valence levels, the situation is considerably more complex than for atoms. This is so because no simple scheme yet exists by which the continuum wave function can be solved for from a one-electron equation, except for simple homonuclear diatomics [71]. For larger molecules, one must, then, resort to approximate wave functions, the OPW's being the simplest which take a proper angular variation into account [64, 65]. Calculations based on this approach have been performed for a number of molecules, and β-values and cross-sections that compare reasonably well with experiment were obtained [65]. If reliable calculations of β-values are available, measurements of β by, for instance, the method mentioned above using polarized radiation, may lead to assignment of the levels. As an example of how valence levels may be distinguished by means of measurement of β-values, Fig. 15 shows a valence electron band in benzene containing three levels [69]. Above the spectrum (which is shown recorded in the "magic angle") the β-values measured at intervals of 0.1 eV are shown. It is clear that the three orbitals involved have distinctly different β-values, which in principle could be used to assign the spectrum. The "β-spectrum" can also be used as an aid in deciding the width of the levels involved.

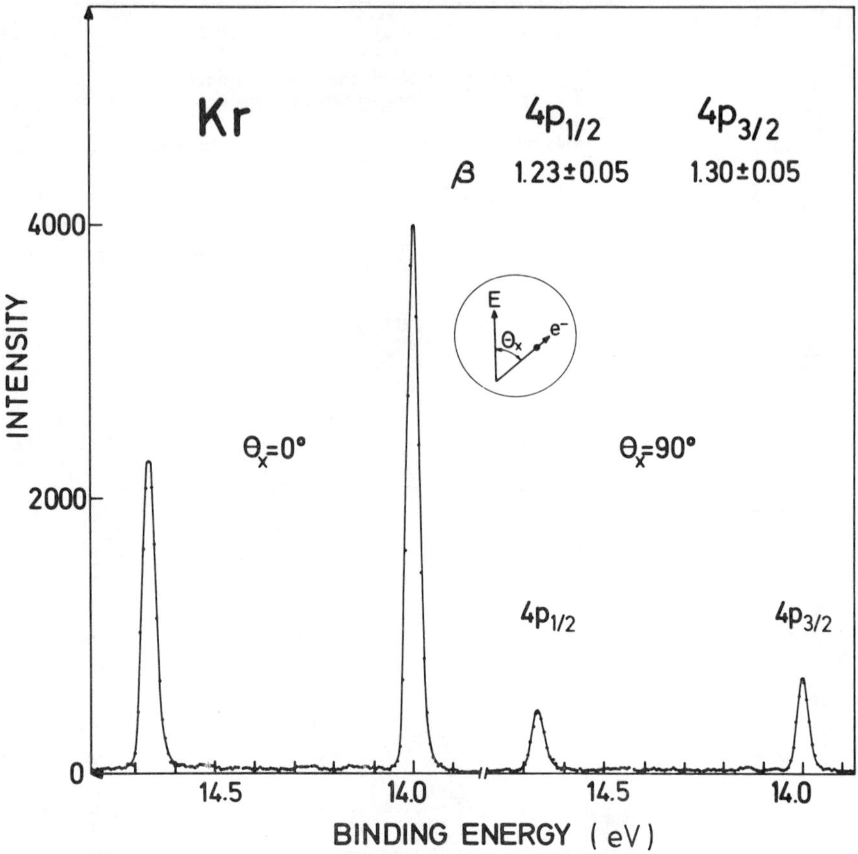

FIG.13. Intensity variation of Kr 4p lines for two different angles between polarization and outgoing electron directions.

In the high photon-energy limit, studies of cross-sections of valence levels may be interpreted in a fashion which leads to some useful consequences [66]. One can see from Eq. (43) that the cross-section is a function of the overlaps between the plane wave and the bound wave functions. It is clear that this in general will not be a simple function, since the bound orbitals are extended over several centres and the contribution to the integrals will thus come from large volumes of space. However, if the wavelength of the outgoing electron is short enough, the contributions to the overlap integrals from the regions of space, where the molecular orbitals are slowly varying, i.e. between the atomic centres, will diminish substantially. The significant contributions to the cross-section will then come from those parts of space where the molecular orbital oscillates rapidly, which is close to the nuclei. This suggests, then, that the cross-section from a particular molecular orbital, j, using X-radiation, may be written as the sum of atomic terms:

$$\sigma_j^{MO} = \sum_A \sigma_{Aj} \tag{47}$$

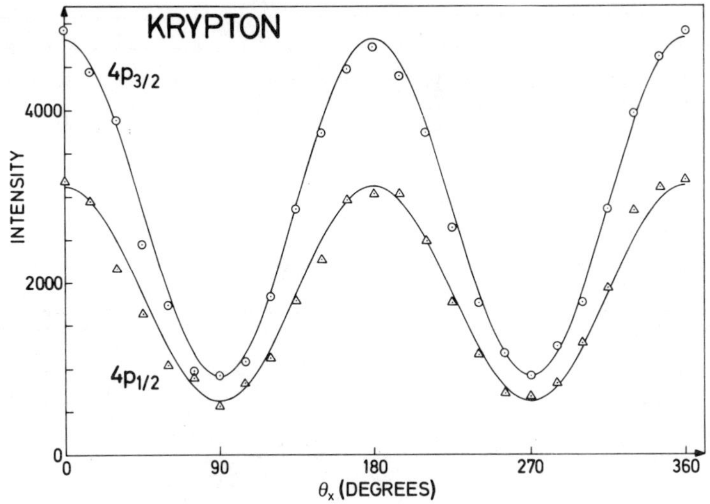

FIG.14. Intensity variation over complete angular interval between polarization and outgoing electron directions for Kr 4p lines.

In terms of an MO-LCAO formalism, the molecular orbitals are described by a superposition of atomic wave functions:

$$\varphi_j = \sum_{A,\lambda} C_{A\lambda j}\, \phi_{A\lambda} \tag{48}$$

This is a good approximation close to the nuclei, where the field experienced by the electrons is essentially atomic. One can therefore expect that the atomic parts of the cross-section, σ_{Aj}, will themselves be composed of subshell cross-sections weighted by the proper populations of the molecular orbital:

$$\sigma_{Aj} = \sum_{\lambda} P_{A\lambda j}\, \sigma_{A\lambda} \tag{49}$$

where $P_{A\lambda j}$ is the population on atom A of atomic symmetry λ and $\sigma_{A\lambda}$ the corresponding cross-section. Equation (47) then becomes:

$$\sigma_j^{MO} = \sum_{A\lambda} P_{A\lambda j}\, \sigma_{A\lambda} \tag{50}$$

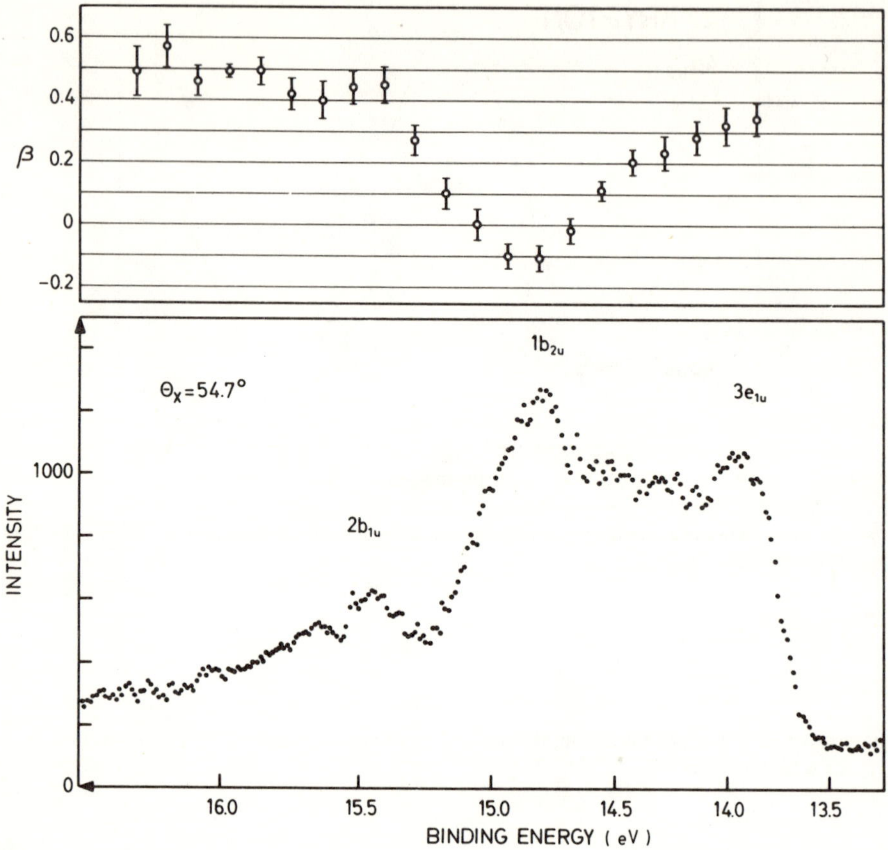

FIG.15. Valence electron lines in benzene recorded in the "magic angle" and variation of the asymmetry parameter β.

The differential cross-section is then written, using cross-sections relative to an atomic symmetry rather than absolute ones:

$$\left(\frac{d\sigma}{d\Omega}\right)_{Rel} = \frac{1}{4\pi} \sum_{A\lambda} P_{A\lambda j} \frac{\sigma_{A\lambda}}{\sigma_{A_0\lambda_0}} \left[1 - \tfrac{1}{2}\beta(E_{fi})\, P_2 (\cos\theta)\right] \tag{51}$$

This is a very useful expression in that it allows the calculation of differential cross-sections of *molecular* photoionization from *atomic* quantities (subshell cross-sections) provided populations and β-values are known. It has its value in that it may be used as an independent tool to assign symmetry to valence electron levels from electron line intensities. The other, more commonly used ways to achieve this are first the comparison of the experimentally observed energies with those calculated theoretically, obtained either under the assumption of equal relaxation energies (i.e. comparison with orbital energies) or from differences in total energies between initial and final states. Secondly, as previously mentioned, the resolution in u.v. excited electron spectra is in general such that the vibrational structure in the electron lines may be recorded. The frequencies excited in a particular hole state are closely related to orbital symmetry and localization, so that careful studies of the observed progressions may lead to assignments of the spectrum. However, the intensities of the electron lines in u.v. excitation are not easily related to orbital parentage and will in general vary quite strongly with u.v. photon energy. This is easy to understand from a simple consideration of overlap between the continuum state (which for u.v. excitations has a wavelength comparable with the molecular dimensions) and the initial state molecular orbital. In the region around 10–25 eV of binding energy one very often encounters a number of closely spaced levels such that, from an energy point of view, several assignments may be possible. The additional information deduced from intensity considerations in the X-ray excited spectrum may then be combined with the vibrational assignments to obtain a more conclusive assignment.

As an example of the sort of agreement obtainable by (51) and experiment, a study of vinyl chloride taken from an investigation of chloroethenes [72] is shown in Fig. 16. The experiments were performed with non-polarized Al Kα-radiation, with an angle between photon and electron directions of 90°. The three theoretical spectra have been calculated by somewhat different means. The uppermost one has been calculated by using CNDO/2 molecular wave functions and theoretical atomic subshell cross-sections. The two lower spectra were obtained from wave functions using another semiempirical method, the extended Hückel (EHT). Also, in the bottom spectrum, a different set of atomic cross-sections is employed, namely those obtained from empirical fits to experiments on other small molecules using the atomic cross-sections as parameters. In the two bottom spectra, the difference in intensities using two different definitions of charge is also compared. The overall agreement with experiment is seen to be good, the best results being obtained by EHT wave functions and net atomic populations. The question of the values of β is a source of possible error in an analysis of this type. In general, as noted above, one must expect these to be different for different symmetries and orbitals. However, experience from experiment and theory seems to indicate [64] that, in the high photon energy limit, β-values for the same symmetry in closely related molecules tend to fall within limits that are reasonably small. As long as this is the case the use of average values instead of the true values will not influence the calculated intensities significantly.

4.2. Photoelectric cross-sections and angular distributions from solids

Photoionization in solids may be treated in a fashion similar to that used for atoms and molecules with band theory as a basis. However, the additional feature of conservation laws of crystal momentum of the one-particle Bloch states leads to a mode of treatment which is different from the atomic and molecular cases. Analogously to these cases one may write down the one-particle expression for the photoelectron intensity in a Golden Rule from:

$$I(E_f) \propto \sum_{i,f} |\langle i|\vec{p}\cdot\vec{A}|f\rangle|^2 \cdot n(E_i) \cdot \delta(E_f - E_i - h\nu) \qquad (52)$$

where i and f represent initial and final one-electron states in the photoelectric process, respectively, and $n(E_i)$ is the occupation function of the initial state. The transitions are assumed to be direct, i.e. crystal momentum is conserved in the photoelectric process

$$\vec{k}_f = \vec{k}_i + \vec{k}_{ph} + \vec{G}$$

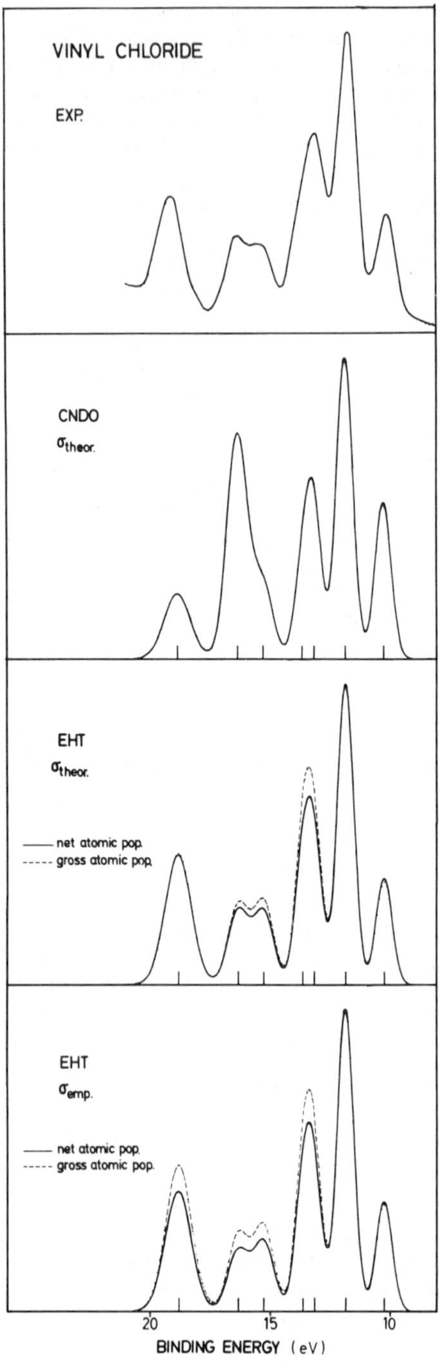

FIG.16. Theoretical valence electron spectra of vinyl chloride obtained by means of Eq. (51), using different semiempirical methods.

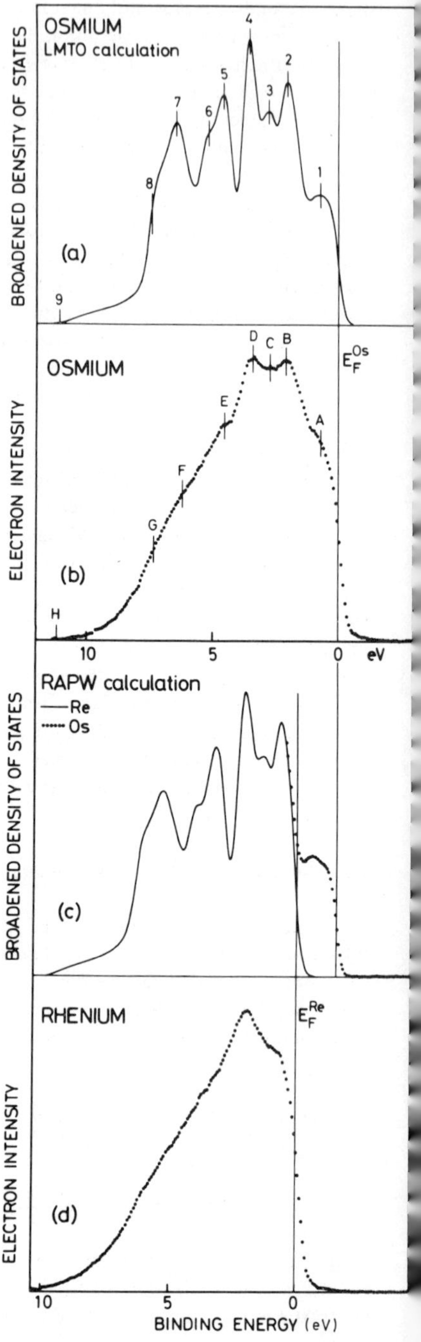

FIG.17. Valence bands of Os and Re ((b) and (d)) compared to broadened density of states curves obtained from (a) LMTO and (c) RAPW calculations.

The expression implies summation over initial and final states satisfying the δ-function requirement. On account of this summation the photoelectron intensity observed will not in general reflect the initial density of states only but the final density of states as well, i.e. the joint density of states. This is then also weighted with the transition matrix element. Due to the restrictions imposed by the conservation laws of energy and momentum, the spectra from valence bands in solids will be rather heavily dependent on the photon energy used to excite the spectra. This is true in particular for u.v. photon energies, where the final state momentum is rather low so that the selection rules in $\vec{k}$-space play an important role. Moreover, the transition matrix elements for the various symmetries from which the band is constructed will, according to what has been said above for molecules, be rather strong functions of energy in the low-energy region. In the high-energy limit, the k-selection will be such that all electrons of a given initial energy in a band will have about the same probability of being excited, since a great number of direct transitions become allowed. Furthermore, the high energy implies that the final density of states may be considered constant. This leads to an appearence of the X-ray excited electron spectrum which closely reflects the initial density of states for each symmetry. The conclusions stated previously for molecular cross-sections in the X-ray regime, Eqs (47)–(51), indicate, in addition, that the cross-sections for bands containing different symmetries may be computed from band structure calculations.

Figure 17 shows an example of a valence band-structure study using X-ray excited electron spectra [73]. Parts (a) and (d) of the figure are the experimentally obtained electron spectra for Os (Z = 76) and Re (Z = 75) using Al Kα-radiation. The (a) and (c) figures are initial density of states (DOS) curves for osmium and rhenium, obtained in band-structure calculations by two methods [74, 75], LMTO (Linear combination of Muffin Tin Orbitals) and RAPW (Relativistic Argumented Plane Wave). The initial DOS curves have been broadened by the electron spectrometer function. It is clear that the experimental spectra are very well represented by the DOS curves; there is an almost perfect one-to-one correspondence between structures experimentally observed and those calculated. The resolution of the structures in the electron spectra is not as good as could be expected from a simple consideration of the spectrometer function. This indicates additional broadening mechanisms, like many electron excitations and volume effects caused by the finite electron escape depth. In Fig. 17(c) the DOS curve of osmium has been obtained from the RAPW calculation on rhenium by means of a rigid band approximation. Obviously this is a model which seems to apply quite accurately to these metals.

The angular distributions observed for the photoelectrons from single crystals will be mappings of the selection of $\vec{k}$-values in reciprocal space of the outgoing waves. The distributions are characterized on the one hand by the symmetry of the band from which the electrons are excited. (The angular variation in this respect will not, of course, follow the relations (45) or (46), because the atoms or molecules in the crystal are not randomly oriented.) A recent study by McFeely et al. [76] on the valence bands of gold and silver by means of Al Kα-excitation showed that the T_{2g} and E_g symmetry states of the bands could be distinguished by means of observation in different crystal directions. This is then a $\vec{k}$-selection originating from vanishing matrix elements in (52). On the other hand, the crystal symmetry gives rise to diffraction phenomena in a fashion analogous to LEED, but with different boundary conditions. Figure 18 shows an early study of the angular variation of the photoelectrons from a sodium chloride single crystal [77]. The angle θ is the angle between the surface normal and the outgoing electrons. The angle between the incoming photons and the electrons was kept constant at 90°. The diffraction due to the low-index planes is prominent in the observed patterns and there is a clear energy-dependence of the main structure when increasing the electron energy by a factor of two for the Na 1s ionization. The narrowing of the 110-reflections follows the decrease in Bragg angle, 13.2° and 8.7°, respectively. Included in the figure is also the variation observed for the Na KLL($^1D_2$) Auger peak. It is interesting to note that the appearance of this distribution very closely resembles that obtained for the Na 2s photoelectron peak, in spite of the fact that they correspond to completely different transitions with different types of matrix elements.

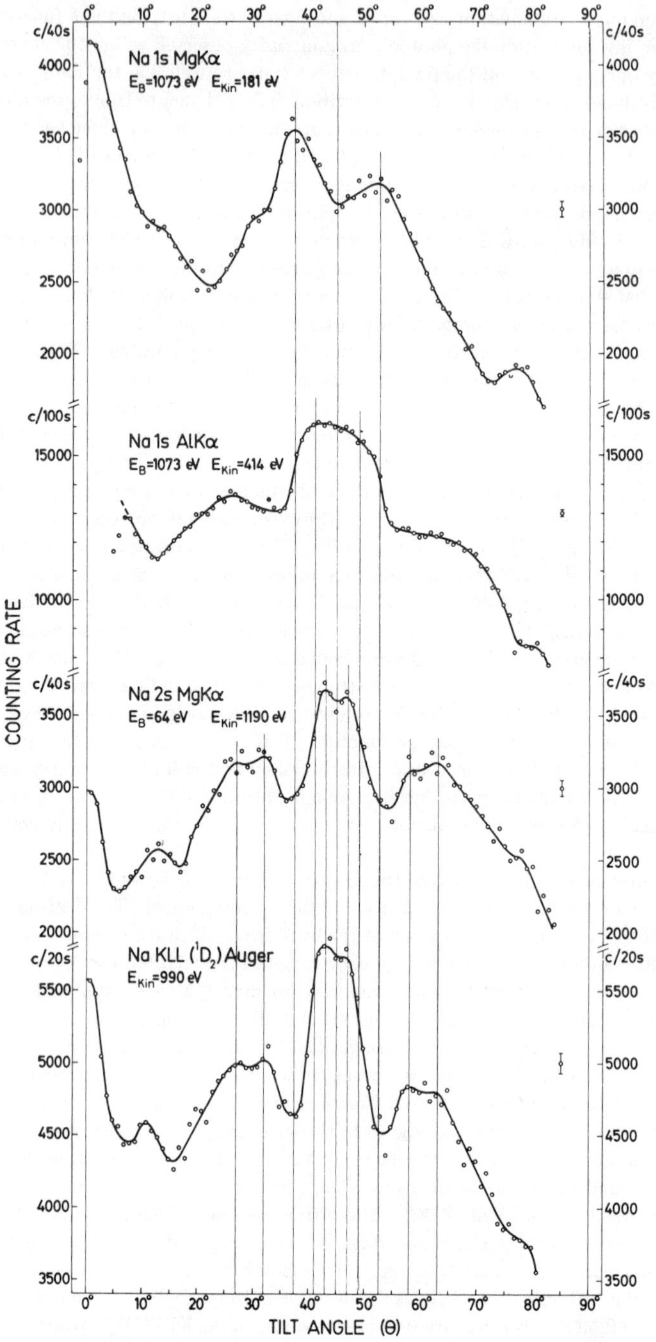

FIG.18. Angular variation of photoelectrons and Auger electrons from a NaCl single crystal. θ is the angle between the ⟨100⟩ direction and the direction of the emitted electrons.

The theory of these diffraction phenomena in the photoelectric process in solids is still in an incomplete state. A treatment of angular distributions of photoelectrons from adsorbate levels has appeared [78], which is a field where electron spectroscopy has a high potential. This is so since it is possible to decide immediately, by selection of the proper electron line, that the primary electron waves originate in the adsorbate atoms. Thus the photoelectron angular distributions yield direct information on the geometrical structure of the overlayer relative to the substrate [78].

It may be mentioned that the study of these angular distributions is closely related to the study of EXAFS (Extended X-ray Absorption Fine Structure). In these X-ray absorption experiments one studies the Kronig fine structure on the high-energy side of the absorption edge from ~ 40 eV up to a few hundreds of electronvolts. The fine structure is due to the diffraction of the outgoing electron with the atomic environment of the atom being studied, which leads to oscillations in the photoelectric cross-section. The evaluation of the Fourier transform of the EXAFS yields accurate values for the distance from the excited atom to the surrounding centres [79]. One would expect to see the same variation with photon energy of the photoelectrons in experiments with polycrystalline samples. This is a field which is being developed at present in connection with the use of synchrotron radiation for electron spectroscopic studies.

Angular distributions in u.v. excited electron spectra are, although complicated, valuable sources of information on the detailed band structure. As mentioned, final-state effects are important in these studies and careful analysis is called for in order to relate observed variations to initial-state band structure. Studies by Smith et al. [80] on $TaSe_2$ single crystals showed structure that could be correlated to valence-electron distributions.

5. AUGER AND AUTO-IONIZATION SPECTRA

5.1. General aspects of Auger transitions

As previously mentioned, de-excitation processes by means of electron emission will be normal ingredients in electron spectra. Although Auger and autoionization are formally identical processes, one has chosen to distinguish between those that occur in initially ionized species (Auger) and those in neutral (autoionization). The Auger processes were first discovered in 1925 by P. Auger in cloud chamber experiments on X-ray absorption [81]. He found that the kinetic energies of certain electron tracks were independent of the incoming photon energy and concluded that these were due to internal conversion processes in the ions. Specifically, the Auger transition occurs between a discrete ion hole state with a hole in an inner shell to a state with two vacancies in outer shells and one electron in the continuum. The rate of such a transition is given by the matrix element:

$$P \propto |\langle \Psi(YZ)\,|H - E_X|\,\Psi(X)\rangle|^2 \qquad (53)$$

were $\Psi(YZ)$ designates the final state with vacancies in Y and Z and $\Psi(X)$ the initial state with vacancy in X. In general terminology, the transition is referred to as an XYZ-transition (KLL, LMM etc.). Special types of transitions are those in which one of the final vacancies is present in the same shell as the primary vacancy. These are referred to as Coster-Kronig transitions and are, in general, characterized by large transition rates. For a non-zero rate the initial and final states (including the continuum electron) must have the same symmetry. This introduces selection rules in the spectra, which for atoms amount to conservation of total angular momentum, its z-component and parity. In pure LS-coupling the same conservation applies to spin and orbital angular momenta. For KLL-spectra, which are the Auger spectra most extensively studied, this leads to a nine-line spectrum in intermediate coupling, seven in LS-coupling and six in jj-coupling. Figure 19 shows the KLL Auger spectrum of magnesium and magnesium oxide obtained at different stages of oxidation of the sample. The spectra are compared to the neon spectrum and are seen to be similar both to

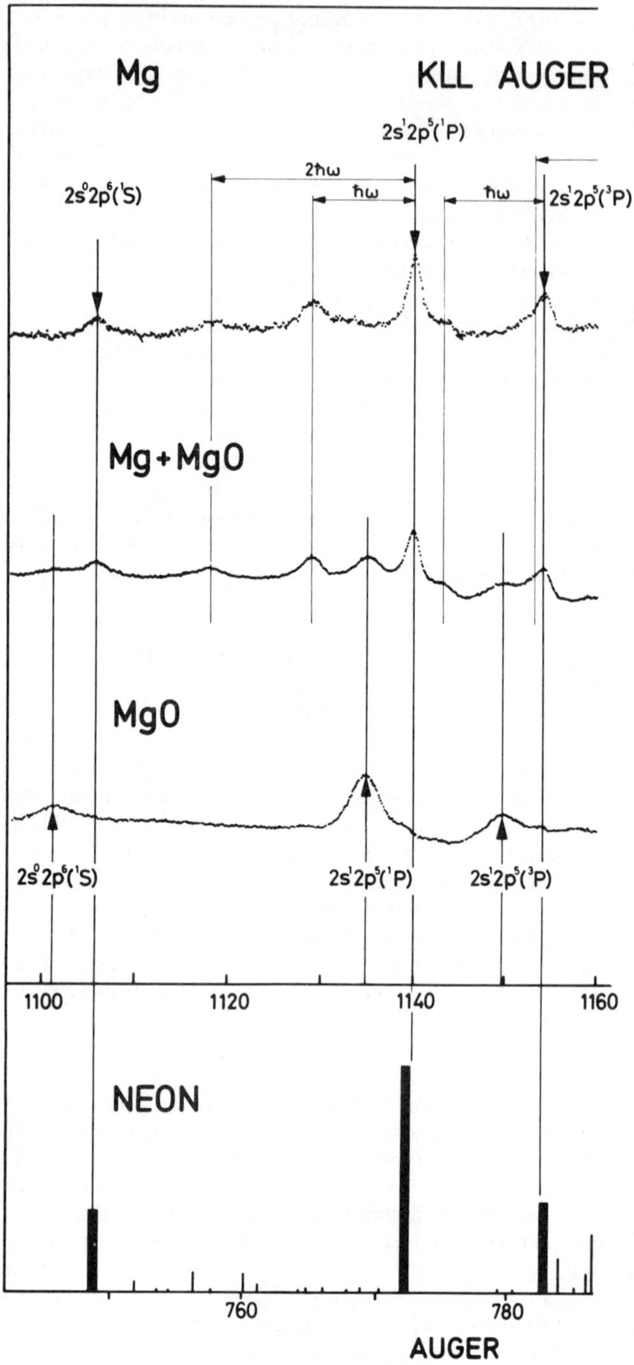

FIG.19. KLL Auger electron spectra of Mg and MgO at different stages of oxidation.

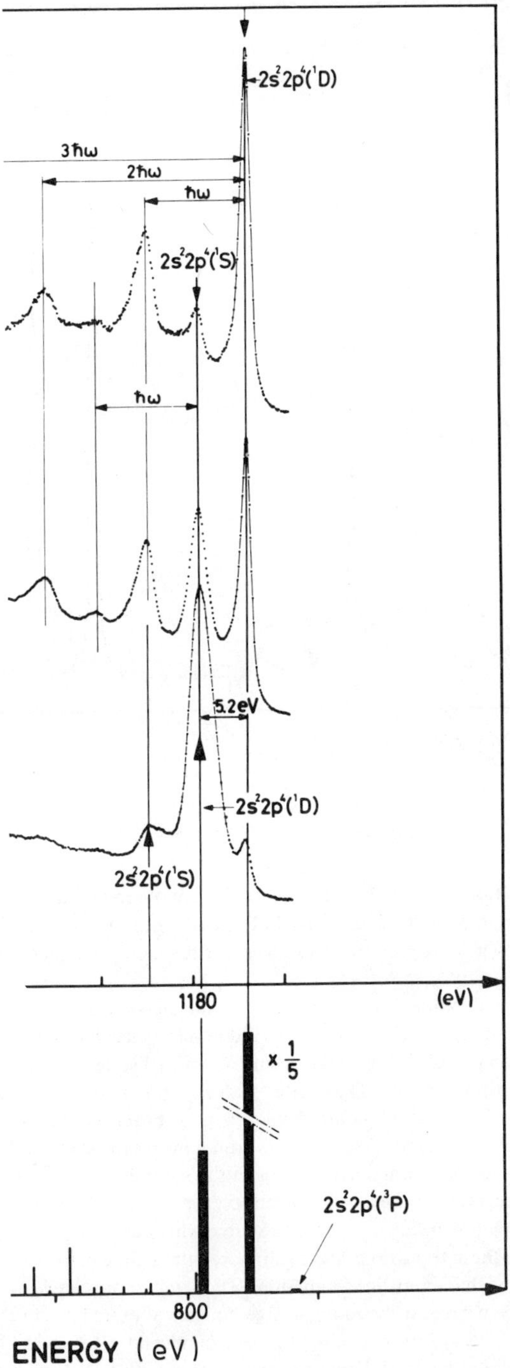

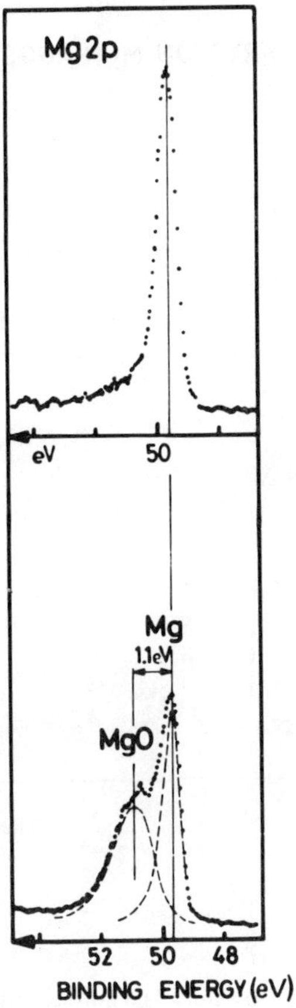

Bottom spectrum is a bar diagram of KLL spectrum of neon [39].

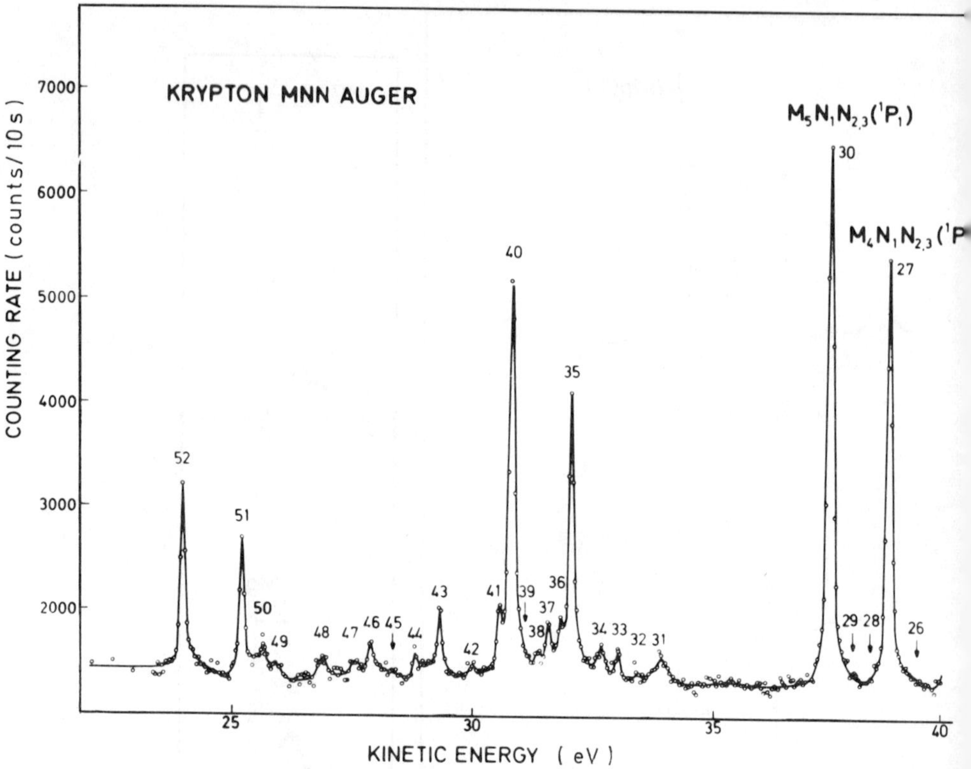

each other and to neon. The normal transitions expected from LS-coupling are observed. In addition to these lines, a number of satellite lines are observed. These arise in this case from the excitation of volume plasmons in the Auger process. They are seen to be strongest for the metal and successively decrease as the oxidation proceeds.

Satellite transitions may occur through a number of processes. First, one may have excitations in the ejection of the Auger electron in a fashion similar to that observed in photoelectron spectra (plasmons, shake-up, shake-off, etc.) Secondly, transitions may occur from initial states that are excited above the hole ground state or multiply ionized. The excited states are populated through shake-up in the primary ionization process. The multiply-ionized states may be produced either by shake-off in the primary ionization or through de-excitation processes from lower-lying single-hole states. In addition to this, one may get transitions from excited states in the neutral species -- autoionization. The intensity of these latter transitions is highly dependent on the mode of excitation. To hit an autoionizing level in the excitation one needs, in general, a source which has a more or less continuous energy distribution. Thus, these transitions are much more intense in electron-excited Auger spectra than those excited by characteristic X-radiation. The satellite processes usually introduce, due to the many possible degrees of freedom, a large number of extra lines in the Auger spectra. Figure 20 shows the Kr $M_{4,5}$NN-spectrum excited by electron impact. It is clear, that the number of normal Auger processes will be much larger for Auger decay with initial vacancies in shells of higher main quantum numbers, which is apparent in this spectrum. Also, more than half

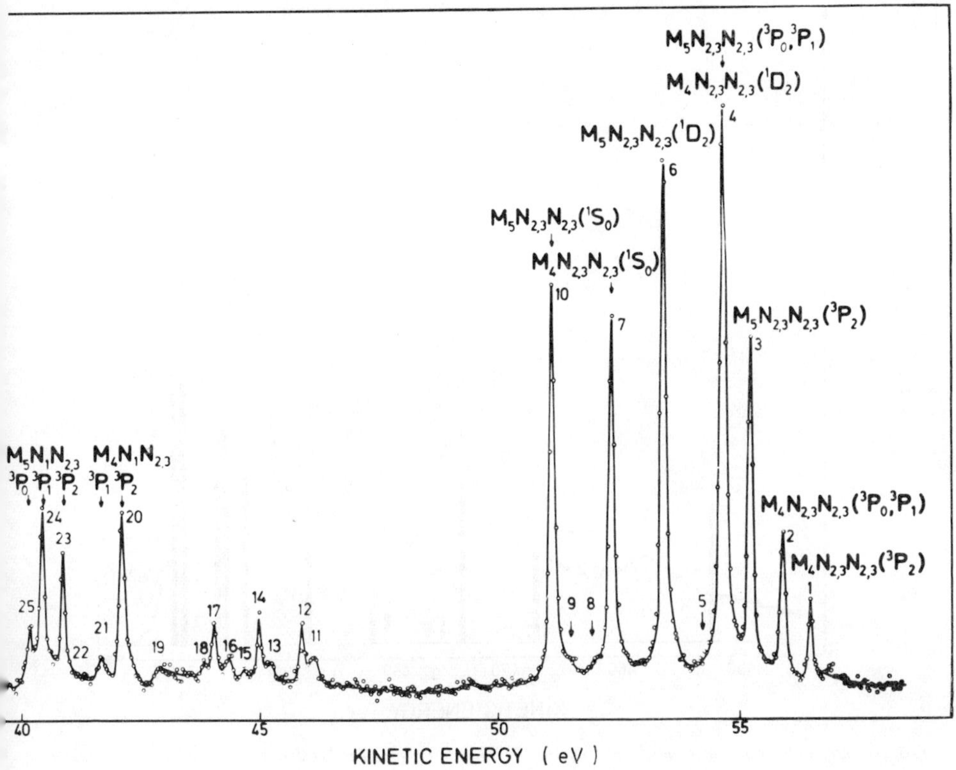

FIG.20. Krypton MNN Auger electron spectrum.

of the lines observed originate from satellite processes. The intensities of the satellites observed are in many cases comparable to the intensities of the ordinary Auger transitions. Theoretical development in the calculation of Auger decay for atoms has undergone considerable development during recent years, both regarding energies and partial and total transition rates [82–86]. This has led to improved possibilities for assigning the extensive and often complex spectra observed for the noble gases.

5.2. Auger transitions with final vacancies in the valence shell

The Auger spectra from molecules and solids constitute a field where many of the effects observed lack a proper treatment. This, of course, is due to the fact that the analysis becomes complicated for these transitions when the effects of final-state couplings and electron polarization have to be considered in multi-atom aggregates. Thus, one finds, for instance, for metallic copper that the line width in the $L_3M_{4,5}M_{4,5}$ transitions is narrower than the conduction electron band, although the latter is involved in the transitions [87–89]. From a first simple consideration one ought to expect that the Auger transitions should be selfconvolutions of the conduction band which would give approximately twice the width of the conduction band. This surprising result still awaits a satisfactory explanation although tentative suggestions have been made [90–92].

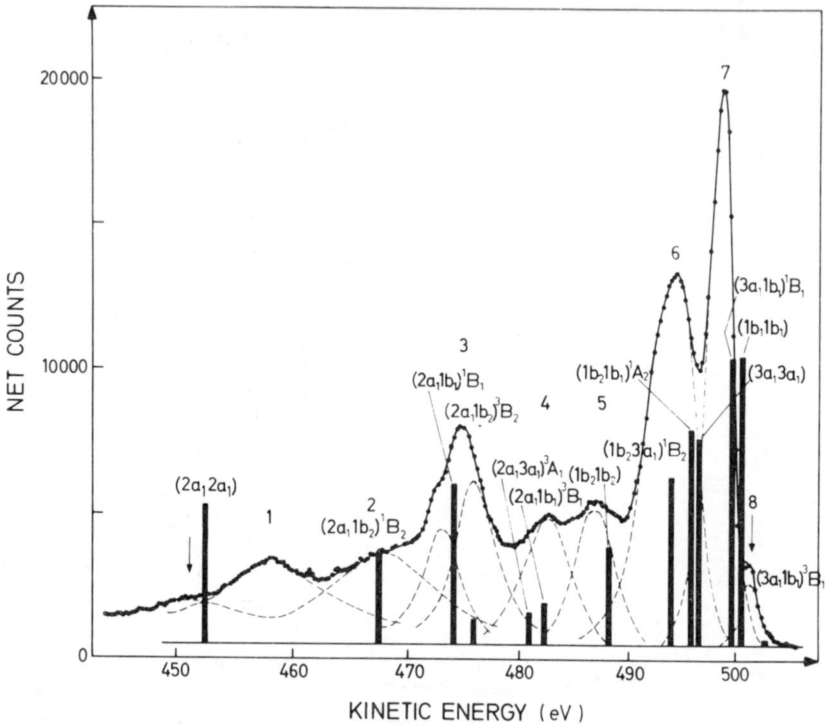

FIG.21. Auger electron spectrum of water excited by photon impact. The bar diagram is theoretically calculated using Eq. (61) for the intensities and a CI for the energies.

For transitions in molecules, when the final vacancies are among the valence electron levels, the situation is simpler, though still complicated enough. A number of small molecules have been investigated both by electron and photon impact, and it has been possible to give an overall interpretation of the main structures [9, 93–97]. One comprehensive study that has recently been made is an investigation of the water spectrum [96]. The spectrum obtained by photon impact is shown in Fig. 21. Although the final states are strongly influenced by chemical bonding, for this particular case the spectrum does have a certain resemblance to a normal atom-like KLL Auger spectrum, which is seen by comparing it with the magnesium and neon spectra, Fig. 19. The four peaks of major intensity in the water spectrum have both relative energies and intensities which are reminiscent of these spectra. This crude resemblance can be traced back to the shapes of the molecular orbitals in water, which indicate that most of the charge density in the water molecule is located on the oxygen. Roughly speaking then, one may expect that in this context of Auger transitions the molecular behaviour will be characteristic of that of an oxygen atom or ion, as seems to be the case. In view of this observation, a simple model was suggested for the calculation of the Auger transition rates in this and similar molecules [96]. One starts from expression (53) for the Auger rate and neglects relaxation in the final state, that is one assumes that the set of orbitals from which the states are constructed are common to initial and final states (restricted HF-determinants combined to give the correct symmetry). For the water molecule, there is no spatial degeneracy and for each combination of molecular orbital hole states either a singlet or triplet spin state may occur –

except when the two final vacancies are in the same molecular orbital, when only a singlet is possible. The intensities for these various possibilities are given by (holes in orbitals φ_i and φ_j):

$$\left.\begin{array}{ll} i \neq j & \begin{cases} I_{ij}(\text{singlet}) \propto |J+K|^2 \\ I_{ij}(\text{triplet}) \propto 3|J-K|^2 \end{cases} \\ i = j & I_{ii} \propto 2|J|^2 \end{array}\right\} \tag{54}$$

where $J = \left\langle \varphi_i \varphi_j \left| \frac{1}{r_{12}} \right| 1s\, \varphi_\epsilon \right\rangle$ and $K = \left\langle \varphi_j \varphi_i \left| \frac{1}{r_{12}} \right| 1s\, \varphi_\epsilon \right\rangle$

For the calculation of J and K the molecular orbitals are expanded in LCAO-form:

$$\begin{aligned} \varphi_i &= \sum_s c_s^i \phi_s \\ \varphi_j &= \sum_r c_r^j \phi_r \end{aligned} \tag{55}$$

where ϕ_s are atomic basis functions. This gives for J:

$$J = \sum_{r,s} c_s^{i*} c_r^{j*} \left\langle \phi_s \phi_r \left| \frac{1}{r_{12}} \right| 1s\, \varphi_\epsilon \right\rangle \tag{56}$$

In this expression one neglects cross-transition elements, i.e. one only considers those matrix elements in (56) non-zero if ϕ_s and ϕ_r are situated on the atom with the primary 1s-vacancy. Further, J becomes a one-centre quantity if one assumes that one can approximate the continuum orbital φ_ϵ by a spherical wave centred on the atom with the primary vacancy:

$$\varphi_\epsilon = R_\ell(\epsilon)\, Y_{\ell m} \tag{57}$$

One then obtains:

$$\begin{aligned} J_{\ell m} = \sum_{\ell' m'} \sum_{\ell'' m''} & c_{\ell' m'}^{i*} c_{\ell'' m''}^{j*} \\ & \times \left\langle R_{n\ell'} Y_{\ell' m'} R_{n\ell''} Y_{\ell'' m''} \left| \frac{1}{r_{12}} \right| 1s\, R_\ell(\epsilon)\, Y_{\ell m} \right\rangle \end{aligned} \tag{58}$$

if the basis functions are given by:

$$\phi_{n\ell m} = R_{n\ell}\, Y_{\ell m} \tag{59}$$

This may be transferred to ordinary atomic radial integrals:

$$J_{\ell m} = \sum_{\ell' m'} \sum_{\ell'' m''} c^{i*}_{\ell' m'} c^{j*}_{\ell'' m''} \, \delta(m' + m'', m)$$

$$\times \sum_{k} c^{k}(\ell' m', 00) \cdot c^{k}(\ell m, \ell'' m'') \tag{60}$$

$$\times R^{k}(n\ell n\ell'', 10 \, \epsilon \, \ell)$$

To get the total intensity one must sum over all possible ℓ, m channels for the ejected electron:

$$\left.\begin{array}{ll} i \neq j & \left\{ \begin{array}{l} I_{ij}(\text{singlet}) \propto \sum_{\ell m} |J_{\ell m} + K_{\ell m}|^2 \\ \\ I_{ij}(\text{triplet}) \propto \sum_{\ell m} 3|J_{\ell m} - K_{\ell m}|^2 \end{array} \right. \\ \\ i = j & I_{ii} \propto 2 \sum_{\ell m} |J_{\ell m}|^2 \end{array} \right\} \tag{61}$$

The expression (60) is easily evaluated provided one has access to the molecular orbital coefficients (which may be obtained from a semiempirical calculation) and the radial integrals. The latter quantities may, to a fair approximation, be taken from those of the atomic case. Strictly, these have to be calculated for the relevant energy of the outgoing electron but at the high energies usually dealt with in this context (450–500 eV for oxygen Auger) the continuum orbitals and hence the radial integrals are not expected to be particularly energy-sensitive. One also has to assume that the basis functions used are reasonably similar to those of the atomic calculation. The relative intensities were calculated according to (61) and inserted as the vertical bars in Fig. 21. The energies of the transitions were calculated by means of an openshell Hartree-Fock program including a limited configuration interaction (CI) employing an extended basis set of Gaussian functions [98]. It is seen that the experimentally observed spectrum is well accounted for by this line approach in terms of normal Auger processes. One notes, however, that satellite processes add intensity, which is difficult to assign in any reliable fashion in this case. Although this molecule is simple, it is possible that studies of this type will add to the possibilities of interpreting such spectra for other molecules.

5.3. Auger transitions between core-like levels

The Auger transitions in molecules and solids that are simplest to interpret are those that occur between levels which can be considered essentially core-like. The spectrum will then be very similar to an atomic Auger spectrum between the same levels for the normal transitions. The satellite spectrum will in general be different, since it depends on the valence levels (occupied and unoccupied). The entire spectrum will, however, be shifted with respect to the atomic spectrum and also with respect to spectra obtained for the same atom in different chemical environments. This Auger shift

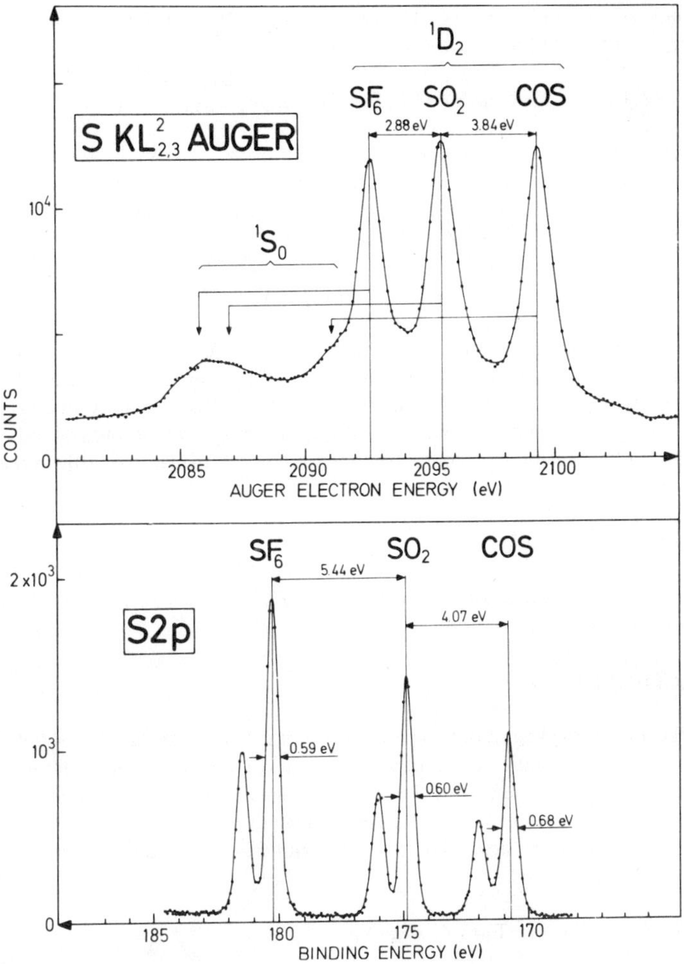

FIG.22. Sulphur KLL Auger electron lines ($^1D_2$) for SF_6, SO_2 and COS compared to the corresponding S2p photoelectron lines.

is analogous to the ordinary binding-energy shift, but has a somewhat different background, which I shall analyse below. As an example of this, Fig. 22 shows a comparison of results on SF_6, SO_2 and COS recently obtained at this laboratory [99]. The Auger shifts observed in the $^1D_2$-transition are 3.84 eV (COS-SO_2) and 2.88 eV ($SO_2 - SF_6$)· This is to be compared with the binding energy shifts in S 2p of 4.07 eV (COS-SO_2) and 5.44 eV ($SO_2 - SF_6$). It can be seen, then, that there are considerable differences between these two types of shifts for these cases. In a lowest order consideration one would expect the Auger shift to be the negative of the binding energy shift (the Auger shift is referred to *kinetic* energies) since the Auger energy is given by:

$$E_{Aug}(XYZ, \Gamma) = E_X - E_{YZ,\Gamma} \qquad (62)$$

where Γ denotes final state symmetry. This expression may be rewritten in terms of one-electron binding energies instead of the total energies in (62):

$$E_{Aug}(XYZ, \Gamma) = E_B(X) - E_B(Y) - E_B^{\ddagger}(Z) + E_\Gamma(YZ) \qquad (63)$$

where $E_B(X)$ and $E_B(Y)$ are neutral binding energies of shells X and Y. $E_B^{\ddagger}(Z)$ is the binding energy of the Z electron corrected for the absence of the Y electron, i.e. taking into account that relaxation has occurred in the system when the first electron was removed. The final term is the multiplet coupling energy for the YZ double-hole state. The Auger shift is:

$$\Delta E_{Aug}(XYZ, \Gamma) = \Delta[E_B(X) - E_B(Y)] - \Delta E_B^{\ddagger}(Z) + \Delta E_\Gamma(YZ) \qquad (64)$$

The first term is nothing but the shift in the X-ray emission spectrum, which is known to be a very small quantity when only core levels are involved since all core levels shift by almost exactly the same amount. Also, since it is observed that the entire Auger spectrum shifts in a rigid fashion, for instance in Fig. 19 for Mg and MgO, we may put the last term to zero in this context. This is expected, since it depends only on core wave functions, which change very little in shape upon chemical changes. Equation (64) thus reduces to:

$$\Delta E_{Aug}(XYZ, \Gamma) \approx -\Delta E_B^{\ddagger}(Z) \qquad (65)$$

Now, the neglect of all the relaxation contributions in $E_B^{\ddagger}(Z)$, which amounts to a totally frozen level of approximation, leads to:

$$\Delta E_{Aug}(XYZ, \Gamma) \approx \Delta\epsilon_Z \qquad (66)$$

which suggests that in the lowest-order approximation the binding energy shift should be equal to the negative of the Auger shift. Apparently, this is a line of approach which is far too simple for any quantitative estimates, considering the results obtained for sulphur and magnesium. One has therefore to include the effects of relaxation in the analysis. In order to do this we shall adopt the transition operator formalism described above. Rewriting Eq.(63) in another form involving a two-electron binding energy yields:

$$E_{Aug}(XYZ, \Gamma) = E_B(X) - E_B(YZ) + E_\Gamma(YZ) \qquad (67)$$

Now consider the sum:

$$\Delta(E_{Aug}(XYZ, \Gamma) + E_B(X)) = 2\Delta E_B(X) - \Delta E_B(YZ) + \Delta E_\Gamma(YZ) \qquad (68)$$

which is zero in the simplest approximation. According to the above, one neglects the last term and introduces transition potentials to evaluate the expression:

$$\Delta(E_{Aug} + E_B) = 2\Delta(\epsilon^{N,N-1} - \epsilon^{N,N-2}) \qquad (69)$$

where the shell designation is dropped since it is assumed that the shifts are the same for all core levels. The first term on the right-hand side designates transition orbital energies from ground state to one core hole (i.e. ordinary transition orbital energy described above) and the second transition orbital energy from ground state to a double-hole state. In order to introduce the charge-potential description of the type

$$-\epsilon = kq + V + l$$

it turns out to be necessary to impose the consistency conditions on the potentials [100]:

$$\epsilon^{N,N-1} = \tfrac{1}{2}(\epsilon^{N-1} + \epsilon^{N})$$
$$\epsilon^{N,N-2} = \tfrac{1}{2}(\epsilon^{N,N-1} + \epsilon^{N-1,N-2}) \qquad (70)$$

These conditions are equivalent to the Lieberman and Hedin-Johansson result for core ionization [101, 102]. If, in addition, it is assumed that the charges and potentials, V, follow the same sort of linear behaviour:

$$\begin{cases} q^{N,N-1} = \tfrac{1}{2}(q^{N-1} + q^{N}) \\ q^{N-1} = \tfrac{1}{2}(q^{N,N-1} + q^{N-1,N-2}) \end{cases} \qquad (71)$$

$$\begin{cases} V^{N,N-1} = \tfrac{1}{2}(V^{N-1} + V^{N}) \\ V^{N-1} = \tfrac{1}{2}(V^{N,N-1} + V^{N-1,N-2}) \end{cases} \qquad (72)$$

the expression (69) may be straightforwardly calculated in terms of quantities of the initial ionization. An assumption of this type is necessary in order to be able to relate the second-step Auger ionization to the first-step ionization in the photoelectric process. In other words, it is necessary to assume that the polarization in the second step may be inferred from the first step. The result is:

$$\Delta[E_{Aug} + E_B] = 2\Delta E_{Relax}^{N,N-1} + 6(k^{N} - k^{N,N-1})\,\Delta Q \qquad (73)$$

where $Q = q^{N,N-1} - q^{N}$ and the other quantities have been previously defined. Equation (73) constitutes a very interesting result. It tells us that any shift in the sum of Auger and ionization energy is closely related to the relaxation properties of the system. The first term is simply understood from the fact that the initial state potential for the Auger process will be more repulsive than that of the photoionization process by $2E_{Relax}$ due to the relaxation of the valence electrons. The last term then accounts for the difference in relaxation energies between the first and second ionization steps. The explicit form of this term may be traced back to the general expression for the relaxation energy in Eq. (32). The 'contraction term' (35) will be different for the two steps since the initial state populations are different by the screening charge of the first step. The first part of the 'flow term' (36) will also be different since the atom is more attractive in the second step than in the first, which is reflected in the increase of the value of k. Our linear approximations (71) and (72) imply, however, that the second part of the 'flow term' will be the same for the two steps. Therefore, in this approximation, the difference in relaxation energy between the Auger and the photoionization process will be strictly a one-centre quantity depending only on the amount of charge which flows onto the ionized atom. A detailed derivation shows that it is proportional to ΔQ by a factor $6(k^{N} - k^{N,N-1})$ as indicated in Eq. (73). The above model is applicable to molecules and clusters and whenever a transition potential model is a valid starting point. Preliminary calculations [100] indicate that in, for instance, sulphur molecules the differences between binding-energy shifts and the negative of the KLL Auger energy shifts may be substantial. In particular, it is noted that the last term in Eq. (73) may be quite important in the analysis. Neglect of this term for the shift in the sum, Eq. (73), between CS_2 and SF_6, leads to a value of about − 1 eV. Inclusion of the term, however, results in a value of about 1.5 eV. Experimentally one obtains 1.4 eV for this quantity [99].

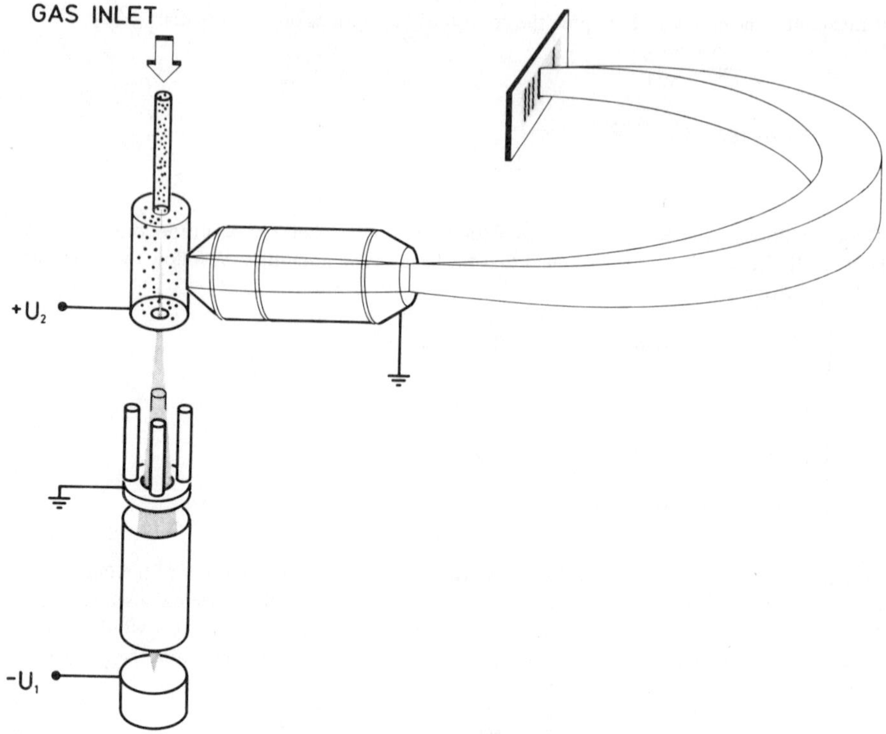

FIG.23. Schematic view of the experimental set-up for recording Auger electron spectra.

The experimental equipment used to obtain the sulphur KLL spectra in Fig. 22 has been recently built at this laboratory and is shown schematically in Fig. 23. It employs a fine-focus electron gun to excite the Auger electrons which, before entering the electron analyser (30 cm magnetic electron spectrometer), are preretarded in an electron lens down to 20% of their initial kinetic energy. By this means the spectrometer contribution to the widths of the peaks is diminished, which is otherwise of major importance in deciding the resolution in the Auger spectra when the kinetic energies become larger than say, 1500 eV. Figure 24 shows a correlation between the measured Auger electron energies and the binding energies for a series of sulphur-containing molecules. The broken line with unity slope has been arbitrarily referred to the CS_2 point. The lengths of the vertical lines connecting each point to this line are then equal to the left-hand side in Eq. (73) referred to CS_2. It is seen that the sum is negative for all molecules except SF_6. Calculations indicate that this is due to the larger relaxation energy of CS_2 as compared to these molecules; the ΔQ-term does not seem to be important for these cases [103]. This, in turn, is an effect of ΔV in Eq. (32) for $E^{N,N-1}_{Relax}$. The screening charge in CS_2 is taken through the π-system mainly from the other sulphur atom, which results in a high relaxation energy. The same possibilities for relaxation do not exist for the other molecules. For SF_6, ΔQ seems to be of importance in deciding the magnitude of the sum, as indicated above.

A similar plot is shown in Fig. 25, which summarizes a series of measurements on methyl-substituted silicon chlorides. In this case, one notices the interesting trend that the sum of Auger and binding-energy shifts seems to change linearly with increasing number of methyl ligands. This

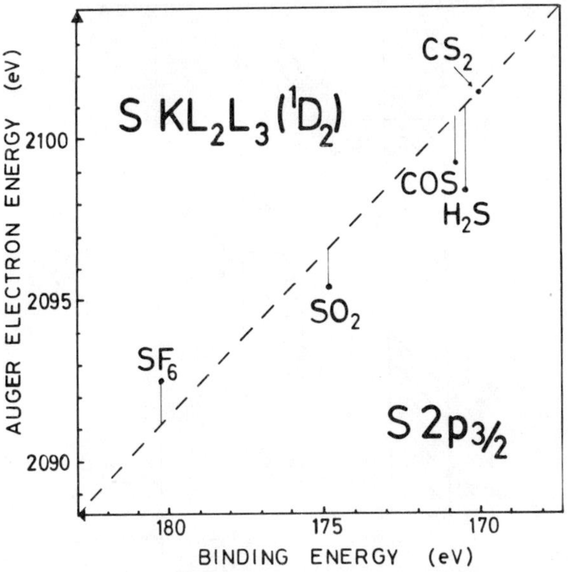

FIG.24. Auger electron energies versus photoelectron binding energies for a series of sulphur molecules. Solid lines connecting the experimental points to the broken lines correspond to the left-hand side quantity in Eq. (73).

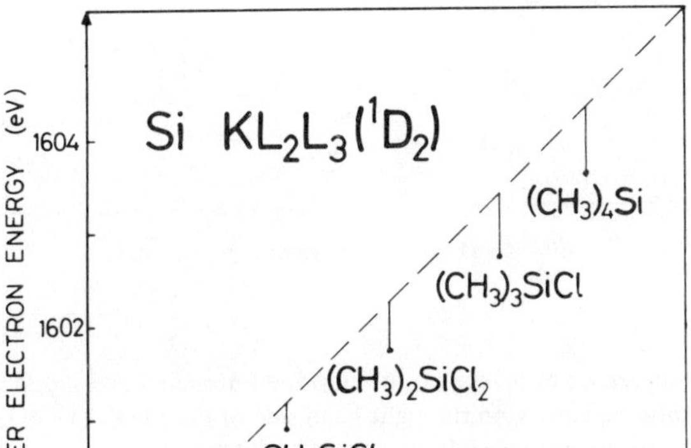

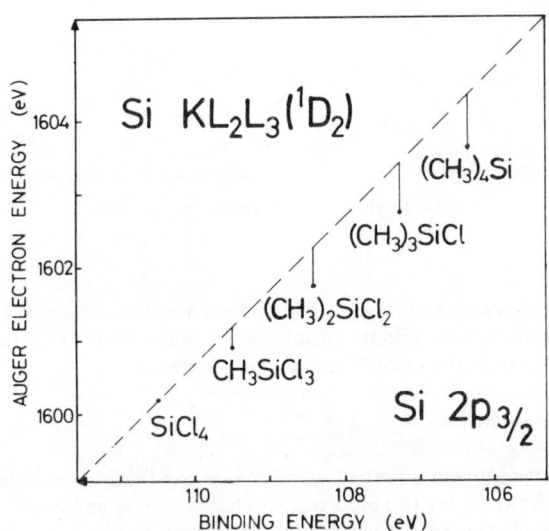

FIG.25. Auger electron energies versus photoelectron binding energies for a series of methyl-substituted silicon chlorides.

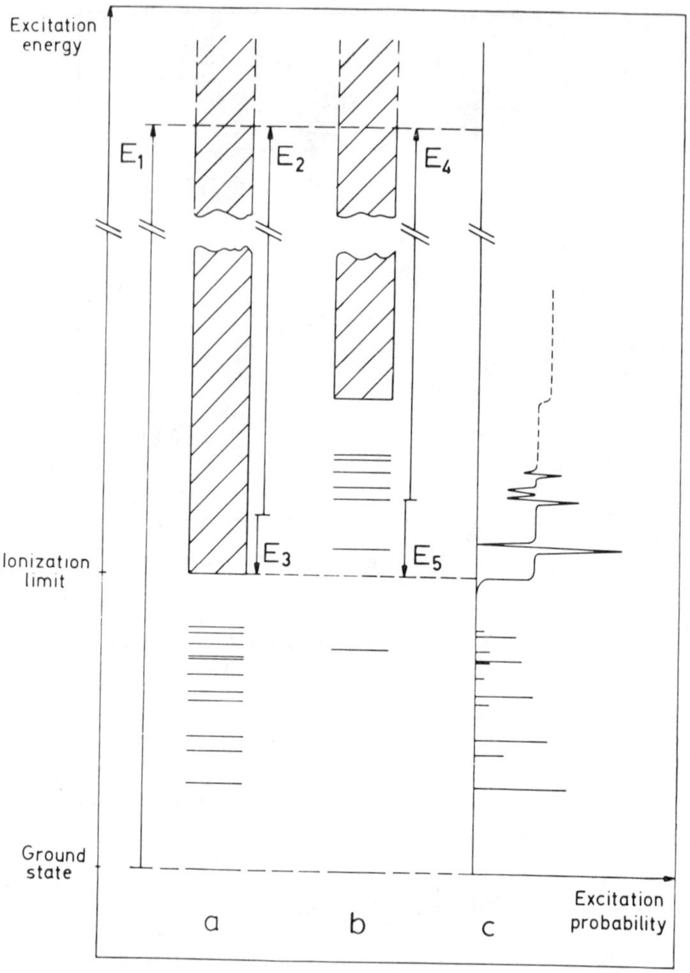

FIG.26. Principle of autoionization (see text).

implies that for each successive exchange of one chlorine for one methyl group a constant contribution is added to the relaxation quantity on the right-hand side of Eq. (73). The detailed implications of this observation will have to await calculations on these systems.

5.4. Auto-ionization spectra

As mentioned, the autoionization and Auger process constitute in principle the same type of process. In general, however, the two processes have been observed under quite different experimental conditions. In Auger electron spectra one usually employs a high-energy excitation source such as electrons or X-radiation and monitors the high kinetic electron energies for the spectra. In autoionization, the prominent features are to be found in a fairly low excitation energy range of the system. Moreover, different aspects of autoionization may be studied either by the direct observation of the autoionized electrons or by measuring the photon absorption or emission spectrum. The general characteristics of autoionization are indicated in Fig. 26. Part (a) of the figure represents the excited levels of a system which goes to an ionization limit at a certain excitation energy. Part (b) represents another excitation spectrum of lower-lying levels of the system. The excitation probability of the system is indicated in part (c) of the figure. When the excitation energy is such as to exceed the first ionization potential, the discrete states in level system (b) may interact with the continuum in (a). The resulting excitation probability function in the vicinity of this level will be that characteristic of a resonance as seen in (c). The shape of this resonance will be a reflection of the interaction between the discrete state and the continuum as well as the mode of excitation (photons, electrons or ions). The mean lifetime of any of the autoionizing states is given by

$$\tau = \frac{h}{4\pi^2 |V_E|^2}$$

where V_E is equal to the interaction integral between the discrete state and the continuum. Theory and experiment indicate that it is generally of the order of 10^{-14} to 10^{-15} seconds [104]. Since decay of autoionizing states in general is very strongly dominated by electron emission, the direct observation of the electron spectrum will give a picture of the excitation probability function. Also, by selecting the mode of excitation, different series of autoionizing states may be excited due to selection rules. Thus, electron excitation will give a larger number of possible series than photon impact studies. For the assignment of states the two modes will then give complementary information.

Figure 27 shows an autoionization spectrum obtained from helium by electron impact where a Rydberg series of asymmetric lines may be identified. Autoionization in helium is of special interest, since the transitions must occur from a state where two electrons are excited. There are two possible series of configurations of the initial state of ^1P symmetry, namely 2s np and 2p ns. These series have a common lowest level (n = 2) and converge to the same energy (2s and 2p in He^+ are degenerate). Due to this, there will be strong configuration interaction effects in the initial state and one will have, roughly, the pair of functions:

$$\Psi(2n\pm) = [|2s\,np| \pm |2p\,ns|](1/\sqrt{2}) \quad (74)$$

describing the initial state [105, 106]. A closer study [106] indicates that the intense spectra are observed from the + states of these pairs. These are indicated in Fig. 27.

Studies of this type give very valuable information on excited states that is not easily obtained by other means. So far, studies of autoionization have been most extensive for the noble gases [104] but measurements have also been made on a number of simple molecules, such as CO [9] and benzene [107].

6. LINE SHAPES AND WIDTHS

6.1. Shapes and widths of unperturbed single levels

When there are no close-lying states of the same symmetry with which the hole state may appreciably interact, first-order perturbation theory implies that the electron line will be a Lorentzian shaped line (disregarding instrumental contributions) with a width given by the decay rate of the final state. The decay rate is usually divided into the decay of the final state by means of Auger electron emission, Coster-Kronig electron emission and radiative transitions. This division into two classes of electron emission channels is made since, as mentioned before, the Coster-Kronig transitions are characterized by very large matrix elements and are thus, when energetically possible, much faster than the ordinary Auger transitions. Figure 28 shows the result of a compilation of

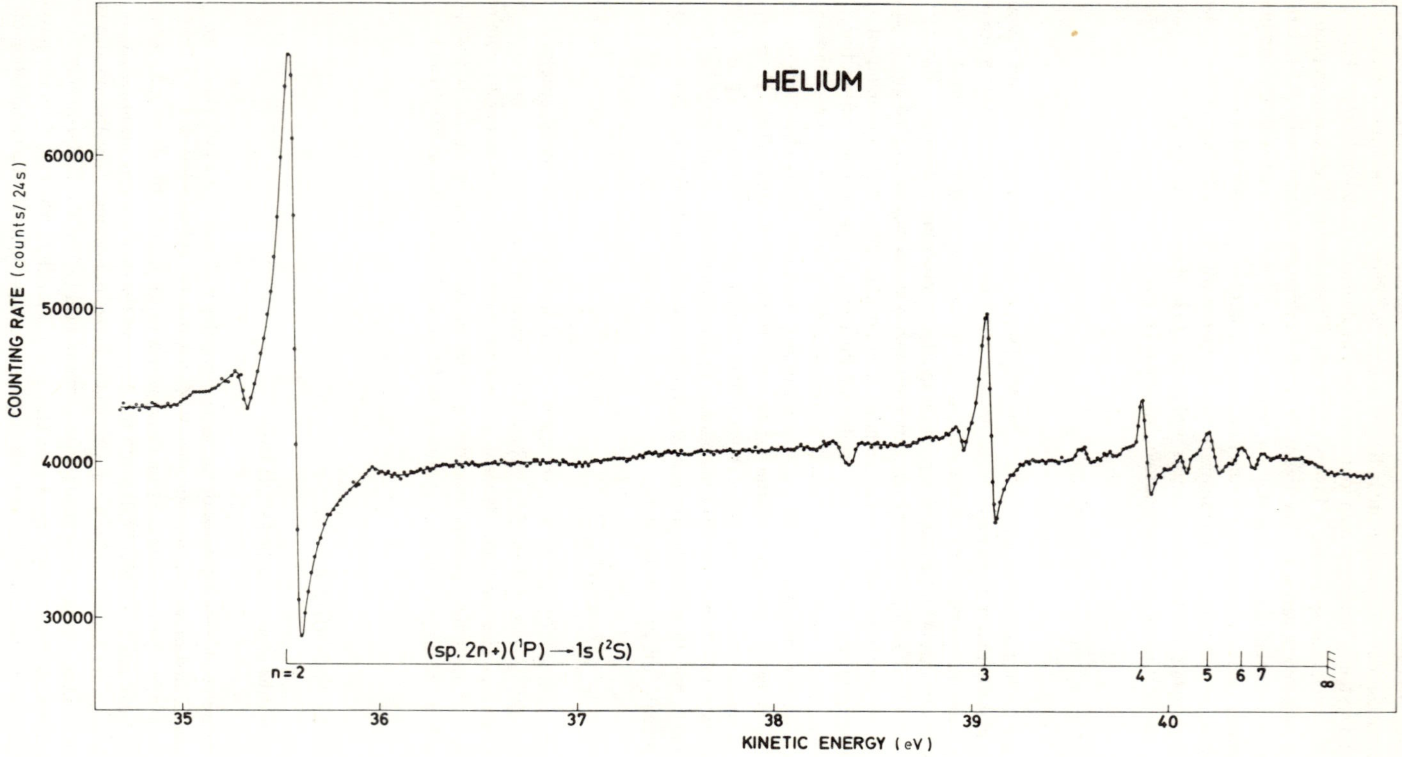

FIG.27. Autoionization spectrum obtained from helium (for explanation of designation of lines, see text).

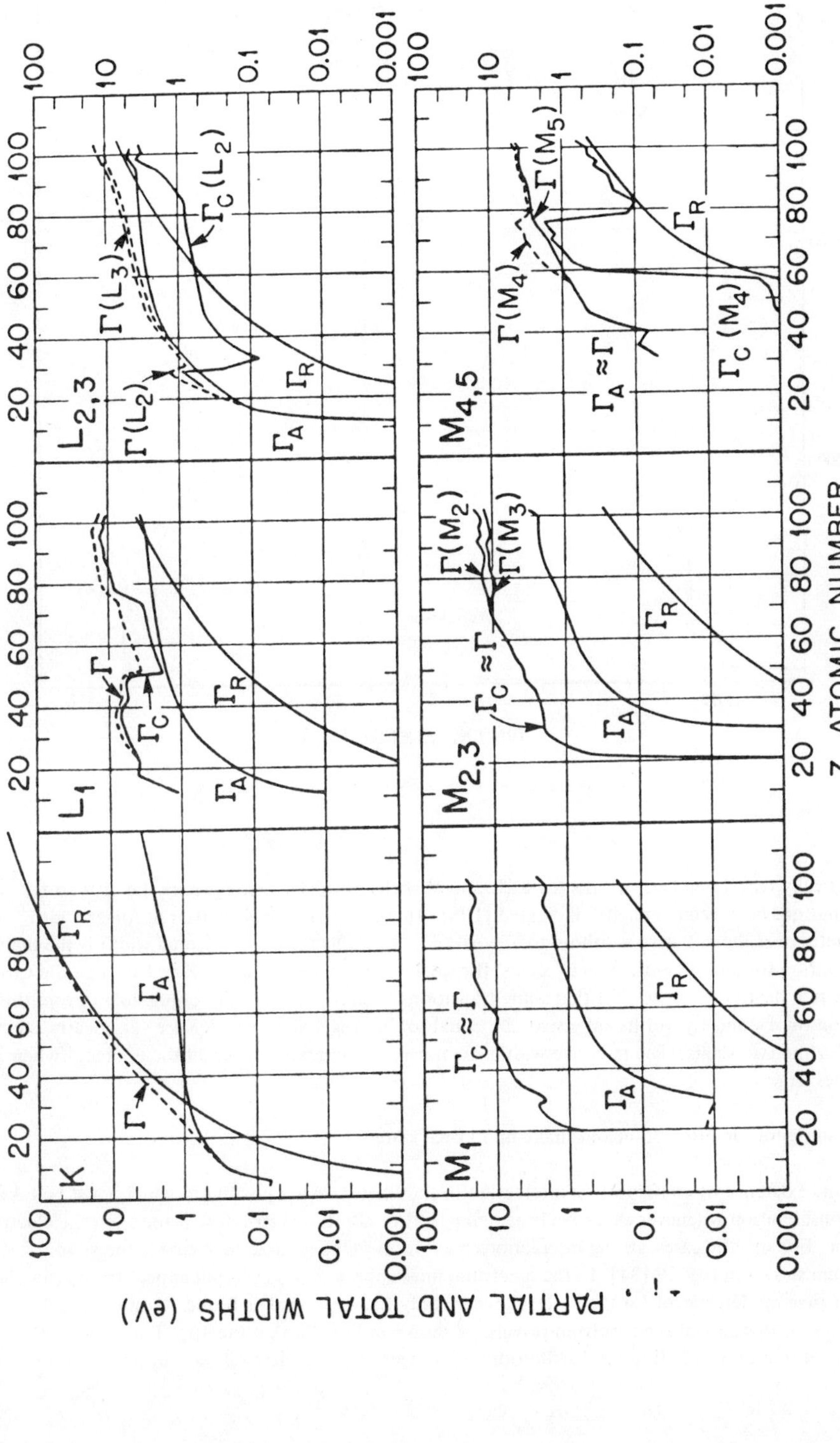

FIG.28. Compilation of theoretically calculated level widths for the elements, and the Auger, Coster-Kronig and radiative contributions to the widths (from Ref. [108]).

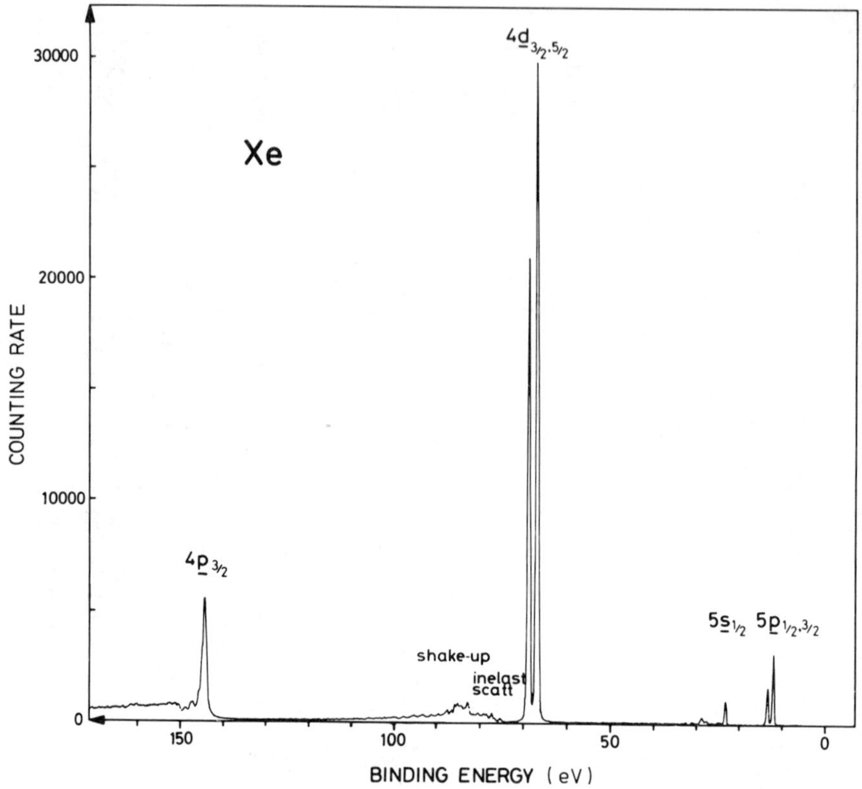

FIG.29. Electron spectrum of xenon obtained by AlKα excitation.

theoretically calculated level widths for K, L and M shells of the elements, where the different contributions have been explicitly indicated [108]. It is seen that, for K-shells, the Auger width dominates over the radiative width up to $Z \sim 30$. For L_1 shells the Coster-Kronig width is the major contribution for all elements. For L_2 shells, the same seems to be true up to $Z \sim 30$, when the Coster-Kronig rate drops due to the fact that some transitions become energetically impossible. From this point on the L_2 and L_3 widths are essentially equal to the Auger widths which are very nearly the same for the two shells. The radiative widths are normally unimportant for all shells except when Z becomes large.

6.2. Configuration interaction and shake-up in final state

The existence of configuration interaction in the final state is a possibility which must be taken into consideration whenever excessive broadening and/or additional structure is found in the electron spectra. For specific cases, strong interactions may occur. One notable such case is the xenon spectrum shown in Fig. 29 [34]. In the spectrum, lines appear that one would expect from a simple one-particle model, except for the 4p level, where only the $4p_{3/2}$ peak is found. A more detailed recording of this part of the spectrum reveals, as shown in Fig. 30, that the $4p_{1/2}$ line seems to be smeared into a broad continuous distribution. Moreover, the $4p_{3/2}$ line displays upon closer inspection

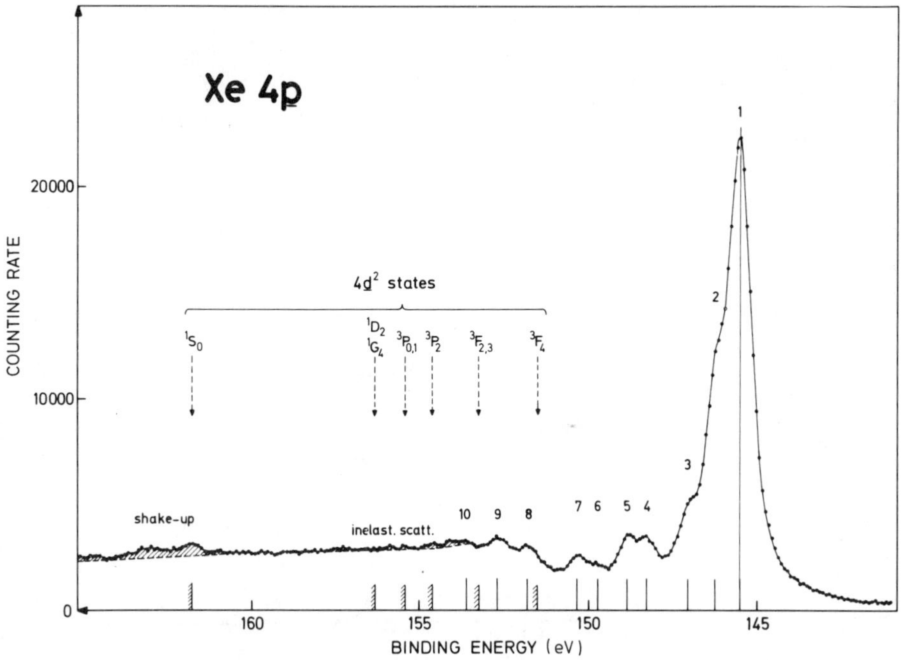

FIG.30. Detail of xenon spectrum showing the Xe 4p region.

a structure, indicating the existence of several close-lying discrete states. The explanation for this appearance of the spectrum is configuration interaction with both discrete states and continuum states. A consideration of the level diagram of xenon shows that almost exactly the same energy is required to remove one 4p-electron as to remove two 4d-electrons. In fact, the energy is slightly higher (6.1 eV) than the binding energy of the main $4p_{3/2}$ peak. The calculated spin-orbit splitting (from SCF calculations) for Xe 4p is 12.6 eV, and it is thus clear that by the same energy needed to remove a $4p_{1/2}$ electron one just barely removes two 4d-electrons. As mentioned, Coster-Kronig transitions are usually very fast. The case of a hole in $4p_{1/2}$ represents a situation where a Coster-Kronig transition may occur such that *both* the final vacancies occur in the same shell as the initial vacancy. This is an example of a so-called "super Coster-Kronig" transition, which is so fast that the primary ionization and the decay must be considered processes on the same time scale. The two outgoing electrons will then share the available kinetic energy. This explains the broad distribution obtained instead of the $4p_{1/2}$ line. As for the appearance of the $4p_{3/2}$ line, when the excitation energy is lower than the super Coster-Kronig threshold, excitations may occur in a 4d-hole state ion from 4d to discrete states close to the continuum, with which the $4p_{3/2}$ hole state may interact. It is found that the ΔSCF value for the binding energy of the $4p_{3/2}$ electron is approximately 10 eV higher than that obtained experimentally. This indicates substantial configuration interaction effects lowering the energy of the final state. The dynamics of these processes in xenon and other elements have been treated by Lundqvist and Wendin [109].

When any of the valence electrons is ionized the situation described seems more apt to occur, due to the existence of more close-lying levels. The situation may be described as the interaction of single hole-states with the shake-up states of other primary holes. Such effects have indeed been found for several cases and are probably of great importance in deciding the intensities of satellite

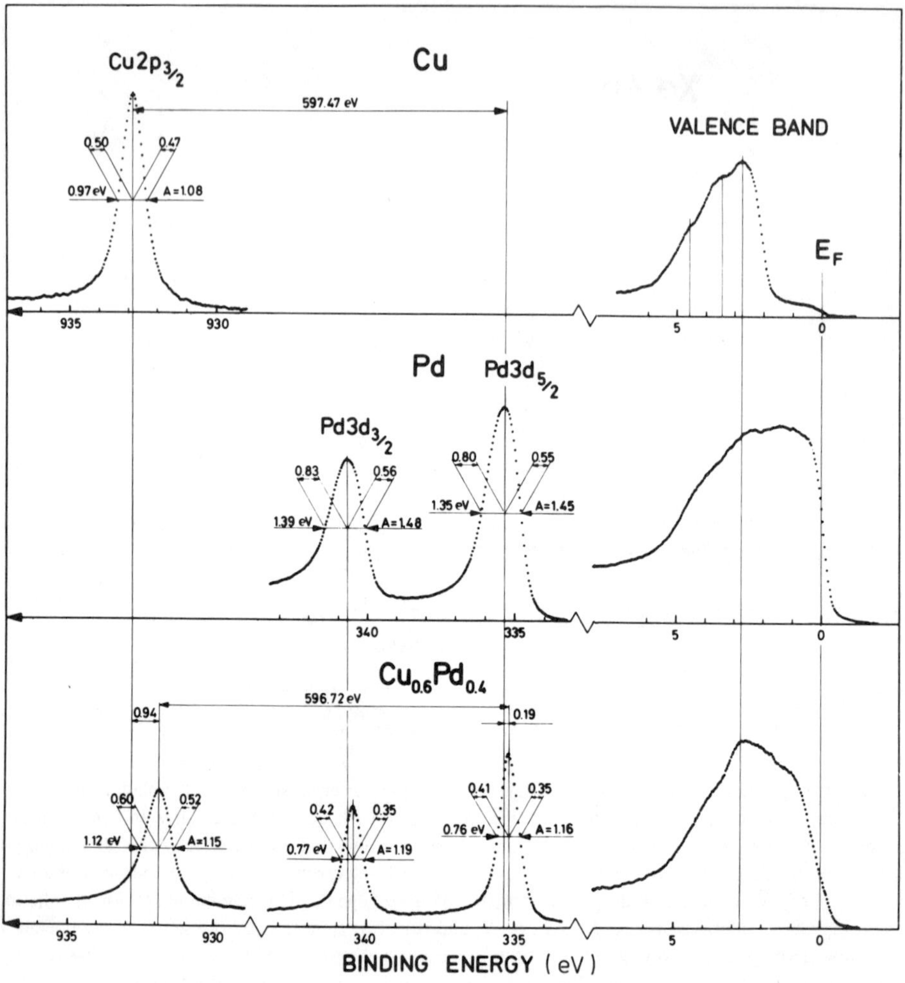

FIG.31. Core electron lines and valence electron bands of metallic copper and palladium and $Cu_{0.6}Pd_{0.4}$ alloy.

structures in valence-electron spectra. In particular CI effects are likely to occur for the inner, core-like valence levels, which are usually found to be very broad and often asymmetric structures.

Another interesting phenomenon observed in core electron spectra is the creation of electron-hole pairs in the conduction band of metals. This may roughly be described as shake-up in the conduction band and gives rise to a marked asymmetry of the electron line. This is seen in Fig. 31 which shows an investigation of a Cu-Pd alloy [39]. The top spectrum shows the Cu $2p_{3/2}$ line and the valence band of pure copper. The Cu $2p_{3/2}$ line is seen to be almost symmetric. The core Pd 3d-lines from pure palladium are, however, rather strongly asymmetric due to shake-up in the conduction band, which for this metal has a high density of states at the Fermi level. In the alloy $Cu_{0.6}Pd_{0.4}$, whose spectrum is shown on the bottom of the figure, the Pd 3d-lines have become almost entirely symmetric as an effect of the change in the density of states at the Fermi level and thus shake-up probability. A similar effect is observed in the electron spectrum if a palladium surface is made amorphous by means of sputtering, which also reduces the density of states at the Fermi level [110].

6.3. Vibrational line broadening in molecules

The effect which is normally the most important in deciding the linewidths for molecules is vibrational excitation. In valence electron spectra this is clearly so, the vibrational progressions are prominent features in u.v. excited spectra, as seen in Fig. 32, which shows the spectrum of ammonia. The outermost orbital, $3a_1$, has a strong progression, with the fundamental frequency of the inversion mode. This frequency is mainly excited, because the $3a_1$-orbital is the lone-pair orbital on the nitrogen and removal of one of its electrons yields a planar ionic state for equilibrium configuration of the molecule. Obviously the vibrational excitations upon valence electron ionization occur because the valence electrons are more or less involved in the bonding of the molecule. For this reason, it was generally believed that vibrational excitation in core electron ionization was not very likely to occur, since core electrons in this terminology are non-bonding. However, it was found in recent experiments at this laboratory under high resolution in the electron spectrum that vibrational excitations are also important in core electron spectra [111]. Figure 33 shows the case of methane, whose C 1s peak has a rather broad, asymmetric structure. It was possible to resolve three vibrational components in the line with relative intensities 61%, 33% and 6%, respectively. The fundamental frequency was found to be 0.43 eV, which is an increase of 0.07 eV with respect to the ground state, indicating a narrowing of the potential curve of the ion. A large number of other cases have been found, for which excessive broadening must be attributed to vibrational structure [34]. An extreme such case is seen in Fig. 34, which shows the S 2p spectrum from trithiapentalene, whose structure is shown in the figure. The broad structure interpreted as the signal from the $S_{2,3}$ positions is vibrationally broadened. In the ground state the middle sulphur atom is rather free to move, that is, it is represented by a rather flat potential function as indicated in the schematic diagram of Fig. 35. Ionization of the middle sulphur is not likely to change this situation very much, since the symmetry of the system is retained. If the positions S_2 to S_3 are ionized, however, the S_1–S_2, S_3 bond length will change and the ionized state may be represented by the more narrow, strongly shifted potential curve in the figure. Obviously, this latter ionization will give rise to considerable vibrational structure, which is indeed observed experimentally.

Other cases that may probably be attributed to vibrational or rather dissociative broadening are the generally broad F 1s lines. This may also be a possible contribution to the large widths of inner valence electron lines apart from configuration interactions as mentioned above.

6.4. Line-broadening effects resulting from condensation of matter

When molecules aggregate and form a condensed phase, liquid or solid, interactions occur among the valence levels which result in band-broadening effects. Thus, the comparison between free-molecule electron spectra and spectra from the condensed phase may yield information on these interactions and the structure of the condensed matter. As an example of this we choose a study made in the liquid phase [112]. The liquid studied was formamide, $HOCNH_2$, which has a low vapour pressure at room temperature, which indicates the presence of hydrogen-bonded structures in the liquid; These are of N-H---O character. The valence levels were obtained for the gaseous phase by X-ray excitation and u.v. excitation and for the liquid by X-ray excitation. The combined use of X-ray and u.v. excitation yielded an assignment of the levels in the formamide molecule. The gaseous and liquid spectra are compared in Fig. 36 and the designations of the levels

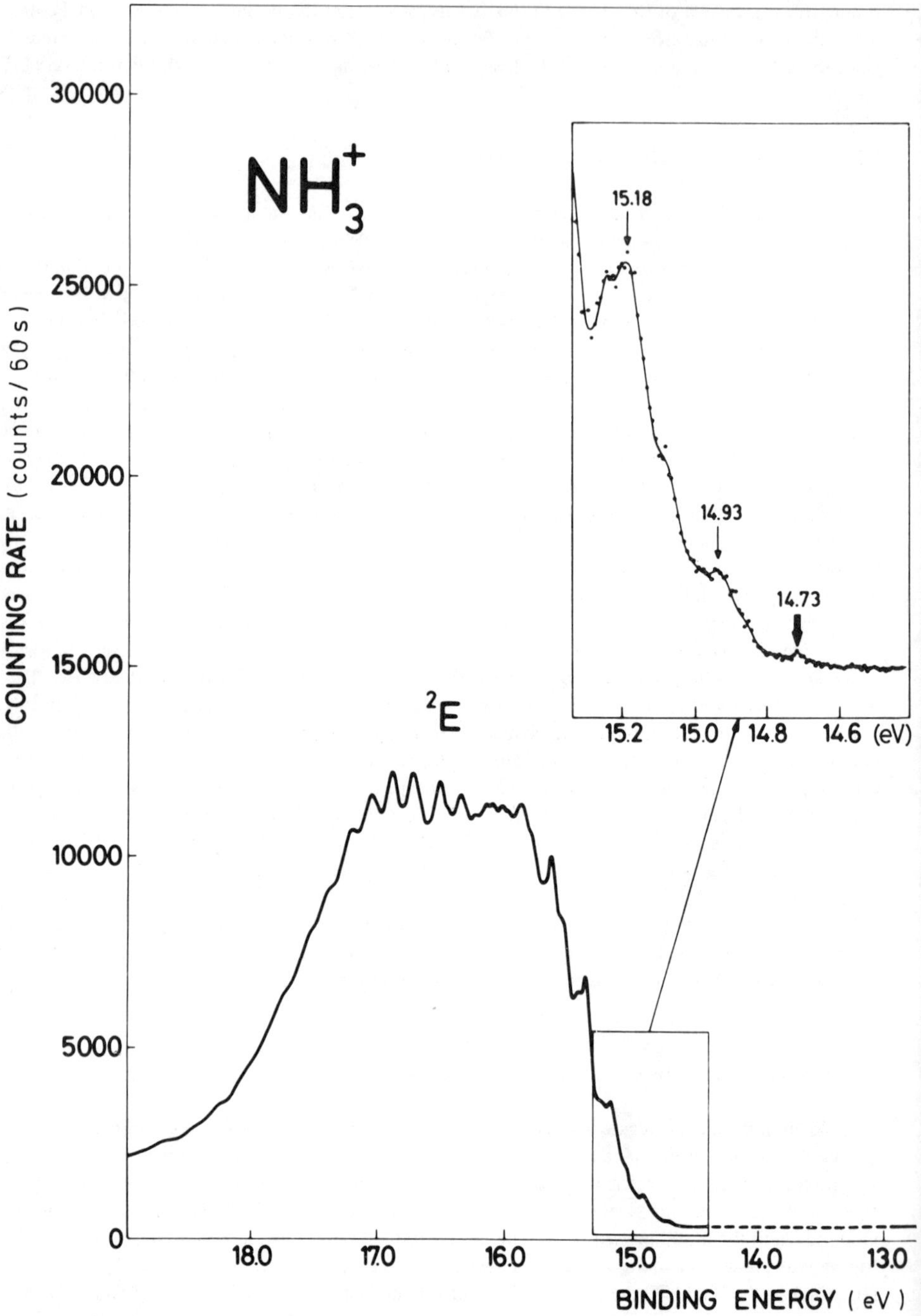

FIG.32. Valence electron spectrum of the ammonia molecule. The $^2A_1$-band shows an extensive vibrational

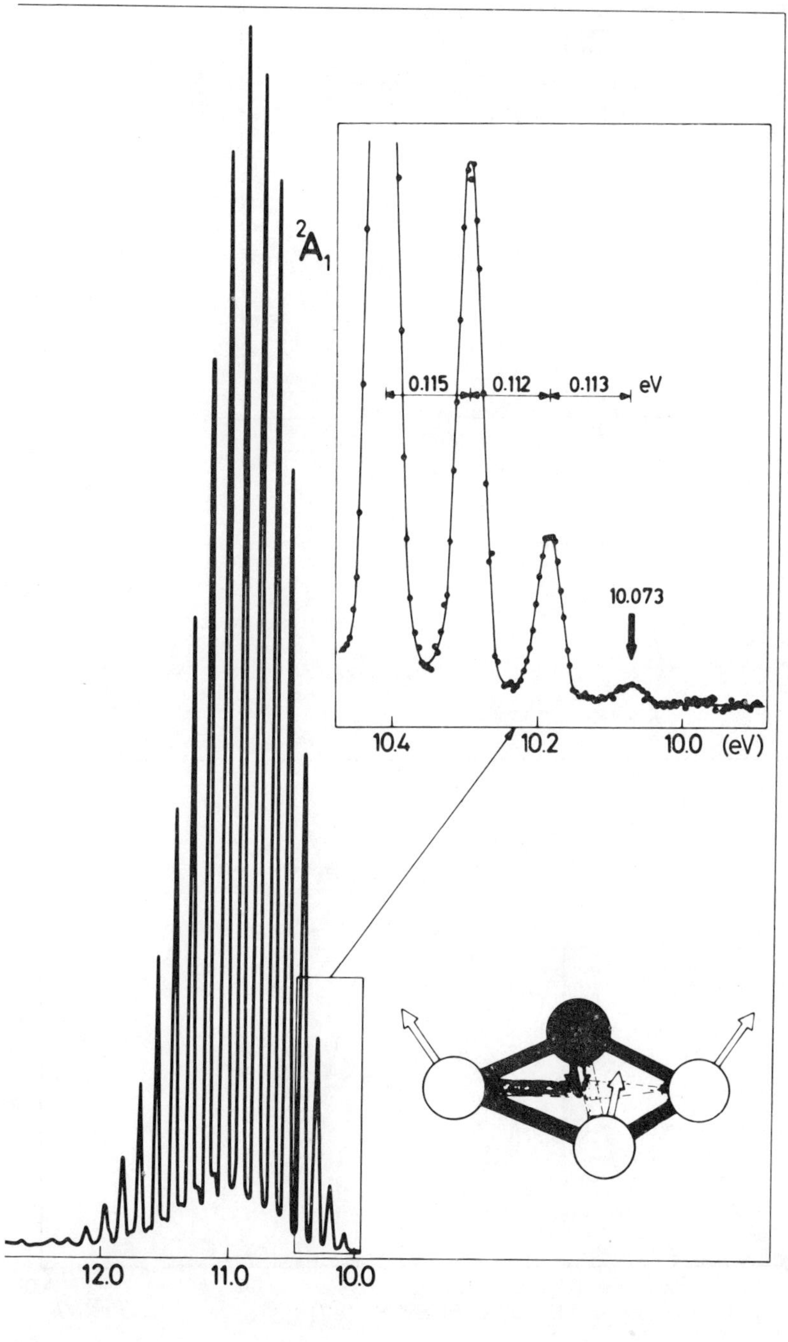

progression due to the inversion mode.

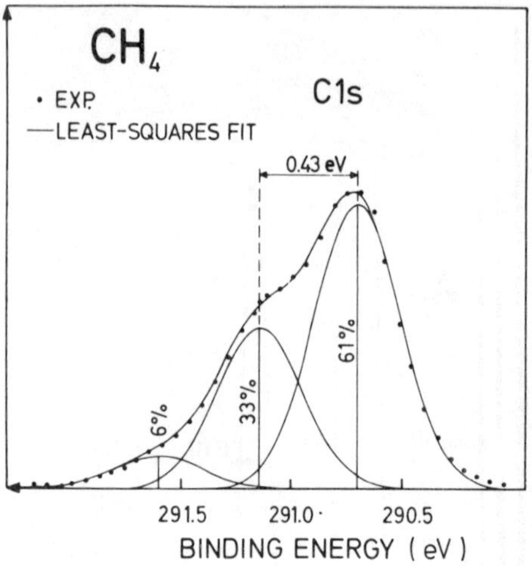

FIG.33. C1s electron line in methane showing vibrational structure.

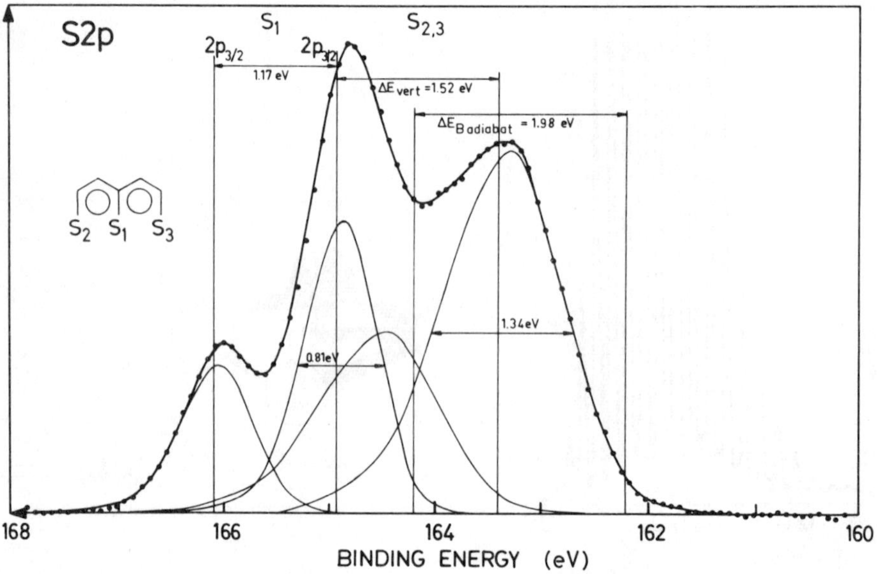

FIG.34. S2p electron line from trithiapentalene. The $S_{2,3}$ electron lines are vibrationally broadened.

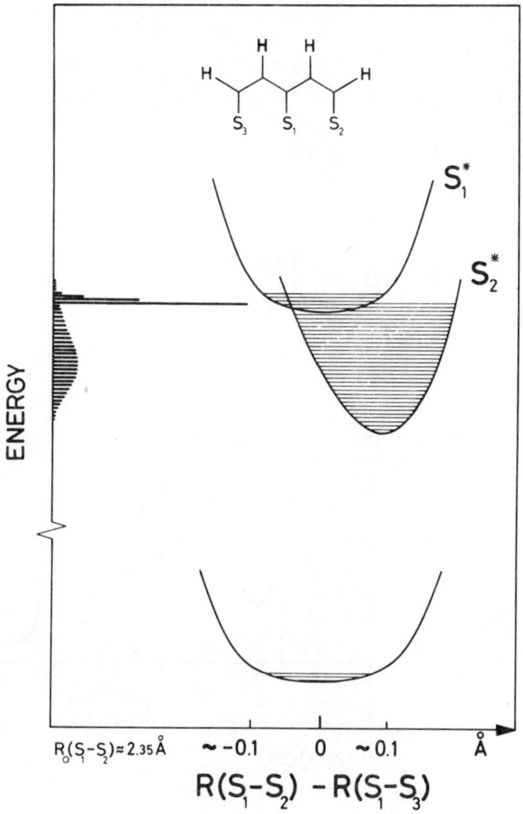

FIG.35. Schematic potential curves of trithiapentalene in the ground state and in the S_1 and $S_{2,3}$ ionized species. The expected vibrational envelope is indicated on the left side.

indicated. The spectra are similar except for the structure involving the 8σ, 9σ and 1π orbitals. An analysis of this band showed that this change in shape was due to excessive broadening of the 9σ level in going to the liquid phase. From the solid-phase crystal structure it may readily be seen that two possible basic structures in the liquid can occur. These are shown in Fig. 37. Calculations were performed on these two model structures to decide the nature of the interactions among the valence levels. It was found that the structure I gave a significant broadening to the 9σ level, whereas structure II gave broadening only to the 8σ level. It could thus be concluded that the structure I was predominantly present in the liquid, or rather chains of several molecules. This conclusion is supported by the high dielectric constant of formamide.

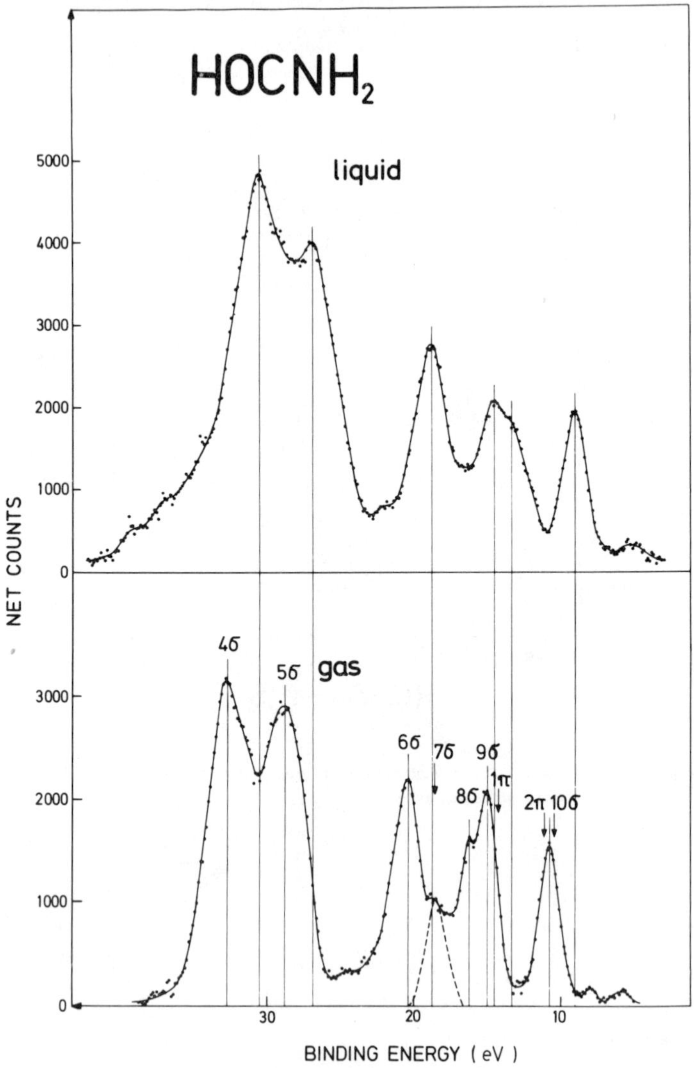

FIG.36. Electron spectra from gaseous and liquid formamide, $HOCNH_2$. Orbital designations are indicated. The 8σ, 9σ, 1π-band shows a marked change in shape from the gaseous to the liquid phase.

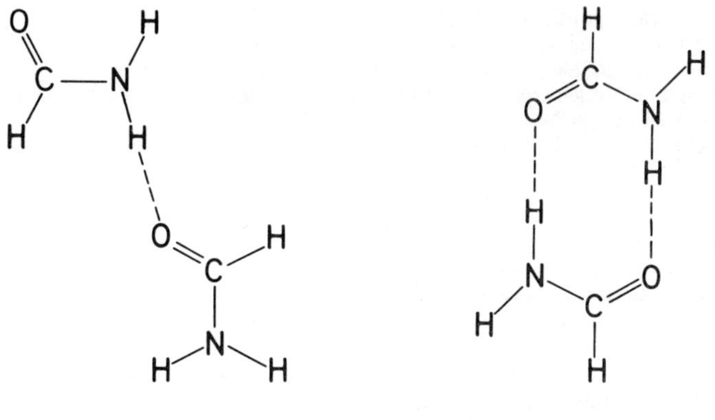

FIG.37. Possible basic structures in liquid formamide expected from the crystal structure in the solid phase.

REFERENCES

[1] SIEGBAHN, K., Ed., Beta- and Gamma-Ray Spectroscopy, North-Holland, Amsterdam (1955; 2nd ed. 1965).

[2] NORDLING, C., SOKOLOWSKI, E., SIEGBAHN, K., Phys. Rev. **105** (1957) 1676.

[3] SOKOLOWSKI, E., NORDLING, C., SIEGBAHN, K., Ark. Fys. **12** (1957) 301.

[4] SIEGBAHN, K., NORDLING, C., FAHLMAN, A., NORDBERG, R., HAMRIN, K., HEDMAN, J., JOHANSSON, G., BERGMARK, T., KARLSSON, S.-E., LINDGREN, I., LINDBERG, B., ESCA-Atomic, Molecular and Solid State Structure Studied by Means of Electron Spectroscopy, Nova Acta Reg. Soc. Sci. Upsaliensis, Ser. IV, **20** Uppsala (1967).

[5] FAHLMAN, A., HAMRIN, K., NORDBERG, R., NORDLING, C., SIEGBAHN, K., Phys. Rev. Lett. **14** (1965) 127.

[6] SIEGBAHN, K., NORDLING, C., SOKOLOWSKI, E., Proc. Rehovoth Conf. Nuclear Structure 291, North-Holland, Amsterdam (1957).

[7] NORDLING, C., SOKOLOWSKI, E., SIEGBAHN, K., Ark. Fys. **13** (1958) 483.

[8] SOKOLOWSKI, E., NORDLING, C., SIEGBAHN, K., Phys. Rev. **110** (1958) 776.

[9] SIEGBAHN, K., NORDLING, C., JOHANSSON, G., HEDMAN, J., HEDEN, P.F., HAMRIN, K., GELIUS, U., BERGMARK, T., WERME, L.O., MANNE, R., BAER, Y., ESCA Applied to Free Molecules, North-Holland, Amsterdam (1969).

[10] SIEGBAHN, K., ALLISON, D.A., ALLISON, J.H., in CRC Handbook of Spectroscopy, Vol. 1 (ROBINSON, J.W., Ed.), CRC Press, Cleveland, Ohio (1974) 257.

[11] TURNER, D.W., AL-JOBURY, M.J., J. Chem. Phys. **37** (1962) 3007.

[12] TURNER, D.W., BAKER, C., BAKER, A.D., BRUNDLE, C.R., Molecular Photoelectron Spectroscopy, Wiley – Interscience, London (1970).

[13] BERGLUND, C.N., SPICER, W.E., Phys. Rev., A **136** (1964) 130; Phys. Rev., A **136** (1964) 1044.

[14] EASTMAN, D.E., J. Appl. Phys. **40** (1969) 1387.

[15] ASAAD, W.N., BURHOP, E.H.S., Proc. Phys. Soc. **71** (1958) 369.

[16] BERGSTRÖM, I., NORDLING, C., in Alpha-, Beta- Gamma-Ray Spectroscopy (SIEGBAHN, K., Ed.), North-Holland, Amsterdam (1965) 1523.

[17] MELHORN, W., Z. Phys. **187** (1965) 21.

[18] KÖRBER, H., MELHORN, W., Z. Phys. **191** (1966) 217.

[19] MELHORN, W., STALHERM, D., Z. Phys. **217** (1968) 295.

[20] KRAUSE, M.O., Phys. Lett. **19** (1965) 14.

[21] MODDEMAN, W.E., CARLSON, T.A., KRAUSE, M.O., PULLEN, B.P., J. Chem. Phys. **55** (1971) 2317.

[22] MADDEN, R.P., CODLING, C.K., Phys. Rev. Lett. **10** (1963) 516.

[23] FANO, U., COOPER, J.W., Rev. Mod. Phys. **40** (1968) 441.
[24] TEMPKIN, A., Ed., Autoionization, Mono Book Corp., Baltimore (1966).
[25] SHIRLEY, D.A., Ed., Electron Spectroscopy, North-Holland, Amsterdam (1972).
[26] DEKEYSER, W., FIERMANS, A., VANDERKELEN, G., VENNIK, J., Eds, Electron emission spectroscopy, D. Reidel Publ. Co. (1973).
[27] CAUDANO, R., VERBIST, J., Eds, Electron Spectroscopy, Elsevier, Amsterdam (1974).
[28] BAKER, A.D., BETTERIDGE, D., Photoelectron Spectroscopy, Pergamon Press (1972).
[29] ELAND, J.H.D., Photoelectron Spectroscopy, Butterworths, London (1974).
[30] CARLSON, T.A., Photoelectron and Auger Spectroscopy, Plenum Press, New York (1975).
[31] EINSTEIN, A., Ann. Phys. **17** (1905) 132.
[32] LUNDQVIST, B.I., Phys. Kondens. Mater. **9** (1969) 236.
[33] MANNE, R., ÅBERG, T., Chem. Phys. Lett. **7** (1970) 282.
[34] GELIUS, U., in Electron Spectroscopy (CAUDANO, R., VERBIST, J., Eds), Elsevier, Amsterdam (1974) 985.
[35] MARTIN, R.L., SHIRLEY, D.A., Phys. Rev., A **13** (1976) 1475.
[36] SVENSSON, S., ÅGREN, H., WAHLGREN, U.I., Chem. Phys. Lett. **38** (1976) 1.
[37] MARTIN, R.L., SHIRLEY, D.A., J. Chem. Phys. **64** (1976) 3685.
[38] MARTIN, R.L., MILLS, B.E., SHIRLEY, D.A., J. Chem. Phys. **64** (1976) 3690.
[39] SIEGBAHN, K., J. Electron Spectrosc. Relat. Phenom. **5** (1974) 3.
[40] BRUNDLE, C.R., ROBERTS, M.W., Proc. R. Soc. (London), Ser. A **331** (1972) 383.
[41] HERCULES, D.M., COX, L.E., ONISICK, S., NICHOLS, G.D., CARVER, J.M., Anal. Chem. **45** (1973) 1973.
[42] BRINEN, J.S., McCLURE, J.E., Anal. Lett. **5** (1972) 737.
[43] WATSON, R.E., Phys. Rev. **118** (1960) 1036.
[44] ALLISON, D.A., JOHANSSON, G., ALLAN, C.J., GELIUS, U., SIEGBAHN, H., ALLISON, J., SIEGBAHN, K., J. Electron Spectrosc. Relat. Phenom. **1** (1972/73) 269.
[45] PERRY, W.B., JOLLY, W.L., Inorg. Chem. **13** (1974) 1211.
[46] CLARK, D.T., KILCAST, D., ADAMS, D.B., MUSGRAVE, W.K.R., J. Electron Spectrosc. Relat. Phenom. **1** (1972/73) 227.
[47] SCHWARTZ, W.E., HERCULES, D.M., Anal. Chem. **43** (1971) 729.
[48] BLACKBURN, J.R., NORDBERG, R., STEVIE, F., ALBRIDGE, R.G., JONES, M.M., Inorg. Chem. **9** (1970) 2374.
[49] STEC, W.J., MODDEMAN, W.E., ALBRIDGE, R.G., WAZER, J.R., J. Phys. Chem. **75** (1971) 3975.
[50] STUCKY, G.D., MATTHEWS, D.A., HEDMAN, J., KLASSON, M., NORDLING, C., J. Am. Chem. Soc. **94** (1972) 8009.
[51] CLARK, D.T., in Electron Emission Spectroscopy (DEKEYSER, W., FIERMANS, L., VANDERKELEN, G., VENNIK, J., Eds), D. Reidel Publ. Co. (1973).
[52] GOSCINSKI, O., PICKUP, B.T., PURVIS, G., Chem. Phys. Lett. **30** (1975) 87.
[53] GOSCINSKI, O., HOWAT, G., ÅBERG, T., J. Phys., B (London) **8** (1975) 11.
[54] GOSCINSKI, O., HEHENBERGER, M., ROOS, B., SIEGBAHN, P., Chem. Phys. Lett. **33** (1975) 427.
[55] HOWAT, G., GOSCINSKI, O., Chem. Phys. Lett. **30** (1975) 87.
[56] SIEGBAHN, H., MEDEIROS, R., GOSCINSKI, O., J. Electron Spectrosc. Relat. Phenom. **8** (1976) 149.
[57] YANG, C.N., Phys. Rev. **74** (1948) 764.
[58] CARLSON, T.A., Phys. Rev. **156** (1967) 142.
[59] WUILLEUMIER, F., KRAUSE, M.O., Phys. Rev., A **10** (1974) 242.
[60] SAMSON, J.A.R., Philos. Trans. R. Soc. London, Ser. A (1970) 141.
[61] COOPER, J., ZARE, R.N., in Lectures in Theoretical Physics (GELTMAN, S., MAHANTHAPPA, K., BRITTIN, W., Eds), Vol. 11, Gordon and Breach, New York (1969) 317.
[62] KENNEDY, D.J., MANSON, S.T., Phys. Rev., A **5** (1972) 227.
[63] McGUIRE, E.J., Phys. Rev. **175** (1968) 20.
[64] LOHR, L., in Electron Spectroscopy (SHIRLEY, D.A., Ed.), North-Holland, Amsterdam (1972) 245.
[65] RABALAIS, J.W., DEBIES, T.P., BERKOSKY, J.L., HUANG, J.-T.J., ELLISON, F.O., J. Chem. Phys. **61** (1974) 516.
[66] GELIUS, U., in Electron Spectroscopy (SHIRLEY, D.A., Ed.), North-Holland, Amsterdam (1972) 311.
[67] LECKEY, R.C.G., Phys. Rev. A **13** (1976) 1043.
[68] SIEGBAHN, K., in Electron Spectroscopy (CAUDANO, R., VERBIST, J., Eds), Elsevier, Amsterdam (1974) 3.
[69] KARLSSON, L., MATTSSON, L., JADRNY, R., SIEGBAHN, K., THIMM, K., Phys. Lett., A **58** (1976) 381.
[70] HANCOCK, W.H., SAMSON, J.A.R., J. Electron Spectrosc. Relat. Phenom. **9** (1976) 211.
[71] TULLY, J.C., BERRY, R.S., DALTON, B.J., Phys. Rev. **176** (1968) 95.
[72] BERNDTSSON, A., BASILIER, E., GELIUS, U., HEDMAN, J., KLASSON, M., NILSSON, R., NORDLING, C., SVENSSON, S., Phys. Scr. **12** (1975) 235.

[73] NILSSON, R., BERNDTSSON, A., MÅRTENSSON, N., NYHOLM, R., HEDMAN, J., Phys. Status Solidi **75** (1976) 197.
[74] JEPSEN, O., ANDERSEN, O.K., MACKINTOSH, A.R., Phys. Rev. B **12** (1975).
[75] MATTHEISS, L.F., Phys. Rev. **151** (1966) 450.
[76] McFEELY, F.R., STÖHR, J., APAI, G., WEHNER, P.S., SHIRLEY, D.A., Univ. of California, Lawrence Berkeley Laboratory Rep. LBL-4325 (1975).
[77] SIEGBAHN, K., GELIUS, U., SIEGBAHN, H., OLSON, E., Phys. Scr. **1** (1970) 272.
[78] LIEBSCH, A., Phys. Rev., B **13** (1976) 544.
[79] STERN, E.A., Phys. Rev., B **10** (1974) 3027.
[80] SMITH, N.V., TRAUM, M.M., DiSALVO, F.J., Solid State Commun. **15** (1974) 211.
[81] AUGER, P., J. Phys. Radium **6** (1925) 205.
[82] WALTERS, D.L., BHALLA, C.P., Phys. Rev., A **3** (1971) 1919.
[83] McGUIRE, E.J., Phys. Rev., A **3** (1971) 1801.
[84] YIN, L.I., ADLER, I., CHEN, M.H., CRASEMANN, B., Phys. Rev., A **7** (1973) 897.
[85] McGUIRE, E.J., Phys. Rev., A **5** (1972) 1052.
[86] BURHOP, E.H.S., ASAAD, W.N., Adv. At. Mol. Phys. 8 (1972) 163.
[87] POWELL, C.J., Phys. Rev. Lett. **30** (1973) 1179.
[88] SCHÖN, G., Surf. Sci. **35** (1973) 96.
[89] BASSET, P.J., GALLON, T.E., MATTHEW, J.A.D., PRUTTON, M., Surf. Sci. **35** (1973) 63.
[90] YIN, L., ADLER, I., TSANG, T., CHEN, M.H., CRASEMANN, B., Phys. Lett., A **46** (1973) 113.
[91] CRASEMANN, B., private communication.
[92] MELHORN, W., private communication.
[93] STALHERM, D., CLEFF, B., HILLIG, H., MELHORN, W., Z. Natuforsch., A **24** (1969) 1728.
[94] MODDEMAN, W.E., CARLSON, T.A., KRAUSE, M.O., PULLEN, B.P., BULL, W.E., SCHWEITZER, G.K., J. Chem. Phys. **55** (1971) 2317.
[95] KARLSSON, L., WERME, L.O., BERGMARK, T., SIEGBAHN, K., J. Electron Spectrosc. Relat. Phenom. **3** (1974) 181.
[96] SIEGBAHN, H., ASPLUND, L., KELFVE, P., Chem. Phys. Lett. **35** (1975) 330.
[97] CONNOR, J.A., HILLIER, I.H., KENDRICK, J., BARBER, M., BARRIE, A., J. Chem. Phys. **64** (1976) 3325.
[98] ÅGREN, H., SVENSSON, S., WAHLGREN, U.I., Chem. Phys. Lett. **35** (1975) 336.
[99] ASPLUND, L., KELFVE, P., SIEGBAHN, H., GOSCINSKI, O., FELLNER-FELDEGG, H., HAMRIN, K., BLOMSTER, B., SIEGBAHN, K., Chem. Phys. Lett. **40** (1976) 353.
[100] SIEGBAHN, H., GOSCINSKI, O., Phys. Scr. **13** (1976) 225.
[101] LIBERMAN, D., Bull. Am. Phys. Soc. **9** (1964) 731.
[102] HEDIN, L., JOHANSSON, G., J. Phys., B (London) **2** (1969) 1336.
[103] SIEGBAHN, H., GOSCINSKI, O., unpublished results.
[104] BERGMARK, T., Uppsala Univ. Inst. Phys. Rep. UUIP-820 (Feb. 1974) (Review).
[105] MADDEN, R.P., CODLING, K., Phys. Rev. Lett. **10** (1963) 516.
[106] COOPER, J.W., FANO, U., PRATS, F., Phys. Rev. Lett. **10** (1963) 518.
[107] KOCH, E.E., OTTO, A., Chem. Phys. Lett. **12** (1972) 476.
[108] KESKI-RAHKONEN, O., KRAUSE, M.O., At. Data Nucl. Data Tables **14** (1974) 139.
[109] LUNDQVIST, S., WENDIN, G., J. Electron Spectrosc. Relat. Phenom. **5** (1974) 513.
[110] HÜFNER, S., WERTHEIM, G.K., BUCHANAN, D.N.E., Chem. Phys. Lett. **24** (1974) 527.
[111] GELIUS, U., SVENSSON, S., SIEGBAHN, H., BASILIER, E., FAXÄLV, Å., SIEGBAHN, K., Chem. Phys. Lett. **28** (1974) 1.
[112] SIEGBAHN, H., ASPLUND, L., KELFVE, P., HAMRIN, K., KARLSSON, L., SIEGBAHN, K., J. Electron Spectrosc. Relat. Phenom. **5** (1974) 1059.

Added in proof

[113] GELIUS, U., HEDÉN, P.F., HEDMAN, J., LINDBERG, B.J., MANNE, R., NORDBERG, R., NORDLING, C., SIEGBAHN, K., Phys. Scr. **2** (1970) 70.
[114] LINDBERG, B.J., HEDMAN, J., Chem. Scr. **7** (1975) 155.

SOLITONS

R.K. BULLOUGH
Department of Mathematics,
University of Manchester
Institute of Science and Technology
Manchester,
United Kingdom

Abstract

SOLITONS.

Solitons are mathematical objects which arise as solutions of certain non-linear dispersive wave equations like the Korteweg-de Vries (KdV), the non-linear Schrödinger (NLS) and the self-induced transparency (SIT) and sine-Gordon (s-G) equations. These govern, respectively, ion acoustic waves in plasmas, the self-steepening of optical pulses and the formation of optical filaments by intense laser light, and the propagation of $\lesssim 10^{-9}$ s optical pulses in resonant media, for example. The equations and applications are very different, yet solitons have many features in common: they collide like particles and, for example, the break-up of coherent 10^{-9} s optical pulses of 'area' 6π into three 2π pulses is a break-up into three solitons. The KdV, NLS and s-G equations are introduced and some single and multi-soliton solutions displayed. As one example of an application in non-linear physics the KdV equation is derived in detail for ion acoustic waves. Next the relevance of the KdV to recurrence phenomena in non-linear lattices (the Fermi-Pasta-Ulam problem) is noted. The theory of SIT in non-degenerate media is developed and used as a physical example of the s-G equation. A double s-G is then derived for SIT in degenerate media. It is shown that soliton-like behaviour is now established by 'wobbling' 4π pulses. SIT for the $^2S_{\frac{1}{2}}(F = 2) \rightarrow {}^2P_{\frac{1}{2}}(F = 1,2)\ D_1$ transitions in sodium vapour is treated. Applications of solitons to Josephson junctions, to optical filaments and to other non-linear physics (plasmas, lattices, particle physics) are briefly sketched.

1. Introduction

What is a "soliton" and what has it to do with the interaction of radiation and condensed matter? In his paper in General Theory (these Proceedings, paper IAEA-SMR-20/16), F.T. Arecchi notes that solitons are involved in the propagation of short optical pulses, 10^{-9} s or shorter, in resonant media and especially with the phenomenon of self-induced transparency (SIT). We shall also look at solitons in SIT in these lectures because they prove to be excellent examples of the soliton phenomena. However these are particular examples, and one of the purposes of these lectures is to show that solitons occur in a much wider context than that of non-linear optics.

In this general context they are simply particle-like solutions of various non-linear wave equations which show certain features in common. These common features are:-

(i) Pulsed solutions of these equations break up into trains of single, usually distortionless bell shaped, pulses which we call solitons.

(ii) The speed of an individual soliton often depends on the amplitude of the soliton: in particular a frequently occurring situation is that large amplitude solitons travel faster (BIGGER PULSES GO FASTER).

(iii) Individual solitons may collide and pass through each other without change of shape or velocity but with a small shift of phase.

The situation is exemplified in SIT where the single soliton is the 2π hyperbolic secant pulse [1]:

$$\varepsilon(x, t) = E_0 \operatorname{sech} \tfrac{1}{2}p\hbar^{-1} E_0(t - xv^{-1}) . \tag{1.1}$$

The amplitude of the electric field envelope $\varepsilon(x, t)$ is E_o, p is the matrix element of the resonant transition and the velocity v depends on E_o:

$$\frac{c}{v} - 1 \approx \alpha(\tfrac{1}{2}p\hbar^{-1} E_o)^{-2} . \qquad (1.2)$$

Note that v approaches c only if E_o is large. Ratios $c/v \sim 10^3$ have been observed in atomic vapours where the atom number density $n \sim 10^{11}$ cm^{-3} only (α is a parameter linear in n).

We shall see later that optical pulses break up into trains of pulses like (1.1). This exemplifies the property (i). We shall see that such pulses could also collide and pass through each other. But this has not yet been seen in the laboratory because of the difficulty of the experiment. The "phase shift" referred to in (iii) is this: after collision the argument $t - xv^{-1}$ of a typical soliton (1.1) is replaced by $t - xv^{-1} + \delta$; the number δ is the (small) phase shift. The phase shift δ_i for the ith soliton is actually a linear sum of pairwise shifts δ_{ij} $(j \neq i)$

$$\delta_i = \sum_{j \neq i} \delta_{ij}$$

and the total phase shift in a multi-soliton collision is conserved:

$$\sum_i \delta_i = \sum_i \sum_{j \neq i} \delta_{ij} = 0 .$$

It is a feature of most of present soliton theory that there are many other conserved quantities associated with solitons.

These introductory remarks on solitons in SIT show that soliton theory has a proper place in this volume. However solitons are important in a vastly greater range of physics than is treated here. Solitons already have some significance in the following areas of non-linear physics:

(i) Theory of water waves in both deep and shallow water.
(ii) Theory of plasmas and the interaction of radiation with plasmas.
(iii) Theory of Josephson junctions.
(iv) Resonant and non-resonant non-linear optics (not just in SIT).
(v) Non-linear crystal physics: theory of dislocations; anharmonic crystals; recurrence phenomena in thermal transport; ferrodistortive phase transitions in crystals in the displacive regime.
(vi) The A-phase of liquid ^{3}He below 2.6 mK where spin waves may propagate as solitons, and likewise in the B-phase.
(vii) Ferromagnetics: Bloch wall motion.
(viii) Theory of fundamental particles.

Reference [2] is a useful review and its references describe some of these applications.

Perhaps the three most important non-linear wave equations with soliton solutions are:

(I) The "Korteweg-de Vries" equation in one space variable x and one time variable t

$$u_t + 6uu_x + u_{xxx} = 0 . \qquad (1.3)$$

(II) The "non-linear Schrödinger" equation in x, t

$$i\, u_t + 2u|u|^2 + u_{xx} = 0 . \qquad (1.4)$$

(III) The "sine-Gordon" equation

$$u_{xx} - u_{tt} = \sin u \qquad (1.5a)$$

in Lorentz invariant form which is also

$$u_{xt} = \sin u \qquad (1.5b)$$

in "characteristic" form.

Equation (I) arises in water wave and plasma theory as well as in anharmonic lattice theory [3, 4, 5]: it is the first order development describing a weakly non-linear weakly dispersive wave system. The corresponding equation for a dissipative system is the Burgers equation

$$u_t + 6uu_x + bu_{xx} = 0 , \qquad b < 0 \qquad (1.6)$$

Equation (II) arises in non-linear optical self-focussing and self-phase modulation [6] as well as in plasma theory [7] and deep water wave theory [8]. In the optical context u is the electric field envelope ε and the factor $|u|^2 = |\varepsilon|^2$ is a contribution from the non-linear "refractive index" - see the section below and see the paper by McLean in General Theory (these Proceedings, paper IAEA-SMR-20/4).

Equation (III) arises in many contexts where rotations of large amplitude are possible: the situation is typified by a system of coupled pendulums in which the pendulums can swing through large angles [9]. Whole twists of 2π, 4π and so on can propagate through this system. Equivalent pendulums arise as real spins (spin waves) or as "pseudo-spins". as we shall see a resonant atom has a pseudo-spin description in which the spin rotates as the excitation oscillates up and down. This is the situation in SIT. Spin one-half systems are two-state systems. Another two-state system is the Josephson junction. Here charge oscillates across the two sides of the junction in the presence of a d.c. voltage across the junction. This is why equation (III) governs the propagation of "fluxons" in a Josephson junction. Notice that the sine-Gordon equation (III) is non-linear in the driving term $\sin u$. Its linearised form is the Klein-Gordon equation - hence the label!

All these equations are given in dimensionless units: the factor 6, for example, in (I) is conventional and can be removed by replacing u by $6u$. Physical constants are absorbed in the scales of the space (x) and time (t) variables.

If the dispersive-dissipative terms are neglected both the Korteweg-de Vries and Burgers equations reduce to the equation of a "simple wave"

$$u_t + 6uu_x = 0 . \qquad (1.7a)$$

This is an example of the more general simple wave

$$u_t + v(u)\, u_x = 0 \qquad (1.7b)$$

in which $v(u)$ is a velocity depending on u. The solution of (1.7b) is the implicit one

$$u = u(v(u)\, t - x)$$

i.e. u is any function of $v(u)\, t - x$. Evidently (1.7a) has $v(u) \equiv 6u$. Thus in those regions where u is large the velocity $v(u)$ is also large and the disturbance in these regions will overtake the disturbance in regions where u is small. The front of a pulse therefore steepens and the wave breaks like the wave on a sea shore as illustrated in Fig. 1.

In contrast in the KdV equation (I) the dispersive term u_{xxx} spreads the wave out. The result is a balance between dispersion and non-linearity in the form of the solitary wave

$$u = 2\xi^2 \operatorname{sech}^2(\xi x - 4\xi^3 t) . \qquad (1.8)$$

A solitary wave is one which travels without change of shape (a distortionless wave = wave of permanent profile). In this case it proves to be of bell shaped form, its velocity $v = 4\xi^2$ and increases with amplitude ξ and in fact this solitary wave is actually a soliton as we shall see.

Not all solitary waves are solitons: for example nerve fibres transmit solitary wave electric impulses but these transmit by electrical discharge through the sheath of the nerve fibre and "eat" themselves - very much as a flame travelling down a candle eats itself. Two flames at either end of a candle annihilate each other when they meet. Such flames travel at a velocity v determined by the chemical energy stored per unit

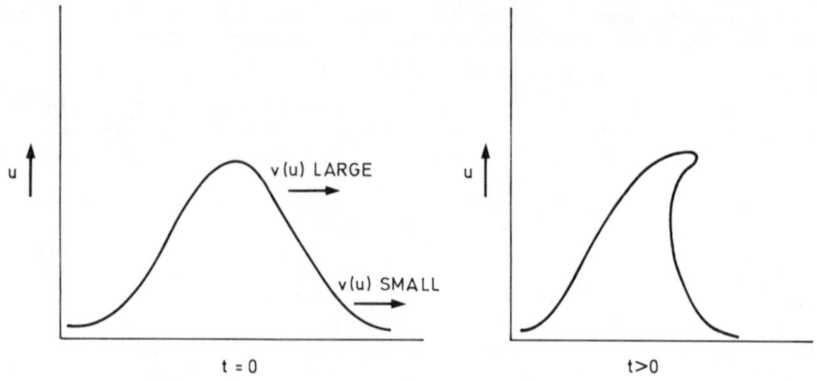

FIG.1. Shock and breakdown of $u(v(u)\,t - x)$.

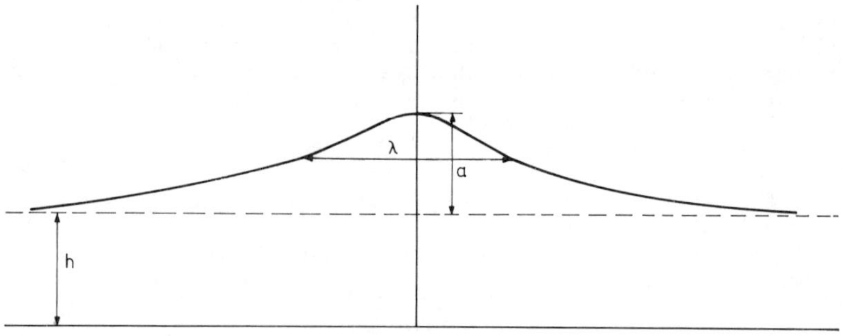

FIG.2. Russell's solitary wave. h is the depth of undisturbed water, a is the amplitude of the solitary wave and λ, the width at half height, is a measure of its length. This solitary wave is a soliton.

length E and the power P required to sustain the flame. Solitary waves of this character do not have the collision property of solitons: they satisfy the local conservation law [10]

$$\varepsilon_t + P_z = 0$$

where ε is the energy density and P the energy flux (power per sq. cm.).

John Scott Russell [11] reported the first recorded observation of a soliton. He was still classifying waves at this time and was led to distinguish the "Wave of Translation" or "giant solitary wave". He reports "I believe I shall best illustrate the phenomena by describing the circumstances of my first acquaintance with it. I was observing the motion of a boat which was rapidly drawn along a narrow channel by a pair of horses, when the boat suddenly stopped - not so the mass of water in the channel which it had put in motion; it accumulated round the prow of the vessel in a state of violent agitation, then suddenly leaving it behind rolled forward with great velocity, assuming the form of a large solitary elevation, a rounded, smooth and well-defined heap of water, which continued its course along the channel apparently without change of form or diminution of speed. I followed it on horseback, and overtook it still rolling on at a rate of some eight or nine miles an hour, preserving its original figure some thirty feet long and a foot to a foot and a half in height. Its height gradually diminished, and after a chase of one or two miles I lost it in the windings of the channel. Such in the month of August 1834 was my first chance interview with the singular and beautiful phenomenon which I have called the Wave of Translation."

What did Russell observe? We now know he saw the wave profile displayed in Fig. 2.

The total displacement y in this Fig. 2 is given by

$$y = h + a \operatorname{sech}^2 \frac{1}{2b}(x - ct)$$

where

$$c^2 = g(h + a)$$
$$b^2 = \frac{h^3}{3a} \quad .$$

Note that the wave is distortionless and that its velocity depends on the pulse amplitude a (BIGGER PULSES GO FASTER). This is a solution $u = y-h$ of Boussinesq's equation [12]

$$u_{tt} = gh \frac{\partial^2}{\partial x^2} \left\{ u\left(1 + \frac{3u}{2h} + \frac{h^2}{3u} u_{xx}\right) \right\} \tag{1.9a}$$

which in dimensionless units takes the form

$$u_{tt} = \frac{\partial}{\partial x} \left\{ u_x + 6uu_x + u_{xxx} \right\} \quad . \tag{1.9b}$$

The Korteweg-de Vries equation (I) can be derived from (1.9b) by restricting the motion to waves travelling one way only. We shall exemplify a restriction of this type (and substantiate its physical validity) in the treatment of SIT later. In this case the electric field envelope is the sech function (1.1): the energy $\propto \operatorname{sech}^2$.

The Fig. 3, taken from [11] shows Russell's observations on the propagation of water waves in shallow troughs about six inches deep. Solitons were created by withdrawing or moving diaphragms. Note the double humped pulse created in the "Fig. 6" which breaks up into two solitons. Recent data on the break-up of water waves governed by the Korteweg-de Vries equation is given in a report on "Tsunamis" [13]. Ref. [4] ("Fig. 3") shows the collision of ion acoustic wave solitons in a plasma. As an indication of how the Korteweg-de Vries equation arises in a physical context we give a derivation of this equation for ion acoustic waves next.

2. Soliton theory of ion acoustic waves

We assume a continuum description for the plasma;[1] we neglect electron inertia ($m_e = 0$) and adopt the isothermal equation of state $p_e = n_e k_B T_e$ for the electrons. With $m_e = 0$ Newton's equation in one space variable x is

$$0 = e \frac{\partial \phi}{\partial x} - \frac{k_B T_e}{n_e} \frac{\partial n_e}{\partial x} \tag{2.1}$$

in which ϕ is the e.s. potential. This has the integral $n_e = n_o \exp(e\phi/k_B T_e)$ where n_o is the uniform background electron density. Poisson's equation is therefore

$$\phi_{xx} = 4\pi e \left\{ n_o \left[\exp(e\phi/k_B T_e) \right] - n_i \right\} \tag{2.2}$$

where n_i is the ion density. Conservation of ion density and momentum yield

$$\frac{\partial n_i}{\partial t} + \frac{\partial n_i v_i}{\partial x} = 0 \quad , \tag{2.3}$$

$$\frac{\partial v_i}{\partial t} + v_i \frac{\partial v_i}{\partial x} = - \frac{e}{m_i} \frac{\partial \phi}{\partial x} \quad . \tag{2.4}$$

[1] See Ref. [14].

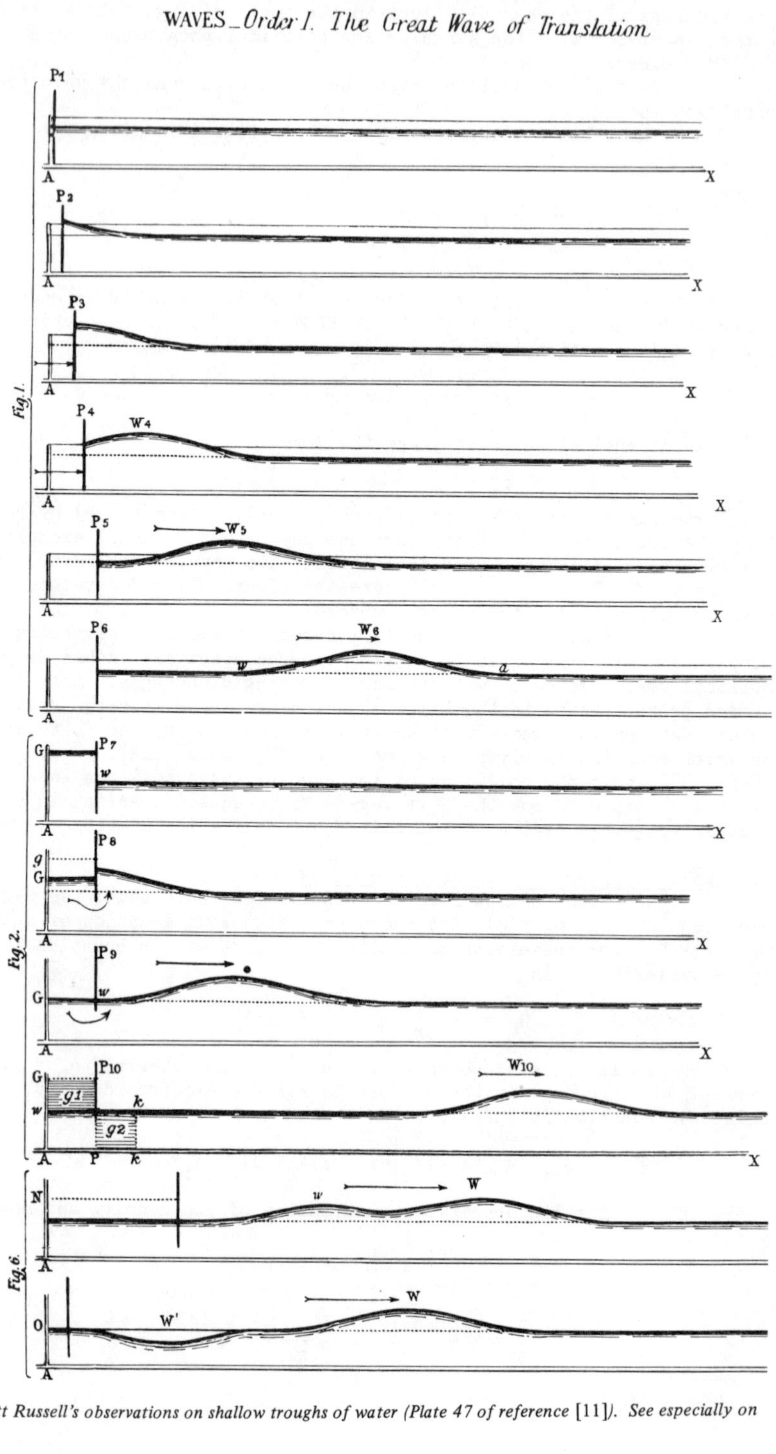

FIG.3. Scott Russell's observations on shallow troughs of water (Plate 47 of reference [11]). See especially on

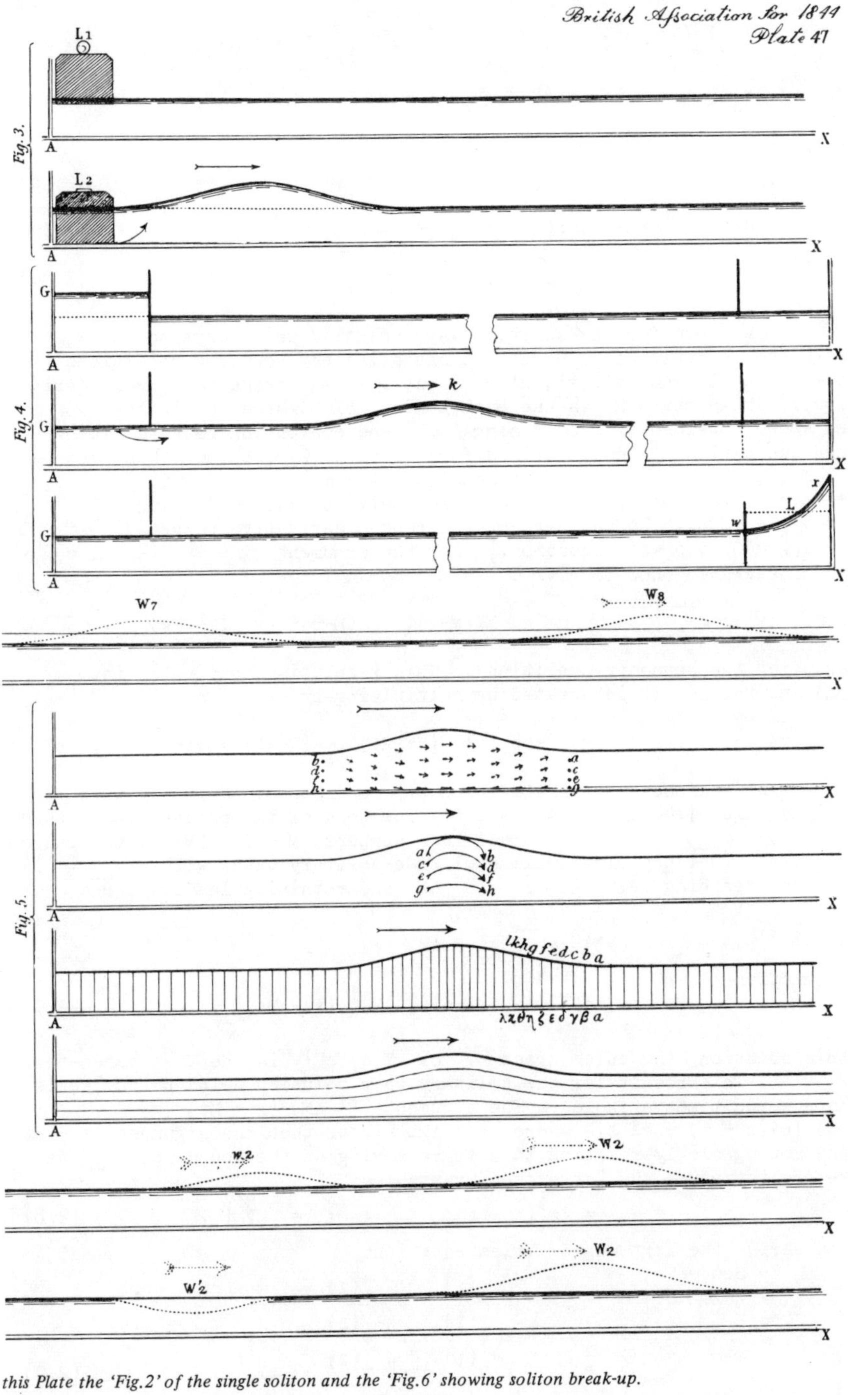

this Plate the 'Fig.2' of the single soliton and the 'Fig.6' showing soliton break-up.

Set [14]

$$(k_B T_e/4\pi n_o e^2)^{-\frac{1}{2}} x \to x, \quad t\,(\frac{4\pi n_o e^2}{m_i})^{\frac{1}{2}} \to t, \quad (\frac{k_B T_e}{e})^{-1}\phi \to \phi,$$

$$n_i n_o^{-1} \to n, \quad v_i (k_B T_e/m_e)^{-\frac{1}{2}} \to v.$$

Then the dimensionless equations are

$$\phi_{xx} = \exp\phi - n \tag{2.5a}$$

$$n_t + (nv)_x = 0 \tag{2.5b}$$

$$v_t + vv_x = -\phi_x \,. \tag{2.5c}$$

We look for the solitary wave solution which depends on x, t through the combination $x - Mt$ and satisfies the boundary conditions $\phi, v \to 0$, $n \to 1$ and $\phi', v', n' \to 0$ as $x \to \pm\infty$ (prime here means derivative). The number M is the Mach number V/c where c is the ion sound speed obtained by linearising (2.5) and converting back to dimensioned quantities. [The linearised form of (2.5) is obtained thus: set $n = 1 + \varepsilon\rho$, $v = \varepsilon u$, $\phi = \varepsilon\theta$, then $\theta_{xx} = \theta - \rho$, $\rho_t + u_x = 0$, $u_t + \theta_x = 0$ to first order in ε. From this follows $\theta_{xx} - \rho_{tt} = 0$, $\theta_{xx} = \rho_{xx} + \theta_{xxxx}$. We can neglect the fourth derivative to get $\rho_{xx} - \rho_{tt} = 0$ the wave equation.] For the argument $x - Mt$ (2.5b, c) can be integrated once to give

$$n = M/(M-v), \quad (M-v)^2 = (M^2-2\phi)$$

after using the boundary conditions. This result is substituted into (2.5a) and the result integrated by multiplying by ϕ'. One finds

$$\tfrac{1}{2}(\phi')^2 = \left[\exp\phi + M(M^2-2\phi)^{\frac{1}{2}} - (M^2+1) + \text{const.}\right] \,. \tag{2.6}$$

The constant vanishes through the boundary conditions. Solitary wave solutions exist [14] for $1 < M \lesssim 1.6$. The peak of the potential ϕ then lies in $0 < \phi_{max} \lesssim 1.3$. For small Mach numbers $M - 1 \equiv \Delta M$ in the range $0 < \Delta M << 1$ (which yields small amplitude solitary waves with $\phi_{max} << 1$) we get by expanding (2.6) in ϕ and ΔM and retaining leading terms

$$(\phi')^2 = \frac{2}{3}\phi^2(3\,\Delta M - \phi)$$

with solution

$$\phi = 3\,\Delta M\,\text{sech}^2\left[(\tfrac{1}{2}\,\Delta M)^{\frac{1}{2}}\,(x - Mt)\right] \,. \tag{2.7}$$

In this solution the pulse speed V or $M \equiv Vc^{-1}$ is the free parameter.

This result for the solitary wave solution helps us to "scale" the space and time variables. The argument $(\frac{1}{2}\,\Delta M)^{\frac{1}{2}}(x - Mt) = (1/\sqrt{2})\,[\varepsilon^{\frac{1}{2}}(x - t) - \varepsilon^{\frac{3}{2}} t]$ where $\varepsilon \equiv \Delta M$. [Note that the argument $x - Mt$ is now moving slowly relative to a frame moving at the sound speed.] We therefore introduce the "stretched" variables

$$z = \varepsilon^{\frac{1}{2}}(x - t), \qquad \tau = \varepsilon^{\frac{3}{2}} t \,. \tag{2.8}$$

We now derive the Korteweg-de Vries equation.

Set

$$\begin{aligned} n &= 1 + \varepsilon\, n^{(1)} + \varepsilon^2\, n^{(2)} + \dots \\ \phi &= \varepsilon\, \phi^{(1)} + \varepsilon^2\, \phi^{(2)} + \dots \\ v &= \varepsilon\, v^{(1)} + \varepsilon^2\, v^{(2)} + \dots \end{aligned} \tag{2.9}$$

Use

$$\partial/\partial x = \varepsilon^{\frac{1}{2}}\,\partial/\partial z, \qquad \partial/\partial t = \varepsilon^{\frac{3}{2}}(\partial/\partial\tau) - \varepsilon^{\frac{1}{2}}(\partial/\partial z).$$

Then to lowest order (2.5) is

$$\phi^{(1)} = n^{(1)}, \quad n_z^{(1)} = v_z^{(1)}, \quad v_z^{(1)} = \phi_z^{(1)}$$

with solution

$$\phi^{(1)} = n^{(1)} = v^{(1)}.$$

To next order we get

$$\phi_{zz}^{(1)} = \phi^{(2)} + \tfrac{1}{2}(\phi^{(1)})^2 - n^{(2)}$$

$$-n_z^{(2)} + n_\tau^{(1)} + (n^{(1)}v^{(1)})_z + v_z^{(2)} = 0$$

$$-v_z^{(2)} + v_\tau^{(1)} + v^{(1)}v^{(1)}{}_z = -\phi_z^{(2)}. \qquad (2.10)$$

The quantities $\phi^{(2)}$ and $n^{(2)}$ can be eliminated from (2.10) and using $n^{(1)} = \phi^{(1)} = v^{(1)}$ we find

$$n_\tau^{(1)} + n^{(1)}n^{(1)}{}_z + \tfrac{1}{2}\,n_{zzz}^{(1)} = 0.$$

If we set $n^{(1)} = 3u$, replace z by x, and rescale $\tau = 2t$, we get the KdV equation (I) for u. One can check that (1.8) is the solitary wave solution (2.7).

3. The Fermi-Pasta-Ulam problem

The word "Soliton" was introduced by Zabusky and Kruskal [15]. They realised for the first time the remarkable particle-like properties of the solitary wave solutions of the KdV equation.

They were solving the KdV equation with periodic boundary conditions numerically. We now know that to do this analytically is a much harder problem than on the infinite line $-\infty < x < \infty$ [16, 17]. The periodic unit was of length 2. The initial condition was

$$u(x, 0) = \cos \pi x, \qquad u_t(x, 0) = 0.$$

Their KdV was $u_t + uu_x + \delta\, u_{xxx} = 0$ with $\delta = 0.022$. They observed the remarkable behaviour shown in Fig. 4. The cosine developed spikes of roughly sech^2 form. These travelled as solitons (Fig. 5). Note the phase shifts signified by the steps in the trajectories.

The periodic boundary conditions have a remarkable consequence. Although the spikes (solitons) collide and pass through each other without change of shape they can move into the next unit cell and recreate what is essentially the initial condition - the cosine wave. The phenomenon is as we now know periodic in space and, in general, almost periodic in time [16].

The recurrence phenomenon has a bearing on the FPU problem [18]. Following Debye it was believed that the initial excitation of a single phonon mode in a non-linear lattice would result in ergodic behaviour - the sharing of the energy in the single mode with all the other modes. The time constant for this energy sharing process would be a measure of the thermal transport properties of the lattice. Pasta and Ulam took a one-dimensional lattice with cubic and quartic anharmonicities. The behaviour they observed numerically (in the case of cubic anharmonicity) is shown in Fig. 6. Energy is shared only amongst the small group of modes labelled 2, 3, 4 and after 158 "linear periods" the energy is almost entirely back in the first mode. We now know that solitons improve the capacity of lattices to transport energy and in one dimension anharmonic features do not lead to the Fourier law of heat conduction [19]. We return to the problem of solitons in lattices later. For references on the FPU problem see [2, 19].

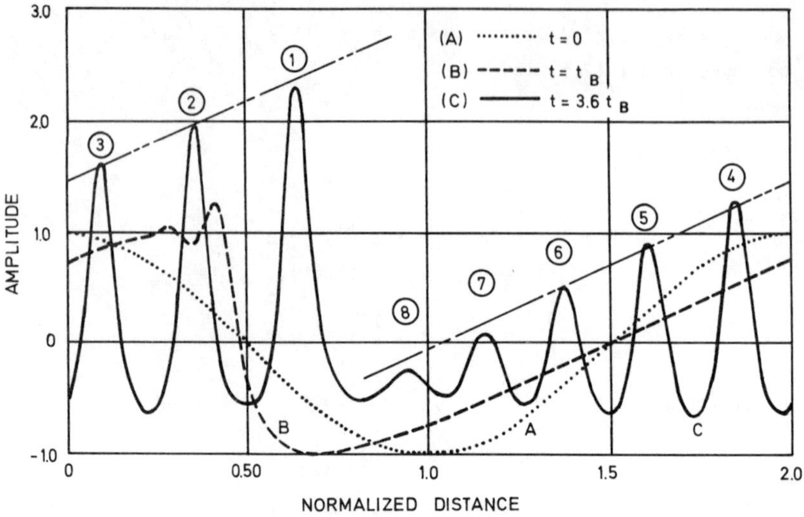

FIG.4. Solutions of the KdV equation $u_t + uu_x + \delta u_{xxx} = 0$ with $\delta = 0.022$ at three different times (reproduced from Fig.6 of Ref. [5]).

These two sections 1 and 2 complete a brief introduction to the properties of solitons and the properties of the Korteweg-de Vries equation (I). The KdV equation is important not only because of its physical applications, but because it was the first equation for which the so called "inverse scattering method" of analytical solution was developed [20, 2]. The inverse scattering method can even be used to solve the initial value problem for this non-linear equation. Analytically the method is a generalisation of the method of Fourier transforms for linear equations [21] and has now been applied to a number of other equations, notably the non-linear Schrödinger equation (II) [6] and the sine-Gordon equation (III) [22, 23, 24, 21].

We turn now to the properties of the s-G equation by considering the theory of optical self-induced transparency.

4. Solitons in SIT - the sine-Gordon equation

We first recall the features observed in SIT (see the paper by Arecchi in General Theory (these Proceedings, paper IAEA-SMR-20/16)). The Fig. 7 shows observations (left) and computer simulations (right) of short resonant pulse break-up in ^{87}Rb vapour [25]. The source was a ^{202}Hg laser and tuning was achieved via a magnetic field. The pulses are in the 25 Wcm^{-2} range (a peak power). Since time is the horizontal axis higher spikes are to the left (bigger spikes emerge first). By comparison with the break-up of water and plasma ion acoustic waves it is hard to avoid the conclusion that this is another example of soliton break-up.

The high transparency in SIT is shown in Fig. 8. Note the successive peaks in transparency with increasing intensity. We define a pulse "area" $\theta(x)$ for a pulse envelope $\varepsilon(x, t)$ by

$$\theta(x) = p\hbar^{-1} \int_{-\infty}^{\infty} \varepsilon(x, t)\, dt \; ; \qquad (4.1)$$

p is the dipole matrix element of the resonant transition. Peaks in transparency occur at $\theta = 2\pi, 4\pi, 6\pi, \ldots$ The "area theorem" has already been

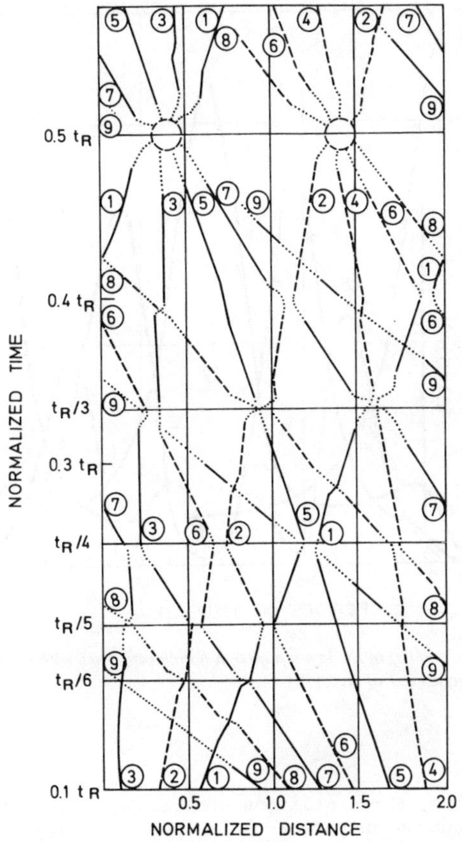

FIG.5. Soliton trajectories on a space time diagram for the data of Fig.4 and beginning at $t = 0.1\ t_R = 3.04\ t_B$ (reproduced from Fig.7 of Ref. [5]).

mentioned (cf. Arecchi's paper, these Proceedings). It was first derived in [1] and states that[2]

$$\frac{d\theta}{dx} = \pm\ \alpha \sin\theta \quad . \tag{4.2a}$$

The signs are (+) for amplifiers and (-) for attenuators. The attenuator is a resonant medium initially unexcited: the amplifier is one initially inverted. We mostly consider attenuators. For weak fields θ is small and (4.2) can be approximated by

$$\frac{d\theta}{dx} = \pm\ \alpha\ \theta\ . \tag{4.2b}$$

The solution is $\theta = \theta_0 \exp \pm \alpha x$. For a fixed length L of sample $\theta(L) = \theta_0 \exp \pm\ \alpha L$. Since the intensity $\propto \varepsilon^2 \propto \theta^2$ for small ε, αL is a measure of the linear attenuation (attenuators) or linear gain (amplifiers). For larger intensities solutions of (4.2a) are θ = constant (amongst other solutions) where $\theta = 2\pi,\ 4\pi,\ 6\pi, \ldots$ We shall find that this is just when one

---

[2] The parameter α used here is defined in terms of the so-called "inhomogeneous broadening" as explained on page 397. The different α used in Eq. (1.2) is the α defined below Eq. (4.17) on page 397.

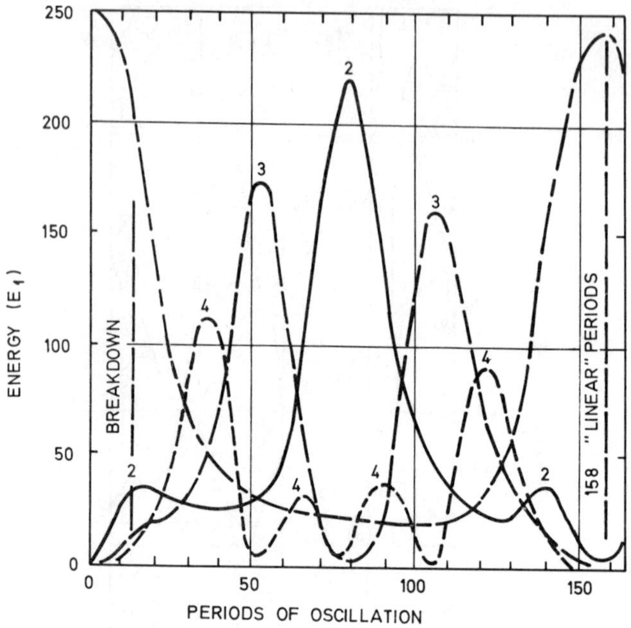

FIG.6. The energy (arbitrary units) in the low modes of a non-linear lattice with a cubic non-linearity in the Hamiltonian (reproduced from Fig.1 of Ref. [18]*).*

soliton, two solitons, three solitons contribute to the optical pulse, there is no other contribution and break-up is into one pulse (= no change), two soliton pulses, three soliton pulses. More generally pulse areas evolve towards 2π, 4π, 6π and pulses reshape until they are pure soliton pulses. Fig. 9 illustrates the evolution of pulse area according to the area theorem (4.2a).

In order to construct a theory we consider a dielectric made up of non-degenerate 2-level atoms. Although atoms have many levels we can expect that close to resonance only the two levels with the resonant frequency spacing need be considered. The Fig. 10 shows how Gibbs and Slusher [25] created a non-degenerate 2-level ^{87}Rb atom. The $5s^2S_{\frac{1}{2}} \rightarrow 5p^2P_{\frac{1}{2}}$ D_1 line at 7947.64 Å had the hyperfine $F = 1, 2$ degeneracies removed by a strong magnetic field ($B = 74.5 \times 10^3$ gauss). Good quantum numbers become J, I, M_J and M_I. At $B = 74.5 \times 10^3$ gauss)the $\Delta J = 0$ $\Delta M_J = 1$ transition $M_J = -\frac{1}{2} \rightarrow M_J = +\frac{1}{2}$ (rotating field) was on resonance with the ^{202}Hg laser line. The question of what happens when the hyperfine levels are left degenerate is a very interesting one and we look at this problem in §7 below.

The 2-level atom is relatively easy to handle. It is a two state quantum mechanical system and has a pseudo-spin (spin one half) description [26]. Fig. 11 shows the equivalent spin description. It is clear that the free atom has Hamiltonian

$$H_0 \equiv \tfrac{1}{2}\hbar\omega_s\,\sigma_z\,, \qquad \sigma_z = \begin{bmatrix} 1 & 0 \\ 0 & -1 \end{bmatrix}. \qquad (4.3)$$

The eigenvalues are $\pm\frac{1}{2}\hbar\omega_s$ and the spacing is $\hbar\omega_s$: ω_s is the resonant frequency of this 2-level transition. In Dirac ket notation if the states are $|0\rangle$ (lower) and $|s\rangle$ (upper) the dipole operator must be essentially

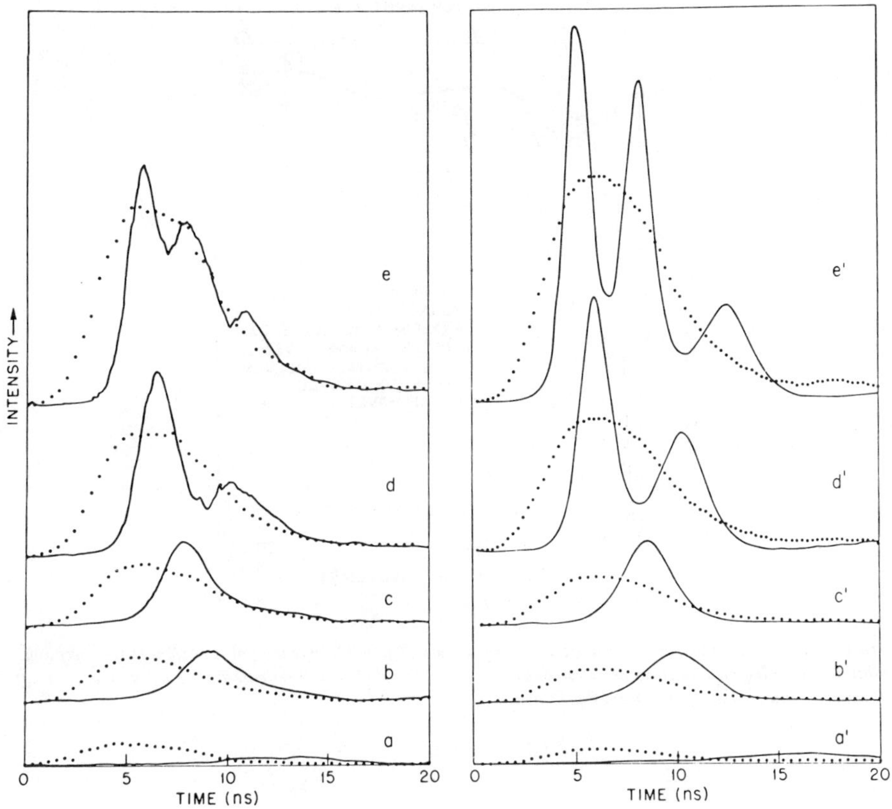

FIG. 7. Input pulses (dotted) and output pulses (hard lines) for $\alpha L \approx 5$ in the SIT experiments of Ref. [25]. *The left shows the observations and the right shows computer simulations based on the homogeneously broadened SIT equations (see below): (a) shows a pulse of area less than π which attenuates; (b) shows a $> 2\pi$ pulse which reshapes to a sech; (c) shows peaking to a sharper sech; and (d) and (e) show break-up of pulses with input areas just less than 3π and 5π (for more complete information see Fig.3 of Ref.* [25]*).*

$p\{|0\rangle\langle s| + |s\rangle\langle 0|\}$ where p is the matrix element of the transition ($p = \int \psi_s^* e\, z\, \psi_0 dz = e\langle s|z|0\rangle$). Dipole allowed transitions are assumed. The interaction Hamiltonian for plane polarised light is therefore

$$H_{int} = -p\,\sigma_x\,\hat{u}\cdot\vec{E}(t) \quad . \tag{4.4}$$

Here σ_x is the usual Pauli operator replacing $|0\rangle\langle s| + |s\rangle\langle 0|$, $\hat{u}$ is a unit vector in the z-direction[3] and $\vec{E}(t)$ is a semi-classical (c-number) field. Heisenberg's equations of motion yield the equations

$$\dot{\vec{\sigma}} = \vec{\omega}\times\vec{\sigma}$$

$$\vec{\omega} \equiv \left(-\frac{2p}{\hbar}E(t),\ 0,\ \omega_s\right) \tag{4.5}$$

[3] We consistently use x as the propagation direction and z is transverse to this. The labels x, y, z on σ_x, σ_y, σ_z are simply the usual ones in *spin* space – distinct from ordinary x, y, z space. For rotating fields, (4.4) involves *spin* co-ordinates σ_x and σ_y in the *space* directions z and y.

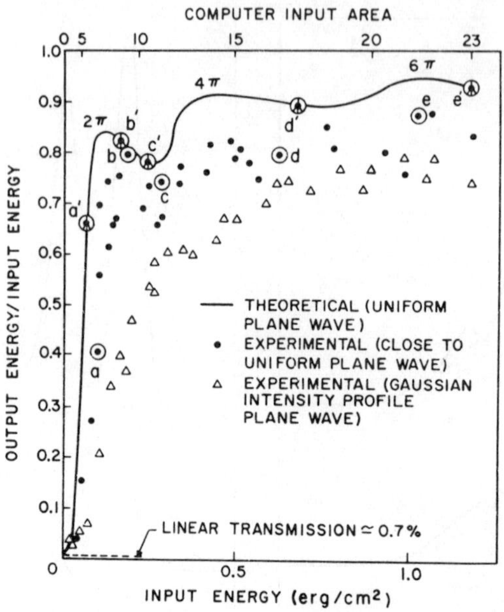

FIG.8. Self-induced transparent transmission in Rb vapour. The solid curve is a uniform plane-wave computer solution. Note the high transparency at areas of 2π, 4π and 6π where pulses reshape at constant area to trains of 1, 2 and 3 solitons (reproduced from Fig.13 of Ref. [25]).

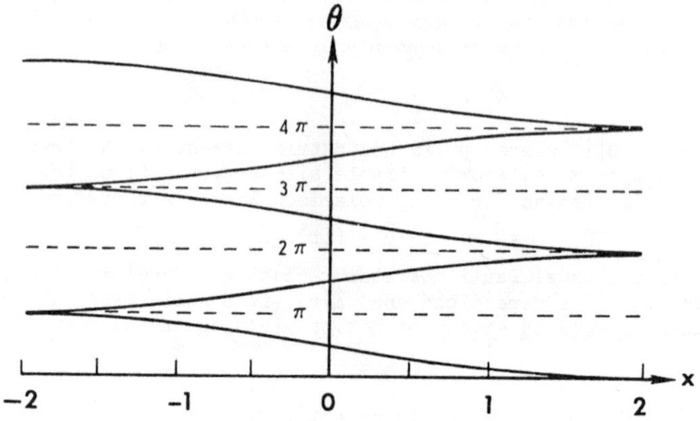

FIG.9. Evolution of $\theta(x)$ (area against distance). Note that for example, if $\pi < \theta(x) < 3\pi$ at $x = 0$, $\theta(x) \to 2\pi$. At the same time the pulse reshapes to a 2π sech envelope pulse – a single soliton. The diagram is to be read from right to left for amplifiers: small area pulses grow to area π.

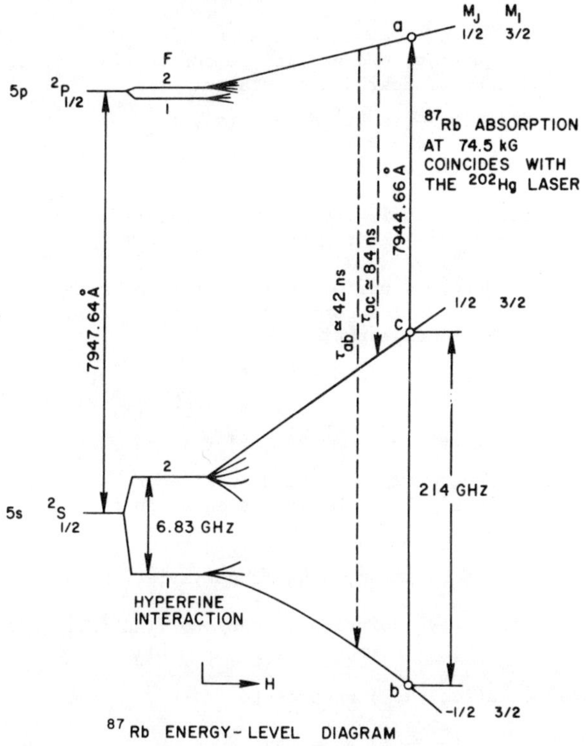

FIG.10. Diagram of the hyperfine structure of ^{87}Rb D_1 transitions as a function of magnetic field strength. The Zeeman interaction at 74.5 kG lifts the low field degeneracy and tunes the absorption to coincide with the ^{202}Hg laser line.

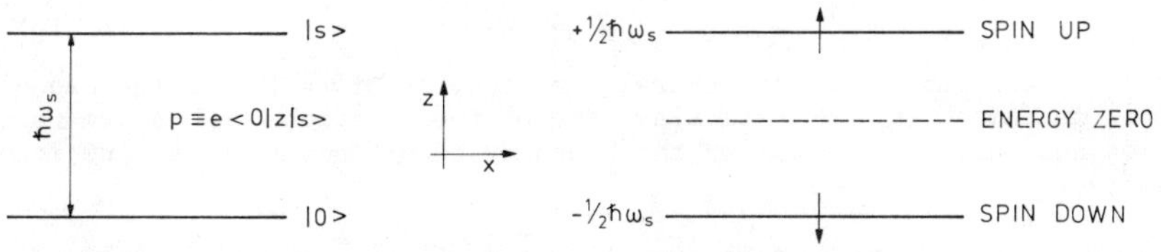

FIG.11. Equivalent spin description for the 2-level atom. The direction of propagation is x; z is transverse to x.

with $\vec{\sigma} = (\sigma_x, \sigma_y, \sigma_z)$. This equation for the pseudo-spin is exactly the equation of motion of a "real" spin in a magnetic field $(B(t), 0, B_o)$ with

$$\frac{\gamma e\hbar}{m_e c} B_o \leftrightarrow \hbar\omega_s .$$

For this reason it is called a Bloch equation.

The choice of the c-number field means that the motions of the atoms are solely coherent motions (spontaneous emission is ignored). The expectation value of the spin operator $\vec{\sigma}$ is consequently very like the behaviour of the operator. The expectation value

$$\langle\vec{\sigma}\rangle \equiv \vec{r}(t) = \text{"Bloch vector"}$$

satisfies

$$\dot{\vec{r}} = \vec{\omega} \times \vec{r} \tag{4.6}$$

the c-number Bloch equation. We assign the components $(u, v, w) \equiv \vec{r}$. It is clear that, in terms of Schrödinger states,

$$|\underline{\psi}(t)\rangle = c_s(t)|s\rangle + c_o(t)|0\rangle,$$

$$w(t) \equiv \langle|\sigma_z(t)|\rangle = \langle\underline{\psi}(t)|\sigma_z|\underline{\psi}(t)\rangle = |c_s|^2 - |c_o|^2 . \tag{4.7}$$

Hence $w(t)$ measures the atomic inversion. Similarly $p\,u$ is the dipole moment $p(c_s{}^* c_o + c_o{}^* c_s)$: the component $v = i(c_s{}^*c_o - c_o{}^*c_s)$ carries phase information.

We now suppose a uniform distribution of 2-level atoms in the medium: the density is n. The components u, v, w will depend on position x as well as t. The Bloch equation becomes

$$\begin{aligned} u_t(x, t) &= - \omega_s\, v(x, t) \\ v_t(x, t) &= + \omega_s\, u(x, t) + \frac{2p}{\hbar} E(x, t)\, w(x, t) \\ w_t(x, t) &= - \frac{2p}{\hbar} E(x, t)\, v(x, t) \end{aligned} \tag{4.8}$$

in component form. Each atom is coupled to all others through the fields radiated by the atomic dipoles. Maxwell's equation is

$$E_{xx} - c^{-2} E_{tt} = 4\pi\, n\, c^{-2}\, p\, u_{tt} \tag{4.9}$$

(we assume plane wave conditions, that is no variation in the y, z directions).

A solitary wave solution is easy to find [27]. Set (compare §2)

$$E = E(t - x\, v_o{}^{-1}) \quad . \tag{4.10}$$

Then the Maxwell equation (4.9) reduces to

$$\left(\frac{c^2}{v_o{}^2} - 1\right) E = 4\pi\, n\, p\, u$$

so that

$$E \propto u \; . \tag{4.11}$$

Substitution in the Bloch equation (4.8) then yields a non-linear ordinary differential equation of the form

$$u'' + A\,u + B\,u^3 = 0 \tag{4.12}$$

(A and B are numbers). In general the solution of (4.12) is the Jacobian elliptic function cn. A particular case of the function cn is the sech function and only this satisfies the boundary conditions that E and its derivative $\to 0$ as $x \to \pm\infty$.

We find

$$E = E_o \operatorname{sech}\left\{\frac{pE_o}{\hbar}\left(t - \frac{x}{v_o}\right)\right\} \tag{4.13a}$$

$$\frac{c^2}{v_o{}^2} - 1 = \frac{8\pi\, n\, p^2\, \omega_s}{\omega_s{}^2 + p^2 E_o{}^2 \hbar^{-2}} \; . \tag{4.13b}$$

This wave satisfies the condition "bigger pulses go faster". However, the numerical evidence does not support the view that this is a soliton. The Fig. 12 shows what happens when a positive pulse collides with a negative one (we call this a "kink anti-kink" collision later). Furthermore (4.13a) is not a resonant solution which is the condition for the validity of the 2-level atom model.

For a physically acceptable approximate solution we look for a slowly varying field envelope modulating a high frequency carrier:

$$E(x, t) = \varepsilon(x, t) \cos\left\{\omega t - kx + \phi(x, t)\right\} \; ; \tag{4.14}$$

$\omega \sim ck \sim 10^{15}$ radians per second at optical frequencies. For 10^{-9} s pulses the envelope ε and phase ϕ both vary on the scales 10^9 radians per second and $3\times10^{10} \times 10^{-9} \simeq 10$ cm. (strong retardation changes the length scale to about 10^{-2} cm).

At exact resonance[4] the dipole $p\,u(z, t)$ has an "in phase" and "out of phase" component:

$$\text{("in phase")} \qquad u(x, t) = Q(x, t) \cos\left\{\omega_s t - k_s x + \phi(x, t)\right\}$$
$$+ \text{("out of phase")} \qquad + P(x, t) \sin\left\{\omega_s t - k_s x + \phi(x, t)\right\}. \tag{4.15}$$

It is convenient to relabel the inversion as $N(x, t)$

$$N(x, t) \equiv w(x, t) . \tag{4.16}$$

By substituting (4.14) - (4.16) into the coupled Maxwell-Bloch equations and equating coefficients of cosine and sine terms exploiting the fact that ε, Q, P, N vary slowly compared with ω^{-1} and k^{-1} we reach the "SIT equations" [28]

$$\varepsilon_x + c^{-1}\varepsilon_t = c^{-1}\alpha\langle P(x, t, \Delta\omega)\rangle$$
$$P_t(x, t, \Delta\omega) = \varepsilon N + (\Delta\omega + \phi_t)Q$$
$$N_t(x, t, \Delta\omega) = - \varepsilon P$$
$$\phi_t(x, t, \Delta\omega) = - (\Delta\omega + \phi_t)P$$
$$\varepsilon(x,t)(\phi_x(x, t) + c^{-1}\phi_t(x, t)) = -c^{-1}\alpha\langle Q(x, t, \Delta\omega)\rangle \tag{4.17}$$

where $\alpha = 2\pi\, n\, p^2\, \omega_s\, \hbar^{-1}$ and $\Delta\omega = \omega_s' - \omega_s$. (The field envelope is scaled to a quantity with the dimensions of frequency by $E(x, t) = \hbar\, p^{-1}\varepsilon(x, t) \cos\{k_s x - \omega_s t + \phi(x, t)\}$).

The SIT equations are "inhomogeneously broadened". The broadening function is $g(\Delta\omega)$ where $\Delta\omega \equiv \omega_s' - \omega$ is the off-set from resonance.[5] In a vapour the inhomogeneous broadening is primarily Doppler motion which causes the atomic resonance to be shifted. In consequence even at exact resonance (ω coincident with the maximum ω_s of $g(\Delta\omega)$ for a symmetric $g(\Delta\omega)$) individual groups of atoms are not on resonance in general. One of the remarkable features of SIT is that atomic dipoles resonant or not couple coherently to permit solitons to propagate.

The inhomogeneously broadened one-soliton solution was first found by McCall and Hahn [29, 1]. It is (cf. equation (1.1))

$$\varepsilon(x, t) = \frac{p}{\hbar} E_o \operatorname{sech}\left\{\frac{\frac{1}{2}p\, E_o}{\hbar}\left(t - \frac{x}{v}\right)\right\}$$

with

$$\frac{c}{v} - 1 = \frac{2\pi\, n\, p^2\, \omega}{\hbar} \int_o^\infty \frac{g(\Delta\omega)\, d\,\Delta\omega}{\Delta\omega^2 + \frac{1}{4}p^2 E_o^2 \hbar^{-2}} \tag{4.18}$$

and remains a solution whether the carrier wave frequency ω is resonant (i.e. $\omega = \omega_s$) or not (if not $\Delta\omega = \omega_s' - \omega$ in (4.18)).

For present purposes we simplify the situation and consider "sharp line SIT". This is achieved in practice by eliminating Doppler broadening using an atomic beam [30]. On exact resonance $\Delta\omega = 0$ and the Bloch equation reduces to

$$P_t = \varepsilon N$$
$$N_t = - P \varepsilon \tag{4.19}$$

[4] One moves off resonance by including terms $\omega - \omega_s$ in ϕ_t and $k - k_s$ in ϕ_x [28].

[5] The parameter α used in (4.17) is not that used in (4.2a). In (4.2a) α is defined by $4\pi^2 g(0) p^2 n\, \omega_s/\hbar c$.

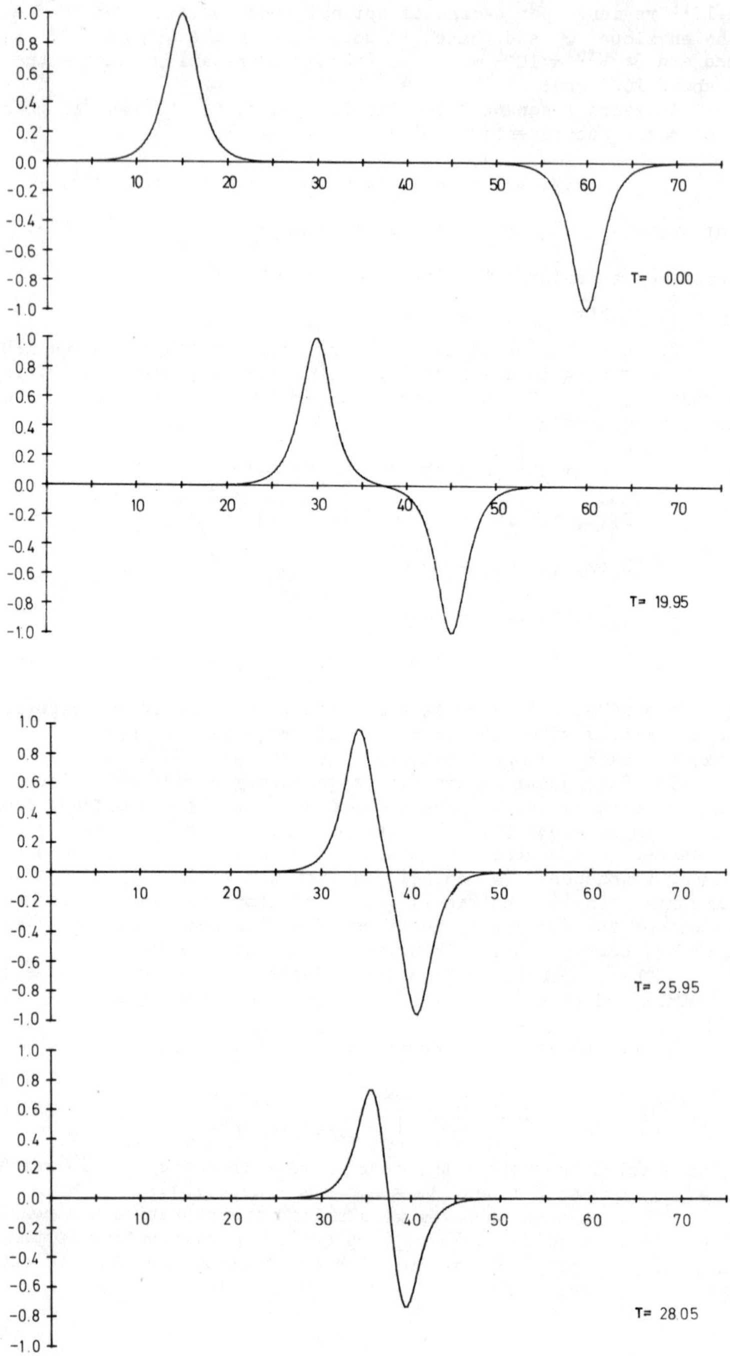
1.0
0.8
0.6
0.4
0.2
0.0
-0.2
-0.4
-0.6
-0.8
-1.0
10
20
30
40
50
60
70
T= 0.00
T= 19.95
T= 25.95
T= 28.05

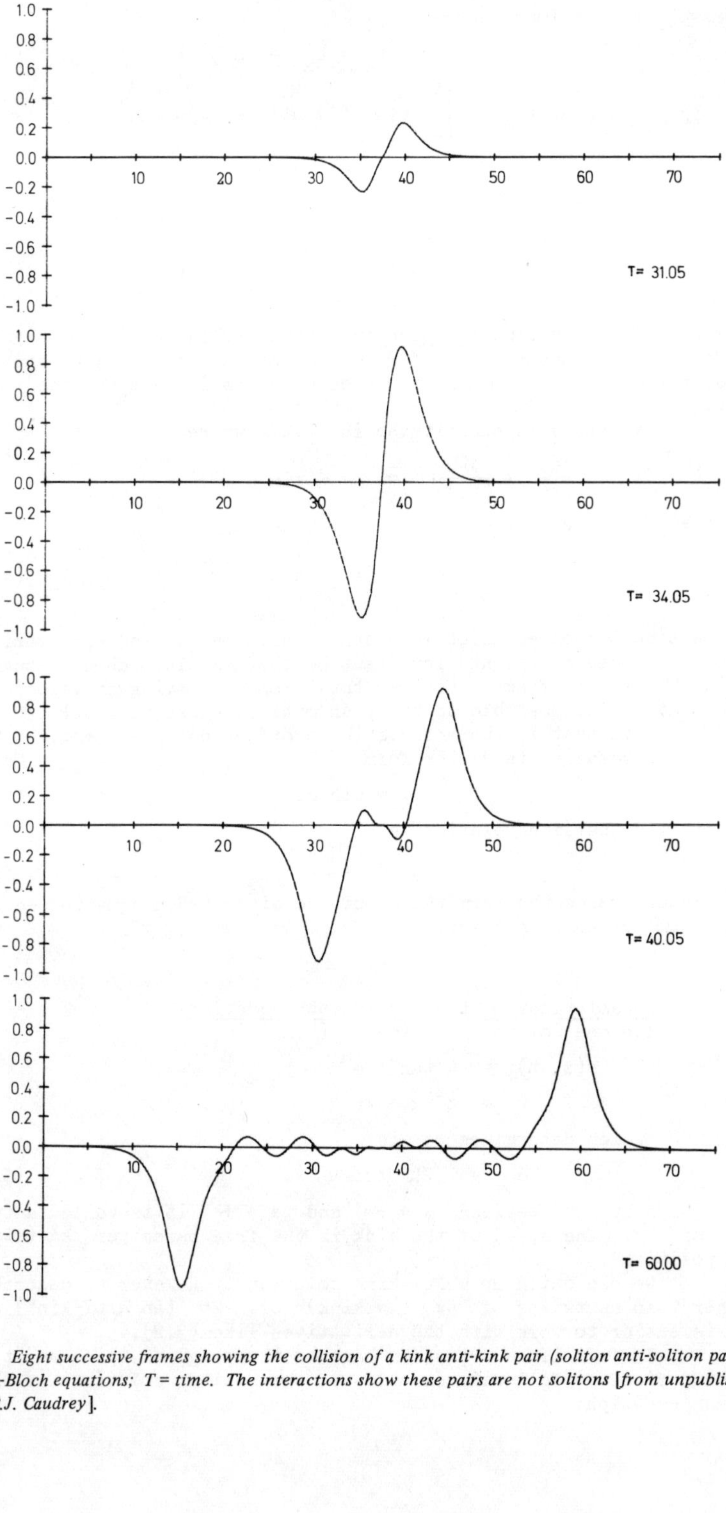

FIG.12. Eight successive frames showing the collision of a kink anti-kink pair (soliton anti-soliton pair) for the Maxwell-Bloch equations; T = time. The interactions show these pairs are not solitons [from unpublished work by Dr. P.J. Caudrey].

whilst Maxwell's equations become

$$c^{-1}\,\varepsilon_t + \varepsilon_x = \alpha c^{-1}\,P \ . \tag{4.20}$$

Set

$$\sigma(x,\,t) = \int_{-\infty}^{t} \varepsilon(x,\,t')\,dt'$$

or

$$\varepsilon(x,\,t) = \sigma_t \tag{4.21}$$

Then

$$P = -\sin\sigma$$
$$N = -\cos\sigma \tag{4.22}$$

satisfies the Bloch equation identically. Moreover the choice of signs is correct for the attenuator. For, for $\sigma \to 0$ as $x \to \pm\infty$, $P \to 0$ and N (the inversion) $\to -1$. The opposite choice of signs is made for the amplifier.

If (4.22) is substituted in (4.20) we reach

$$\frac{1}{c}\frac{\partial^2\sigma}{\partial t^2} + \frac{\partial^2\sigma}{\partial t\,\partial x} = -\frac{\alpha}{c}\sin\sigma \ .$$

Set $\sqrt{\alpha}\ ct = \tau\sqrt{c}$, $\sqrt{\alpha}\ (ct - 2x) = \xi\sqrt{c}$. Then

$$\sigma_{\xi\xi} - \sigma_{\tau\tau} = \sin\sigma \ . \tag{4.23}$$

This is the sine-Gordon equation in a space variable ξ and time variable τ. Notice that this is Lorentz invariant so that a pulse can be brought to rest in the ξ, τ frame. However this frame is moving at velocity $\frac{1}{2}c$ so that it is not possible to bring an optical pulse to rest!

Note that in characteristic co-ordinates $\xi = x + t$, $\tau = x - t$ the sine-Gordon equation takes the form

$$\sigma_{xt} = \sin\sigma. \tag{4.24}$$

The linearised dispersion relation is

$$\omega k = 1$$

and this feature makes the form (4.24) of the sine-Gordon equation easier to handle in terms of soliton theory.

5. The soliton solutions of the sine-Gordon equation

The one-soliton solution of (4.24) is

$$\sigma(x,\,t) = 4\tan^{-1} e^{u}$$
$$u = k^{-1}x + kt \ . \tag{5.1}$$

Evidently σ_t which determines ε is

$$\sigma_t = 2k\,\mathrm{sech}\,u \ . \tag{5.2}$$

Since σ jumps by 2π between $x \to -\infty$ and $x \to +\infty$ it is called a twist or "kink" of 2π. The speed of the kink is the free parameter k^2 (supposed real and positive).

We can build up multi-kink solutions. However to describe these other than as twists of 4π (4π kink) or -2π (an anti-kink) and so on it is easier to work with the derivatives like (5.2).

In fact if we transform back to the original space and time variables we find the following examples of the soliton solutions of the sine-Gordon equation:

1-soliton (2π)

$$\varepsilon(x,t) = E_1 \operatorname{sech} \theta_1, \qquad \theta_1 = \omega_1 t - k_1 x + \delta_1, \quad \omega_1 = \tfrac{1}{2}E_1,$$

$$\frac{k_1}{\omega_1} = 1 + 4p\beta\hbar^{-1}E_1^{-2} \qquad (\beta = \hbar p^{-1}\alpha)$$

$$E_1 = \text{free parameter (real)}$$

2-soliton (4π)

$$\varepsilon(x,t) = \frac{E_1^2 - E_2^2}{E_1^2 + E_2^2}\; \frac{E_1 \operatorname{sech}\theta_1 + E_2 \operatorname{sech}\theta_2}{1 - \frac{2E_1E_2}{E_1^2+E_2^2}(\tanh\theta_1 \tanh\theta_2 - \operatorname{sech}\theta_1 \operatorname{sech}\theta_2)}$$

E_1, E_2 = free parameters (real); θ_1, θ_2 as θ_1 above.

2-soliton (0π)

$$\varepsilon(x,t) = 2E_o \operatorname{sech}\theta_R \left[\frac{\cos\theta_I - \gamma \sin\theta_I \tanh\theta_R}{1 + \gamma^2 \sin^2\theta_I \operatorname{sech}^2\theta_R}\right]$$

$$\gamma \equiv E_1 E_o^{-1}, \qquad \theta_R = \tfrac{1}{2}E_o(t - m_e x) + \delta_R, \qquad \theta_I = \tfrac{1}{2}E_1(t - m_c x) + \delta_c$$

E_o, E_1 = free parameters (real) .

Notice the existence of one-solitons (= solitary waves), two-solitons which break up only asymptotically into two sech pulses, and a two-soliton solution which does not break up. In optics this is called a 0π pulse. In other cases where similar pulses arise they have been called "breathers" (There is no breather solution of the KdV but there is one of the NLS, equation (II) for example, cf. [6]).

The one-soliton rotates the pseudo-spin of successive atoms by 2π (ground state $|0\rangle \to$ excited state $|s\rangle \to |0\rangle$). Physically the atom is stimulated to absorb once and to emit once leaving the atom in its ground state. This is done fast enough to beat spontaneous emission,[6] no energy is lost, and the medium is transparent to 2π pulses. The remarkable feature is that the whole system is kept coherently in phase throughout the passage of the pulse. This is true even when the system is inhomogeneously broadened.

In the case of the 0π pulse the pseudo-spin rotates from the ground state configuration but then rotates back again. No "genuine" 0π pulses of this type have been seen in SIT but the combination of a pair of pulses one phase delayed by π with respect to the other (which is equivalent to negative ε and the kink-anti-kink situation) shows enhanced transparency. The Fig. 13 shows an experimental arrangement used by E.L. Hahn and J.C. Diels on ruby [31]. The Fig. 14 shows an enhanced transparency [31].

The 0π solutions of the sine-Gordon equation act like solitons. They collide and pass through each other. Fig. 15 shows the break-up of a 4π pulse into two 2π sech pulses. Note the two equal time-areas of the small and large amplitude pulses. Fig. 16 shows the collision of two 2π pulses. Fig. 17 shows a 0π pulse. Fig. 18 shows the break-up of a pulse into two 0π's travelling at different speeds. Fig. 19 shows the collision of a 0π pulse (at the back) with a 2π pulse. The 0π overtakes the 2π and ultimately the 2π sech pulse is at the back.

---

[6] For this reason spontaneous emission (so-called homogeneous broadening) is omitted from (4.17).

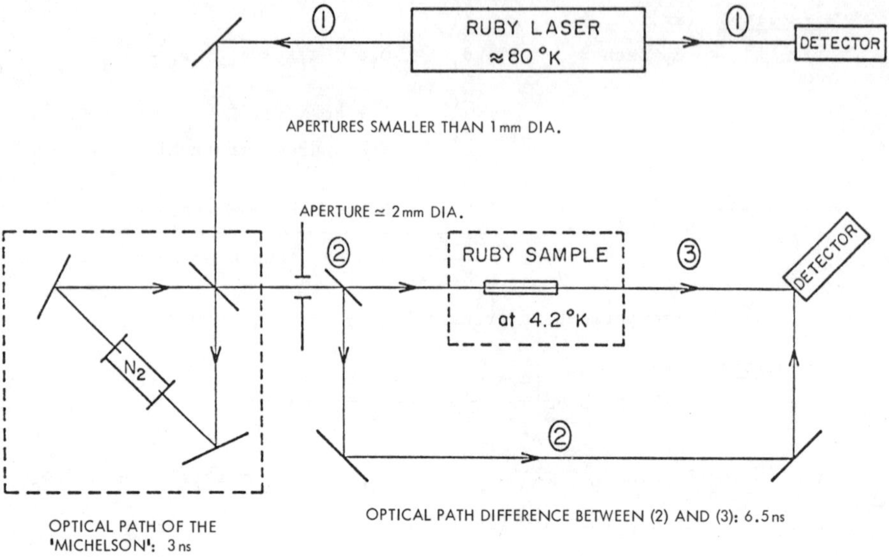

FIG.13. Experimental arrangement used by Diels and Hahn [31] to see enhanced transparency by kink anti-kink transmission in ruby.

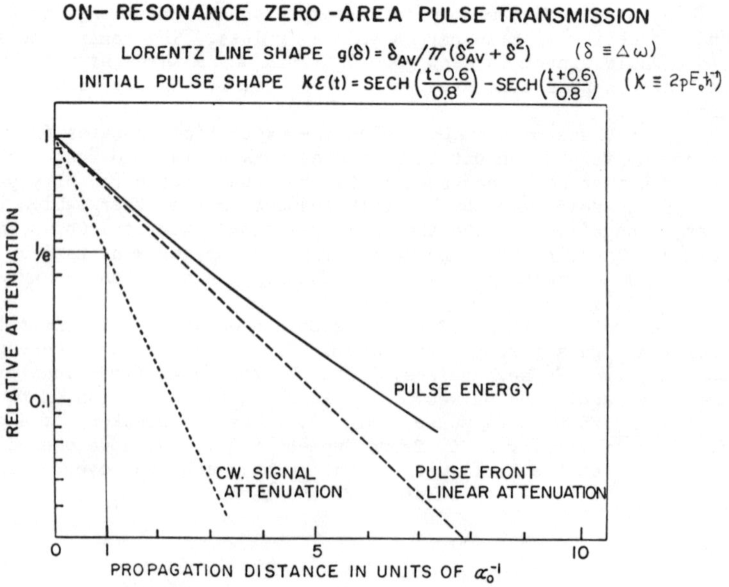

FIG.14. Enhanced transparency calculated by Diels and Hahn for Lorentzian inhomogeneous broadening [31].

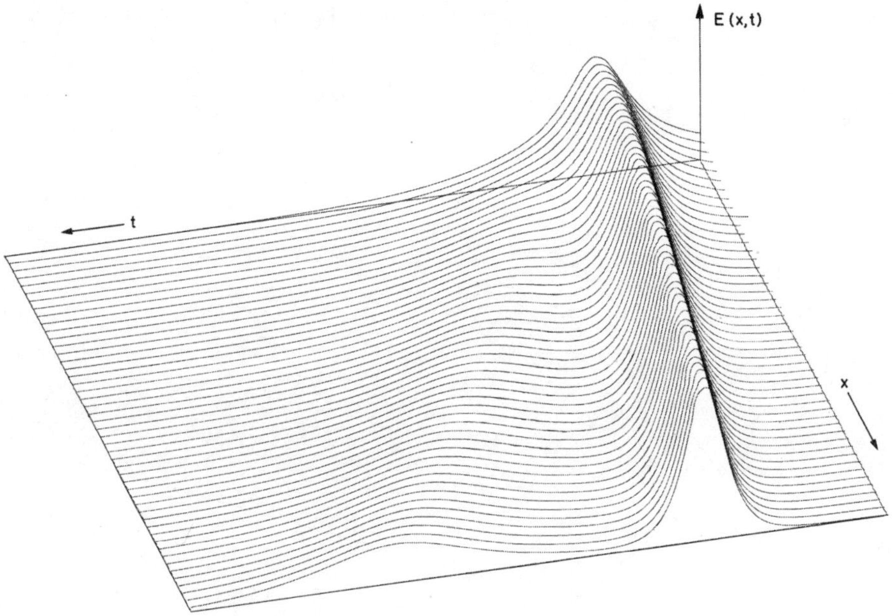

FIG.15. Break-up of a 4π pulse into two 2π sech pulses.

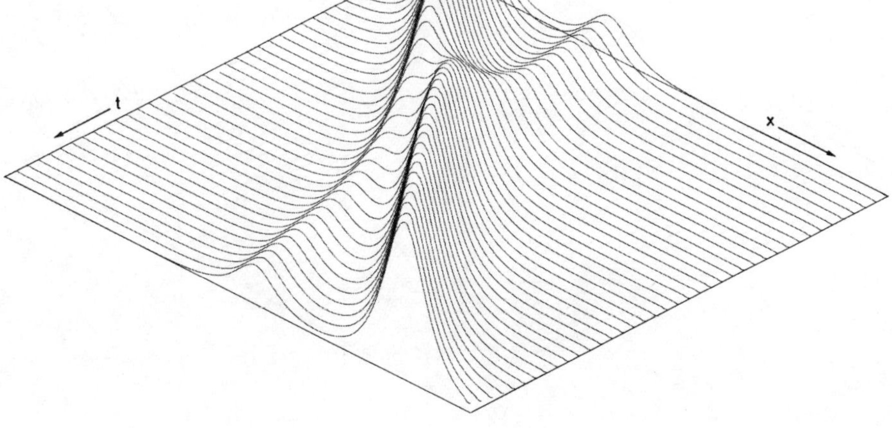

FIG.16. Collision of two 2π pulses.

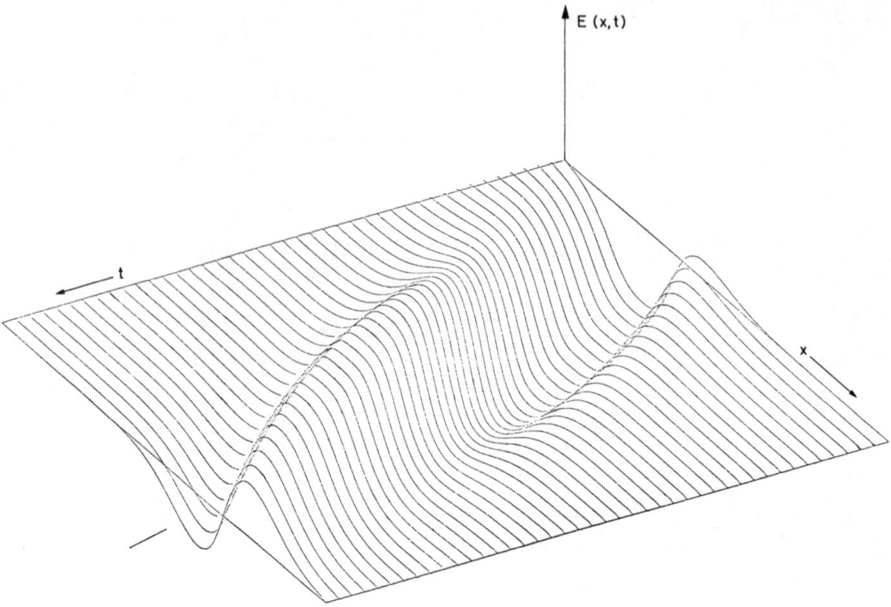

FIG.17. A 0π pulse.

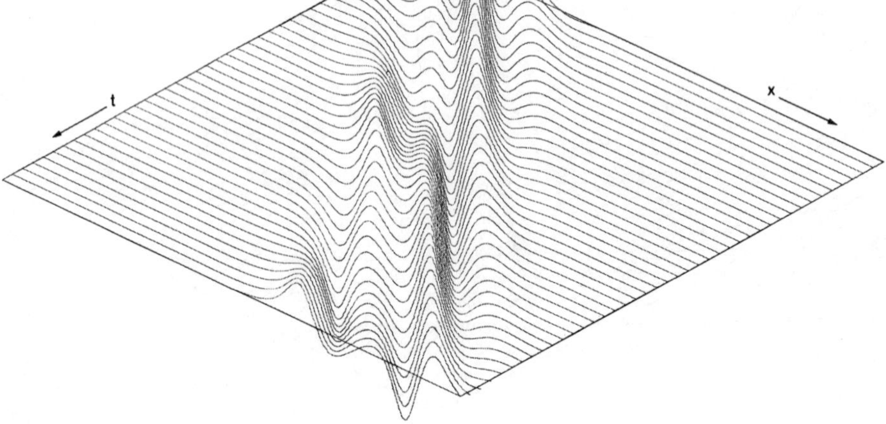

FIG.18. Break-up of an initial pulse into two 0π's.

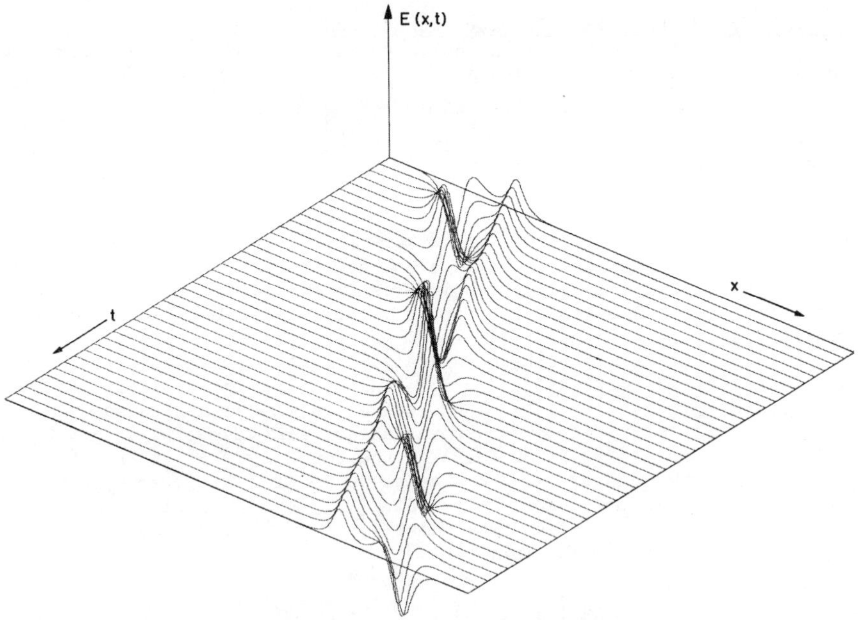

FIG.19. Collision of a 0π pulse and a 2π pulse.

We now know how to find the analytical N-soliton solution of the sine-Gordon equation for any integer N:

Note added in proof: We have notational difficulties! In (4.17) ϵ is the frequency indicated in the parenthesis there. Here ϵ is the field defined in (4.14) then scaled to the frequency called E; whilst β replacing α matches the work in §7. The essential point is that the equations as written have the N-soliton solution as written. The juxtaposition of "N-soliton" and N = N(x,t) here is also unfortunate, but these and the other usages are the usages of the published literature.

The equations $E_x + c^{-1} E_t = \frac{p\beta}{\hbar c} P$ $\quad (\beta = \hbar p^{-1}\alpha)$

$P_t = EN , \quad N_t = - EP \quad (E \equiv p\hbar^{-1}\varepsilon)$

have the N-soliton solution

$$E^2 = 4\frac{\partial^2}{\partial t^2} \ln f \quad , \quad f = \det ||M||$$

$$M_{ij} = \frac{2(E_i E_j)^{\frac{1}{2}}}{E_i + E_j} \{\exp \Theta_i + (-1)^{i+j} \exp -\Theta_j\}$$

$$\Theta_i = \omega_i t - k_i x + \delta_i \quad , \quad \omega_i = \tfrac{1}{2} E_i \quad , \quad \frac{ck_i}{\omega_i} = 1 + \frac{4p\beta}{\hbar E_i^2}$$

The best way of finding this solution is through the inverse scattering method [21, 22, 23, 24]. It was found in Manchester by a different method [32, 33, 34] . J.C. Eilbeck has made a short piece of film[7] which shows the time evolution of some of these analytical multi-soliton solutions. The film also shows the time evolution of certain multi-soliton solutions of the so called Reduced Maxwell-Bloch equations (RMB equations). We look briefly at these.

---

[7] This film can now be purchased with sound track either in English or French. Apply to the author. The cost is about £25 (1976).

6. The Reduced Maxwell-Bloch equations

The RMB equations take the Bloch equations (4.8) with the reduced (one way going) Maxwell equation

$$E_x + c^{-1} E_t = - 2\pi n c^{-1} p u_t . \qquad (6.1)$$

In the linear case the dipole pu at frequency ω is $\alpha(\omega) E_\omega$ so that (6.1) has dispersion relation

$$k - \omega c^{-1} = 2\pi n \omega c^{-1} \alpha(\omega) . \qquad (6.2)$$

The (linear) atomic polarisability is $\alpha(\omega)$.

It is clear that (6.2) is a low density result. Closer analysis[8] shows that (6.1) is also a low density result independent of the strength of the field E. Thus (6.1) applies to vapours irradiated by fields of arbitrary strength.

The RMB equations have an analytic multi-soliton solution. This is very similar in mathematical form to the multi-soliton solution of the sine-Gordon equation. We interchange space and time variables and then again use (x, t) for the switched variables: this allows us to get the linearised dispersion relation (6.2) in the form $\omega = f(k)$ where $f(k)$ is a rational function of k. This means we can use the inverse scattering method in the form due to Ablowitz et al. [23, 21]. With Bloch vector (r, s, u) and μ a constant (essentially the frequency), the RMB equations can be put in the form

$$E_t = s ; \qquad r_x = - \mu s$$
$$s_x = Eu + \mu r ; \qquad u_x = - Es . \qquad (6.3)$$

The N-soliton solution is then [24]

$$E^2 = 4 \frac{\partial^2}{\partial x^2} \ln \det \|M\|$$

$$M_{nm} = \frac{2}{E_m+E_n} \cosh \tfrac{1}{2} (\theta_n + \theta_m)$$

$$\theta_n = \tfrac{1}{2} E_n(x - 4[E_n^2 + 4\mu^2]^{-1} t + \delta_n) . \qquad (6.4)$$

Notice that when $\mu = 0$ equations (6.3) reduce to

$$E_t = s , \qquad s_x = Eu , \qquad u_x = - Es \qquad (6.5)$$

and these become the sine-Gordon equation

$$\sigma_{xt} = \sin \sigma$$

with $E \equiv \sigma_x$.

The last result is not surprising: if we choose $\omega_s = 0$ in the Bloch equations (4.8) we find

$$v_t = \frac{2p}{\hbar} E(x, t) w(x, t)$$

$$w_t = - \frac{2p}{\hbar} E(x, t) v(x, t) . \qquad (6.6)$$

Together with the reduced Maxwell equation (6.1) (with $u_t = -\omega_s v$ and the ω_s here absorbed into a constant α) equations (6.6) are of the same form as (4.19) with (4.20). Hence (6.1) and (6.6) are also equivalent to a sine-Gordon equation.

However, the physics is now very different from that in (4.19) and (4.20): there $\Delta\omega \equiv \omega_s - \omega$ and is certainly zero on resonance whereas in (4.8) ω_s is the optical transition frequency and can never be set equal to zero. Even so there is real physics in the RMB equations. The one-soliton solution corresponds to the solitary wave solution (4.13a) of the MB equations. In the case of the RMB equations this is a one-soliton

---

[8] See Ref. [35].

solution, not just a solitary wave. However it is essentially a d.c. field: there is no carrier wave to justify the 2-level atom model.

The physics lies in the oscillatory solutions of "breather" type we call bions. These look like 0π solutions of the sine-Gordon equation. There is one free parameter like E_0 determining an amplitude and a speed of a field envelope. And there is a second free parameter E_1 which determines a frequency ω of oscillatory trigonometric terms which act like carrier waves. The trick is to choose E_0 for time scales of 10^{-9} s and E_1 for 10^{-15} s [28]. The bion solution then behaves very like [28]

$$E_0 \operatorname{sech}\left\{\tfrac{1}{2}p\hbar^{-1} E_0(t - \tfrac{x}{v})\right\} \cos(\omega t - kx) \; ; \qquad (6.7)$$

this is the McCall-Hahn [1] 2π pulse! Moreover any number of such bions can collide and pass through each other and they need not have the same carrier wave frequencies. Thus via the RMB equations we find a multi-soliton solution which will describe resonant or non-resonant SIT applicable to arbitrary intensities (given the validity of the 2-level atom model) and restricted only by the low density condition. G.L. Lamb [36] has found multi-soliton solutions of the SIT equations themselves applicable both on and off resonance. Off resonance one cannot discard the slowly varying phase ϕ and the pulses "chirp". However the chirp largely represents the fact that the pulses (6.4) have different carrier waves even if they all have the same carrier frequency. This is because k and ω satisfy a dispersion relation which depends on E_0 [28].

The six sections just completed constitute both an introduction to solitons and an introduction to SIT. In the next two sections we choose one particular problem in the theory of non-linear optics for investigation in greater depth. This concerns how degeneracies in the resonant atomic energy levels affect the theory of SIT.

SOLITONS IN NON-LINEAR OPTICS

7. Degenerate SIT - the multiple sine-Gordon equations

I turn to the problem of 2-level atoms which have level degeneracies. Later we look at experiments done by G.J. Salamo, H.M. Gibbs and G.G. Churchill [37] on Na vapour which illustrate some of the rather remarkable results.

Consider the level scheme labelled by angular momentum quantum numbers J, m_J. We take $J = 2$ as example. The Fig. 20 displays the situation. The levels are supposed strictly degenerate and the selection rule imposed is $\Delta J = 0$. For plane polarised light $\Delta m_J = 0$.

In molecular rotation spectroscopy these are the Q-branch transitions. However J is an angular momentum and could be F, the hyperfine level angular momentum in an atom.

The key result obtainable through Clebsch-Gordan coefficient theory is that the matrix elements for the transitions labelled by m_J have matrix element $\frac{m_J}{J}p$ where p is the largest matrix element. In the case $J = 2$ the matrix elements are p, $\frac{1}{2}p$, 0, $-\frac{1}{2}p$, $-p$. As we shall see the ratio $2:1$ of $p:\frac{1}{2}p$ plays a major role in the theory.

Since the atoms are 10-level atoms we are concerned with an SU_{10} Lie algebra. However the selection rules reduce this to the direct sum of 5 SU_2 Lie algebras.[9] This is a 5 spin system in spin language. We therefore expect 5 Bloch equations for each atom: these are independent except only that they are all coupled one to the other via the interatomic field.

We can expect that the sign of the matrix element is unimportant. We therefore expect 2 spins with matrix element p, 2 with matrix element $\frac{1}{2}p$, one with matrix element zero. A zero matrix element means no transitions. We

[9] Without this simplification there are 45 raising and 45 lowering operators as well as 9 independent number operators!

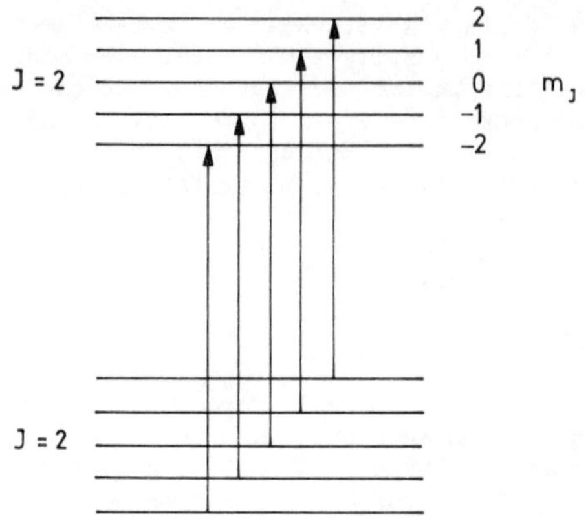

FIG.20. Level scheme for transitions between degenerate levels with selection rules $\Delta J = 0$, $\Delta m_J = 0$ when $J = 2$.

are left then with only two spins with matrix elements p, $\frac{1}{2}p$. This expectation is borne out.

We consider the case of sharpline degenerate SIT with Q(2) symmetry. We assume exact resonance. We therefore set

$$E(x, t) = \varepsilon(x, t) \cos(\omega_s t - k_s x) \tag{7.1}$$

for the electric field and choose;

$$np\left\{\sum_{i=1}^{5} g_i \lambda_i Q^{(i)}(x, t) \cos(\omega_s t - k_s x) + \sum_{i=1}^{5} g_i \lambda_i P^{(i)}(x, t) \sin(\omega_s t - k_s x)\right\} \tag{7.2}$$

for the dipole density. By definition $\lambda_i = p_i/p$ where $\{p_i\} = \{-p, -\frac{1}{2}p, 0, \frac{1}{2}p, p\}$. The g_i are relative occupation numbers of the initial states. It is supposed that the ground state degeneracy can be Zeeman split[10] so that the initial state occupations can be changed. In most of the discussion we assume equal state occupations: $g_i = \frac{1}{5}$ for all i. When we consider the experimental work on Na vapour we need to vary the g_i because we become concerned with "spins" in different atoms.

On exact resonance without inhomogeneous broadening

$$\varepsilon_x + c^{-1}\varepsilon_t = c^{-1}\beta(P^{(1)} + \tfrac{1}{2}\lambda P^{(2)}) \qquad (\beta = \hbar p^{-1}\alpha)$$

$$P_t^{(1)} = \frac{p}{\hbar}\varepsilon N^{(2)}; \qquad N_t^{(1)} = \frac{-p}{\hbar}\varepsilon P^{(1)}$$

$$P_t^{(2)} = \frac{\frac{1}{2}p}{\hbar}\varepsilon N^{(2)}; \qquad N_t^{(2)} = \frac{-\frac{1}{2}p}{h}\varepsilon P^{(2)} \tag{7.3}$$

where

$$\tfrac{1}{2}\lambda = g_2\lambda_2/g_1\lambda_1 = \tfrac{1}{2}\, g_2/g_1 .$$

(The $P^{(i)}$, $N^{(i)}$ with $i = 1, 2$ are relabelled in this index i in an obvious way.)[11] As in the non-degenerate case set

[10] There is no first order Stark shift which requires degeneracies with $\Delta J = \pm 1$.

[11] For example $P^{(5)} = -P^{(1)}$ in (7.2) and $P^{(5)}$ in (7.2) becomes $P^{(1)}$ in (7.3).

$$\sigma = p\hbar^{-1} \int_{-\infty}^{t} \varepsilon(x, t')\, dt' \qquad (4.7)$$

$$P^{(1)} = -\sin\sigma\,, \qquad P^{(2)} = -\sin\tfrac{1}{2}\sigma$$

$$N^{(1)} = -\cos\sigma\,, \qquad N^{(2)} = -\cos\tfrac{1}{2}\sigma\,. \qquad (7.5)$$

Set

$$\xi = \sqrt{\frac{p\beta}{\hbar c}}\,(ct - 2x)\,, \qquad \tau = \sqrt{\frac{p\beta}{\hbar c}}\,ct\,.$$

Then

$$\varepsilon_x + c^{-1}\varepsilon_t = c^{-1}\beta(P^{(1)} + \tfrac{1}{2}\lambda P^{(2)})$$

means

$$\frac{\partial^2\sigma}{\partial\xi^2} - \frac{\partial^2\sigma}{\partial\tau^2} = \sin\sigma + \tfrac{1}{2}\lambda\sin\tfrac{1}{2}\sigma \qquad (7.6a)$$

with

$$\varepsilon(x, t) = \frac{\hbar}{p}\frac{\partial\sigma}{\partial t} = \sqrt{\frac{\beta c\hbar}{p}}\left[\frac{\partial\sigma}{\partial\xi} + \frac{\partial\sigma}{\partial\tau}\right]. \qquad (7.6b)$$

Since λ is an adjustable parameter we reach the non-degenerate problem by setting $\lambda = 0$. In this case (7.6a) is the simple sine-Gordon equation treated in §4. We recall there are two cases

$$\sigma_{\xi\xi} - \sigma_{\tau\tau} = \pm\sin\sigma\,. \qquad (7.7)$$

The positive sign is the attenuator; the negative sign is the amplifier.

The multi-soliton solution of (7.7) with positive sign is described in §4. Unfortunately none of the analytical methods for finding multi-soliton solutions has proved successful on equation (7.6). For example one can prove [38] that the inverse scattering method in the form due to Ablowitz et al. [21] will not work. Indeed it is an open question whether (7.6) has multi-soliton solutions or not. However the numerical integration of (7.6) suggests that some rather unusual multi-soliton solutions exist.

To understand these we look for the solitary wave solutions of (7.6): these will presumably be 1-solitons if there are multi-soliton solutions. To find these solitary waves we use a two-dimensional phase space, the phase plane. To see how this can be used we take the case $\lambda = 0$ in (7.6) namely the simple sine-Gordon (7.7).

The elegant route to distortionless solutions exploits the Lorentz invariance of (7.7):

$$\xi^2 - \tau^2 = \text{invariant} = \xi'^2 - \tau'^2\,.$$

For distortionless solutions $\sigma = \sigma(\xi')$ where

$$\left.\begin{aligned} \xi' &= \frac{\xi - V\tau}{\sqrt{1 - V^2}} \\ \tau' &= \frac{\tau - \xi V}{\sqrt{1 - V^2}} \end{aligned}\right\} V < 1 \quad \text{or} \quad \left.\begin{aligned} \xi' &= \frac{V\tau - \xi}{\sqrt{V^2 - 1}} \\ \tau' &= \frac{V\xi - \tau}{\sqrt{V^2 - 1}} \end{aligned}\right\} V > 1\,. \qquad (7.8)$$

The two conditions $V < 1$, $V > 1$ in dimensionless units are $V < c$, $V > c$ in physical units. Pulse velocities are group velocities so the case $V > c$ needs discussion.

When $V = 0$ equation (7.7) becomes the pendulum equation

$$\frac{d^2\sigma}{d\xi^2} = \pm\sin\sigma\,. \qquad (7.9)$$

A first integral is obtained by multiplying by σ_ξ so that

$$\tfrac{1}{2}\sigma_\xi^{\,2} = \mp\cos\sigma + \text{constant.}$$

The "energy" is $\frac{1}{2}\sigma_\xi^{\,2} \pm \cos\sigma = \text{constant}$. The phase plane plots σ_ξ against σ. The two cases resulting from (7.9) are shown in Fig. 21 (case + sign in (7.9) is uppermost).

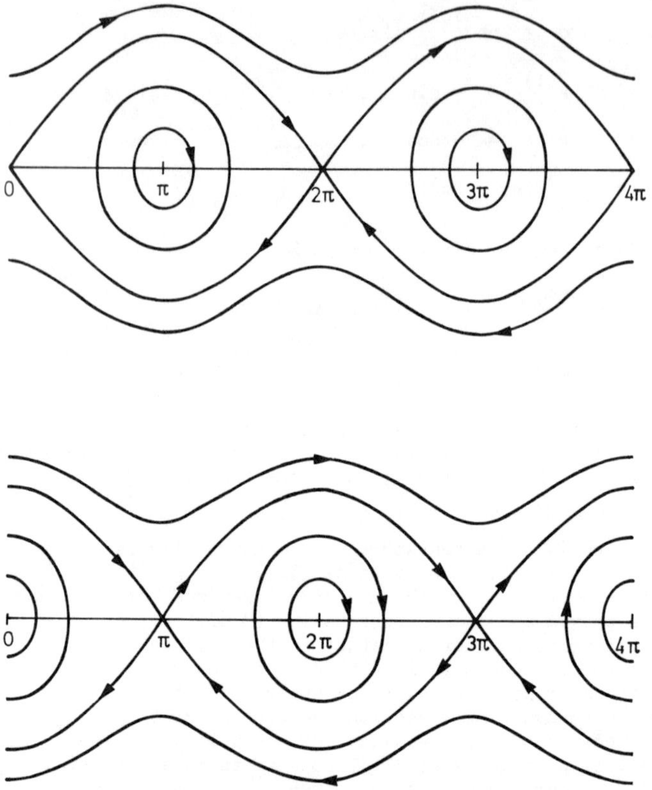

FIG.21. Phase planes for the sine-Gordon equation $\sigma_{\xi\xi} - \sigma_{\tau\tau} = \pm \sin\sigma$. Note the roles of the zeros 0, π of sin σ in the two cases.

The important points are the zeros of sin σ. In the upper figure the zeros 0, 2π are "unstable" singular points: the zero at π is a "stable" singular point. The trajectory $0 \to 2\pi$ is the trajectory of the sech solution (a pendulum standing on end and just displaced from that unstable equilibrium). The conclusion is that when $V = 0$ there is a sech pulse solution in the attenuator. Then by Lorentz transformation there is a sech pulse with $V < 1$ (i.e. $V < c$) in the attenuator.

The amplifier demands the negative sign in (7.9). The singular points are now π, 3π (unstable) and 0, 2π (stable). There is no trajectory from $\sigma = 0$ ($\sigma = 0$ means $N = +1$ since $N = +\cos\sigma$, $P = +\sin\sigma$ in this case). Hence no sech pulse with $V < c$ travels in the amplifier.

However for $V = \infty$ we find

$$\frac{d^2\sigma}{d\tau^2} = \mp \sin\sigma. \qquad (7.10)$$

The signs are opposite to (7.9). The amplifier now has positive sign, its phase plane is the upper one, and the conclusion is that sech pulses travel in amplifiers with speeds $V > c$. This difficulty is resolved by noticing that the sech pulse has an infinite front and tail and cannot be created in the laboratory: the physical pulse piles up on the front travelling at $V = c$, is not distortionless, and evolves [39, 40].

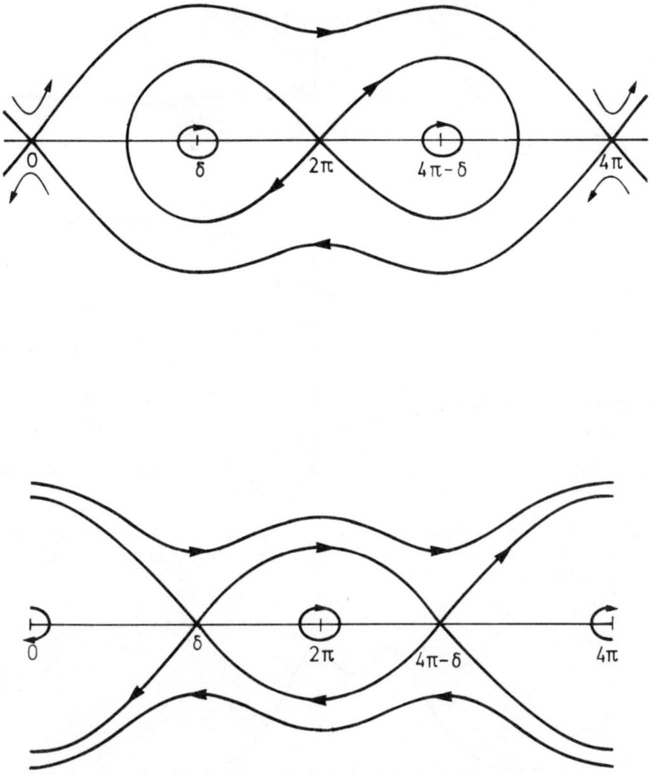

FIG.22. Phase planes for the double sine-Gordon equation $\sigma_{\xi\xi} - \sigma_{\tau\tau} = \pm (\sin\sigma + \frac{1}{2}\sin\frac{1}{2}\sigma)$. The zeros of the right-hand side occur at 0, δ, 2π and 4π−δ where $\delta = 2\cos^{-1}(-\frac{1}{4})$. There are 4 amplifiers and 4 attenuators.

The phase planes for equation (7.6) which we shall now call the "double sine-Gordon" equation are more complicated because there are 4 zeros of $\sin\sigma + \frac{1}{2}\sin\frac{1}{2}\sigma$ (we set $\lambda = 1$) (see Fig. 22). There will be 4 attenuators and 4 amplifiers! These can be characterised by the two spins: we use ↓(↑) for spin with matrix element p down (up); we use ↡(↟) for spin with matrix element $\frac{1}{2}$p down (up).

The "full" attenuator is[12] ↓↡ ($\sigma = 0$)
The "half" attenuator is[12] ↓↟ ($\sigma = 2\pi$)
The "full" amplifier is[13] ↑↟ ($\sigma = 0$)
The "half" amplifier is[13] ↑↡ ($\sigma = 2\pi$)

Note that the phase plane plots σ_ξ (or σ_τ) against σ. Hence the trajectory is rather like the shape of the pulse.

We conclude the following: a double humped 4π pulse with $V < c$ ($V > c$) is the solitary wave solution of the double sine-Gordon equation in the "full" attenuator (amplifier). A zero-π pulse with $V < c$ ($V > c$) is

[12] For these attenuators $N^{(1)} = -\cos\sigma$, $P^{(1)} = -\sin\sigma$;
$N^{(2)} = -\cos\frac{1}{2}\sigma$, $P^{(2)} = -\sin\frac{1}{2}\sigma$.

[13] For these amplifiers $N^{(1)} = +\cos\sigma$, $P^{(1)} = +\sin\sigma$;
$N^{(2)} = +\cos\frac{1}{2}\sigma$, $P^{(2)} = +\sin\frac{1}{2}\sigma$.

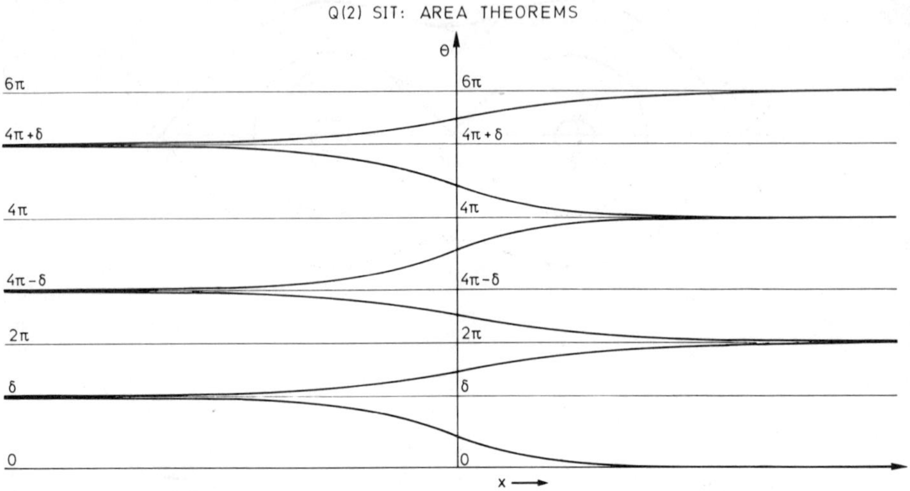

FIG.23. The eight area theorems for the double sine-Gordon equation $\sigma_{\xi\xi} - \sigma_{\tau\tau} = \pm (\sin \sigma + \frac{1}{2} \sin \frac{1}{2} \sigma)$. The $\frac{1}{2}$- and full-amplifiers are read backwards from $\theta = 0$ and 2π and from 2π and 4π, respectively. The $\frac{1}{4}$- and $\frac{3}{4}$-attenuators are read backwards towards δ and $4\pi - \delta$.

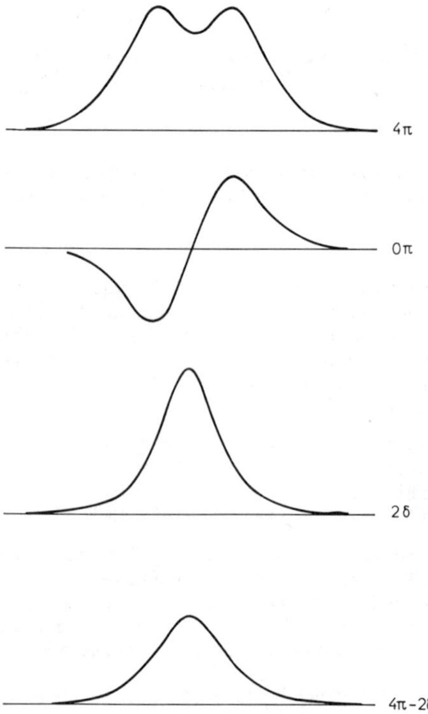

FIG.24. The four distortionless pulses of the double sine-Gordon equation. These travel in the full- and $\frac{1}{2}$-attenuators and the $\frac{1}{4}$- and $\frac{3}{4}$-attenuators, respectively.

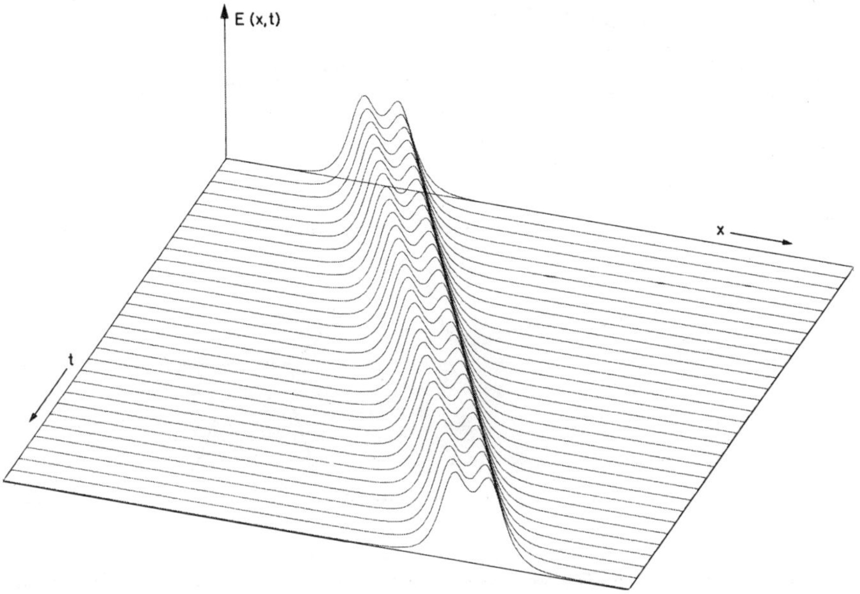

FIG.25. Three-dimensional plot of the distortionless double-humped 4π pulse which travels in the full attenuator.

the solitary wave solution in the "half" attenuator (amplifier). Notice that the excited states (half and full amplifier, half attenuator) are equilibrium states only because spontaneous emission is neglected. Hence they have a lifetime of nano-seconds only.

The places δ and $4\pi-\delta$ ($\delta = 2\cos^{-1}-\frac{1}{4}$) are also zeros of $\sin\sigma + \frac{1}{2}\sin\frac{1}{2}\sigma$. They also represent equilibrium dielectrics since the dipole moment vanishes there and the system cannot radiate. The configuration of the two spins in these equilibrium states is neither up nor down. The second phase plane shows trajectories $\delta \to 4\pi-\delta$ and $4\pi-\delta \to \delta$. These represent single humped pulses of areas $4\pi-2\delta$ and 2δ which propagate through the "quarter attenuator" ($\sigma = \delta$) and "three-quarter attenuator" ($\sigma = 4\pi-\delta$) with speeds $V < c$. For example the 2δ pulse converts the three-quarter attenuator to the quarter attenuator but these are of equal energy so that this is an acceptable pulse (again the lifetime of the quarter, three-quarter states $\sim 10^{-9}$ s because of spontaneous emission). Finally it is now clear that similar pulses are the solitary wave solutions for the quarter and three-quarter amplifiers: the speeds will be $V > c$.

There are 8 area theorems for this dielectric! For example in the full attenuator pulses with areas $\Theta < \delta$ attenuate; if $\delta < \Theta < 4\pi-\delta$ the area $\Theta \to 2\pi$; if $4\pi-\delta < \Theta < 4\pi+\delta$, $\Theta \to 4\pi$. In the quarter attenuator $\Theta < 2\pi-\delta$ attenuates ; if $2\pi-\delta < \Theta < 4\pi-\delta$ then $\Theta \to 4\pi-2\delta$. In the quarter amplifier if $\Theta < 2\pi-\delta$ then $\Theta \to 2\pi-\delta$; similarly if $0 < \Theta < 4\pi-2\delta$, $\Theta \to 2\pi-\delta$, and if $4\pi-2\delta < \Theta < 4\pi$ then $\Theta \to 4\pi-\delta$. The Fig. 23 shows the eight area theorems on a single plot. The half and full amplifiers are read backwards; the quarter and three-quarter attenuators are read backwards.

The equations for the phase plane trajectories of the 4π, 0π, 2δ and $4\pi-2\delta$ pulses can be integrated. The shapes are shown in the Fig. 24 and are very much as expected. The analytical expressions in dimensionless units are shown in the Table 1.

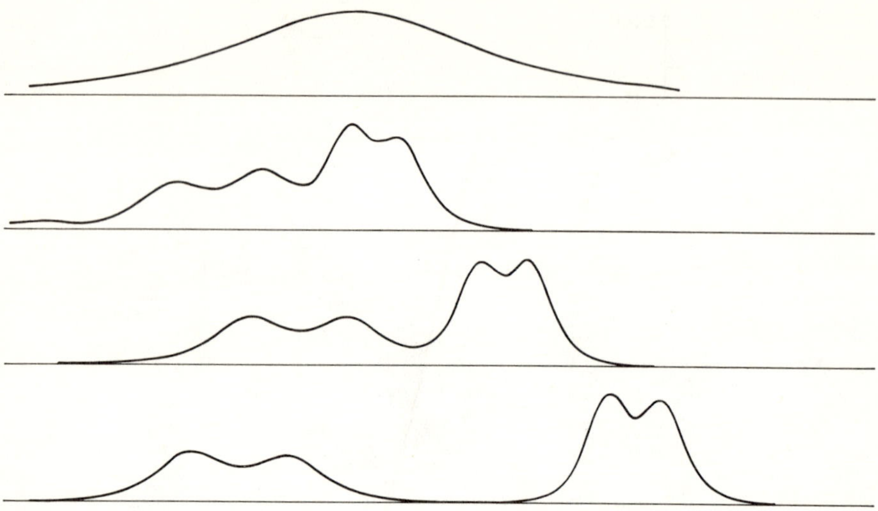

FIG.26. Break-up of a 9π sech pulse into two 4π double-humped pulses. Note the small asymmetry in the leading pulse in the last frame.

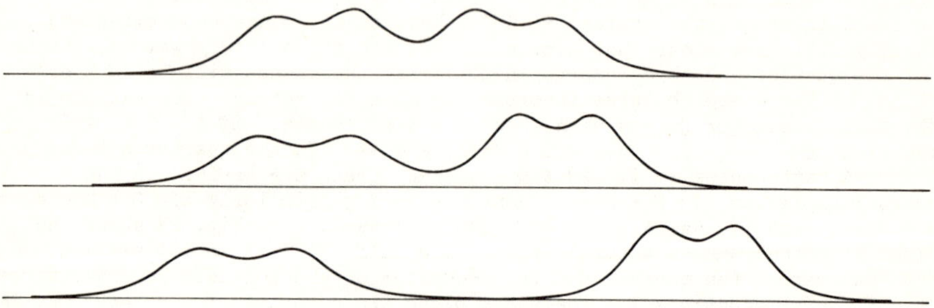

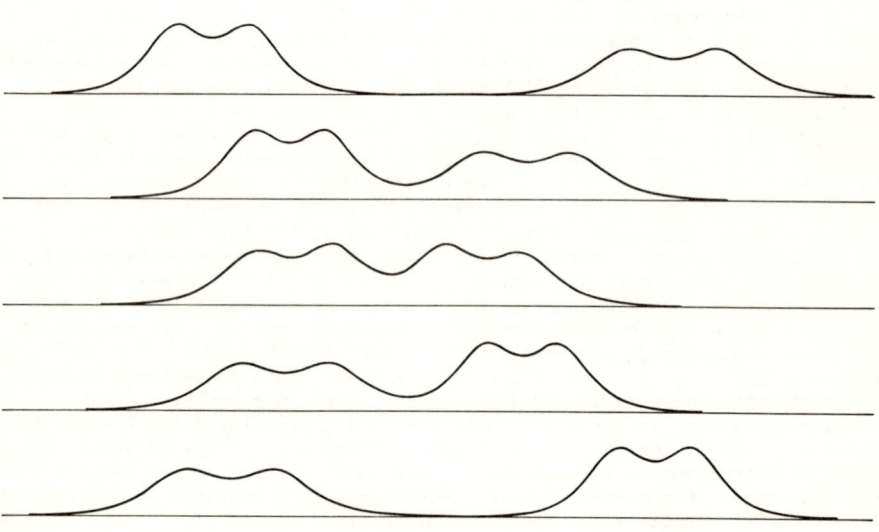

FIG.27. Collision of two double-humped 4π pulses.

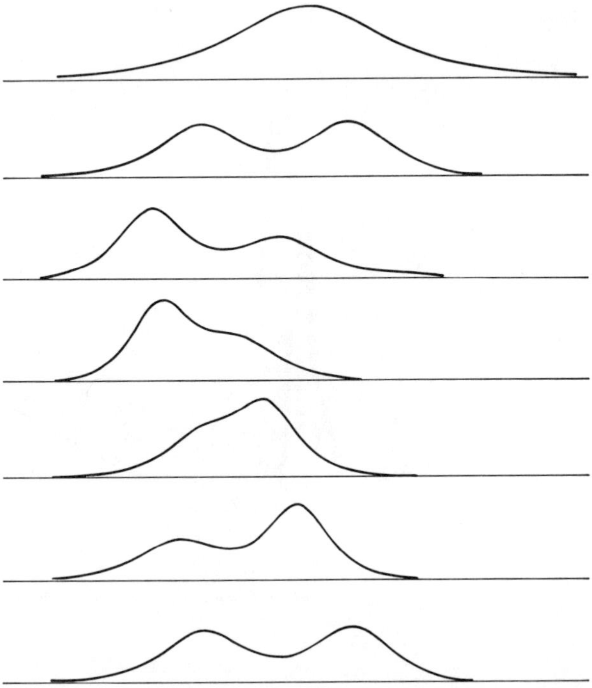

FIG.28. Evolution of a 3.9π sech into a 4π wobbling double-humped pulse.

The Fig. 25 shows the propagation of the 4π double humped pulse through the full attenuator. The following Fig. 26 shows the break-up of an $8-9\pi$ pulse into two 4π double humped pulses in the full attenuator. This is a numerical result and suggests that the 4π-pulses behave like solitons. The next figure, Fig. 27, shows the collision of two 4π pulses.

However the situation is complicated. First note the small asymmetry in the leading 4π pulse in Fig. 26. The following figure, Fig. 28, shows the evolution of a 3.9π pulse towards a double humped 4π pulse. The two humps plainly oscillate! The next two, Figs. 29 and 30, show this oscillation in three dimensions. The second of these figures shows a wobble which takes the pulse through the initial pulse shape.

The situation can be understood by observing that the 4π and 0π solitary wave solutions can be written identically in the following way:

$$4\pi : \quad \varepsilon = 2\omega\left\{\operatorname{sech}(\theta + \Delta) + \operatorname{sech}(\theta - \Delta)\right\} \qquad \cosh^2\Delta = 5$$

$$0\pi : \quad \varepsilon = 2\omega\left\{\operatorname{sech}(\theta + \Delta) - \operatorname{sech}(\theta - \Delta)\right\} \qquad \cosh^2\Delta = 4 .$$

The 4π is a "bound pair" of 2π sech pulses; the 0π is also a bound pair of 2π sech pulses but in the "kink anti-kink" pairing rather than "kink-kink" pairing.

These results suggest that by varying the spacing 2Δ in the pulse input the pulse which propagates will oscillate or "wobble". The Figs. 31a, b show how an initial spacing of 4Δ produces a wobbler.

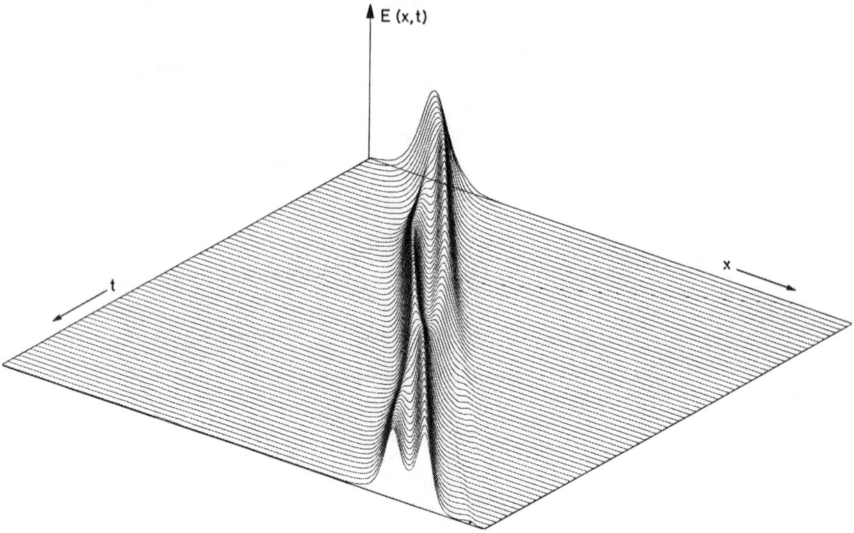

FIG.29. Evolution of a 4π sech pulse into a 4π double-humped wobbler.

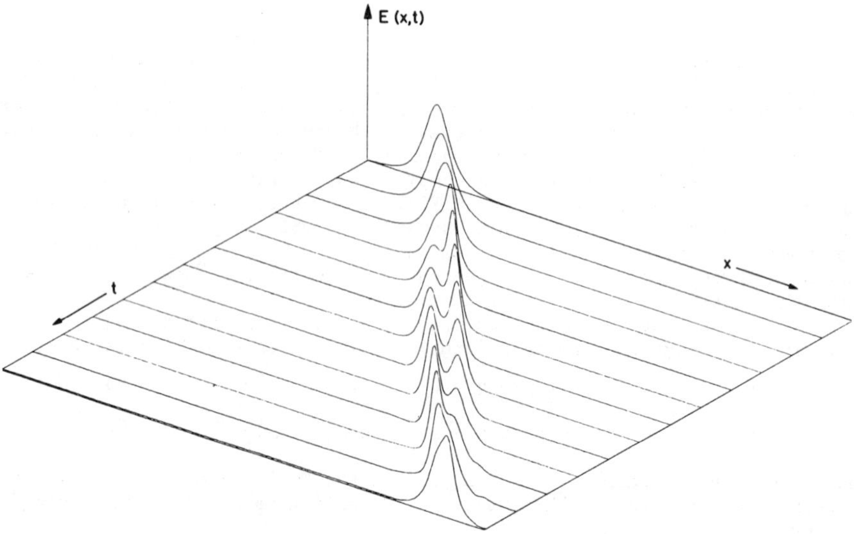

FIG.30. The 4π sech pulse becomes a wobbling two-humped pulse and later wobbles back to roughly the sech form.

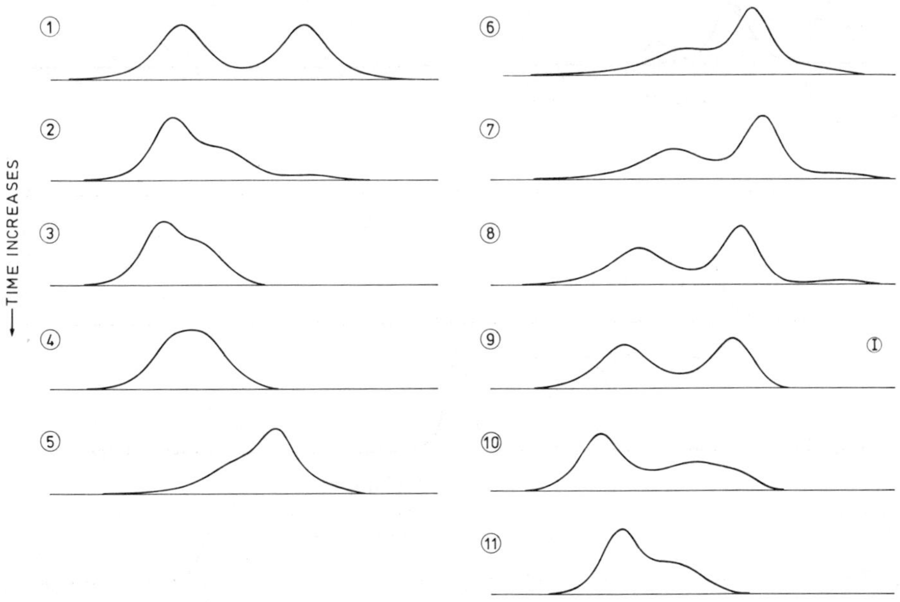

FIG.31a,b. Successive frames showing two 2π sech pulses with spacing 4Δ evolving into a wobbler via a single peaked pulse. Note the initial shape re-emerging in the frame labelled (*I*).

TABLE I. ANALYTICAL EXPRESSIONS FOR ϵ IN DIMENSIONLESS UNITS ($\theta \equiv \omega(t - xV^{-1}) + \delta_0$

Equilibrium dielectric	Field envelope ε	Area Θ	Velocity
att. → att. (amp. → amp.)	$\frac{4\sqrt{5}\omega \operatorname{sech}\Theta}{1 + 4\operatorname{sech}^2\Theta}$	4π	$\frac{c}{V} = 1 \pm \frac{5c\alpha}{4\omega^2}$
$\frac{1}{2}$ att. → $\frac{1}{2}$ att. ($\frac{1}{2}$ amp. → $\frac{1}{2}$ amp.)	$\frac{4\sqrt{3}\omega \sinh\Theta}{3 + \cosh^2\Theta}$	0π	$\frac{c}{V} = 1 \pm \frac{3c\alpha}{4\omega^2}$
$\frac{1}{4}$ att. → $\frac{3}{4}$ att. ($\frac{1}{4}$ amp. → $\frac{3}{4}$ amp.)	$\frac{4\sqrt{15}\omega}{3\sinh^2\Theta + 5\cosh^2\Theta}$	$4\pi - 2\delta$	$\frac{c}{V} = 1 \pm \frac{15c\alpha}{16\omega^2}$
$\frac{3}{4}$ att. → $\frac{1}{4}$ att. ($\frac{3}{4}$ amp. → $\frac{1}{4}$ amp.)	$\frac{4\sqrt{15}\omega}{3\cosh^2\Theta + 5\sinh^2\Theta}$	2δ	$\frac{c}{V} = 1 \pm \frac{15c\alpha}{16\omega^2}$

The most remarkable situation occurs when Δ is large. In this case the 4π pulse entering the attenuator consists of two well separated 2π pulses. The first of these enters the attenuator and will rotate the spin with element p right round (through 2π). But the Rabi frequency[14] for the other spin with matrix element $\frac{1}{2}$p is one half of that for p. This spin is rotated half way round (through π). Thus the first pulse converts the

[14] Rabi frequencies are defined from matrix elements p and fields ϵ by $p\epsilon\hbar^{-1}$.

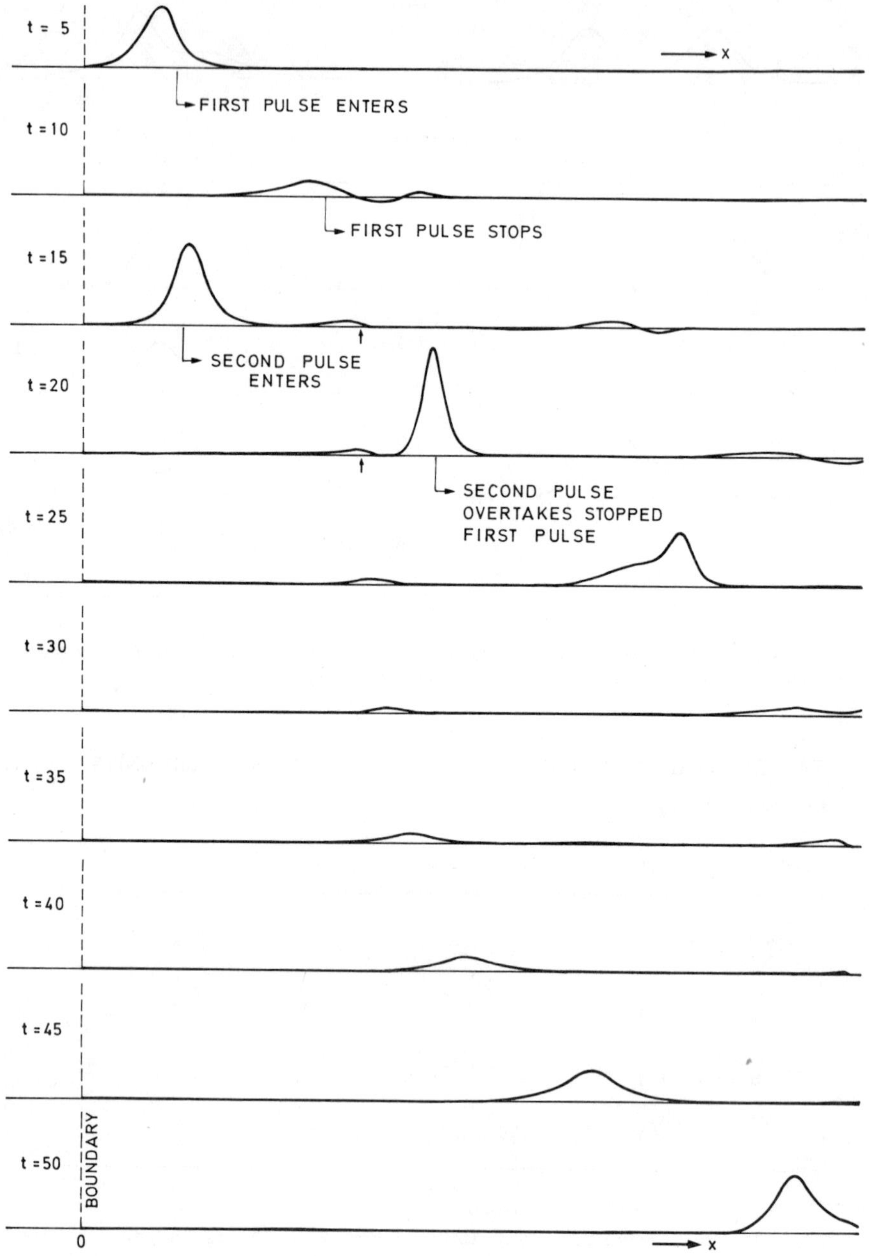

FIG.32. Two well-separated 2π sech pulses entering the full-attenuator. Each frame, taken after uniform intervals of time t (in arbitrary units) shows the spatial profile of the two pulses as they enter the Q(2) attenuator medium from vacuo (boundary indicated by broken line). The arrows indicate the stopping point of the first pulse.

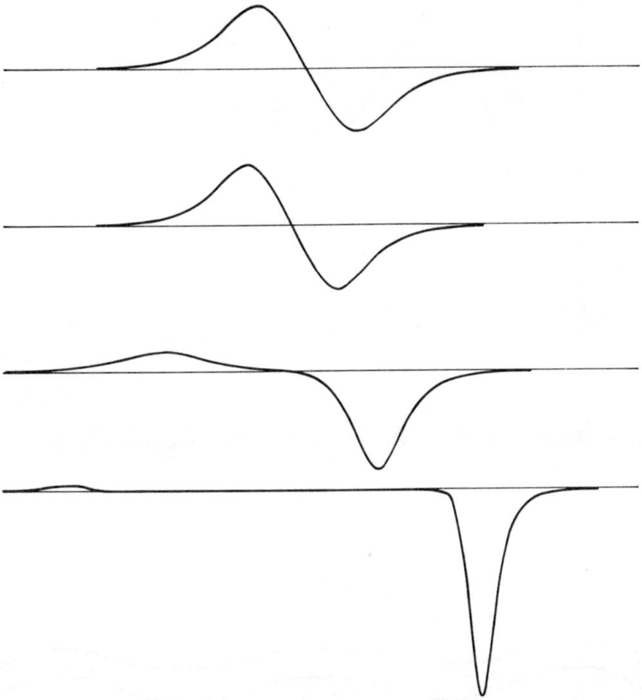

FIG.33. Instability of the 0π pulse in the ½-attenuator.

dielectric to the half attenuator (↓↟). This is an excited state: the pulse loses energy but conserves area (at 2π). It therefore distorts in shape, loses amplitude and spreads out. Its speed therefore drops and it slows down towards a stopping point. (This is an optical pulse being actually brought to rest!)

The second 2π pulse enters the ½-attenuating medium prepared by the first pulse: it restores the ½-attenuator to the full attenuator (↓↡) gains energy from the medium, spikes in amplitude at constant area 2π, and speeds up. It overtakes the stopped pulse.

It now sees before it the full attenuator (↓↡). It flattens at constant area, slows down and stops. The pulse previously stopped now sees before it the ½-attenuator. Its coherence information, spread over an area of 2π of essentially zero amplitude, is preserved. It speeds up and spikes at constant area and will overtake the stopped pulse in front of it. This "leap frog" behaviour (without spontaneous emission) will continue through the resonant medium. The Fig. 32 illustrates this remarkable behaviour.

We now "understand" why the 4π pulse is a <u>bound</u> pair of 2π pulses. The leading pulse tends to stop, the trailing pulse to accelerate, and in general the two pulses wobble through each other even when the peak spacing is small.

To confirm the analysis we consider the 0π solitary wave pulse in the ½-attenuator. It is the difference of two 2π sechs. The leading 2π sech accelerates and the trailing 2π stops: the pulse is therefore unstable. The Fig. 33 illustrates this instability.

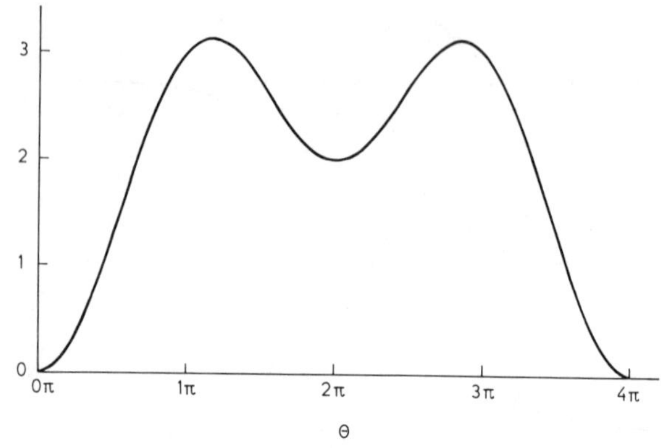

FIG.34. Form of potential in which the 2π and -2π pulses move in the $\frac{1}{2}$-attenuator.

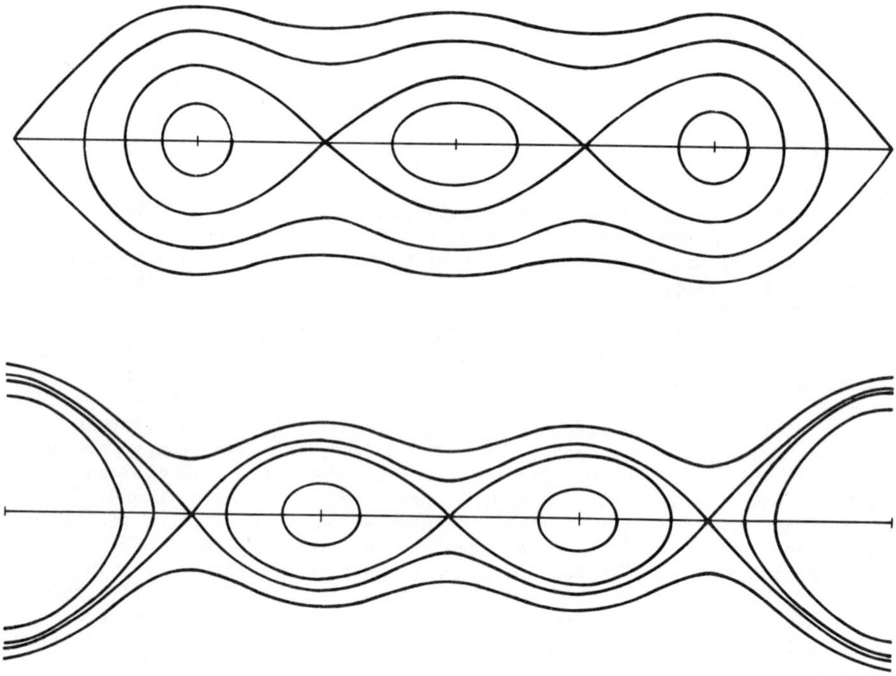

FIG.35. Phase planes for the triple sine-Gordon equation $\sigma_{\xi\xi} - \sigma_{\tau\tau} = \pm\, (\sin \sigma + \frac{1}{3} \sin \frac{1}{3} \sigma + \frac{2}{3} \sin \frac{2}{3} \sigma)$. Zeros occur at 0 and 3π, at $\sigma_1 = 3 \cos^{-1}\{\frac{1}{6} (-1 + \sqrt{7})\}$ and $6\pi - \sigma_1$, and at $\sigma_2 = 3 \cos^{-1} \{\frac{1}{6} (-1-\sqrt{7})\}$ and $6\pi - \sigma_2$.

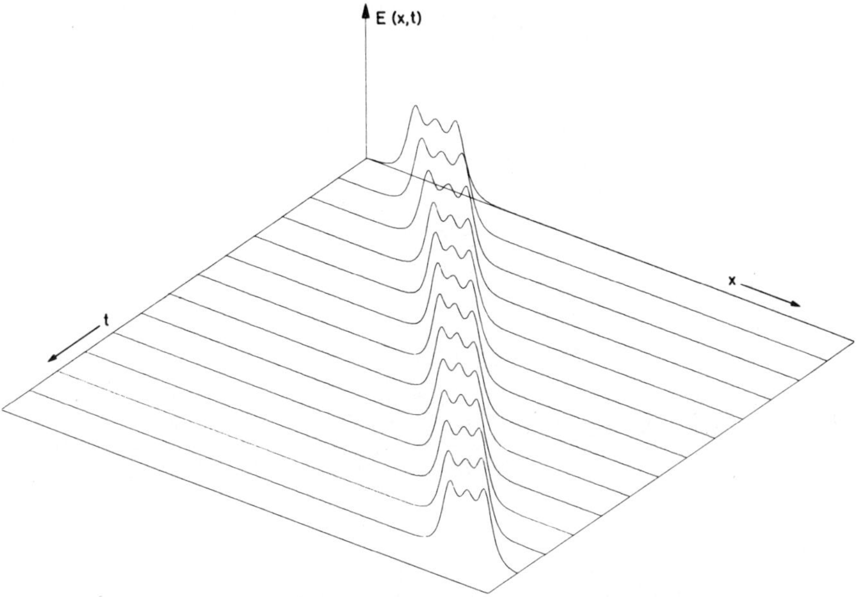

FIG.36. Propagation of a 3-peaked 6π pulse in the full-attenuator when J = 3.

However from the phase plane one sees that the trajectories also represent the energy left in the atoms. Consider the full attenuator described by

$$\frac{d^2\sigma}{d\xi^2} = \sin\sigma + \tfrac{1}{2}\sin\tfrac{1}{2}\sigma$$

with velocity $V = 0$. Then

$$\tfrac{1}{2}\sigma_\xi{}^2 = -\cos\sigma - \cos\tfrac{1}{2}\sigma + \text{constant}$$
$$= N^{(1)} + N^{(2)} + \text{constant} .$$

The energy in the atoms is $\frac{1}{2}\hbar\omega_s(N^{(1)} + N^{(2)})$. One can guess from this and the actual trajectories where $\sigma \sim 2\pi$ that the bound pair of 2π pulses move in a potential of the form shown in Fig. 34. For $\Delta < \Delta_0$ we might expect the 0π pulse to propagate without break-up in the $\frac{1}{2}$-attenuator. The numerical results [41] show it does (though it is certainly not clear why Δ does not drop to zero).

We conclude this section on the multiple sine-Gordon equations by considering the case when $J = 3$. The degeneracy $2J + 1 = 7$ and the problem is a 3-spin problem. The Fig. 35 shows the phase planes. One can deduce immediately that a 3-peaked 6π pulse propagates in the full attenuator (↓↓↓). The Fig. 36 shows it does. By varying the peak spacing wobblers are produced as the Fig. 37 shows. The next figure, Fig. 38, showing the evolution of a 6π sech shows that this becomes a 3-peaked wobbler. The Fig. 39 shows a 12π 3-peaked pulse breaking up into two 6π wobblers.

The remaining figure, Fig. 40, shows the single peaked pulse of area $2\sigma_1$ (see the phase plane) which propagates in the $\frac{1}{6}$ attenuator. This pulse "bumps" the double peaked pulse of area $6\pi - 2\sigma_1$ as described in a recent Letter [38].

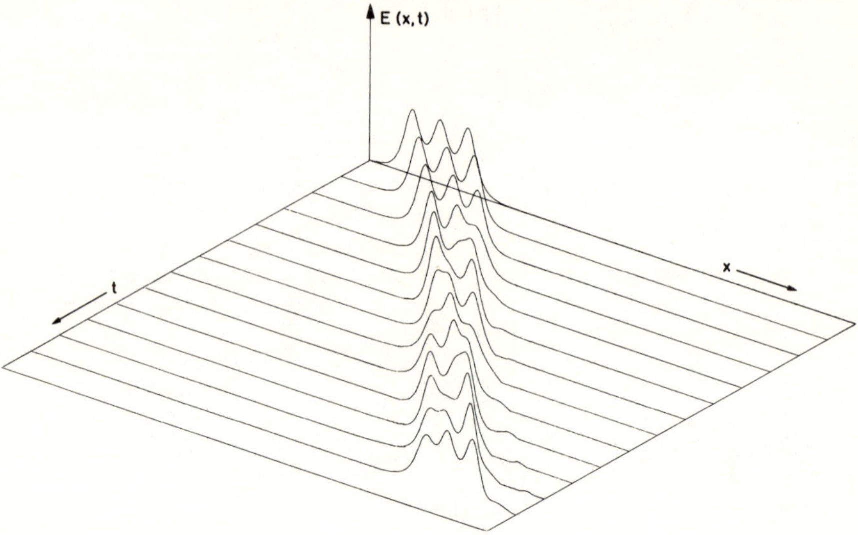

FIG.37. A 3-peaked wobbling 6π pulse in the Q(3) attenuator.

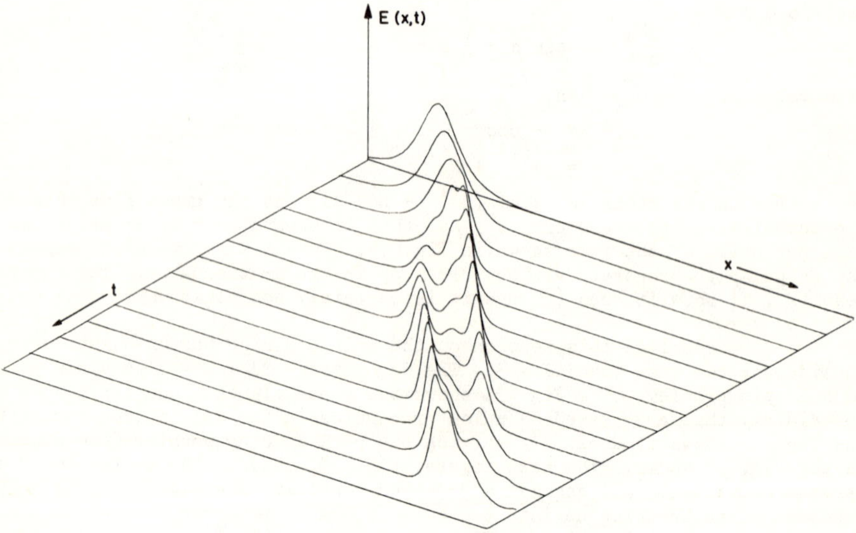

FIG.38. Evolution of a 6π sech into a wobbling 3-peaked 6π pulse in the Q(3) attenuator.

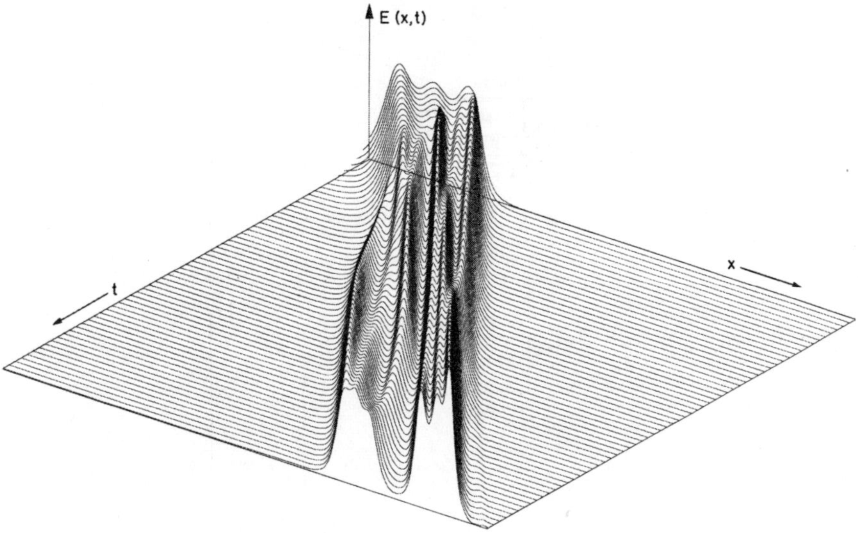

FIG.39. A 12π 3-peaked pulse breaks up into two 6π 3-peaked wobblers in the Q(3) attenuator.

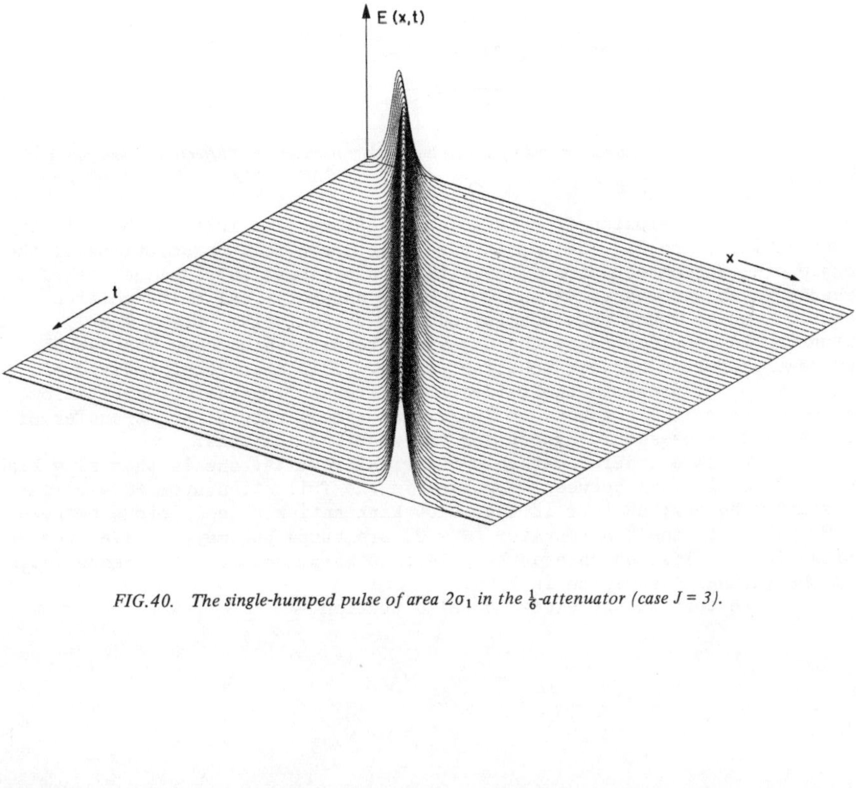

FIG.40. The single-humped pulse of area $2\sigma_1$ in the $\frac{1}{6}$-attenuator (case J = 3).

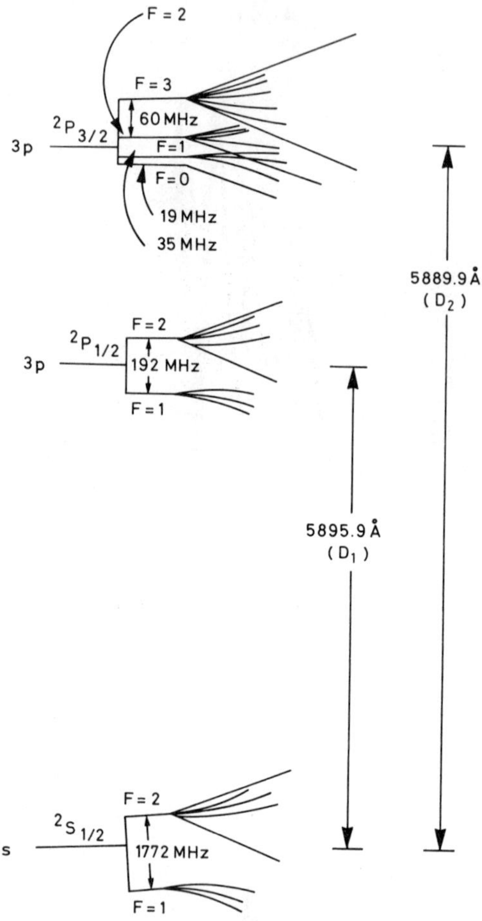

FIG.41. The hyperfine structure of the Na D_1 and D_2 lines (taken from Salamo "Effects of degeneracies on SIT" [42]*).*

The conclusion from these numerical results is that the solitary waves exhibited in Fig. 24 and in the Table 1 are soliton solutions of the double sine-Gordon equation (7.6) with an additional internal degree of freedom in the two cases 4π and 0π. Until an analytical multi-soliton solution can be found this conclusion must be tentative only. The situation is complicated by the numerical results which show that wobble can be transferred between wobbling pulses. This means that individual wobblers do not preserve their identity. Even so transfer of excitation by inelastic collisions between "particles" seems natural and transfer of wobble is what we should expect for the wobbling solitons.

We have other remarkable numerical results: one is that slow kink anti-kink collisions between 4π pulses in the full attenuator (J = 2) may generate wobblers; another is that slow kink anti-kink collisions between $4\pi-2\delta$ pulses in the $\frac{3}{4}$-attenuator (J = 2) are bumps but may involve breathers, whilst fast collisions change the pulses to 2δ pulses (See reference [41] and the further discussion in §10).

We turn to the experimental situation.

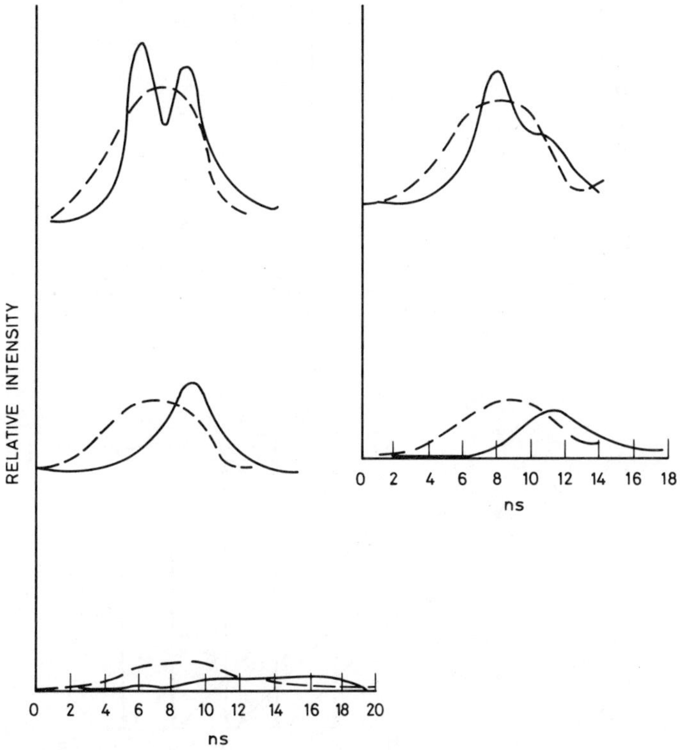

FIG.42. Experimental output pulse shapes as seen for the D_1 transitions. Linearly polarised light is used. A large magnetic field is imposed to remove the degeneracy (taken from [42]*).*

8. Degenerate SIT in Sodium vapour

The Fig. 41 shows the hyperfine structure of the Na D_1 and D_2 lines. The Fig. 42 taken from a thesis by G.V. Salamo, "Effects of degeneracies on SIT", [42] shows observations on the D_1 lines in a strong magnetic field. Here the hyperfine degeneracy is lifted and normal SIT is seen (compare the work on ^{87}Rb vapour in §5).

The Fig. 43 shows observations of pulse input and output from a cell of Na vapour in a weak magnetic field. The nuclear spin $\vec{I}$ and the total electron a.m. $\vec{J} = \vec{L} + \vec{s}$ interact through a term $\gamma\vec{I}\cdot\vec{J}$. The magnetic field interaction is weaker than this so that F and M_F are good quantum numbers $(\vec{F} = \vec{I} + \vec{J})$. The Rabi frequency $p\,E_0/\hbar \sim 10^9\ s^{-1}$ and the dipole interaction $\vec{p}\cdot\vec{E} < \gamma\vec{I}\cdot\vec{J}$ - that is the pulse lifetimes are long enough so that their spectral widths are narrow enough to resolve the hyperfine spacing. Hence we can expect to resolve the $F = 2 \rightarrow F' = 2$ transition. This is just a Q(2) transition (see Fig. 43(b)): the matrix elements are $-p, -\frac{1}{2}p, 0, \frac{1}{2}p, p$.

This situation should mean that a 4π wobbling pulse should in general propagate through the Na vapour. Unfortunately the data (Fig. 43(a)) are not good enough to see a wobble. Indeed Salamo (and see [37]) had no reason to interpret the result in Fig. 43a as anything but "normal SIT": the double peak appears to be the beginnings of normal break-up into two 2π

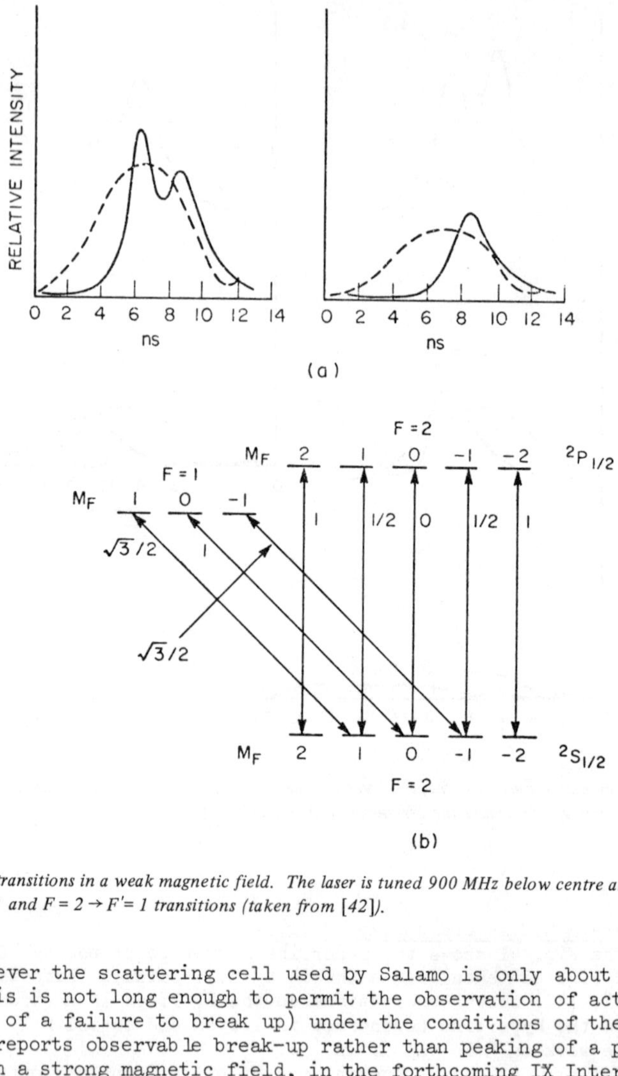

FIG.43. The D_1 transitions in a weak magnetic field. The laser is tuned 900 MHz below centre and picks up both $F = 2 \rightarrow F' = 2$ and $F = 2 \rightarrow F' = 1$ transitions (taken from [42]).

pulses. However the scattering cell used by Salamo is only about 2 mm. long; and this is not long enough to permit the observation of actual break-up (or of a failure to break up) under the conditions of the experiment (Gibbs reports observable break-up rather than peaking of a pulse, presumably in a strong magnetic field, in the forthcoming IX International Quantum Electronics Conference [43]).

The situation obtaining in practice here is very much complicated by Doppler broadening: the Doppler width is 1700 MHz at 200°C and the $F = 2 \rightarrow F' = 1$ transitions (192 MHz below the $F = 2 \rightarrow F' = 2$) are not resolved by this Doppler width. We can argue that different atoms with different thermal velocities are simultaneously on resonance through $F = 2 \rightarrow F' = 2$ and $F = 2 \rightarrow F' = 1$ transitions. Hence we need to add three spins in the proper ratios associated with the matrix elements p, $\frac{1}{2}p$ $(F = 2 \rightarrow F' = 2)$ and p, $\frac{\sqrt{3}}{2}p$ $(F = 2 \rightarrow F' = 1)$.

Counting spins equally the weightings are 3:2:2 for p, $\frac{\sqrt{3}}{2}p$, $\frac{1}{2}p$. The irrational ratio of spins now means that no pulse with an area equal to an integral number of 2π will travel without distortion. We can try to use the phase plane as we did in §7 but we shall find that this is not periodic and there are no closed trajectories.

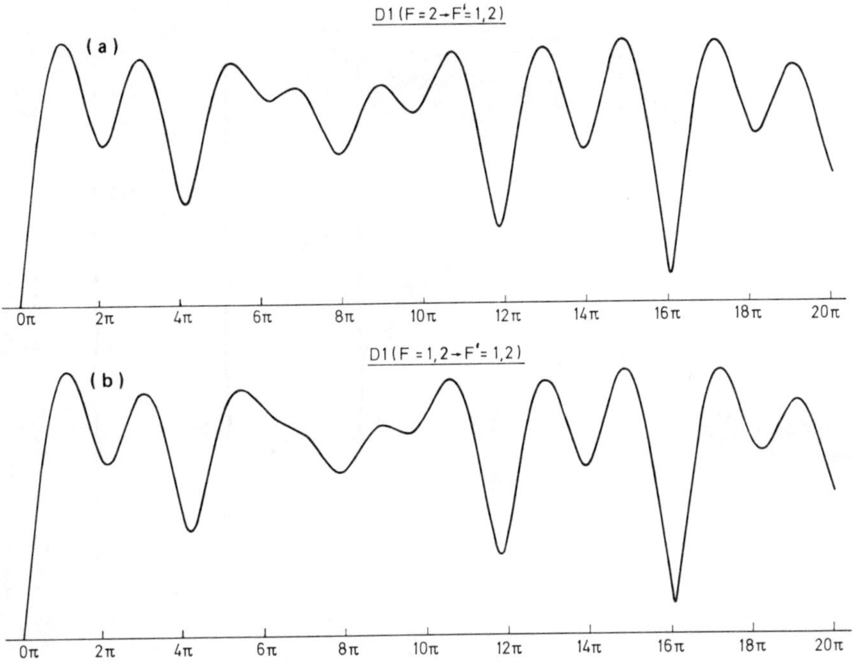

FIG.44. Phase-plane trajectories for Doppler broadened transitions. The matrix elements are p, $\frac{1}{2}$ p and $\frac{\sqrt{3}}{2}$ p for both (a) D_1 transitions F = 2 → F' = 1, 2 and (b) D_1 transitions F = 1, 2 → F = 1, 2 (taken from [41]*).*

The Fig. 44(a) shows the phase plane trajectory from $\sigma = 0$: a pulse of this shape (and infinite area) would propagate without distortion. However a shorter pulse must adapt. It does this as follows. We saw in §7 that the phase plane describes both pulse shapes and the energy left as excitation in the atoms. A closed trajectory as for the double humped pulse in the Q(2) attenuator leaves zero energy in the atoms. Evidently the deep minimum at $\sigma \sim 4.2\pi$ in Fig. 44(a) means that a pulse reshaping to a double humped 4.2π pulse leaves minimum energy in the atoms. The Fig. 45 shows how such a pulse develops. Notice that it looks like Salamo's "break-up" Fig. 43(a), but in fact it is a wobbler! Thus it seems that with a longer scattering cell Salamo will see a "wobbler".

Notice now the weak minimum at about 6.2π in Fig. 44(a). A 6.2π pulse breaks up into a 2π pulse which slows and stops and a 4.2π wobbler which continues (cf. Fig. 46. Figs. 44, 45, 46 come from S. Duckworth's Ph.D. thesis [41]). Duckworth finds also that a 7.8π pulse breaks up into a 4.2π and a 3.6π, that an 11.8π becomes three double peaked wobblers, and that a 16π (probably) becomes four double peaked wobblers. All these results can be guessed from the phase plane trajectory in Fig. 44(a). The 6.2π pulse is perhaps the most interesting: it shows that the energy given to the medium to create the high minimum at 6.2π is best achieved by throwing off the 2π which then slows down and stops.

The Fig. 47 shows the results (a) and the transitions (b) when the laser is tuned to line centre. Both the F = 2 → F'= 1 and F = 1 → F'= 2 transitions are included with the F = 2 → F'= 2. The former is 192 MHz below the F = 2 → F'= 2; the latter is 1772 MHz above. The matrix elements are still p, $\frac{\sqrt{3}}{2}$ p, $\frac{1}{2}$p but the weighting is 6:4:2. The phase plane is not

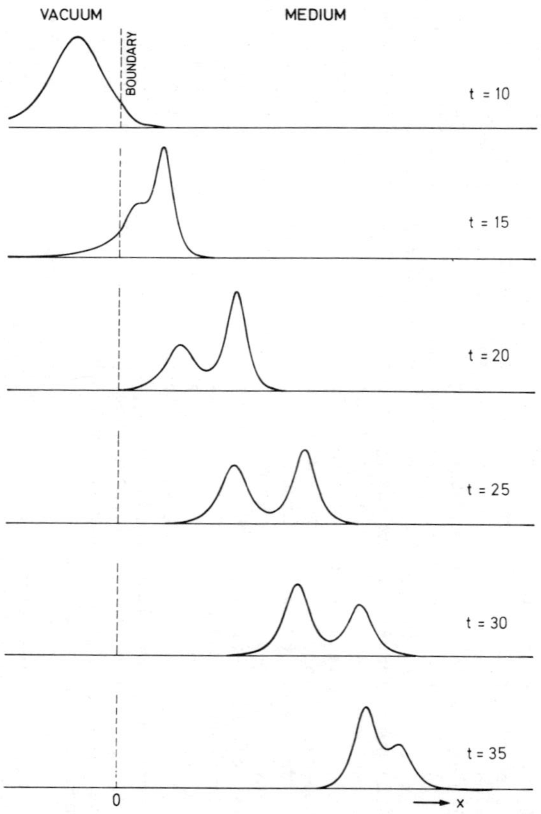

FIG.45. Evolution of a 4.2π sech pulse in a Na D_1 ($F = 2 \rightarrow F' = 1, 2$) level system (taken from [41]).

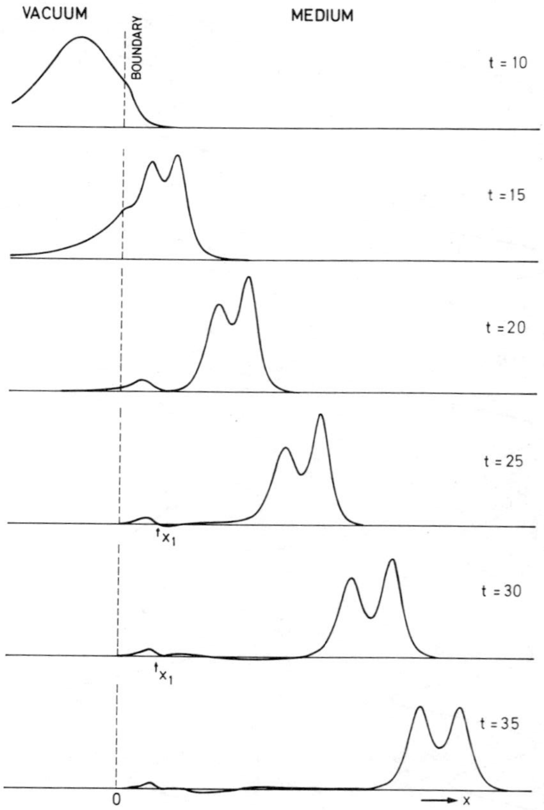

FIG.46. Evolution of a 6.2π sech pulse in a Na D_1 ($F = 2 \rightarrow F' = 1, 2$) degenerate level system (taken from [41]).

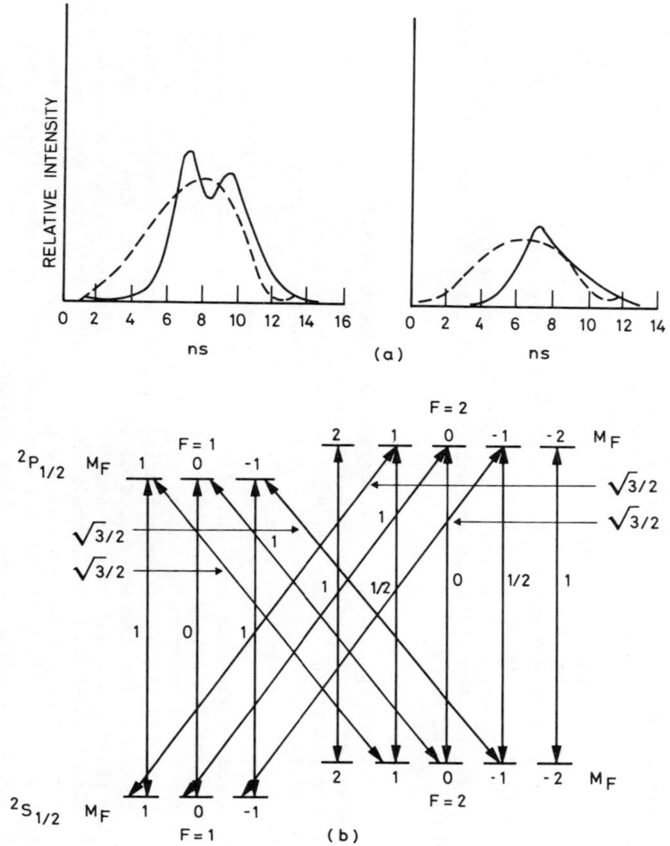

FIG. 47. Laser tuned to centre in a weak magnetic field: F = 1, 2 → F' = 1, 2 transitions are picked up (taken from [42]).

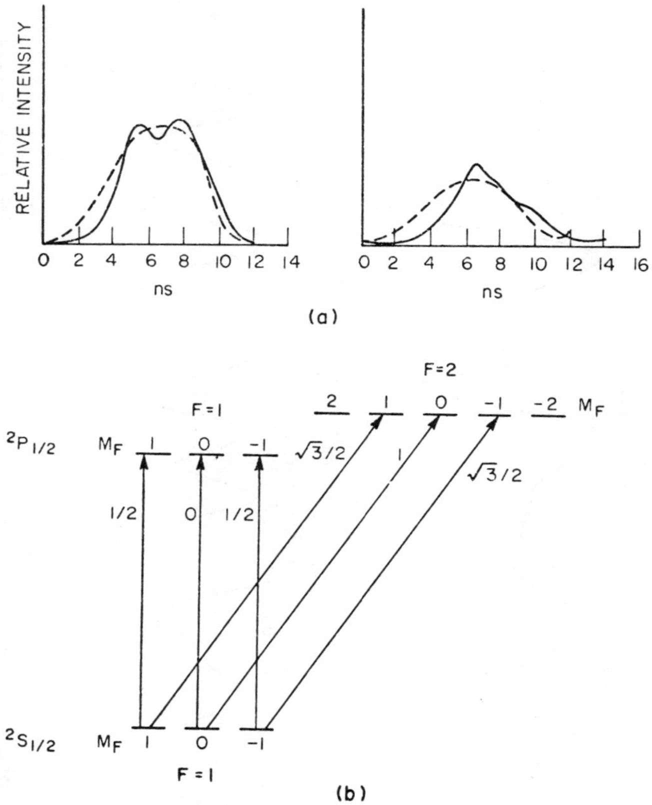

FIG. 48. Laser tuned to 900 MHz above centre: F = 1 → F' = 1, 2 transitions are picked up (taken from [42]).

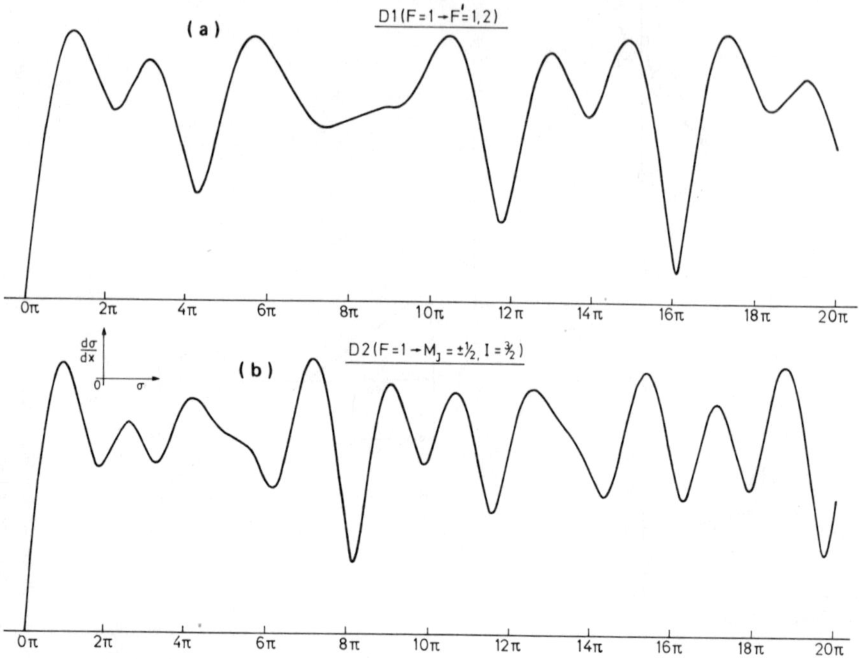

FIG.49. Phase-plane trajectories for Doppler broadened transitions in the cases (a) D_1 $(F = 1 \rightarrow F' = 1, 2)$ and (b) D_2 $(F = 1 \rightarrow m_J = \pm\frac{1}{2}, I = \frac{3}{2})$.

changed much from that in Fig. 44(a) and the observations are also little changed. The phase plane for this case is shown in Fig. 44(b).

In Fig. 48 the laser is tuned 900 MHz above centre. The $F = 1 \rightarrow F' = 1, 2$ transitions are picked up. The matrix elements are still p, $\frac{\sqrt{3}}{2}$ p, $\frac{1}{2}$p but the weightings are 1:2:2. A 4.2π pulse is still the preferred pulse and it proves to be a wobbler. Salamo interprets his shallow double peaked pulse (Fig. 48) as one about to break up.

The D_2 transitions are different: the largest hyperfine splitting of the $^2P_{3/2}$ excited states is now only ~ 60 MHz, and for nano second pulses with areas up to 6π

$$\underset{\sim}{p} \cdot \underset{\sim}{E} > h \, \Delta \, \nu_{HF}$$

($\Delta\nu_{HF}$ = hyperfine splitting). However a weak magnetic field of 60 to 150 Gauss decouples I and J so that good quantum numbers are J, M_J, I, M_I for the excited state. The $M_J = \pm \frac{1}{2}$ states are separated by 110 - 220 MHz and may be considered distinct. Fig. 50 shows the situation in a weak magnetic field with the laser tuned 900 MHz on the high frequency side of centre. (This picks up the F = 1 level of the $^2S_{\frac{1}{2}}$) The system is a three spin one: matrix elements are p, $1/\sqrt{2}$ p, $\sqrt{3}/2$ p; weightings are 1:1:1. The phase plane trajectory Fig. 49(b) is consistent with the double peaked pulse (area ~ 3π) observed by Salamo (Fig. 50).

The Fig. 51 shows the situation when the laser is tuned 1800 MHz off centre on the low frequency side of the D_2 transition. The Fig. 52 shows the phase plane and less pronounced double peaking at area about 3π. This is also consistent with Salamo's observations. Note that this case is a four spin one: matrix elements are $\sqrt{3}/2$ p, $\sqrt{2}$ p, p, $1/\sqrt{2}$ p and the weights are 2:2:2:2. Once more the numerical work [41] shows that pulses wobble rather than break up over large distances.

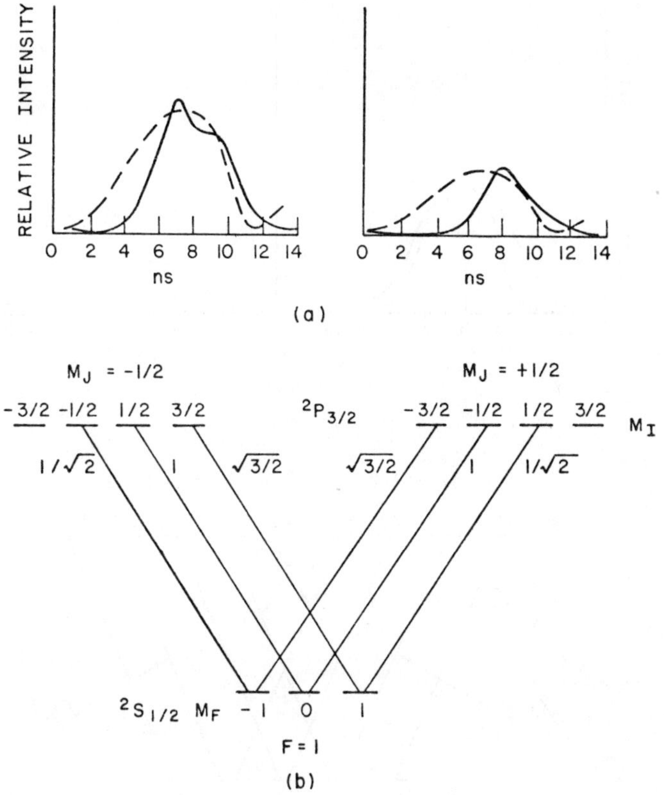

FIG.50. The laser is tuned 900 MHz above centre for the D_2 transition in a field of 60 − 150 Gauss (taken from [42]*).*

In conclusion: degeneracy leads to a rich field for the experimental observation of unusual soliton behaviour. Such observations will be possible only if scattering cells longer than those used so far on the metal vapours (~ 2 mm) can be developed. We calculate for Na vapour at $n = 10^{12}$ cm^{-3} that a 2π pulse stops in about 5 cm. and a leap-frog occurs in about 15 cm. However Fig. 53 (also taken from reference [41]) shows some agreement with Salamo's data for computed distances ~ 12 cm. (compare (b) of Fig. 53 with Salamo's results in Fig. 43). The numerical work is sharp line (no inhomogeneous broadening) and the length scales may need adjusting for this reason.

The final figure, Fig. 54, of this §8 shows the 8π pulse reshaping in time at particular points in space into two 4π's. This is the experimental situation: a pulse emerges at the end of a fixed length of scattering cell - although the effective length of the cell can be varied in successive experiments by varying the atomic density.[15] Only the first 4π can be seen in the output. It is a wobbler but note how it compares (say) with the data in Fig. 44.

We now turn to other physical problems where solitons play a role.

---

[15] For a length L of cell, the key quantity is αL, where α is the linear absorption given by $4\pi^2 g(0) p^2 n\, \omega_s/\hbar c$, as in equation (4.2a).

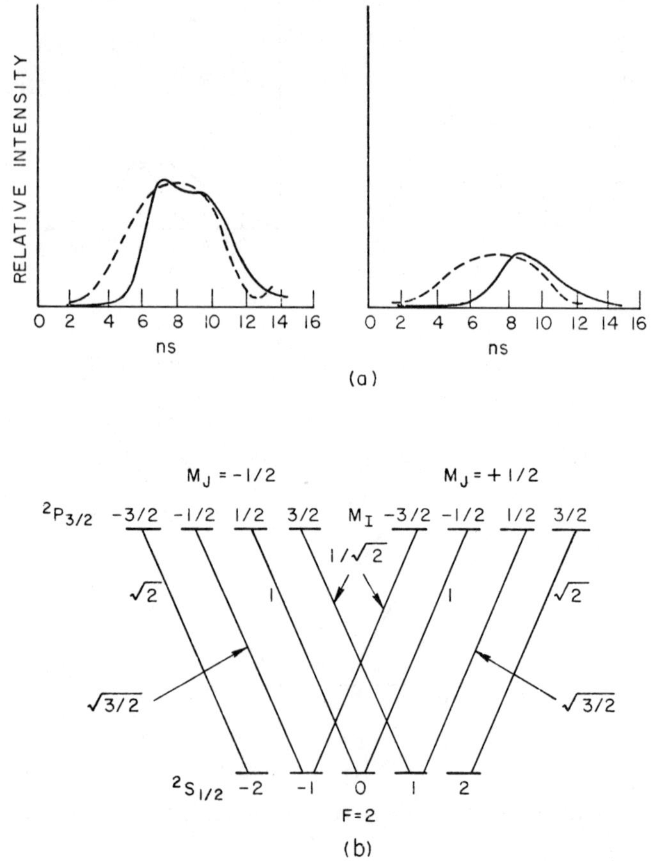

FIG.51. D_2 transitions in a weak magnetic field. The laser is tuned 1800 MHz below centre (taken from [42]*).*

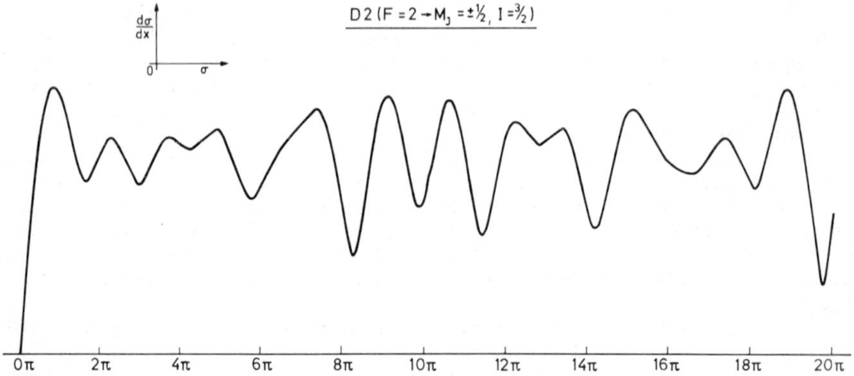

FIG.52. Phase-plane trajectory for the Doppler broadened transitions D_2 ($F = 2 \rightarrow m_J = \pm \frac{1}{2}$, $I = \frac{3}{2}$).

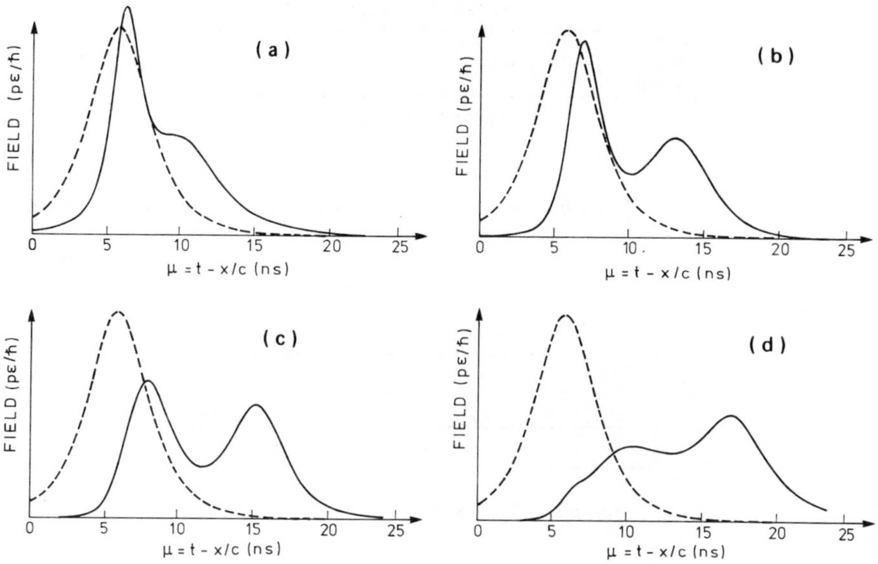

FIG.53. Computed outputs for different sample lengths of a homogeneously broadened Q(2) degenerate system. The input pulse (broken line) is a 5 ns 4π sech pulse: (a) shows the output pulse (solid line) for a sample length x ≏ 6 cm; (b) that for x ≏ 12 cm; (c) that for x ≏ 18 cm; and (d) that for x ≏ 24 cm.

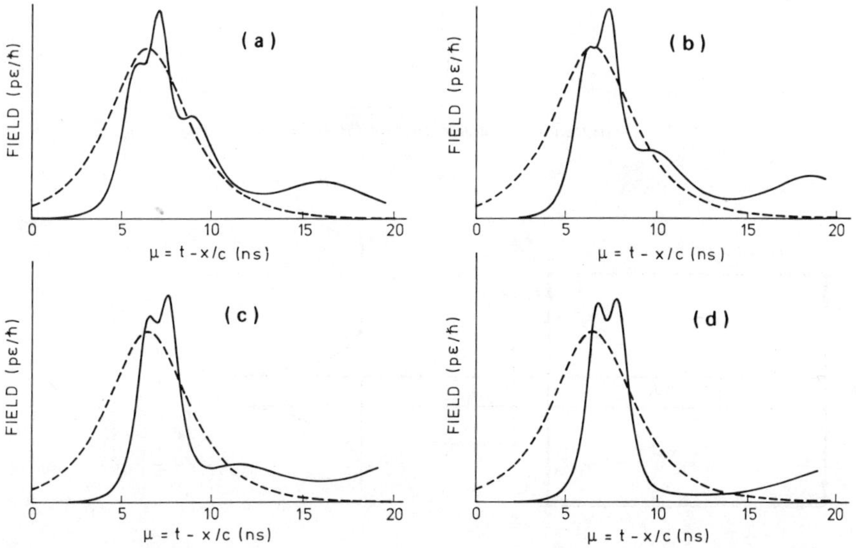

FIG.54. As in Fig.53, but the input is an 5 ns 8π sech pulse and the sample lengths are: (a) x ≏ 12 cm; (b) x ≏ 18 cm; (c) x ≏ 24 cm; (d) x ≏ 30 cm.

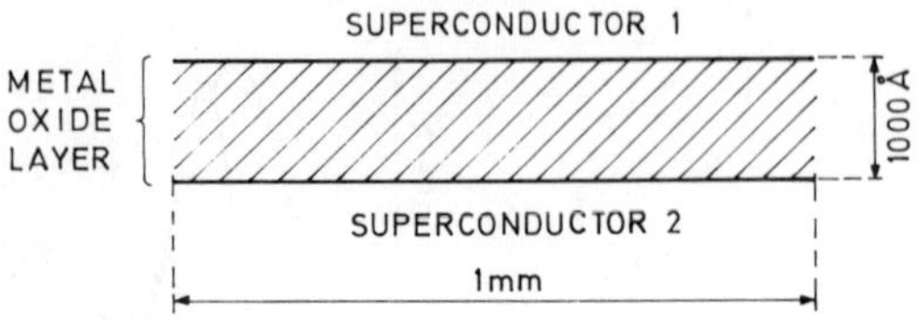

FIG.55. Sketch of a large area Josephson junction with a metal oxide layer between the two superconductors.

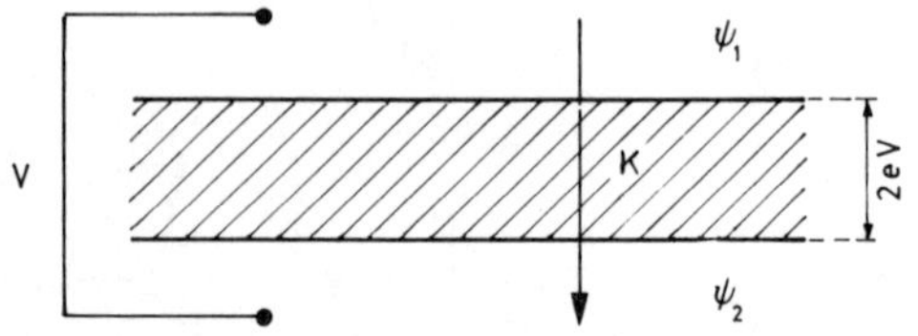

FIG.56. "Wave functions" ψ_1 and ψ_2 for a Josephson junction.

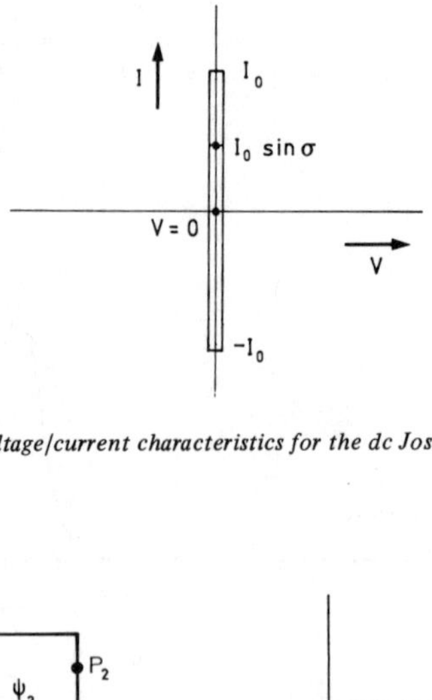

FIG.57. Voltage/current characteristics for the dc Josephson effect.

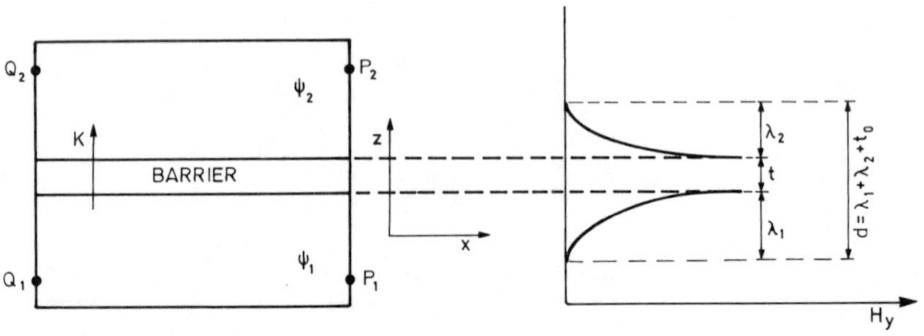

FIG.58. Notation and geometry for the application of Maxwell's equations to the case of the large-area Josephson junction.

SOLITONS IN OTHER NON-LINEAR PHYSICS

9. Josephson fluxons: solitons in large area Josephson junctions

9.1 In this §9 I shall show an exact parallelism between the quantised units of flux, the fluxons, in large area Josephson junctions, the kinks and anti-kinks of the sine-Gordon equation, and self-induced transparency in a resonant optical medium.

A Josephson junction is typically a "sandwich" of two superconductors embracing a gap or a metal oxide layer. In the second case the oxide is 500-1000 Å thick. Fig. 55 is a sketch. The dimension (1 mm) is that of a "large area" junction. Most work has been done on small area junctions. Often these are simply point contacts. We treat this case first.

The two sides of the small junction can be assigned "wave functions" (order parameters) ψ_1 and ψ_2. Fig. 56 illustrates this. The system is a two state quantum mechanical system and will have a pseudo-spin description. Indeed we can see it as a giant 2-level atom with energy spacing 2eV. The voltage V is applied across the junction as shown. The charge 2e rather than e represents the fact that Cooper electron pairs tunnel across the gap. The parameter K (here supposed a real number) is a tunnelling parameter coupling the two sides of the junction.

In Volume 3 of his Lectures on Physics, R.P. Feynman writes down the equations of his two state system in the form:

$$\begin{aligned}\psi &= a_1\psi_1 + a_2\psi_2\\ i\hbar\,\dot{a}_1 &= eVa_1 + Ka_2\\ i\hbar\,\dot{a}_2 &= Ka_1 - eVa_2 \quad .\end{aligned} \tag{9.1}$$

The quantities a_1 and a_2 depend on time t and are probability amplitudes for the pair 2e to be on one side ("one") or the other ("two") of the junction. The equations are phenomenological. One can adopt a deeper phenomenological starting point, namely in terms of a tunnelling Hamiltonian. For example, Rogovin and Scully [44] show from such a starting point that a Josephson junction is a giant 2-level atom with however a non-local coupling instead of the number K.

In order to see the spin description define a Bloch vector $\underline{r}(t) = (r_1, r_2, r_3)$ by

$$\begin{aligned}r_3 &\equiv |a_1|^2 - |a_2|^2\\ r_2 &\equiv i^{-1}(a_1{}^*a_2 - a_1a_2{}^*)\\ r_1 &\equiv (a_1{}^*a_2 + a_1a_2{}^*) \quad .\end{aligned} \tag{9.2}$$

The Feynman equations (9.1) now imply the Bloch equations of motion

$$\frac{dr_1}{dt} = -\frac{2eV}{\hbar}r_2 \tag{9.3a}$$

$$\frac{dr_2}{dt} = \frac{2eV}{\hbar}r_1 + \frac{2K}{\hbar}r_3 \tag{9.3b}$$

$$\frac{dr_3}{dt} = -\frac{2K}{\hbar}r_2 \ . \tag{9.3c}$$

Notice that these are exactly the Bloch equations for a 2-level atom with energy spacing 2eV and "applied field" $p^{-1}K!$ (compare (9.3) with (4.8)). However for mathematical purposes it is more fruitful to note that the voltage V may depend on time (the energy spacing $\hbar\omega_s$ of the 2-level atom does not) whilst K is time independent (the field E(t) on the atom usually depends on time). Mathematically the equations now go into correspondence by the simple switch $r_1 \leftrightarrow r_3$, $r_2 \leftrightarrow r_2$. We use this later.

Notice that if no voltage is applied r_1 = constant and

$$\ddot{r}_3 + 4K^2\hbar^{-2}r_3 = 0 \quad . \tag{9.4}$$

Thus

$$r_3(t) = \cos(2K\hbar^{-1}t + \delta) \tag{9.5a}$$

$$\dot{r}_3(t) = -2K\hbar^{-1}\sin(2K\hbar^{-1}t + \delta) \quad . \tag{9.5b}$$

The Bloch equations have the constant of the motion

$$r_1^2 + r_2^2 + r_3^2 \quad (= |a_1|^2 + |a_2|^2 = 1) \tag{9.6 }$$

so $|r_3| \leqslant 1$ and the amplitude in (9.5a) is unity. Further (9.6) and (9.2) together mean that

$$|a_1|^2 = \tfrac{1}{2}(1 + r_3) \tag{9.7a}$$

and

$$\frac{d|a_1|^2}{dt} = \frac{1}{2}\frac{dr_3}{dt} \quad . \tag{9.7b}$$

But the current across the junction is

$$2e\,\frac{d|a_1|^2}{dt}$$

(for charges 2e). Hence the current across the junction is

$$I(t) = I_o \sin(2K\hbar^{-1}t + \delta) \quad . \tag{9.8}$$

Josephson currents are extremely weak and the usual situation has the two sides of the junction close to balance: $r_3 \approx 0$. In so far as (9.4) still applies $r_3 = 0$ implies $\ddot{r}_3 = 0$ so that $\dot{r}_3$ = constant and from (9.3c) r_2 = constant. Then since $|r_2| \leqslant 1$, $I(t) = I_o \sin\sigma$ = constant. This is the celebrated d.c. Josephson effect. Fig. 57 shows the unusual voltage/current characteristic.

A better argument sets $r_3 = 0$ in (9.3). Then

$$\frac{dr_2}{dt} = \frac{2eV}{\hbar}\,r_1$$

$$\frac{dr_1}{dt} = -\frac{2eV}{\hbar}\,r_2 \quad . \tag{9.9}$$

A solution is

$$r_1 = -\cos\sigma \tag{9.10a}$$

$$r_2 = -\sin\sigma \tag{9.10b}$$

with

$$\frac{2eV}{\hbar} = \frac{d\sigma}{dt} \quad . \tag{9.10c}$$

Equation (9.10c) is one of Josephson's equations: the quantity σ which now emerges can be identified with the phase difference across the junction which is what (9.10a, b) already show. From (9.3c) follows the second of Josephson's equations

$$I(t) = I_o \sin\sigma. \tag{9.11}$$

The result above now follows when $V = 0$ so that σ = constant.

For the a.c. Josephson effect choose V = constant. Then $\sigma = 2eV\hbar^{-1}t + \delta$ from (9.10c) and

$$I(t) = I_o \sin(2eV\hbar^{-1}t + \delta). \tag{9.12}$$

9.2 These results were all for small area Josephson junctions. In the case of a "large" junction we need to consider the variation of Josephson current across the width of the junction. This is governed by Maxwell's equations.

The Fig. 58 illustrates the notation and geometry: x, y, z form a right handed system with z normal to the plane of the junction. Faraday's law of electro-magnetic induction

$$\oint \vec{E} \cdot d\vec{s} = -\frac{1}{c}\frac{d}{dt}\int_S \vec{B} \cdot d\vec{S}$$

yields for the difference between the voltage difference between point pairs P and Q

$$-V(P) + V(Q) = -\frac{1}{c}\frac{d}{dt}\int_Q^P dx\left[\lambda_1 + \lambda_2 + t_o\right] H_y(x, t)$$

$$= -\frac{d}{c}\frac{d}{dt}\int_Q^P dx\, H_y(x, t) \qquad (9.13)$$

where d is the effective thickness of the barrier including the penetration distances λ_1, λ_2 of the magnetic field. From this follows the differential relationship

$$\frac{\partial V}{\partial x} = \frac{d}{c}\frac{\partial H_y}{\partial t} \qquad (9.14a)$$

whilst Ampère's theorem is

$$\frac{\partial H_y}{\partial x} = \frac{4\pi}{c} j_z + \frac{1}{c}\frac{\partial D_z}{\partial t} = \frac{4\pi}{c}\left[j_z + C\frac{\partial V}{\partial t}\right] \; ; \qquad (9.14b)$$

C is the capacity of the junction per sq. cm. $(C = \kappa/4\pi d)$ and $C\,\partial V/\partial t$ is the displacement current.

If we set

$$j_z(x, t) = 2\beta\frac{\partial}{\partial t}[a_1(x, t)]^2 = \beta\frac{\partial r_3}{\partial t}(x, t) \qquad (9.15)$$

where in the second relation the spatially dependent equivalent of (9.7a) has been used, Maxwell's equations (9.14) become

$$\frac{\partial^2 V}{\partial x^2} - \frac{\kappa}{c^2}\frac{\partial^2 V}{\partial t^2} = \frac{4\pi}{c^2} d\,\beta\frac{\partial^2 r_3}{\partial t^2} \qquad (9.16a)$$

or

$$\frac{\partial^2 E}{\partial x^2} - \frac{\kappa}{c^2}\frac{\partial^2 E}{\partial t^2} = \frac{4\pi}{c^2}\beta\frac{\partial^2 r_3}{\partial t^2} \; . \qquad (9.16b)$$

The electric field E is in the z-direction and is defined by $E \equiv Vd^{-1}$. Typical figures are $V \sim 10\mu V$, $d \sim 4 \times 10^{-6}$ cm, $E \sim 2.5$ V $cm^{-1} \sim 8.5 \times 10^{-6}$ e.s.u. cm.$^{-1}$. For junctions $(j_z)_{max} \sim 1$ A·cm^{-2} and $\partial r_3/\partial t \sim 10^9\ s^{-1}$ means $\beta \sim 5$ e.s.u. cm^{-2}.

Since the field E varies in space as well as time the components of the Bloch vector (r_1, r_2, r_3) also vary in space as well as time. There is a remarkable correspondence now between the Maxwell-Bloch equations for the 2-level atom dielectric and those for the large Josephson junction. One problem maps to the other by $r_1 \leftrightarrow r_3$, $r_2 \leftrightarrow r_2$ although the numbers involved are very different. The Fig. 59 displays this correspondence.

In order to find soliton solutions it is tempting to approximate the Maxwell-Bloch equations for the Josephson junctions by the RMB equations (§ 6). Only the Maxwell equation (9.15) is changed - namely to

$$\frac{\partial E}{\partial x} + \frac{1}{\bar{c}}\frac{\partial E}{\partial t} = -\frac{2\pi}{\bar{c}}\frac{\beta}{\kappa}\frac{\partial r_3}{\partial t}$$

$$= \frac{4\pi}{\bar{c}}\frac{K\beta}{\hbar\kappa} r_2$$

in which $\bar{c} \equiv \kappa^{-\frac{1}{2}}c$. This equation cannot really be justified for the junction however since $\beta\kappa^{-1}$ is not small: compare $\beta \sim 5$ e.s.u. cm^{-2} against $np \sim 10^{-8}$ e.s.u. cm^{-2}.

If $r_3 \approx 0$ everywhere however we again set

$$r_1 = -\cos\sigma , \qquad r_2 = -\sin\sigma \; .$$

The Maxwell-Bloch Equations

Josephson Junction	2-level atom dielectric
$\frac{\partial^2 E}{\partial x^2} - \frac{\kappa}{c^2}\frac{\partial^2 E}{\partial t^2} = \frac{4\pi}{c^2}\beta\frac{\partial^2 r_{3-}}{\partial t^2}$	$\frac{\partial^2 E}{\partial x^2} - \frac{1}{c^2}\frac{\partial^2 E}{\partial t^2} = \frac{4\pi}{c^2} np\frac{\partial^2 r_{1-}}{\partial t^2}$
$\frac{\partial r_{1-}}{\partial t} = -\frac{2ed}{\hbar} r_2$	$\frac{\partial r_1}{\partial t} = -\omega_s r_2$
$\frac{\partial r_{2-}}{\partial t} = \frac{2K}{\hbar} r_3 + \frac{2ed}{\hbar} E r_1$	$\frac{\partial r_{2-}}{\partial t} = \omega_s r_1 + \frac{2p}{\hbar} E r_3$
$\frac{\partial r_{3-}}{\partial t} = -\frac{2K}{\hbar} r_2$	$\frac{\partial r_{3-}}{\partial t} = -\frac{2p}{\hbar} E r_2$
$ed \sim 10^{-17}$, $\beta \sim 5$, $2K\hbar^{-1} \sim 10^9$	$p \sim 10^{-18}$, $10^{-8} \lesssim np \lesssim 10^{-2}$, $\omega_s \sim 10^{15}$

Correspondence is

$r_1 \longleftrightarrow r_3$, $r_2 \longleftrightarrow r_2$

FIG.59. A comparison of the Maxwell–Bloch equations for the large-area Josephson junction (left) and the non-degenerate 2-level atom dielectric (right). One maps to the other under $r_1 \leftrightarrow r_3$, $r_2 \leftrightarrow r_2$.

The two Josephson relations are now

$$\frac{\partial\sigma}{\partial t} = \frac{2eV}{\hbar} , \qquad j_z = j_{zo}\sin\sigma \quad . \tag{9.17}$$

From (9.15a) it follows by one integration in time and without change of independent variables (compare § 4) that

$$\frac{\partial^2\sigma}{\partial x^2} - \frac{1}{\bar{c}^2}\frac{\partial^2\sigma}{\partial t^2} = \frac{4\pi}{\bar{c}^2}\frac{\beta}{\kappa}\frac{2K}{\hbar}\sin\sigma . \tag{9.18}$$

This is just the Lorentz invariant form of the sine-Gordon equation. It has the static one-kink solution

$$\sigma = 2\sin^{-1}\mathrm{sech}[(x - x_o)/\lambda_o] \tag{9.19}$$

where

$$\lambda_o^{-2} = \frac{4\pi}{\bar{c}^2}\frac{\beta}{\kappa}\frac{2K}{\hbar} . \tag{9.20}$$

This is a physical kink brought to rest and contrasts with the optical pulse which cannot be brought to rest (compare (4.23)). The derivative

$$\sigma_x = (2/\lambda_o)\,\mathrm{sech}[(x - x_o)/\lambda_o] \tag{9.21}$$

is recognizable as the single soliton. Since

$$\sigma_{tx} = \sigma_{xt} = \frac{2ed}{\hbar} V_x = \frac{2ed}{\hbar c} H_{y,t} \tag{9.22a}$$

$$\sigma_x \equiv \frac{\partial\sigma}{\partial x} = \frac{2ed}{\hbar c} H_y \tag{9.22b}$$

and

$$\sigma(\infty) = \frac{2ed}{\hbar c}\int_{-\infty}^{\infty} dx\, H_y = \frac{2e}{\hbar c}\int B\, dS$$

$$= 2\pi \qquad (\text{for } (9.19)) \quad . \qquad (9.23)$$

This is a single quantised unit of flux - the "fluxon". Notice that this quantisation of flux is achieved by the "quantisation" into solitons.

If the kink (9.19) is moving with velocity u, Lorentz transformation shows that

$$\frac{x_o - x}{\lambda_o} \to \frac{ut - (x - x_o)}{\lambda_o(1 - u^2\bar{c}^{-2})^{\frac{1}{2}}}$$

$$\sigma_t = \frac{2u}{\lambda}\,\text{sech}\left[\frac{u}{\lambda}\left(t - \frac{x - x_o}{u}\right)\right]$$

$$\lambda = \lambda_o(1 - u^2\bar{c}^{-2})^{\frac{1}{2}} .$$

Then from (9.17)

$$V(x, t) = V_o\,\text{sech}\left[\frac{u}{\lambda}\left(t - \frac{x - x_o}{u}\right)\right] \qquad (9.24)$$

where $V_o = \frac{\hbar u}{e\lambda}$. Thus

$$\left(\frac{\bar{c}^2}{u^2} - 1\right) = \frac{\hbar^2\bar{c}^2}{e^2\lambda_o^{\,2}V_o^{\,2}} . \qquad (9.25)$$

Evidently (9.24) is a voltage pulse travelling up the x-axis with a velocity u given by (9.25): bigger pulses go faster. Associated with this voltage pulse is the electric field $E = Vd^{-1}$ and the magnetic field H_y determined by (9.22a).

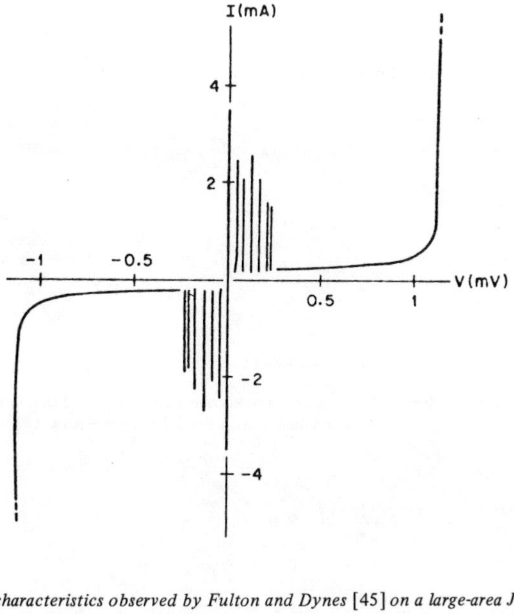

FIG.60. Voltage/current characteristics observed by Fulton and Dynes [45] on a large-area Josephson junction.

The area[16]

$$\Theta = \frac{2e}{\hbar} \int_{-\infty}^{\infty} V(x, t)dt \qquad (9.26)$$

formally satisfies the area theorem

$$\frac{d\Theta}{dx} = \pm \gamma \sin \Theta$$

by analogy with area theorem for non-degenerate SIT (§4, eqns. (4.2)). The quantity γ is formally

$$\gamma = \frac{8\pi^2 \beta K}{\hbar \bar{c} \kappa} g(o)$$

The quantity $g(o)$ is the value of the broadening function at the origin. Since the situation is necessarily sharp line $g(o) = \delta(o)$ and the area theorem takes the form

$$\sin \Theta = 0 . \qquad (9.27)$$

The relevant solutions are $\Theta = 2\pi, 4\pi, 6\pi, \ldots.$

Implicitly all disturbances are moving so that at any point x all pulses pass by in the infinite time between $t = -\infty$ and $t = +\infty$. There is therefore no inconsistency with the definition

$$\Theta = \frac{2ed}{\hbar c} \int_{-\infty}^{\infty} dx\, Hy = \frac{2e}{\hbar c} \int B\, d\, S$$

[16] The situation is actually more complicated than this because the variables x, t here are not the variables x, t of the area theorem (4.2a). It is easy to see that if one assumes the boundary conditions $\sigma(-\infty, t) = 2\nu_1 \pi$, $\sigma(+\infty, t) = 2\nu_2\pi$ in terms of the present variables x, t at $x = \pm\infty$ that

$$\int_{-\infty}^{\infty} \sigma_x dx = 2\nu\pi \; (\nu = \nu_2 - \nu_1 \text{ and the } \nu\text{'s are integers, positive, negative or zero}).$$

Hence

$$2\nu\pi = \int_{-\infty}^{\infty} \sigma_x\, dx = \frac{2ed}{\hbar c} \int_{-\infty}^{\infty} H dx = \frac{2e}{\hbar c} \int BdS$$

and the flux is $\nu hc/2e$. From (4.2) one sees that the area can be taken to be, for example,

$$\theta = \int_{-\infty-2\xi,\, t=-\infty}^{+\infty-2\xi} \sigma_x\, dx + \int_{-\infty,\, x=+\infty-2\xi}^{\infty} \sigma_t\, dt$$

and formally

$$\frac{d\theta}{d\xi} = \pm \gamma \sin \theta.$$

The argument of the text shows that $\sin \theta = 0$. One can then choose to have a fixed flux initially threading the junction at $t = -\infty$ or to permit a given sequence of voltage pulses to cross the point $x = +\infty$ between $t = -\infty$ and $t = +\infty$. In practice the alternative definitions

$$\theta = \frac{2e}{\hbar c} \int BdS \quad \text{and} \quad \theta = \frac{2e}{\hbar} \int_{-\infty}^{\infty} V\, dt$$

are equally acceptable and each has value $2\nu\pi$.

and the magnetic flux is quantised into units of

$$\frac{2\pi \hbar c}{2e} = \frac{hc}{2e} .$$

This is the more fundamental statement and admits the possibility of one fluxon at rest within the junction: there can be no more since the multi-soliton solutions of the sine-Gordon equation admit only one 2π kink at rest. Any number of breathers (zero π type) can be brought to rest but these bring zero flux into the junction. They are probably unimportant because damping is an important factor in a real junction and the 0π's are unstable and self annihilating when damping is present as a sine-Gordon model constructed with pins stuck into a rubber band easily shows.[17]

Fulton and Dynes [45 and cf. 46, 47, 48] make a remarkable inference concerning the interaction of a large area Josephson junction with radiation by considering a mechanical model [48] for the sine-Gordon equation. If a d.c. voltage V is applied to the junction (9.12) shows there is an associated frequency $\omega = 2eV\hbar^{-1}$. If the single fluxon with voltage V given by (9.24) traverses the junction there is associated with it the kink or twist of magnitude 2π. If a twist of 2π travelling along x reaches a fixed end of a mechanical model it can only untwist travelling back along $-x$. It therefore returns as an anti-kink although the hand of the twist about the direction of motion is unchanged. But if the end is an open end $V \propto \sigma_t$ is unchanged in sign and an anti-twist is returned along $-x$. If the single soliton is taken as the natural mode of the open ended cavity of finite length L along x formed by any real junction the soliton must double pass the cavity before completing a full cycle. Fulton and Dynes infer that the fundamental mode frequency $\bar{c}/2L$ must be replaced by the fundamental even mode frequency $\bar{c}/L$ and only the even modes are observable. The Fig. 60 shows the voltage-current curves observed by Fulton and Dynes [45]. The several branches are separated by voltages $V = 2n\hbar\omega/2e$ and the effective charge $2e$ of the Cooper pair is halved! The number n is supposedly the number of solitons involved.

The mathematical problem of solitons in a one-dimensional box with open or closed ends has not been solved. The comparable problem for periodic boundary conditions is solved for the KdV equation only [16, 17, 49]. The relationship between the hierarchy of solutions obtained by Lax [16] and solitons on $-\infty < x < +\infty$ has not been established. The numerical data obtained by Hyams in ref. [16] can look different from the multi-soliton solutions on $-\infty < x < +\infty$.

In §10 we look briefly at two recent applications of the sine-Gordon equations to spin waves in liquid ^{3}He at temperatures below 2.6 mK [50, 51, 52].

10. Spin waves in the A and B phases of liquid ^{3}He

Below 50 mK and above 2.6 mK liquid ^{3}He behaves as a "weakly interacting" Fermi liquid. Landau's theory of normal Fermi liquids [53] solves the paradox of the strong interparticle interactions. Below 2.65 mK liquid ^{3}He displays at least two phases, the A-phase and the B-phase. Fig. 61 shows the phase diagram as it is presently known.

Since ^{3}He is a weakly interacting Fermi system, it is natural to postulate some form of superconductivity. Normal superconductivity is described by Cooper electron pairs with spin singlet pairing and no orbital angular momentum ($\ell = 0$). It is isotropic. In ^{3}He overlap of hard cores would occur for $\ell = 0$; $\ell \neq 0$ prevents the close approach of hard cores and attraction can be maintained by van der Waals forces. Spin pairing must be

---

[17] This model is easily made: take pins about one inch long with heavy heads and stick them at equal spacings of about one eighth of an inch into a $\frac{1}{4}$ in flat rubber strip, say about one foot long.

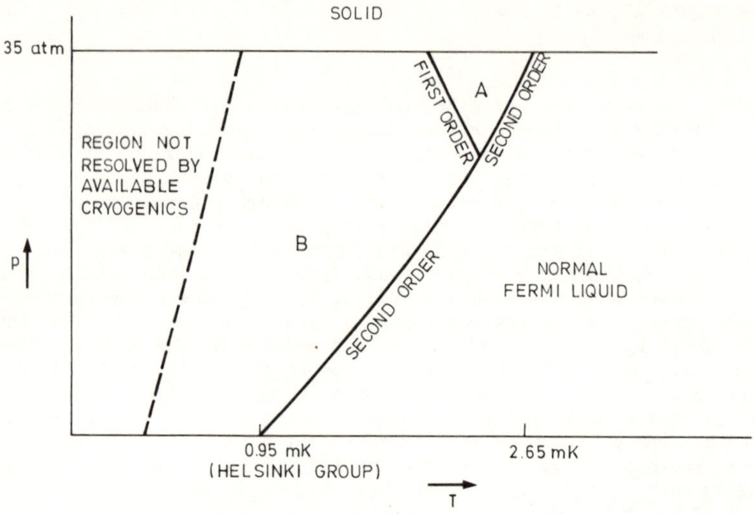

FIG.61. Phase diagram for ^3He below 3 mK.

triplet pairing and ℓ is odd. It is usually assumed $\ell = 1$. The system is anisotropic but not necessarily macroscopically anisotropic and is classified as an anisotropic liquid [54].

The pair wave function is complicated: there are 18 degrees of freedom (from (roughly) the product of a spin vector and a momentum vector[18] $= 3 \times 3$ and elements complex). Two particular states have been discussed: the ABM state [55, 56] and the BW state [57]. There are many others (see e.g. [54]). In the ABM state $S_z = 0$ is excluded. The liquid has two interpenetrating components labelled by "up spins" and "down spins" ($S_z = \pm 1$) respectively. Each component has relative angular momentum ℓ and there is net angular momentum. The two spin components are not coupled to first approximation. There is a very weak coupling through spin dipole forces however. This coupling breaks the symmetry and is important for that reason [54]. In particular spin waves are possible. In the BW state all 3 spin states are formed. There is no net angular momentum since e.g. $\ell_{\uparrow\uparrow} = -\ell_{\downarrow\downarrow}$ and the state is isotropic. Of the many possible $\ell = 1$ states it is the most stable in BCS approximation. The ABM state is unstable in BCS approximation but is supposedly stabilised by fluctuations.

Possible candidates for the two phases are: A-phase - ABM state; B-phase - BW state. We assume these assignments in what follows. Evidently the ABM state is a two-state system with weak coupling and approximates to a Josephson junction [54]. We sketch a theory of this.

If the system is homogeneous (no spatial variations) we identify $S_z(t)$ with $r_3(t)$ and can expect to use the two Josephson equations ((9.3c) with (9.10b))

$$\frac{dr_3}{dt} = \frac{2K}{\hbar} \sin \sigma \tag{10.1}$$

and an extended form of (9.10c)

$$\frac{d\sigma}{dt} = \frac{2eV}{\hbar} + 4\, I\, r_3 \quad . \tag{10.2}$$

[18] An *angular* momentum vector. The ABM state involves the outer product of a real unit vector in spin space and a complex unit vector in orbital (angular momentum) space; the BW state involves a bilinear combination of real spin and real orbital unit vectors and is invariant under the joint rotation of both spaces.

There is no voltage and the quantity "2eV" is the more fundamental quantity $\mu_\uparrow - \mu_\downarrow$, the difference between the chemical potentials of the up and down spin components. This assumes the components separately achieve thermal equilibrium. The component r_3 of the Bloch vector is $r_3 \equiv |c_\uparrow|^2 - |c_\downarrow|^2$; K is now a constant which can be interpreted in terms of the weak dipole interaction [50]. The chemical potential term in (10.2) is augmented by a spin exchange term $4\,I\,r_3$ [50].

In the presence of a magnetic field B (10.2) becomes [50]

$$\frac{d\sigma}{dt} = 2\omega_L - 4(1-\bar{I})\ r_3[N(o)]^{-1} \tag{10.3}$$

in which $\omega_L = e\gamma/m_pc\ B$ is the Larmor frequency[19] and N(o) the density of states at the Fermi level: $\bar{I} \equiv I\,N(o)$. Except for spin exchange effects[20] the analogy with Josephson junction theory is complete: $\gamma e\hbar B/m_pc \leftrightarrow eV$ as comparison below (4.5) would have indicated.

Together (10.1) and (10.3) mean

$$\sigma_{tt} = -\Omega_\ell^2 \sin\sigma$$

$$\Omega_\ell^2 = [(1-\bar{I})\,6\pi\,\gamma^2\,\Delta^2(T)/5g^2\,N(o)] \tag{10.4}$$

in which Ω_ℓ is the longitudinal NMR resonance frequency obtained by Legett [58]. The notation is given in [50, 51]. γ is the gyroscopic ratio of the ^{3}He atom, g the pairing interaction constant, $\Delta(T)$ the temperature dependent order parameter: Ω_ℓ is real.

Equation (10.3) is the pendulum equation. We expect the sine-Gordon equation in the spatially inhomogeneous case. Equation (10.1) is extended by inclusion of a spin current $\vec{J}$ [51]:

$$r_{3,t} + \nabla\cdot\vec{J} = 2K\hbar^{-1}\sin\sigma\ .$$

We are concerned with longitudinal excitations and this is a new feature compared with the transverse theory involved in §9. The current $\vec{J}$ is related to the velocity fields $\nabla\sigma_{\uparrow,\downarrow}$ of the two spin systems: Maki and Tsuneto [51] show that $\vec{J} \propto \nabla\sigma;\ \sigma \equiv \sigma_\uparrow - \sigma_\downarrow$. In one spatial dimension they then find

$$\sigma_{tt} - c_A^2\,\sigma_{xx} = -\Omega_\ell^2 \sin\sigma \tag{10.5}$$

where

$$c_A^2 = \tfrac{1}{3}(1-\bar{I})\,v_F^2\rho_{xx}\,\rho^{-1}:$$

v_F is the Fermi velocity, ρ_{xx} is the xx-component of the superfluid density tensor $\overleftrightarrow{\rho}_S$. Equation (10.5) scales to the sine-Gordon equation as already discussed. In the infinite system with phase difference an integral multiple of 2π at $x = \pm\infty$ spin waves are sine-Gordon solitons.

---

[19] Subsequently we use γ for $e\gamma/m_pc$ and we set $\hbar = 1$.

[20] The spin exchange term in (10.2) which is therefore the term $\bar{I}$ in (10.3) is less than crucial to the theory. What *is* crucial is the term $4r_3[N(o)]^{-1}$ in (10.3). This term represents the change in r_3 induced by B, a description of the magnetisation induced by B. In the steady state the magnetisation is

$$(e\gamma/m_pc)r_3 = \tfrac{1}{2}N(o)(1-\bar{I})^{-1}(e\gamma/m_pc)^2\,B = \chi B$$

where χ is the static susceptibility (with spin exchange effects included as a mean field correction). See the work based on an adiabatic Hamiltonian in the review by A.J. Leggett (Rev. Mod. Phys. **47** (1975) 331); and see also a paper "Bumping spin waves in the B-phase of liquid ^{3}He" by P.J. Caudrey and the writer (in preparation, 1976) which considerably amplifies the necessarily condensed treatment of spin waves in ^{3}He given in this §10.

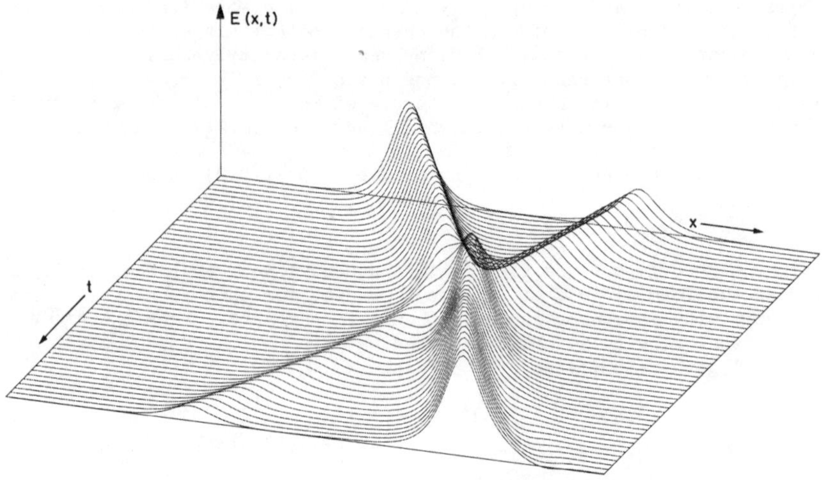

FIG. 62. Computer simulation showing spin waves colliding and bumping off each other in the B-phase of liquid ^3He. For comparison with the optical pulses in a degenerate medium displayed in § 7 the derivatives of the kinks have been plotted (taken from [41]*).*

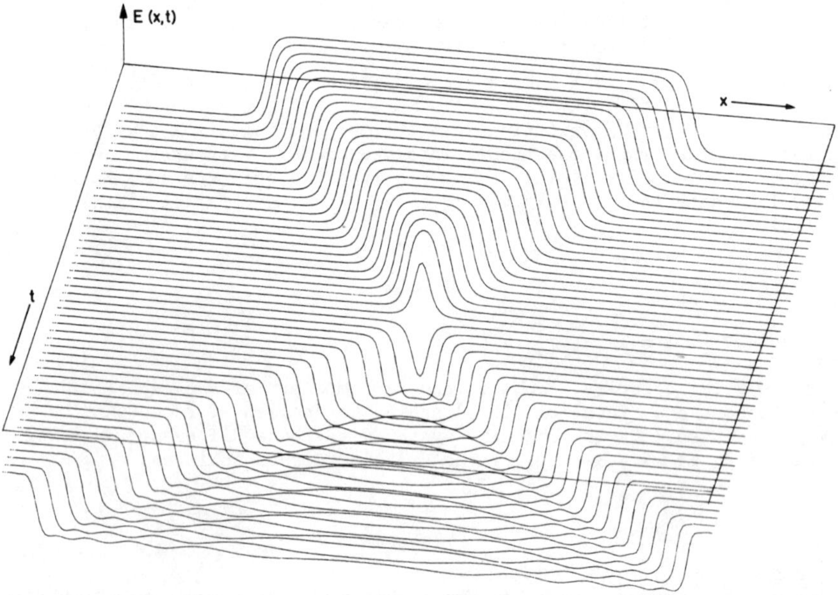

FIG. 63. Spin waves in the B-phase of liquid ^3He. The kink anti-kink sequence 2δ with -2δ emerges as the anti-kink kink sequence $2\delta-4\pi$ with $4\pi-2\delta$. The quantity σ is plotted since oscillations mask the form of the derivative and the collision is not perfect. This collision is an example of particle pair annihilation and creation. The rest energy of the 2δ pulse exceeds that of the $4\pi-2\delta$ pulse. Hence there is a relative velocity threshold for the $2\delta-4\pi$ with $4\pi-2\delta$ sequence below which particle annihilation and creation cannot occur. Below threshold the $2\delta-4\pi$ with $4\pi-2\delta$ sequence bumps but the detailed motion may be complicated by oscillations. On the other hand the 2δ with -2δ sequence will always annihilate to create the $2\delta-4\pi$ with $4\pi-2\delta$ sequence with finite relative velocity. Bumping does not occur in this case. In the figure the frame is chosen so that the observer travels with the mid-point between a kink and its anti-kink [*from unpublished work by Dr. P.J. Caudrey*].

The B-phase is particularly remarkable: in the A-phase the sign of the dipole source term $\sin\sigma$ in the sine-Gordon equation is determined by the dipole interaction H_d in the effective Hamiltonian

$$H_o + H_I + H_d \; ;$$

H_I is the exchange interaction. In ref. [51]

$$E_d = \langle H_d \rangle = -(\pi\,\gamma^2/20g^2)\,\Delta^2(T)\,[1 + 3\cos(\sigma_\uparrow - \sigma_\downarrow)] \; . \qquad (10.6)$$

In the B-phase [52]

$$E_d = \frac{2}{15}\gamma^{-2}\,\chi_B\,\Omega_\ell^{\,2}\left[\left\{(1+\cos\beta)\,[1+\cos(\alpha+\gamma)] - \tfrac{3}{2}\right\}^2 - \tfrac{5}{4}\right] \qquad (10.7)$$

in which χ_B is the static susceptibility in the BW state and Ω_ℓ is the longitudinal resonance in the B-phase: α, β, γ are Eulerian angles specifying an orientation of the spin axes.

The dipole interaction splits the degeneracies of the non-interacting system, and stable equilibrium states minimise the dipole interaction even though this is weak. The interaction (10.7) is minimised by $\beta = \gamma = 0$, $\alpha = \cos^{-1}(-\tfrac{1}{4})$. The axis for the rotation α is arbitrary, but reasons have been advanced [54] why this should be about the z-axis; this is the direction of the magnetic field. If $\beta = \gamma = 0$ and $\alpha = \tfrac{1}{2}\sigma$

$$E_d = \frac{4}{15}\gamma^{-2}\,\chi_B\,\Omega_\ell^{\,2}\left[\cos\sigma + \cos\tfrac{1}{2}\sigma + \tfrac{1}{2}\right] \; . \qquad (10.8)$$

The dipole energy is minimised at the root $\delta = 2\cos^{-1}(-\tfrac{1}{4})$ which is less than 2π and at $4\pi-\delta$. These are the roots introduced in §7 for the double sine-Gordon equation. Comparison of (10.8) with (10.6) and reference to (10.5) shows that

$$\sigma_{tt} - c_B^{\,2}\sigma_{xx} = \frac{16}{15}\,\Omega_\ell^{\,2}(\sin\sigma + \tfrac{1}{2}\sin\tfrac{1}{2}\sigma) \qquad (10.9a)$$

which scales to

$$\sigma_{xx} - \sigma_{tt} = -\sin\sigma - \tfrac{1}{2}\sin\tfrac{1}{2}\sigma \; . \qquad (10.9b)$$

Equation (10.9b) is the equation of the $\tfrac{1}{4}$- and $\tfrac{3}{4}$-attenuators mentioned only briefly in §7. The relevant phase plane is the lower one in Fig. 22. The "solitons" are the 2δ and $4\pi-2\delta$ pulses shown in Fig. 24. Two such pulses "bump" each other rather than pass through each other as noted in §7 (and [38]). The lower phase plane in Fig. 22 shows why: a 2δ spin-wave pulse traversing B-phase liquid ^{3}He in the state $\sigma = -\delta$ takes it to the state $\sigma = +\delta$; a $4\pi-2\delta$ $+\delta$ to $\sigma = 4\pi-\delta$ $(= -\delta)$. Total pulse area is conserved at 4π whatever the order of the pulses; but if the pulses come in reverse order equilibrium states are not taken to equilibrium states. Fig. 62 illustrates the collision of $4\pi-2\delta$ and 2δ pulses in ^{3}He in the state $\sigma = +\delta$; σ_t is plotted.[21] By the same arguments the order of pulses following the break-up of arbitrary spin waves into 2δ and $4\pi-2\delta$ pulses is determined by the state of the system into which break-up occurs [38].

Both the kink anti-kink pair 2δ with -2δ and the pair $4\pi-2\delta$ and $2\delta-4\pi$ have total area zero. A 2δ takes the state $\sigma = -\delta$ to $\sigma = +\delta$: the -2δ takes $\sigma = +\delta$ to $-\delta$. If the two pulses collide a $2\delta-4\pi$ can take $\sigma = -\delta$ to $\sigma = +\delta$ and a $4\pi-2\delta$ can take $\sigma = +\delta$ to $\sigma = -\delta$. Thus the kink anti-kink sequence 2δ with -2δ can emerge as the anti-kink kink sequence $2\delta-4\pi$ with $4\pi-2\delta$. This occurs (see Fig. 63). The collision is imperfect and the derivative σ_t develops oscillations sufficient to mask the pulse shapes. At lower relative velocities of

[21] For variables ξ, τ (see § 7) σ_t is plotted. For variables x, t as in (10.9) $\sigma_x + \sigma_t$ is plotted.

collision a $4\pi-2\delta$, $2\delta-4\pi$ kink anti-kink pair appears to retain its order and the two kinks bump; but there is a period in which they tend to bind (as a breather type solution) and σ_t subsequently shows oscillations. Apparently similar results are reported by Maki and Kumar [59]. For soliton theory the situation is unclear [38]. For liquid ^{3}He the results are characteristic of the double rather than single sine-Gordon equation. Studies of spin waves in the B-phase of liquid ^{3}He may therefore help with the identification of the state as a BW state with the "best" axis of rotation in the sense of Leggett [54].

This completes the discussion of solitons in Josephson junctions and in liquid ^{3}He. In §11 we turn to optical self-focussing and optical filament formation, the non-linear Schrödinger equation (equation (II) in (1.4) of §1) and some applications of this equation in plasma theory.

11. Solitons in plasma theory and the self-focussing of light in dielectrics

11.1 Ion acoustic waves in a plasma satisfy the KdV equation. This is already treated in §2.

11.2 Self-modulation of optical pulses; self-focussing of optical beams and optical filament formation.

We consider first a dielectric made up of neutral atoms or molecules. The frequency of the light is small in energy compared with the ionisation potential of these atoms or molecules and is well separated from any resonance. In this case a perturbation development is possible. The atomic dipole moment P induced in a centrosymmetric atom will depend on odd powers of the applied field E:

$$P = \alpha E + \alpha_{NL} E^3 + \ldots..$$

We suppose the linear polarisability α and the first non-linear polarisability α_{NL} are both real. These dipoles drive the field through Maxwell's equations:

$$(\nabla^2 - c^{-2}\partial^2/\partial t^2)E = 4\pi n c^{-2} \partial^2 P/\partial t^2$$

$$= \frac{4\pi n\alpha}{c^2}\frac{\partial^2 E}{\partial t^2} + \frac{4\pi n}{c^2}\alpha_{NL}\frac{\partial^2}{\partial t^2}(E^3) \qquad (11.1)$$

in which n is the number of atoms (or molecules) per unit volume.

Set

$$E = \varepsilon(x, y, z, t)\, e^{i(\omega t-kz)} + \text{c.c.} \qquad (11.2)$$

in which $\varepsilon(x, y, z, t)$ is a slowly varying complex amplitude. We shall keep first harmonics only: in principle we should keep all terms which generate first harmonics as well but these are supposed to provide only small corrections. When (11.2) is inserted into (11.1) we find

$$(\omega^2c^{-2} - k^2 + 4\pi\, n\, \alpha\, \omega^2c^{-2})\varepsilon$$
$$- 2ik\, \varepsilon_z + 2i\omega\, c^{-2}\, \varepsilon_t + 4\pi\, n\alpha\, 2i\omega\, c^{-2}\, \varepsilon_t$$
$$+ \varepsilon_{xx} + \varepsilon_{yy} + 4\pi\, n\alpha_{NL}\, \omega^2\, c^{-2}\, 3|\varepsilon|^2\varepsilon = 0 . \qquad (11.3)$$

We consider the linearised problem: in this case $\varepsilon = E_o$ = constant and $|\varepsilon|^2\varepsilon$ is supposed small. Then

$$\omega^2c^{-2} - k^2 + 4\pi\, n\alpha\, \omega^2c^{-2} = 0 . \qquad (11.4)$$

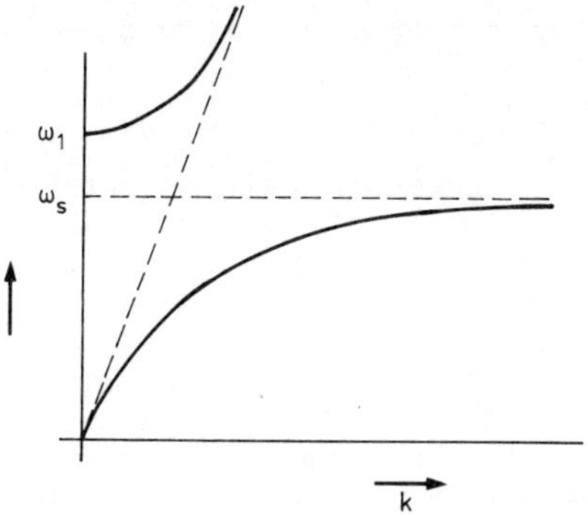

FIG.64. Dispersion relation for electromagnetic waves in a linearised 2-level atom dielectric.

This is the linear dispersion relation. For two-level atoms in particular

$$\alpha(\omega) = \frac{e^2}{m_e}\frac{1}{\omega_s^2-\omega^2} \tag{11.5}$$

and the dispersion relation takes the familiar form shown in Fig. 64. The intersept ω_1 of the upper branch on the line $k = 0$ is given by[22]

$$\omega_1^2 = \omega_s^2 + \omega_p^2 \tag{11.6}$$

where

$$\omega_p^2 = \frac{4\pi\ n\ e^2}{m_e}$$

is the plasma frequency for n electrons cm^{-3}. Formally the linearised dispersion relation for a plasma is obtained by reducing the discrete energy spacing $h\omega_s$ to zero. For $\omega_s \to 0$

$$\alpha = -\frac{e^2}{m_e\omega^2}$$

and

$$\omega^2c^{-2} - k^2 = \omega_p^2\ c^{-2}. \tag{11.7}$$

The lower branch in Fig. 64 is now eliminated.

In the weakly non-linear region we can choose ω and k to satisfy the linear dispersion relation. If (11.4) holds equation (11.3) becomes

$$\varepsilon_{xx} + \varepsilon_{yy} + \omega^2c^{-2}12\pi\ n\ \alpha_{NL}\ |\varepsilon|^2\varepsilon + 2i\ c^{-2}\left[\omega\varepsilon_t(1 + 4\pi\ n\ \alpha) - c^2\ k\ \varepsilon_z\right] = 0\ . \tag{11.8}$$

There are two cases of interest:

Case 1. ε does not depend on t (steady state result). In this case

$$\varepsilon_{xx} + \varepsilon_{yy} + \omega^2c^{-2}\ 12\pi\ n\ \alpha_{NL}\ |\varepsilon|^2\varepsilon - 2ik\ \varepsilon_z = 0$$

and ε and z can be rescaled so that

$$\varepsilon_{xx} + \varepsilon_{yy} + 2|\varepsilon|^2\varepsilon - i\ \varepsilon_z = 0\ . \tag{11.9}$$

---

[22] The argument omits the Lorentz field correction which lowers both ω_1 and the asymptote of the lower branch [60].

Case 2. Either ε does not depend on z or a new "time" variable is introduced so that after scaling

$$\varepsilon_{xx} + \varepsilon_{yy} + 2|\varepsilon|^2\varepsilon + i\,\varepsilon_t = 0\,. \tag{11.10}$$

Equation (11.9) describes steady state optical self-focussing and optical filament formation. Equation (11.10) describes self-phase modulation.

To understand self-focussing we can see from (11.3) that if ε were a constant amplitude E_o there would be a non-linear dispersion relation

$$k^2 = \omega^2 c^{-2}\left[1 + 4\pi\, n(\alpha + 3\,\alpha_{NL}\,|E_o|^2)\right]. \tag{11.11}$$

In both plasmas and neutral dielectrics providing $\alpha_{NL} > 0$ the "refractive index" $m \equiv ck\omega^{-1}$ increases with the intensity $|E_o|^2$. Thus if the light beam shows slow variations in E_o in the plane normal to the propagation direction z the light turns into the high field regions focussing and forming filaments.

An analytical theory is available only in one space dimension x (or y). In this case (11.9) is the NLS equation

$$-\,i\,\varepsilon_z + 2|\varepsilon|^2\varepsilon + \varepsilon_{xx} = 0 \tag{11.12}$$

which has the 1-soliton solution [6]

$$\varepsilon(x, z) = \frac{2\eta\,\exp\{+\,4i(\xi^2-\eta^2)z + 2i\xi x + i\delta\}}{\cosh\left[2\eta(x-x_o) - 8\eta\xi z\right]}\,. \tag{11.13}$$

Free parameters are the amplitude 2η, the carrier phase δ, the envelope "speed" 4ξ, and the envelope phase x_o. In the case $\xi = 0$ the electric field (11.2) is

$$E(x, z, t) = 2\eta\,\mathrm{sech}\{2\eta(x-x_o)\}\;\times$$
$$\times\;\exp i\left[\omega t - (k + 4\eta^2)z + \delta\right]\,. \tag{11.14}$$

The intensity profile is $4\eta^2\ \mathrm{sech}^2\{2\eta(x-x_o)\}$ and the speed V of the carrier is determined from[23]

$$\frac{1}{V} = \frac{k}{\omega} + \frac{4\eta^2}{\omega}\,.$$

The linearised dispersion relation (11.4) is again corrected by a term in the field intensity $(\propto 4\eta^2)$.

It will be seen that the soliton is acting as a wave guide channel along the z-axis. The intensity profile reshapes until for large enough z it takes on the sech^2 form. In general the wave guide channel is at angle $\theta = -\tan^{-1} 4\xi$ to the z-axis. The NLS equation has multi-soliton solutions [6]. A laser pulse of arbitrary transverse intensity profile breaks up into distinct wave guide channels at angles $\theta_i = -\tan^{-1} 4\xi_i$ to the z-axis. These are the optical filaments (in one dimension). For large z each filament has an intensity profile $\varepsilon^2 \propto \mathrm{sech}^2$ normal to the direction making an angle $\theta_i = -\tan^{-1} 4\xi_i$ to the z-axis.

The time dependent form (11.10) of the NLS displays self phase modulation in the following way. Comparison with (11.9) and (11.13) already shows that the carrier phase $\omega t - kz$ in (10.2) gains a term in the intensity $(\propto \eta^2)$. However coupling between the envelope of a pulse and its phase in the general case will always occur and may be complicated.

[23] The soliton solutions of the NLS equation do not exemplify the rule 'Bigger pulses go faster'. The envelope speed ξ and the amplitude η are independent and are connected only in that they occur combined as the complex number $\xi + i\eta$ in the theory.

In (11.10) set

$$\varepsilon(x, t) = \mathcal{A}(x, t)e^{i\phi(x,t)} . \tag{11.15}$$

Then by equating real and imaginary parts

$$\begin{aligned} 2\mathcal{A}^3 + \mathcal{A}_{xx} - (\phi_x)^2\mathcal{A} + \phi_t &= 0 \\ \mathcal{A}_t + \phi_{xx} + 2\mathcal{A}_x \phi_x &= 0 . \end{aligned} \tag{11.16}$$

The amplitude therefore drives the phase and vice versa.

11.3 The ponderomotive force. This force is a possible source of optical filament formation in plasmas. We first show how it arises.

The equation of motion for an electron in electric and magnetic fields $\vec{E}(\vec{x}, t)$, $\vec{B}(\vec{x}, t)$ is

$$m_e \vec{v}_t = - e[\vec{E}(\vec{x}, t) + \vec{v}\, c^{-1} \times \vec{B}(\vec{x}, t)] . \tag{11.17}$$

This is already non-linear in the Lorentz force term. But a second non-linearity arises because in the exact expression (11.17) we evaluate the electric field at the current position of the particle: in order to go to the continuum picture we need to evaluate the force at the electrons mean position $\vec{x}_o$. We suppose

$$\begin{aligned} \vec{E}(\vec{x}, t) &= \varepsilon\, \vec{E}_1(\vec{x}, t) + O(\varepsilon^2) \\ \vec{B}(\vec{x}, t) &= \varepsilon\, \vec{B}_1(\vec{x}, t) + O(\varepsilon^2) \end{aligned} \tag{11.18}$$

and set

$$\vec{E}_1(\vec{x}, t) \equiv \vec{E}_1(\vec{x}) \cos \omega t . \tag{11.19}$$

An electron associated with the point $\vec{x}_o$ has

$$\begin{aligned} \vec{x}(t) &= \vec{x}_o + \varepsilon\, \vec{x}_1(\vec{x}_o, t) + \varepsilon^2\, \vec{x}_2(\vec{x}_o, t) + O(\varepsilon^3) \\ \vec{v}(t) &= \varepsilon\, \vec{v}_1(\vec{x}_o, t) + \varepsilon^2\, \vec{v}_2(\vec{x}_o, t) + O(\varepsilon^3) . \end{aligned} \tag{11.20}$$

To order ε (linearised theory) we can neglect the Lorentz force and

$$\begin{aligned} \vec{v}_{1,t} &= - m_e^{-1}\, e\, \vec{E}_1(\vec{x}) \cos \omega t \\ \vec{v}_1(\vec{x}_o, t) &= - e(m_e\, \omega)^{-1}\, \vec{E}_1(\vec{x}_o) \sin \omega t \\ \vec{x}_1(\vec{x}_o, t) &= e(m_e\, \omega^2)^{-1}\, \vec{E}_1(\vec{x}_o) \cos \omega t . \end{aligned} \tag{11.21}$$

To second order

$$\vec{E}_1(\vec{x}_1) = \vec{E}_1(\vec{x}_o) + \varepsilon\, \vec{x}_1(\vec{x}_o, t)\cdot \nabla \vec{E}_1(\vec{x}_o, t) . \tag{11.22}$$

To this is added the term $c^{-1}\, \vec{v}_1 \times \vec{B}_1$ in (11.17) where, by Faraday's law,

$$\begin{aligned} - \vec{B}_{1,t} &= \nabla \times \vec{E}_1(\vec{x}, t) \\ \vec{B}_1(\vec{x}, t) &= - \omega^{-1}\, \nabla \times \vec{E}_1(\vec{x}) \sin \omega t . \end{aligned} \tag{11.23}$$

The second order part of (11.17) is now

$$m_e\, \vec{v}_{2,t} = - e[(\vec{x}_1\, (\vec{x}_o,t) \cdot \nabla) \times \vec{E}_1(\vec{x}_o, t) + c^{-1}\, \vec{v}_1 \times \vec{B}_1(\vec{x}_o, t)]$$

and if (11.21), (11.19) and (11.23) are inserted appropriately and the resulting expression averaged over t

$$m_e \vec{v}_{2,t} = - e^2(m_e \omega^2)^{-1} \frac{1}{2} \left[(\vec{E}_1(\vec{x}_o) \cdot \nabla) \vec{E}_1(\vec{x}_o) + \vec{E}_1(\vec{x}_o) \times (\nabla \times \vec{E}_1(\vec{x}_o))\right]$$

$$= - \frac{1}{4} e^2(m_e \omega^2)^{-1} \nabla \vec{E}_o^2(\vec{x}_o) . \qquad (11.24)$$

This is the force on one electron so the force per unit volume can finally be expressed as

$$\vec{F} = - (\omega_p^2 \omega^{-2}) \nabla < \vec{E}^2(\vec{x}, t)/8\pi> \qquad (11.25a)$$

$$= (\varepsilon-1) \nabla < \vec{E}^2(\vec{x}, t)/8\pi> . \qquad (11.25b)$$

We use (11.7) for $\varepsilon - 1$ and <....> denotes time average.

The result (11.25a) shows that electrons (and hence plasma) are driven out of regions of intense field (and high field gradient). The expression (11.25b) in terms of $\varepsilon - 1$ also applies to neutral atoms but, since $\varepsilon - 1 > 0$, there is now compressive electrostriction. The compressive action on the atoms is not included in the neutral dielectric theory given in §10.1, but the action on the electrons which largely determines the optical properties of this bound state material is already included in detail there. The corresponding theory for the plasma adopts (11.25a). Both plasma and neutral dielectric self-focus because ε increases with field intensity in both cases.

In the plasma the force $\vec{F}$ provides a secular drift of zero frequency (= d.c.). Charge separation occurs between electrons and ions. This creates a force on the ions to which is added the ponderomotive force on the ions: this is smaller than (11.24) by a factor $m_e m_i^{-1}$. The net force on the plasma is therefore close to (11.25) per unit volume.

To investigate self-focussing further we take the plasma density to be the ion density and take the equations for the ions in the form([61],[24] and compare §2)

$$n_t + \nabla \cdot (n \vec{v}) = 0$$

$$\vec{v}_t + \vec{v} \cdot \nabla \vec{v} = - k_B T_e (m_i n)^{-1} \nabla n + (m_i n_e)^{-1} \vec{F} ; \qquad (11.26)$$

$\vec{F}$ is given by (11.25). To first order in $\vec{v}$ the two equations (11.26) imply

$$m_i n_{tt} - k_B T_e \nabla^2 n = - \nabla \cdot \vec{F} . \qquad (11.27)$$

Maxwell's equation is

$$\vec{E}_{tt} + c^2 (\nabla \times \nabla \times \vec{E}) + n n_e^{-1} \omega_p^2 \vec{E} = \vec{0} . \qquad (11.28)$$

The derivative $\vec{j}_t$ of the electron current density drives this equation:

$$\vec{j}_t = ne \vec{v}_t = - (4\pi n_e)^{-1} n \omega_p^2 \vec{E} \qquad (11.29)$$

and $\vec{v}$ is the electron velocity in (11.21): n is the electron density because it is the ion density. If we look for transverse electric field

[24] Reference [61] treats a two-dimensional problem in which $\vec{E}$ points in the z-direction and does not depend on z.

solutions of the form (11.2) we see from (11.3) that

$$(\omega^2 c^{-2} - k^2 - \omega_p^{\,2} c^{-2})\varepsilon$$

$$2i\,k\,\varepsilon_z - 2i\,\omega c^{-2}\varepsilon_t + \varepsilon_{xx} + \varepsilon_{yy}$$

$$+ (1 - n\,n_e^{-1})\,\omega_p^{\,2} c^{-2}\varepsilon = 0 \ . \qquad (11.30)$$

The term in ε vanishes by the choice (11.7) for the dispersion relation between k and ω. The resulting equation is non-linear because n appears.

The ions respond slowly to the electromagnetic field. Despite the large value of $m_i\, m_e^{-1}$ we can neglect $m_i\, n_{tt}$ in (11.27) and use (11.25) to get

$$-\nabla^2 n = \frac{\omega_p^2}{\omega^2}\,\nabla\cdot\left\{\frac{n}{n_e}\ \frac{\nabla\,|\varepsilon|^2}{16\pi\ k_B T_e}\right\} \qquad (11.31a)$$

with solution

$$n\,n_e^{-1} = \exp -\left\{\frac{\omega_p^2}{\omega^2}\ \frac{|\varepsilon|^2}{16\pi\ k_B T_e}\right\} \ . \qquad (11.31b)$$

The solution (11.31b) shows clearly that plasma is expelled from the high field region.

We can guess now that ε filaments in the x, y plane: (11.30) is

$$-2i\,k\,\varepsilon_z + 2i\,\omega\,c^{-2}\,\varepsilon_t + \varepsilon_{xx} + \varepsilon_{yy}$$

$$+ \left[1 - \exp\left\{-\frac{\omega_p^2}{\omega^2}\ \frac{|\varepsilon|^2}{16\pi\ k_B T_e}\right\}\right]\ \frac{\omega_p^2}{c^2}\ \varepsilon = 0 \qquad (11.32)$$

which by choosing a new z-variable scales to

$$-i\,\varepsilon_z + \varepsilon_{xx} + \varepsilon_{yy} + \left[1 - \exp\left\{-2|\varepsilon|^2\right\}\right]\varepsilon = 0 \ . \qquad (11.33a)$$

In one space dimension x (say) and to first order this is

$$-i\,\varepsilon_z + \varepsilon_{xx} + 2\,|\varepsilon|^2\,\varepsilon = 0 \qquad (11.33b)$$

which is the NLS equation (11.12).

According to (11.31) the filamentary regions with high field inside them drive plasma out: regions of low plasma density trapping (or being trapped by) the high fields inside them have been called "cavitons" [61, 62, 63]. Equation (11.31) suggests the possibility of "photon bubbles" (strictly 3-dimensional regions containing localised high fields and localised drops in plasma density). Two dimensional bubbles in the critical surface are discussed in [61]. Whether these regions have any of the true properties of solitons (the perfect collision property) is an open question since only the NLS equation (11.33b) is integrable for multi soliton solutions at this stage.

The density term in (11.33a) saturates, ie. tends to unity for high fields: in contrast the first order development $2|\varepsilon|^2$ in (11.33b) diverges. Relativistic effects introduce similar terms with the same first order development and which saturate at high fields [64]. First we look briefly at a theory of SIT in semi-conductors [65]. A Hamiltonian

$$H = \left[(m^* c^{*2})^2 + (ec^*\,A/c)^2\right]^{\frac{1}{2}} \qquad (11.34)$$

has been adopted [65] in which particle momenta are suppressed in favour of the vector potential $\vec{A}$ of the field. The induced current density is

$$\vec{j} = - n c \, \delta H/\delta \vec{A} .$$

The (transverse) wave equation is therefore

$$\nabla^2 \vec{A} - \varepsilon_L c^{-2} \partial^2 \vec{A}/\partial t^2 = \varepsilon_L c^{-2} \omega_p^2 \vec{A}[1 + (eA(m/c^* c)^2]^{-\frac{1}{2}}. \quad (11.35)$$

The quantities are: m^* the effective mass of the conduction electrons, n the semi-conductor carrier concentration, and ε_L the lattice dielectric constant: $\omega_p^2 = 4\pi ne^2/m^* \varepsilon_L$. The parameter $c^* = (Eg/m^*)^{\frac{1}{2}}$ where Eg is the energy gap: $c^* < c$ by two orders of magnitude.

The effective plasma frequency is the root of

$$\omega_p^2 \left[1 + (eA/m^* c^* c)^2\right]^{-\frac{1}{2}} \quad (11.36)$$

and is reduced below ω_p^2 falling to zero (like (11.31b)) in strong fields: thus propagation at frequencies below the plasma frequency is possible in semi-conductors. The dielectric constant increases with A and the system self-focusses.

For circularly polarised solutions

$$\vec{A} = \alpha(x, y, z, t) \left[\hat{i} \cos(\omega t - kz + \phi) + \hat{j} \sin(\omega t - kz + \phi)\right] \quad (11.37)$$

equation (11.35) is

$$\nabla^2 \vec{A} - \varepsilon_L c^{-2} \vec{A}_{tt} = \varepsilon_L c^{-2} \omega_p^2 \vec{A} \left[1 + (e\alpha/m^* c^* c)^2\right]^{-\frac{1}{2}} \quad (11.38)$$

and if

$$2\varepsilon = \frac{e\alpha \, e^{i\phi}}{m^* c^* c}$$

$$(\nabla^2 - \varepsilon_L c^{-2} - \partial^2/\partial t^2) \, \varepsilon e^{i(\omega t - kz)} = \frac{\varepsilon_L c^{-2} \omega_p^2 \varepsilon e^{i(\omega t - kz)}}{(1 + 4|\varepsilon|^2)^{\frac{1}{2}}} . \quad (11.39)$$

Equation (11.30) follows with $c \, \varepsilon_L^{-\frac{1}{2}}$ replacing c and $1 - n/n_e$ replaced by $1 - [1 + 4|\varepsilon|^2]^{-\frac{1}{2}}$. The one-dimensional first order development is (11.33b).

A precisely similar argument for the relativistic plasma finds

$$\vec{j} = \frac{-n_e \, e \, \alpha[\hat{i} \cos(\omega t - kz + \phi) + \hat{j} \sin(\omega t - kz + \phi)]}{[1 + (e\alpha/mc^2)^2]^{\frac{1}{2}}} . \quad (11.40)$$

If $2\varepsilon = (e\alpha/mc^2) \exp i\phi$, the same equations follow: in particular (11.33a) becomes

$$- i \varepsilon_z + \varepsilon_{xx} + \varepsilon_{yy} + \left[1 - (1 + 4|\varepsilon|^2)^{-\frac{1}{2}}\right]\varepsilon = 0 . \quad (11.41)$$

In one space dimension x and to first order this is (11.33b). At high powers (11.41) saturates like (11.33a). Peak powers where relativistic effects like this are relevant are $\sim 10^{18}$ W cm^{-2}: these figures will be reached in laser-fusion work.

Tappert [66] has investigated both equation (11.41) and (11.33a) numerically and finds optical filament formation in the plasma. Valeo and Estabrook [61] display contour plots of "bubble" formation in two dimensions.

11.4 The 3-wave interaction. This occurs in neutral dielectrics (see the notes by T.P. McLean, this volume and e.g. [67, 68]), as well as in plasmas [69-73]. The associated system of equations has been solved for soliton solutions [74, 75].

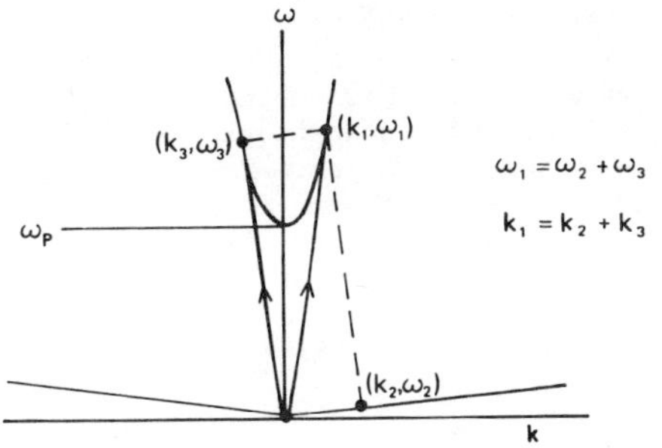

FIG.65. Stimulated 'Brillouin backscattering'. A light wave (k_1, ω_1) excites an ion wave (k_2, ω_2) and another light wave (k_3, ω_3) moving in the opposite direction. The straight lines are ion dispersion relations: the parabola is the dispersion relation (11.7) for light also shown as the upper branch in Fig.64 in the case when $\omega_s \to 0$.

Three wave packets with frequencies ω_1, ω_2 and ω_3 and wave vectors $\vec{k}_1$, $\vec{k}_2$ and $\vec{k}_3$ can interact in two ways - "decay type" interactions and "explosive" type. In the former energy and momentum conservation take the forms

$$\omega_1 = \omega_2 + \omega_3 , \qquad \vec{k}_1 = \vec{k}_2 + \vec{k}_3$$

respectively. The Fig. 65 exemplifies these relationships for stimulated Brillouin back scattering in a plasma: an ion wave (k_2, ω_2) is excited by incident light (k_1, ω_1) and light is back scattered at a shifted frequency ($\omega_1 - \omega_2$), light can similarly excite an ion wave and a backward going plasma (electron) wave. Or it can excite a plasma wave and back scatter as light which has undergone "stimulated Raman scattering". The phraseology is based on the similar processes in solids.

In the decay interaction the complex envelopes ε_1, ε_2 and ε_3 satisfy

$$\begin{aligned} \varepsilon_{1,t} + v_1\, \varepsilon_{1,x} &= iq\, \varepsilon_2\, \varepsilon_3^* \\ \varepsilon_{2,t} + v_2\, \varepsilon_{2,x} &= iq\, \varepsilon_1\, \varepsilon_3 \\ \varepsilon_{3,t} + v_3\, \varepsilon_{3,x} &= iq\, \varepsilon_1^*\, \varepsilon_2 . \end{aligned} \qquad (11.42)$$

The v_i ($i = 1, 2, 3$) are non-dispersive velocities and q is an interaction constant (which can be scaled away). If the three envelopes are well separated

$$\varepsilon_i = \varepsilon_i(t - x\, v_i^{-1}) \qquad (i = 1, 2, 3) \qquad (11.43)$$

so that changes of shape occur only during the interactions between the packets.

"Explosive" interactions are possible only in systems where negative energy waves can exist (but just as in resonant non-linear optics this _is_ possible in the absence of thermal equilibrium [73]). In this case the energy momentum relations are

$$\omega_1 + \omega_2 + \omega_3 = 0 , \qquad \vec{k}_1 + \vec{k}_2 + \vec{k}_3 = \vec{0}$$

and the complex envelopes satisfy

$$\varepsilon_{1,t} + v_1\, \varepsilon_{1,x} = iq\, \varepsilon_2^*\, \varepsilon_3^* \qquad (11.44)$$

together with its obvious permutations. In both cases (11.42) and (11.44) we assume for definiteness that $v_1 < v_2 < v_3$.

Both (11.42) and (11.44) have been treated when ε_1, ε_2 and ε_3 depend only on t or only on x (cf. McLean's notes). When they depend on both x and t a solution has been obtained by an inverse scattering method [74, 75]. The use of the inverse scattering method is unusual in that it involves a 3×3 scattering problem. Problems previously treated this way had involved at most a 2×2 problem [21].

A soliton solution of either (11.42) or (11.44) is [74, 75]

$$V_{23} = \frac{2\eta_3}{D} e^{i\xi_3(x-v_1t-\bar{x}_1)} \left[e^{\eta_1(x-v_3t-x_3)} - \gamma_1\gamma_2 \frac{\bar{\zeta}_1-\zeta_3}{\bar{\zeta}_1{}^*-\zeta_3} e^{-\eta_1(x-v_3t-x_3)} \right] \tag{11.45a}$$

$$V_{31} = \frac{-4\eta_1\eta_3\beta_{13}\gamma_2\gamma_3}{\beta_{12}\beta_{23}(\bar{\zeta}_1-\zeta_3{}^*)} \frac{1}{D} e^{-i\xi_1(x-v_3t-\bar{x}_3)} e^{-i\xi_3(x-v_1t-\bar{x}_1)} \tag{11.45b}$$

$$V_{12} = \gamma_1\gamma_2 \frac{2\eta_1}{D} e^{i\xi_1(x-v_3t-\bar{x}_3)} \left[e^{\eta_3(x-v_1t-x_1)} - \gamma_2\gamma_3 \frac{\bar{\zeta}_1{}^*-\bar{\zeta}_3{}^*}{\bar{\zeta}_1{}^*-\zeta_3} e^{-\eta_3(x-v_1t-x_1)} \right] \tag{11.45c}$$

where

$$D = \left[e^{\eta_1(x-v_3t-x_3)} - \gamma_1\gamma_2 e^{-\eta_1(x-v_3t-x_3)} \right] \left[e^{\eta_3(x-v_1t-x_1)} - \gamma_2\gamma_3 e^{-\eta_3(x-v_1t-x} \right.$$

$$+ \frac{\gamma_1\gamma_3(\bar{\zeta}_1-\bar{\zeta}_1{}^*)(\zeta_3-\zeta_3{}^*)}{(\bar{\zeta}_1-\zeta_3{}^*)(\bar{\zeta}_1{}^*-\zeta_3)} e^{-\eta_1(x-v_3t-x_3)} e^{-\eta_3(x-v_1t-x_1)} . \tag{11.45d}$$

The envelopes ε_i are related to the V_{jk} by

$$V_{23} = \varepsilon_1(\beta_{12}\beta_{13})^{-\frac{1}{2}} \quad \text{etc.} \tag{11.46}$$

where

$$\beta_{ij} = v_i - v_j ;$$

x_1, $\bar{x}_1$, x_3 and $\bar{x}_3$ are arbitrary; the complex numbers $\bar{\zeta}_1$ and ζ_3 are free complex parameters and

$$\beta_{12}\bar{\zeta}_1 \equiv \xi_1 - i\eta_1 \qquad (\eta_1 > 0)$$

$$\beta_{23}\zeta_3 \equiv \xi_3 + i\eta_3 \qquad (\eta_3 > 0) . \tag{11.47}$$

The γ_i satisfy $\gamma_i{}^2 = q^2$ and the choice of signs determines whether the solution satisfies (11.42) or (11.44), that is whether the solutions are of decay or explosive type. For the choice $(\gamma_1, \gamma_2, \gamma_3) = (+q, +q, +q)$ D is not positive definite [75]: in general it has zeros at some finite time t and the wave packets "explode". The only alternative choice proves to be $(\gamma_1, \gamma_2, \gamma_3) = (+q, -q, +q)$ [75], and in this case D is positive definite and the solution is of decay type, satisfying (11.42).

In either case as $t \to -\infty$ (11.45) has the form $\varepsilon_2 = 0$, $\varepsilon_3 = \varepsilon_3(x - v_3t)$, $\varepsilon_1 = \varepsilon_1(x - v_1t)$ with ε_1 to the right of ε_3: in the collision ε_2 is produced and decays back to zero whilst ε_1 and ε_3 separate and move with speeds v_1 and v_3 and unchanged amplitudes but with ε_3 now to the right of ε_1. It is assumed as before that $v_3 > v_1$. The general situation is that if $v_3 > v_2 > v_1$ and ε_2 contains solitons before collision then ε_2 does not contain solitons after collision whilst ε_3 and ε_1 keep their solitons but each gain a number of solitons equal to the number originally in ε_2! Shapes are not preserved after collision. For details the reader is referred especially to [75].

This completes in this §11 a rather selective set of applications of soliton theory to plasmas: there are others (for example Zakharov's treatment of Langmuir turbulence as caviton type collapse [76]). As a final application of the NLS equation we look in the next section at the significance of this equation to waves in deep water.

12. The Non-linear Schrödinger equation for deep water waves

We refer back to Fig. 2 for notation although we are concerned with any pulse of displaced water and not just the KdV soliton displayed there. Two scaling parameters determine the behaviours of waves in water of finite depth: $\varepsilon \equiv ah^{-1}$ and $\delta \equiv h^2\lambda^{-2}$ which scale amplitude and wave length respectively.

In shallow water h is small, δ is small for long wave lengths and ε is large even for moderate values of a. In this case the system is in the non-linear but non-dispersive regime governed by the equation (1.7a) of a "simple wave". Wave solutions break as discussed from (1.7b).

If $\delta = K\varepsilon$ where K is some number and δ and ε are not negligible the system is in the weakly non-linear weakly dispersive regime and motion is governed by the KdV equation (1.3).

If ε is small and δ takes on moderate values the non-linearity is weak but the dispersion is not and the system is governed by the NLS equation. In deep water ε is always very small and δ takes on moderate values in the long wave length limit.

Yuen and Lake [8] derive the equation

$$i\left[A_t + \tfrac{1}{2}(\omega_0/k_0)A_x\right] - (\omega_0/8k_0^2)A_{xx} - \tfrac{1}{2}\,\omega_0 k_0^2|A|^2 A = 0 \qquad (12.1)$$

for the wave envelope A: $A = a\,e^{i\phi}$ and the displacement of the water surface is $a(x,t)\cos\{\omega_0 t - k_0 x + \phi(x,t)\}$; $\omega_0 = \sqrt{gk_0}[1 + O(k_0^2 a_0^2)]$ where a_0 is the amplitude of a. With this scaling the one soliton solution is (compare (11.13))

$$A = a_0 \operatorname{sech} \sqrt{2}\,k_0^2\,a_0\left[(x - X_0) - (\frac{\omega_0}{2k_0} + v_0)t\right]$$
$$\times \exp\left\{-\tfrac{1}{2}k_0^2\,a_0^2\,\omega_0 t - \frac{4i\,k_0^2}{\omega_0}\,v_0\left[(x - X_0) - (\frac{\omega_0}{2k_0} + v_0)t + \theta_0\right]\right\} \qquad (12.2)$$

The speed v_0 is now relative to the linear group velocity $\omega_0/2k_0$: it is independent of the amplitude a_0 and there are two free parameters. The Fig. 66 taken from [8] shows the separate evolution of two wave pulses and (right-hand trace) the interaction of the two pulses: note the break-up of the pulse on the central trace and the soliton collision property exemplified in the right-hand trace.

Yuen and Lake [8] use Whitham's method of the averaged Lagrangian [77] to derive (12.1). It has been derived otherwise (cf. e.g. [78]). The frequency ω and wave number k are slowly varying quantities in Whitham's method and Yuen and Lake find the dispersion relation

$$\omega = \sqrt{gk}\,\left[1 + \tfrac{1}{2}k^2a^2 + a_{xx}/8k^2\right]. \qquad (12.3)$$

The interesting feature is the dispersive term $a_{xx}/8k^2$. Lighthill [79] found similar results but without this term. The system shocks and there are no solitons. The averaged Lagrangian method has been applied to the coupled Bloch-Maxwell equations (4.8) with (4.9) used in resonant non-linear optics [80]: predicted pulse evolutions agree well with observations on Rb vapour for short times [81], but the theory is equivalent to Lighthill's for water waves and optical shocks are predicted over longer times. Dispersive terms are missing however [81] and when these are included [82] results agree closely with those derived from the SIT equations (the equations derived are more complete than the SIT equations (4.17)): the soliton solutions are then the soliton solutions of the SIT equations [82].

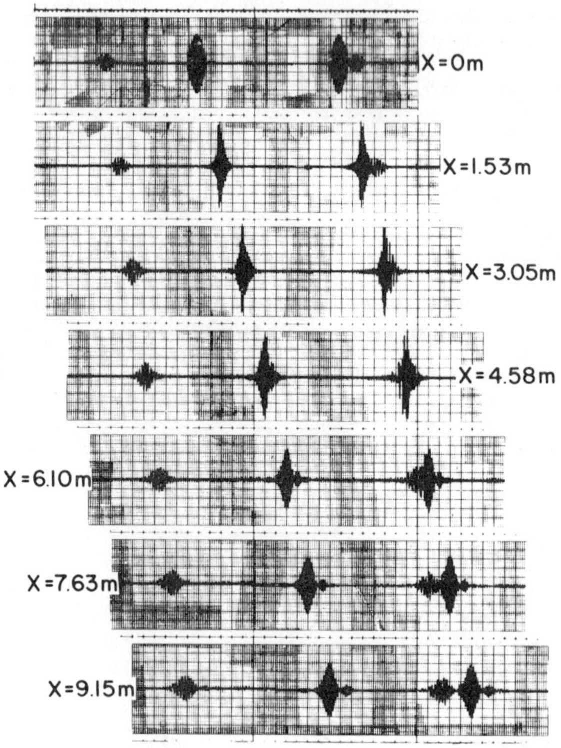

FIG.66. Evolution of two pulses in deep water (left and centre traces) and their soliton-like collision (right-hand trace). The pulse in the centre trace breaks up into two solitons; and this behaviour is not changed by the collision process (right-hand trace) (taken from [8]*).*

This completes a series of physical applications of the NLS equation and we look briefly next at an application of soliton theory to the theory of non-linear lattices.

13. Solitons in lattice theory

13.1 The Fermi-Pasta-Ulam problem was mentioned in §3.

13.2 The Toda Lattice

This beautiful linear lattice is completely integrable. The infinite lattice is exactly soluble by a form of inverse scattering method [83, 84] and has multi-soliton solutions. It is governed by the Hamiltonian

$$H = \sum_{\ell} \left\{ \tfrac{1}{2} p_\ell^{\,2} + [\exp\{-(u_\ell - u_{\ell-1})\} - 1] \right\} \qquad (13.1)$$

in which (p_ℓ, u_ℓ) are the momenta and displacement of the ℓth particle. It may be scaled to encompass the harmonic and hard core limits [2]. The one and two soliton solutions found by Toda [85, 86] approach the corresponding solutions of the KdV equation (1.3) in an appropriate limit; and Toda and Wadati [87] further show that the motion of the infinite lattice approaches the KdV equation itself in that limit. This establishes the connection with the FPU problem as discussed in §3. The properties of the lattice are studied by Toda in an authoritative set of lecture notes [88] and the reader is referred to these.

Krumhansl and Schrieffer in a fundamental study of a strongly anharmonic linear chain [89] found particular solutions of kink-like character defining "clusters". They found that the exactly calculated low temperature static properties could also be derived by assuming that these kinks were elementary excitations on a par with the phonons.[25] The model differs from that of either the FPU problem or the Toda lattice in that phase transitions are possible. The Hamiltonian was of the type

$$H = \sum_{\ell}\left\{\tfrac{1}{2}\,p_{\ell}^{2} + \tfrac{1}{2}\,A\,u_{\ell}^{2} + \tfrac{1}{4}\,B\,u_{\ell}^{4} - C\,u_{\ell}u_{\ell+1}\right\}$$

$$= \sum_{\ell}\left\{\tfrac{1}{2}\,p_{\ell}^{2} + \tfrac{1}{2}(A - 2C)u_{\ell}^{2} + \tfrac{1}{4}\,B\,u_{\ell}^{4} + \tfrac{1}{2}\,C(u_{\ell} - u_{\ell+1})^{2}\right\} . \qquad (13.2)$$

The ℓth particle moves in a local potential

$$\tfrac{1}{2}(A - 2C)u_{\ell}^{2} + \tfrac{1}{4}\,B\,u_{\ell}^{4} \qquad (13.3)$$

superimposed on the harmonic forces provided by first neighbours. The local potential has a minimum at the roots of

$$(A - 2C)u + B\,u^{3} = 0 \; . \qquad (13.4)$$

If $B > 0$ and $(A - 2C) > 0$ this minimum is at $u = 0$. But if $B > 0$ and $(A - 2C) < 0$ there are two minima at

$$u = \pm\left[(2C-A)/B\right]^{\frac{1}{2}} \equiv \pm\,u_{o}$$

and the system can undergo a "displacive" phase transition: $2C - A = 0$ is the displacive limit.

A continuum approximation to the system is

$$c_{o}^{2}\,u_{xx} - u_{tt} = \left\{(A - 2C)u + B\,u^{3}\right\} \qquad (13.5)$$

where $c_{o}^{2} = a^{2}C$ (a is the lattice spacing). The zeros of the right side occur at the roots of (13.4). If $(A - 2C) < 0$, a kink-like solution is

$$u = +\,u_{o}\tanh\left[(x - vt)/\sqrt{2}\xi\right] \qquad (13.6a)$$

where

$$\xi^{2} = (c_{o}^{2} - v^{2})/|A - 2C| \; . \qquad (13.6b)$$

This kink takes u from $+u_o$ to $-u_o$: an anti-kink (change the sign of u) takes u from $-u_o$ to $+u_o$. Between $\pm u_o$, u is displaced positively from $u = 0$ initially and becomes displaced negatively as u approaches $-u_o$. The kink thus forms a boundary between regions where $u > 0$ and where $u < 0$. These regions are the clusters. Kinks and anti-kinks form boundaries between clusters.

Evidently a kink must be followed by an anti-kink. Hence these kinks cannot have the collision property of solitons. Numerical evidence supports this conclusion [90].[26] In the similar situation of a 2-dimensional Ferrodistortive XY model ([91] and see below) Schneider and Stoll derive a sine-Gordon equation rather than (13.5). This indicates that solitons themselves are relevant to cluster formation in non-linear lattices in the

[25] Further work (by Bishop, A.R., Currie, J.R., Trullinger, S.E., Krumhansl, J.A.) to be presented at the Manchester Solid State meeting (January 1977) has both generalised and rigorised this first work by Krumhansl and Schrieffer (see also KOEHLER, T.R., BISHOP, A.R., KRUMHANSL, J.A., SCHRIEFFER, J.R., Solid State Commun. **17** (1975) 1515, and AUBRY, S., J. Chem. Phys. **62** (1975) 3217.).

[26] Dr. A.R. Bishop tells me that bumps (like the double sine-Gordon bumps) have been seen in numerical studies of the collisions between kinks. For other numerical work on (13.5) (on a pseudo-breather solution) see KUDRYAVTSEV, JETP Lett. **22** (1975) 82.

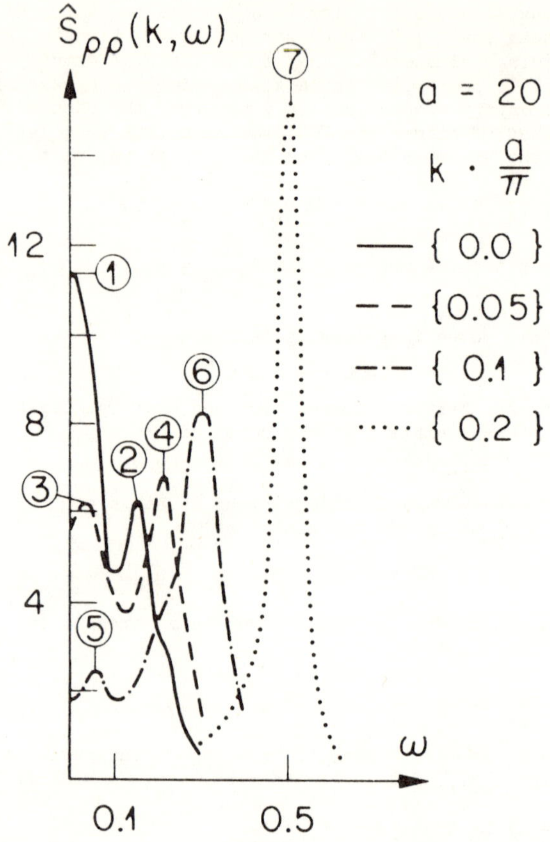

FIG.67. The dynamical structure factor $\hat{S}_{\rho\rho}(k,\omega)$ plotted for some fixed k-values for the Hamiltonian (13.2) in the displacive regime (taken from [92]).

displacive regime. The kinks (13.6) are in some sense "approximate" solitons.

Before looking briefly at the Ferrodistortive model we look at a numerical study [92] by the same authors based on the 1-dimensional Hamiltonian (13.2). The system apparently undergoes a phase transition at $T = T_c = 0$[27] where the quantity $N^{-1} \sum_{\ell} \sum_{\ell'} \langle u_\ell u_{\ell'} \rangle$ diverges. N is the number of particles and periodic boundary conditions are adopted. The average <....> is evaluated as a time average: the method, the definition of temperature, and the connection with averages within the microcanonical ensemble are discussed in [93]. Results reported in [92] traceable to the existence of clusters include that of a "central peak" at $\omega = 0$ in the dynamical structure factor $S(k, \omega)$ and an extra branch in the (ω, k) dispersion relation near $T = T_c$. The central peak phenomenon is also discussed in [89]. A central peak in $S(k, \omega)$ has been observed in Sr Ti O_3 [94].

[27] It is a result due to Landau that no phase transition with $T_c > 0$ can occur in a 1-dimensional system with a finite range interaction. It is not, therefore, possible to explore behaviour for $T < T_c$ through a 1-dimensional model.

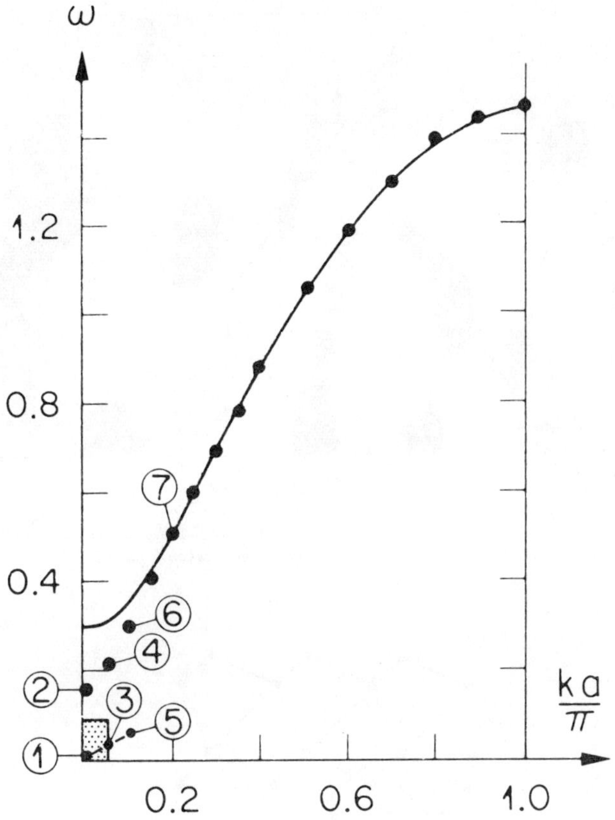

FIG.68. The dispersion relation for the Hamiltonian (13.2) in the displacive regime. Note the new branch of the spectrum connecting the points ①, ② and ③ (taken from [92]).

The dynamical structure factor $S(k, \omega)$ is defined by

$$S_{\rho\rho}(k, \omega) = \int_{-\infty}^{\infty} dt \exp(-i \omega t) \langle \rho(-k, 0) \rho(k, t) \rangle \tag{13.7a}$$

where

$$\rho(k, t) = N^{-1} \sum_{\ell} \exp i k(a\ell + u_{\ell}) \quad . \tag{13.7b}$$

$N = 2000$. The time dependent average $\langle .... \rangle$ is defined in [93]. The Fig. 67 shows the quantity $\hat{S}_{\rho\rho}(k, \omega) \equiv S_{\rho\rho}(k, \omega)/S_{\rho\rho}(k, t = 0)$ plotted at some fixed k-values for the choice $A = 7/8$, $B = 1/3$, $C = 1/2$; $k_B T = 0.0468$. The peak maxima ②, ④, ⑥ are identified as soft mode resonances. In the case when $k = 0$, ② is the soft mode resonance and ① at $\omega = 0$ is the central peak. For $k \neq 0$ the central peak splits into a double peak structure (③ and ⑤). Peaks ①, ③ and ⑤ are evidence for a new branch of the dispersion relation.

The Fig. 68 shows this dispersion relation. The full line is obtained in the self-consistent phonon approximation. The points are peak maxima in $\hat{S}(k, \omega)$. To study the new branch the Fourier transform

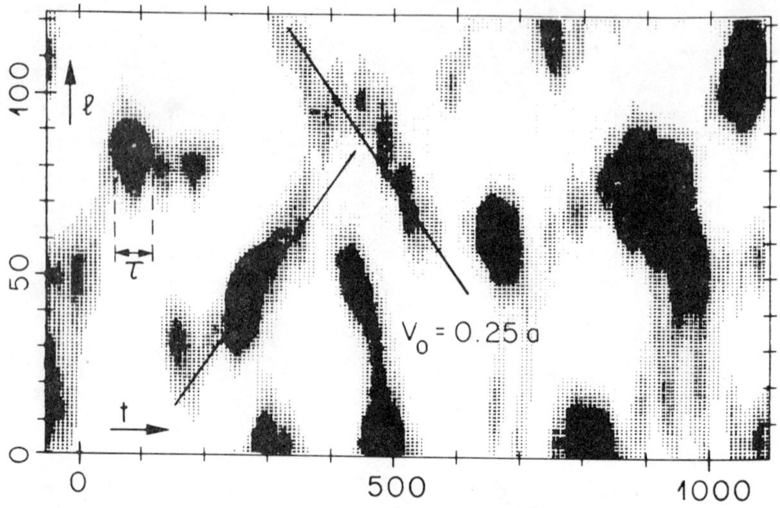

FIG.69. Hypsometric plot of the slowly varying part of $u_\ell(t)$ for the Hamiltonian (13.2) in the displacive regime: v_0 is the mean velocity of cluster waves and τ is the cluster lifetime (taken from [92]).

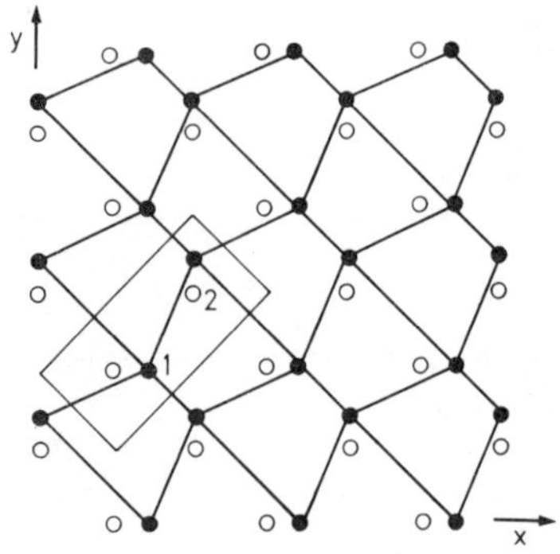

FIG.70. A typical 2-dimensional "ferrodistortive" structure. The open circles mark a primitive square lattice.

$u(k, \omega)$ of $u_\ell(t)$ is Fourier inverted over the set of (k, ω) lying in the shaded region of Fig. 68. This isolates the slowly varying part of $u_\ell(t)$.

The Fig. 69 shows the result: blackening indicates increasing positive $u_\ell(t)$ and negative u_ℓ are suppressed. Evidently clusters of locally ordered u_ℓ exist and move with velocity $v_0 \sim 0.25a$. This is close to the group velocity (0.2a) in the new excitation branch in Fig. 68.

An important result from Fig. 69 is the finite lifetime τ of the clusters: true solitons have infinite lifetime; the continuum kink solution (13.6) also has but its collision properties are uncertain. The finite lifetime means that an approach to an ergodic state is possible in contrast with the results of the FPU model (§3). Schneider and Stoll [92] argue that, for small $\omega = v_0 k \sim 0$, damped travelling wave clusters become over damped and are responsible for the central peak. The inverse lifetime τ^{-1} determines the peak half width. They find in the order-disorder regime where $A < 0$ (e.g. $A = -1$, $B = 1/3$, $C = 1/2$ [92]) that at $k_BT = 1.5$ clusters are long lived and do not propagate whilst at $k_BT = 3.0$ lifetimes are shorter and cluster velocities ($v_o \sim 0.085a$) are finite. These results suggest that τ increases with decreasing T and that the central peak half width vanishes at $T = T_c = 0$.

Similar results are reported [91] for a 2-dimensional "Ferrodistortive XY model".[28] A typical ferrodistortive (2-dimensional) structure is shown in Fig. 70. The points O define a primitive square lattice; but in the distorted structure the unit cell contains two particles as shown. The model used in [91] is defined by the Hamiltonian

$$H = \tfrac{1}{2} \sum_{\ell,\alpha} p^2_{\ell\alpha} + \frac{A}{2} \sum_{\ell,\alpha} u^2_{\ell\alpha} + \frac{B}{4n} \sum_{\ell} \Big(\sum_{\alpha} u^2_{\ell\alpha}\Big)^2 + \frac{B_1}{4} \sum_{\ell,\alpha} u^4_{\ell\alpha} - C \sum_{<\ell\ell'>,\alpha} u_{\ell\alpha}\, u_{\ell'\alpha} \,. \tag{13.8}$$

A, B, B_1, C and n are constants of the model ($n = 2$ [95]). The notation $<\ell\ell'>$ indicates summation over first neighbours (in 2 dimensions); XY models couple $u_{\ell\alpha}$ and $u_{\ell'\alpha}$ but not $u_{\ell\alpha}$ and $u_{\ell'\beta}$; $\alpha = 1, 2$. Other XY models are discussed in [96]. For $C > 0$, $(A - 2C) < 0$, $B_1 < 0$ and $(B/n + B_1) > 0$ the local potential of the ℓth particle has minima lying at points $(u_{\ell 1}, u_{\ell 2}) = (\pm u_o, 0)$ and $(0, \pm u_o)$ where $u_o = [(2C - A)/(B/n + B_1)]^{\frac{1}{2}}$. By choice of parameters the principal motion can be an angular motion at a fixed distance $\rho = u_o$ from $(0,0)$. For a fixed ρ and varying angle ϕ the continuum limit of the angular motion is

$$c_o^2(\phi_{xx} + \phi_{yy}) - \phi_{tt} = -\tfrac{1}{4} B_1 \rho^2 \sin 4\phi \ , \tag{13.9}$$

This is the sine-Gordon equation in two space dimensions.

Distinct zeros of the right side of (13.9) occur at $\phi = 0, \frac{1}{2}\pi, \pi, \frac{3}{2}\pi$. These angular sites are labelled −, O, ● and | in the Fig. 71. This figure shows cluster patterns for the case $A = -1$, $B = 4/5$, $B_1 = -1/15$, $C = 1/4$ and k_BT:- (a) 2.0 (b) 2.5 (c) 3.0 (d) 6.0. Since $k_BT_c \approx 2.72$ temperatures range from below to above T_c. The system is prepared so that $<u_{\ell 1}> \neq 0$ and $<u_{\ell 2}> = 0$ at $T = 0$. Clusters of configurations | , and O appear as T increases: closer to T_c ● configurations also appear.

In the continuum limit the kink solutions

$$\begin{aligned} \phi(x, y = 0, t) &= \tan^{-1} \exp\theta \\ \theta &= \alpha(x - Vt) \\ \alpha^2 &= -\frac{B_1\rho^2}{c_o^2}\Big(1 - \frac{V^2}{c_o^2}\Big)^{-1} \end{aligned} \tag{13.10}$$

travel along x with velocity V and take ϕ from $\phi = \frac{\pi}{2}$ to $\phi = 0$. An N-kink solution can change ϕ by $N\pi/2$. Note that the derivative ϕ_t is $\phi_t = \frac{1}{2}\alpha V \,\mathrm{sech}\,\theta$ which differs from the familiar 1-soliton result (1.1) only by the extra factor $-\frac{1}{4}$. The 1-kink solution (5.1) of (4.24) differs similarly from (13.10) (or compare (9.19)).

The numerical data for the discrete problem (13.8) show central peaks and a new branch of the dispersion relation. The reader is referred to [91] for details. The central peak has finite width and presumably corresponds to finite kink lifetimes. This seems to mean that finite

---

[28] See Ref. [93]. The Hamiltonian in [93] differs from (13.8).

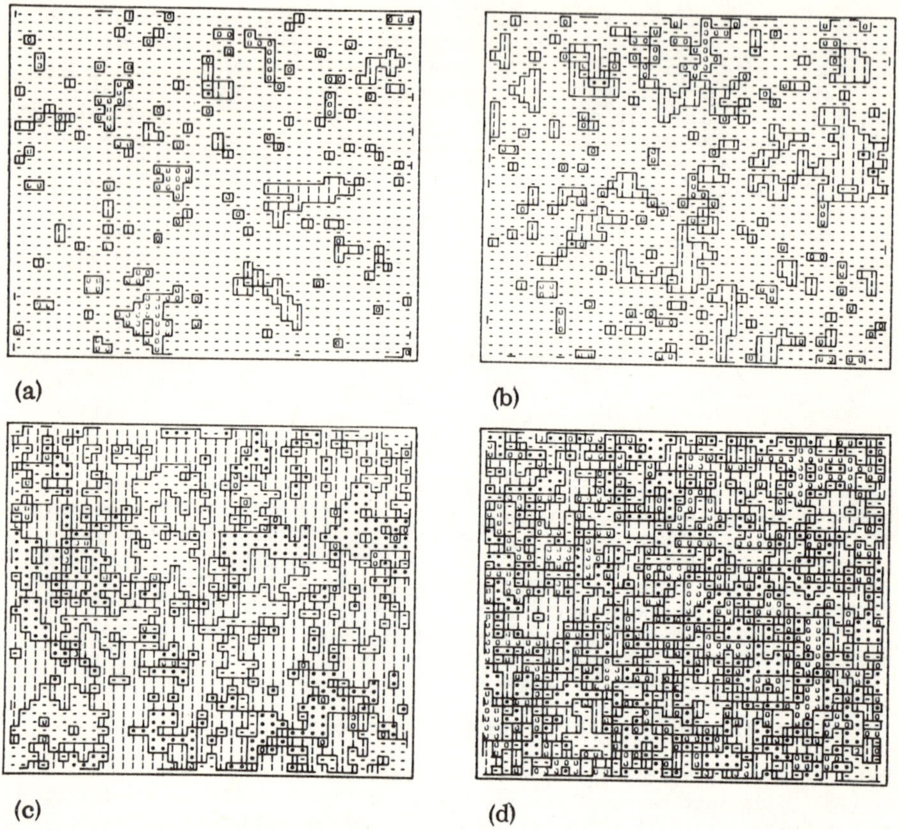

FIG.71. Cluster patterns at different temperatures for the 2-dimensional "ferrodistortive" model with Hamiltonian (13.8). For details see the text near Eq. (13.9) (taken from [91]).

lifetimes can arise simply as a consequence of the discrete rather than continuous nature of the model.[29] For soliton theory a conclusion based on this numerical work is that "approximate" solitons with finite lifetimes will play an important role in physical applications and that true solitons with their infinite lifetimes and exact collision property are a specially elegant sub-class of these physically applicable entities.

We complete these lectures by reference to one further example of the application of solitons in physics. This is to the fundamental particles.

14. Solitons in particle physics

Because of the particle-like properties of solitons, non-linear wave equations with soliton solutions or approximate soliton solutions (§13) are obvious candidates for model field theories. In one space and

[29] Results by Currie, Trullinger, Bishop and Krumhansl ("Numerical simulation of sine-Gordon soliton dynamics in presence of perturbations", to be published) show that, in the displacive limit where the slowly varying nature of the kink makes the continuum approximation acceptable, the correction due to the discrete character of the lattice is very small (this is no longer true in the order-disorder regime). It would seem therefore that the finite lifetime of the kinks is due more to the fact that (13.9) is an approximation based on fixing ρ.

one time dimension the NLS equation (1.4) and the s-G equation (1.5a) are obvious candidates. The latter is Lorentz invariant and was studied as a classical field theory by Skyrme [97, 98]. He found the 1- and 2-kink solutions and the breather (0π). Kinks and anti-kinks are identified as particles and anti-particles: the twists $(2\pi$ or $-2\pi)$ are normalised to ± 1 and identified as charges. The breather is therefore chargeless and is interpreted as a meson. Studies of the phase shifts in collisions show that kink-kink pairs repel but mixed kink anti-kink pairs attract (see especially [99]).

Solitons provide in themselves a natural quantisation (recall the "quantised" fluxons of §9). One can also impose quantisation on the classical fields. Lecture notes by Coleman [100] and Fadeev [101] as well as the papers referred to in e.g. [102] summarize some recent work in this direction.

In the series of papers culminating in reference [102] the authors develop functional integration techniques to quantise a solution of a chosen non-linear classical wave equation. These equations include the "ϕ-four" equation $\phi_{xx} - \phi_{tt} = \phi - \phi^3$ [103], which is the equation (13.5) and the s-G $\phi_{xx} - \phi_{tt} = m^2 \sin\phi$ [102]. In the new variables the kink solution (13.6) is a solution of the ϕ-four which survives under quantisation. It has a finite self-energy and additional quantised excited states can be identified as a mass spectrum. That part of the energy non-analytic in the coupling constant γ appears in the contribution from the classical solution: quantised corrections appear as a power series in γ (γ is so chosen as to absorb the quantum parameter $\hbar$). The same approximate methods applied to the breather solution of the s-G yield results which appear to be exact [102]. The reason is presumably contained in the facts that the ϕ-four does not have exact soliton solutions (cf. §13): the s-G does.

For the significance of this we need to know more about the inverse scattering method for finding soliton solutions. The 2×2 inverse scattering scheme of Ablowitz et al. [21] is referred to in §3. We briefly summarize the "philosophy" of the inverse method here.

Given a field $u(x, t)$ satisfying a non-linear evolution equation $u_t = K[u]$ where $K[u]$ is a non-linear functional of u (e.g. $K[u] \equiv -6uu_x - u_{xxx}$ for the KdV) one wishes to compute from initial data $u(x, 0)$ the quantity $u(x, t)$ at successive times $t > 0$. The inverse scattering method does this by the following route: with $u(x, t)$ associate a "wave function" $\psi(x, t; \lambda)$, a <u>linear</u> scattering problem $L\psi = \lambda\psi$, and a time evolution of ψ $A\psi = \psi_t$. The differential operator L contains $u(x, t)$ linearly as a scattering potential (!) and A is a (non-linear) functional of u (In the case of the KdV $L \equiv -\partial^2/\partial x^2 + u(x, t)$ in a 1×1 scheme and the scattering problem is exactly the Schrödinger problem with u as potential).

Since u evolves in time the eigenvalues λ of the scattering problem depend in general on the time. But under the condition $\lambda_t = 0$ one easily sees, formally at least, that $L_t\psi = [A, L]\psi$. Since L depends linearly on u and on time only through u, the operator relation $L_t = [A, L]$ can be identified as the evolution equation $u_t = K[u]$ if A and L are suitably chosen. This is the procedure!

The initial data $u(x, 0)$ act as a potential in the scattering problem $L\psi = \lambda\psi$ at $t = 0$. This determines certain scattering data: for example the solitons are associated with the bound states of $u(x, t)$ and the scattering data consists of a discrete eigenvalue λ_n of L and a normalisation constant; the latter depends on t. The boundary conditions on u are $u(x, t) \to 0$ as $|x| \to \infty$. By working at large $|x|$ this proves sufficient to allow determination of the time evolution of the the scattering data through $A\psi = \psi_t$.

From the scattering data at time t one wishes to infer $u(x, t)$, the potential which is the source of that scattering data at time t. This is done through the so called Gelfand-Levitan-Marchenko (linear) integral equations. The result is the required $u(x, t)$! For details of this extraordinary and extraordinarily beautiful method see especially [21].

The next step is to recognise as Zakharov and Fadeev [104] did for the KdV that the mapping from the initial data to the scattering data can be interpreted as a canonical transformation of co-ordinates. Then the inverse mapping from scattering data at time t to the "potential" $u(x, t)$ is an inverse canonical transformation.

The key point is that, in terms of the canonical co-ordinates expressed in terms of the scattering data, the Hamiltonian of each of the "integrable" non-linear wave equations is completely integrable. The new canonical co-ordinates are of action-angle type and the Hamiltonian depends <u>only</u> on the action variables. This is not the place to pursue the canonical structure of integrable equations further: the reader is referred in particular to [104, 105, 106] and [107]. We illustrate the remarks above here by quoting from Fadeev [101].

The s-G equation (1.5a) has Lagrangian

$$\mathcal{L} = \frac{1}{\gamma}\int_{-\infty}^{\infty}\left[\tfrac{1}{2}(u_t^{\,2} - u_x^{\,2}) - m^2(1 - \cos u)\right]dx \quad . \tag{14.1}$$

In this m is the mass of the field u ($m = 1$ in (1.5a)) and γ is the coupling constant. The usual Euler equations yield

$$u_{tt} - u_{xx} + m^2 \sin u = 0 \quad . \tag{14.2}$$

The Poisson brackets are

$$\left\{\pi(x, t),\ u(y, t)\right\} = \delta(x - y) \tag{14.3a}$$

where

$$\pi(x, t) = \frac{\delta\mathcal{L}}{\delta u_t} = \gamma^{-1}\, u_t(x, t) \tag{14.3b}$$

is the canonical conjugate of u. The Hamiltonian can be expressed in the two forms

$$H = \gamma^{-1}\int_{-\infty}^{\infty}\left[\tfrac{1}{2}(u_t^{\,2} + u_x^{\,2}) + m^2(1 - \cos u)\right]dx \tag{14.4}$$

$$= \int_{-\infty}^{\infty}\sqrt{p^2+m^2}\,\rho(p)dp + \sum_{a=1}^{A}\sqrt{p_a^{\,2}+M_a^{\,2}} + \sum_{b=1}^{B}\sqrt{p_b^{\,2}+M^2(\alpha_b)} \quad . \tag{14.5}$$

It is a non-trivial step from (14.4) to (14.5). This second form appears in terms of a set of new canonical variables which are obtained from canonical variables expressed in terms of scattering data. Note that H depends only on the momenta and is integrable.

The first term in (14.5) is associated with the continuous part of the spectrum of L. This provides a background contribution to $u(x, t)$ which diffuses away and leaves only the solitons: these are associated with the discrete part of the spectrum of L and determine the remaining terms in (14.5). In the present theory the continuous part of the spectrum provides the energy of a scalar particle of mass m. It has no charge (no twist). The second term associated with the discrete spectrum of L is the contribution of kinks and anti-kinks. Evidently (with $c = 1$) the mass of these is M and this takes the particular value $M = 8m\gamma^{-1}$ in H. The charges are $+1$ (kinks) and -1 (anti-kinks).

The third term is also associated with the discrete spectrum: the eigenvalues come as complex pairs which determine the 0π solutions.

The 0π's are two parameter solutions (cf. §5): they have two degrees of freedom which in (14.5) are associated with centre of mass co-ordinates (p_b, q_b) and two internal degrees of freedom (α_b, β_b). These co-ordinates are defined so that $-\infty < p_b < +\infty$, $-\infty < q_b < +\infty$ whilst [101] $0 < \alpha_b < \pi/2$, $0 < \beta_b < 32\pi\,\gamma^{-1}$.

Quantisation is straightforward: $\hbar = 1$ and e.g. $[p_a, q_{a'}] = -i\,\delta_{aa'}$. The phase space for the internal motion of the breathers is compact however and of volume $16\pi^2\gamma^{-1}$: if N is the number of quantised states in the quantised version this volume is also 2π N. Thus we require $N = 8\pi\,\gamma^{-1}$ which can be approximately true for large N and small γ. If α is quantised its eigenstates α_n, roughly equally spaced, will have eigenvalues $\alpha_n = \pi n/2N$ for $n = 1, 2, \ldots\ldots, N$. In (14.5) $M(\alpha)$ takes the explicit form $M(\alpha) = M \sin\alpha = 16m\gamma^{-1}\sin\alpha$. Thus Fadeev finds the remarkable mass formula

$$M_n = 16m\,\gamma^{-1}\sin(n\gamma/16) \tag{14.6}$$

from the breather (0π) solutions of the s-G equation. The formal limit $\gamma \to 0$, $n = 1$ gives $M_1 = m$: the mass of the lowest excitation is thus the mass m of the field. Dashen, Hasslacher and Neveu [102] first found[30] this mass formula by their very different method. They conjecture it is exact (after a certain renormalisation of γ). The conclusion then is that the soliton and 0π solutions generate the whole mass spectrum.

This field of investigation is little more than one year old: an obvious crucial problem is extension of the arguments to three space and one time dimension. Some results have been achieved [100, 101].

Note the effect of the additional degree of freedom in the 0π's in generating the mass spectrum (14.6). Natural candidates for further investigation in this direction are the wobbling pulse solutions of the multiple s-G equations of §§7 and 8. However, although the numerical data presented in §7 strongly suggest soliton behaviour we know [38] that the multiple s-G's are not integrable by the inverse scattering method in the form due to Ablowitz et al. [21]. No canonical formalism like that connecting (14.4) and (14.5) is therefore available for the multiple s-G's at this time. The WKB technique of Dashen et al. [102] should still be applicable however.

This short §14 on solitons in particle physics completes these lectures on "Solitons". The importance of solitons to non-linear optics should be plain; but I have also tried to indicate the importance of solitons to non-linear physics as a whole.

Note added in proof

The following additional references are of interest: in §2 the first derivation of the KdV equation for ion acoustic waves of small amplitude by Washini and Taniuti [108], the article by Gardner and Morikawa [109] which derives the KdV for a hydromagnetic wave in a cold plasma and Su'sand Gardner's paper [110] which derives the KdV and Burgers equations in a number of different physical contexts; in §4 the early work on SIT by Arecchi and Bonifacio [111], the review article on SIT by Lamb [112] and the work of Hirota [113] on the s-G equation; in §4 the paper [114] where some of the Figures of that section first appeared; in §§7 and 8 the recent Letter [115] reporting the unusual soliton behaviour associated with SIT in Q(2) degenerate systems and a report [116] of the experimental observation of wobbling two peaked pulses in Doppler broadened Na D_1 transitions; in §9 the paper by McLaughlin and Corones [117] which also compares the application of the Maxwell-Bloch equations to SIT and to

[30] Whether Dashen et al. or Fadeev first found (14.6) is by no means clear, but the formula is remarkable enough for "who did what when" scarcely to matter – as the two sets of discoverers also seem to believe.

Josephson junction theory; in §11 (and §4) the prediction and numerical investigation of self-focussing in non-degenerate SIT by Mattar [118, 119] and Mattar and Newstein [120] - both Gibbs (working on Na vapour and Toschek and Krieger (working on Neon) have observed strong enhancement of the central part of SIT pulses of finite transverse aperture [121]; in §§13 and 14 recent work by Luther [122] which relates Coleman's results [123] for the equivalence of the "massive Thirring model" (a model of massive fermions in one space dimension) and the quantised s-G equation to the exactly soluble spin $\frac{1}{2}$ x-y-z chain model (a one-dimensional lattice model of interacting spins) [124, 125]. Luther [122, 126] has used this result to prove that the mass spectrum (14.6) of Fadeev [101] and Dashen et al. [103] is exact. In [126] the theory is applied to a one dimensional electron gas in a study of one-dimensional conductors.

ACKNOWLEDGEMENTS

I am indebted to my colleagues in the Theoretical Physics Group at the University of Manchester Institute of Science and Technology for the collaborative research on which these notes are principally based. R.K. Dodd, J.D. Gibbon and especially P.J. Caudrey have my grateful thanks. Philip Caudrey collaborated on all of the work on the multiple sine-Gordon equations and much of it is his. I am also grateful to S. Duckworth on whose thesis §7 and §8 are based. He did all the computations on the multiple s-G's and all the Figures in §7 and §8 with few exceptions are his. J.C. Eilbeck did the first computations on the sine-Gordon and amongst many other contributions the RMB equations are his. Susan Jackson typed the manuscript and this paper would not exist without her resource and help.

REFERENCES

[1] McCALL, S.L., HAHN, E.L., Phys. Rev. 183 (1969) 457.
[2] SCOTT, A.C., CHU, Y.F., McLAUGHLIN, D.W., Proc. I.E.E.E. 61 (1973) 1443.
[3] KORTEWEG, D.J., DE VRIES, G., Phil. Mag. 39 (1895) 422.
[4] IKEZI, H., TAYLOR, R.J., BAKER, R.D., Phys. Rev. Lett. 25 (1970) 11.
[5] ZABUSKY, N.J., Proc. Sympos. on Non-Linear Partial Differential Equations (Univ. of Delaware, Newark, Del., 1965), Academic Press (New York: 1967) 223.
[6] ZAKHAROV, V.E., SHABAT, A.B., Zh. Eksp. Teor. Fiz. 61 (1971) 118 (Sov. Phys. J.E.T.P. 34 (1972) 62).
[7] VALEO, E.J., KRUER, W.L., Phys. Rev. Lett. 33 (1974) 750.
[8] YUEN, H.C., LAKE, B.M., Phys. Fluids 18 (1975) 956.
[9] BARONE, A., ESPOSITO, F., MAGEE, C.J., SCOTT, A.C., Rivista del Nuovo Cimento 1 (1971).
[10] SCOTT, A.C., Rev. Mod. Phys. 47 (1975) 487.
[11] RUSSELL, J.S., Report on Waves. British Association Reports (1844).
[12] BOUSSINESQ, J., J. Math. Pures Appl. Ser. 2 17 (1872) 55.
[13] HAMMACK, J.L., Jr., Tsunamis - A model of their generation and propagation. Cal. Tech. Report KH-R28 (1972) Chap. 5.
[14] DAVIDSON, R.C., Methods in Non-Linear Plasma Theory. Academic Press (New York: 1972) Chap. 2.
[15] ZABUSKY, N.J., KRUSKAL, M.D., Phys. Rev. Lett. 15 (1965) 240.
[16] LAX, P.D., Comm. Pure and Appl. Maths. 28 (1975) 141.
[17] McKEAN, H.P., VAN MOERBEKE, P., Inventiones math. 30 (1975) 217.
[18] FERMI, E., PASTA, J.R., ULAM, S.M., Studies of non-linear problems, I Los Alamos Rept. LA - 1940 (May 1955) and Collected Works of E. Fermi Vol. II pp. 978-88, Univ. of Chicago Press (1965).

[19] For example, K. Miura, The energy transport properties of one-dimensional anharmonic lattices. Thesis. Department of Computer Science, Univ. of Illinois (1973).
[20] GARDNER, C.S., GREENE, J.M., KRUSKAL, M.D., MIURA, R.M., Phys. Rev. Lett. 19 (1967) 1095.
[21] ABLOWITZ, M.J., KAUP, D.J., NEWELL, A.C., SEGUR, H., Studies in Appl. Math. 53 (1974) 249.
[22] ABLOWITZ, M.J., KAUP, D.J., NEWELL, A.C., SEGUR, H., Phys. Rev. Lett. 30 (1973) 1262.
[23] ABLOWITZ, M.J., KAUP, D.J., NEWELL, A.C., SEGUR, H., Phys. Rev. Lett. 31 (1973) 125.
[24] GIBBON, J.D., CAUDREY, P.J., BULLOUGH, R.K., EILBECK, J.C., Lett. al Nuovo Cimento 8 (1973) 775.
[25] SLUSHER, R.E., GIBBS, H.M., Phys. Rev. A 5 (1972) 1634.
[26] FEYNMAN, R.P., VERNON, F.L., HELLWORTH, R.W., J. Appl. Phys. 28 (1957) 49.
[27] BULLOUGH, R.K., AHMAD, F., Phys. Rev. Lett. 27 (1971) 330.
[28] EILBECK, J.C., CAUDREY, P.J., GIBBON, J.D., BULLOUGH, R.K. 6 (1973) 1337.
[29] McCALL, S.L., HAHN, E.L., Phys. Rev. Lett. 18 (1967) 908.
[30] GIBBS, H.M., SLUSHER, R.E., Phys. Rev. A 6 (1972) 2326.
[31] DIELS, J.C., HAHN, E.L., private communication (1972).
[32] GIBBON, J.D., EILBECK, J.C., J. Phys. A: Gen. Phys. 5 (1972) L122.
[33] CAUDREY, P.J., GIBBON, J.D., EILBECK, J.C., BULLOUGH, R.K., Phys. Rev. Lett. 30 (1973) 237.
[34] CAUDREY, P.J., EILBECK, J.C., GIBBON, J.D., J. Inst. Math. Applics. 14 (1974) 375.
[35] EILBECK, J.C., J. Phys. A: Gen. Phys. 5 (1973) 1355.
[36] LAMB, G.L., Phys. Rev. Lett. (1973) 196.
[37] SALAMO, G.J., GIBBS, H.M., CHURCHILL, G.G., Phys. Rev. Lett. 33 (1974) 273.
[38] DODD, R.K., BULLOUGH, R.K., DUCKWORTH, S., J. Phys. A: Maths. Gen. 8 (1975) L64.
[39] ICSEVGI, A., LAMB, W.E., Jr., Phys. Rev. 185 (1969) 517.
[40] EILBECK, J.C., BULLOUGH, R.K., J. Phys. A: Gen. Phys. 5 (1972) 820.
[41] DUCKWORTH, S., Thesis, Univ. of Manchester (1976).
[42] SALAMO, G.V., Thesis, City Univ. New York (1974).
[43] BÖLGER, B., BAEDE, L., GIBBS, H.M., IX International Quantum Electronics Conference, Amsterdam, June 1976, Paper H.1 (Optics Communications 18 (1976) 67).
[44] ROGOVIN, D., SCULLY, M.O., Does the "Two-level Atom" picture of Josephson junction have a theoretical foundation in B.C.S.? Annals of Physics, November 1974.
[45] FULTON, T.A., DYNES, R.C., Solid State Comm. 12 (1973) 57.
[46] FULTON, T.A., DUNKLEBERGER, L.N., Rev. de Physique Appliquee 9 (1974) 299.
[47] FULTON, T.A., DUNKLEBERGER, L.N., Appl. Phys. Lett. 22 (1973) 232.
[48] FULTON, T.A., DYNES, R.C., ANDERSON, P.W., Proc. I.E.E.E. 61 (1973) 28.
[49] NOVIKOV, S.P., Funkt. Anal. i Ego Prilozh. 8 (1974) 54.
[50] MAKI, K., TSUNETO, T., Prog. Theor. Phys. 52 (1974) 773.
[51] MAKI, K., TSUNETO, T., Phys. Rev. B 11 (1975) 2539.
[52] MAKI, K., Phys. Rev. B 11 (1975 4264.
[53] LANDAU, L.D., Z. Eksper. Teoret. Fiz. 30 (1956) 1058 (Sov. Phys. J.E.T.P. 3 (1957) 920).
[54] LEGGETT, A.J., Annals of Physics 85 (1974) 11.
[55] ANDERSON, P.W., BRINKMAN, W.F., Phys. Rev. Lett. 30 (1973) 1108.
[56] ANDERSON, P.W., MOREL, P., Phys. Rev. 123 (1961) 1911.
[57] BALIAN, R., WERTHAMER, N.R., Phys. Rev. 131 (1963) 1553.

[58] LEGGETT, A.J., Phys. Rev. Lett. 3 (1973) 352.
[59] MAKI, K., KUMAR, P., Creation of Magnetic Solitons in Superfluid ^{3}He (Preprint, 1976).
[60] BULLOUGH, R.K., J. Phys. A: Gen. Phys. 1 (1968) 409.
[61] VALEO, E.J., ESTABROOK, K.G., Phys. Rev. Lett. 34 (1975) 1008.
[62] WILCOX, J.Z., WILCOX, T.J., Phys. Rev. Lett. 34 (1975) 1160.
[63] DONALDSON, T.P., SPALDING, I.J., Phys. Rev. Lett. 36 (1976) 467.
[64] GERSTEN, J.I., TZOAR, N., Phys. Rev. Lett. 35 (1975) 934.
[65] TZOAR, N., GERSTEN, J.I., Phys. Rev. Lett. 28 (1972) 1203.
[66] TAPPERT, F., Private communication (1976).
[67] BLOEMBERGEN, N., Non-linear Optics, Benjamin (New York: 1965).
[68] ARMSTRONG, J.A., JHA, S.S., SHIREN, N.S., J. Quantum Electron. Q E-6 (1970) 123.
[69] SAGDEEV, R.Z., GALEEV, A.A., Non-linear Plasma Theory, Benjamin (New York: 1969).
[70] KADOMTSEV, B.B., MIKHAILOVSKY, A.B., TIMOFEEV, A.V., Sov. Phys. J.E.T.P. 20 (1965) 1517.
[71] NOZAKI, K., TANIUTI, T., J. Phys. Soc. Jap. 34 (1973) 796.
[72] OHSAWA, Y., NOZAKI, K., J. Phys. Soc. Jap. 36 (1974) 591.
[73] ROSENBLUTH, M., COPPI, B., SUDAN, R., Proc. 3rd Intern. Conf. Plasma Physics and Controlled Nuclear Fusion Research, Novosibirsk (1968).
[74] ZAKHAROV, V.E., MANAKOV, S.V., J.E.T.P. Lett. 18 (1973) 243.
[75] KAUP, D.J., Studies in Applied Mathematics 55 (1976) 9.
[76] ZAKHAROV, V.E., Zh. Eksp. Teor. Fiz. 62 (1972) 1745 (Soviet Physics J.E.T.P. 35 (1972) 908).
[77] WHITHAM, G.B., Linear and Non-linear Waves. John Wiley & Sons Inc. (New York: 1974).
[78] HASIMOTO, H., ONO, H., J. Phys. Soc. Japan 33 (1972) 805.
[79] LIGHTHILL, M.J., J. Inst. Math. Applics. 1 (1965) 269.
[80] ARMSTRONG, J.A., Phys. Rev. A 11 (1975) 963.
[81] GRISCHKOWSKY, D., COURTENS, E., ARMSTRONG, J.A., Phys. Rev. Lett. 31 (1973) 422.
[82] ARMSTRONG, J.A., BULLOUGH, R.K., DODD, R.K., JACK, P.M., LEVI, M., to be published (1976).
[83] FLASCHKA, H., Phys. Rev. B 9 (1974) 1924.
[84] FLASCHKA, H., Prog. Theor. Phys. 51 (1974) 703.
[85] TODA, M., J. Phys. Soc. Japan 22 (1967) 431.
[86] TODA, M., Prog. Theor. Phys. Suppl. 45 (1970) 174.
[87] TODA, M., WADATI, M., J. Phys. Soc. Japan 34 (1973) 18.
[88] TODA, M., Arkiv. for Det. Fysiske Seminar i Trondheim, No. 2 (1974).
[89] KRUMHANSL, J.A., SCHRIEFFER, J.R., Phys. Rev. B 11 (1975) 3535.
[90] KRUSKAL, M.D., Lectures in Applied Mathematics 15 (1974) 61 (American Math. Soc. 1974).
[91] SCHNEIDER, T., STOLL, E., Phys. Rev. Lett. 35 (1976) 1501.
[92] SCHNEIDER, T., STOLL, E., Phys. Rev. Lett. 35 (1975) 296.
[93] SCHNEIDER, T., STOLL, E., Phys. Rev. B 13 (1976) 1216.
[94] SCHWABL, F., in Anharmonic Lattices, Structural Transitions and Melting, Ed. T. Riste, Noordhoff, Leiden, Netherlands (1974) p.87.
[95] FISHER, M.E., Rev. Mod. Phys. 46 (1974) 597.
[96] BETTS, D.D., X-Y Model in Phase Transitions and Critical Phenomena Vol. 3, Eds. C. Domb and M.S. Green, Academic Press (1974) p.569.
[97] SKYRME, T.H.R., Proc. Roy. Soc. A 247 (1958) 260.
[98] PERRING, J.K., SKYRME, T.H.R., Nucl. Phys. 31 (1962) 550.
[99] CAUDREY, P.J., EILBECK, J.C., GIBBON, J.D., Il Nuovo Cim. 25 B (1975) 497.
[100] COLEMAN, S., Classical lumps and their quantum descendants. Lectures at the 1975 International School of Subnuclear Physics "Ettore Majorana" (1975).

[101] FADEEV, L.D., Quantisation of solitons. Lectures at Princeton, May 1975 (Copies from Institute for Advanced Study, Princeton, N.J. 08540).
[102] DASHEN, R.F., HASSLACHER, B., NEVEU, A., Phys. Rev. D 11 (1975) 3424.
[103] DASHEN, R.F., HASSLACHER, B., NEVEU, A., Phys. Rev. D 10 (1974) 4130.
[104] ZAKHAROV, V.E., FADEEV, L.D., Funkt. Anal. i Ego Prilozh. 5 (1971) 18.
[105] FLASCHKA, H., NEWELL, A.C., "Integrable systems of non-linear evolution equations" in Dynamical Systems Theory and Applications. Springer Lecture Notes in Physics. Ed. J. Moser, Springer-Verlag (Heidelberg: 1975).
[106] McLAUGHLIN, D.W., J. Math. Phys. 16 (1975) 96.
[107] DODD, R.K., BULLOUGH, R.K. (1976) to be published.
[108] WASHIMI, H., TANIUTI, T. Phys. Rev. Lett. 17 (1966) 996.
[109] GARDNER, C.S., MORIKAWA, G.K., Comm. Pure and Appl. Math. 18 (1965) 35.
[110] SU, C.H., GARDNER, C.S., J. Math. Phys. 10 (1969) 536.
[111] ARECCHI, F.T., BONIFACIO, R., I.E.E.E. Q E-1 (1965) 169.
[112] LAMB, G.L., Rev. Mod. Phys. 43 (1971) 99.
[113] HIROTA, R., J. Phys. Soc. Japan 33 (1972) 1459.
[114] BULLOUGH, R.K., CAUDREY, P.J., EILBECK, J.C., GIBBON, J.D. Opto-electronics 6 (1974) 121.
[115] DUCKWORTH, S., BULLOUGH, R.K., CAUDREY, P.J., GIBBON, J.D., Phys. Lett. 57a (1976) 19.
[116] BULLOUGH, R.K., CAUDREY, P.J., GIBBON, J.D., DUCKWORTH, S., GIBBS, H.M., BÖLGER, B., BAEDE, L. Post deadline paper P7 IX International Quantum Electronics Conference, Amsterdam, June 1976 (Optics Communications 18 (1976) 200).
[117] McLAUGHLIN, D.W., CORONES, J., Phys. Rev. A 10 (1974) 2051.
[118] MATTAR, F.P. Ph.D. Thesis. Polytechnic Institute of New York (1976).
[119] MATTAR, F.P., NEWSTEIN, M.C. Paper H4 IX International Quantum Electronics Conference, Amsterdam, June 1976 (Optics Communications 18 (1976) 70).
[120] MATTAR, F.P., NEWSTEIN, M.C., to be published (1976).
[121] GIBBS, H.M., BÖLGER, B., BAEDE, L. Post deadline paper P6 IX International Quantum Electronics Conference, Amsterdam, June 1976 (Optics Communications 18 (1976) 199).
[122] LUTHER, A. Eigenvalue spectrum of interacting fermions in one dimension. Preprint (April 1976).
[123] COLEMAN, S., Phys. Rev. D 11 (1975) 2088.
[124] BAXTER, R.J., Phys. Rev. Lett. 26 (1971) 832.
[125] JOHNSON, J.D., KRINSKY, S., McCOY, B.M., Phys. Rev. A 8 (1973) 2526.
[126] LUTHER, A. Quantum solitons in one-dimensional conductors. Preprint (August 1976).

SECRETARIAT OF THE WINTER COLLEGE

DIRECTORS

J.M. Ziman	H.H. Wills Physics Laboratory, University of Bristol, Royal Fort, Tyndall Avenue, Bristol BS8 1TL, United Kingdom
G. Chiarotti	Istituto di Fisica "G. Marconi", Università degli Studi, Piazzale delle Scienze 5, Rome, Italy
F. García-Moliner	Departamento de Física, Universidad Autónoma de Madrid, Facultad de Ciencias, C-XII, 6, Canto Blanco, Madrid, Spain
F. Gautier	Laboratoire de structure électronique des solides, Université Louis Pasteur, 4 rue Blaise-Pascal, 67 Strasbourg, France
S. Lundqvist	Chalmers University of Technology, Fack, S-40220 Göteborg, Sweden
N.H. March	Department of Theoretical Physics, Imperial College, Prince Consort Road, London SW7 2BZ, United Kingdom
H. Reik	Fakultät für Physik der Universität Freiburg, Hermann-Herder-Strasse 3, Freiburg, Federal Republic of Germany

EDITOR

L.A. Self	Division of Publications, IAEA, Vienna, Austria

The following conversion table is provided for the convenience of readers and to encourage the use of SI units.

FACTORS FOR CONVERTING UNITS TO SI SYSTEM EQUIVALENTS *

SI base units are the metre (m), kilogram (kg), second (s), ampere (A), kelvin (K), candela (cd) and mole (mol).
[For further information, see International Standards ISO 1000 (1973), and ISO 31/0 (1974) and its several parts]

Multiply		*by*	*to obtain*
Mass			
pound mass (avoirdupois)	1 lbm	= 4.536×10^{-1}	kg
ounce mass (avoirdupois)	1 ozm	= 2.835×10^{1}	g
ton (long) (= 2240 lbm)	1 ton	= 1.016×10^{3}	kg
ton (short) (= 2000 lbm)	1 short ton	= 9.072×10^{2}	kg
tonne (= metric ton)	1 t	= 1.00×10^{3}	kg
Length			
statute mile	1 mile	= 1.609×10^{0}	km
yard	1 yd	= 9.144×10^{-1}	m
foot	1 ft	= 3.048×10^{-1}	m
inch	1 in	= 2.54×10^{-2}	m
mil (= 10^{-3} in)	1 mil	= 2.54×10^{-2}	mm
Area			
hectare	1 ha	= 1.00×10^{4}	m^2
(statute mile)^2	1 mile^2	= 2.590×10^{0}	km^2
acre	1 acre	= 4.047×10^{3}	m^2
yard^2	1 yd^2	= 8.361×10^{-1}	m^2
foot^2	1 ft^2	= 9.290×10^{-2}	m^2
inch^2	1 in^2	= 6.452×10^{2}	mm^2
Volume			
yard^3	1 yd^3	= 7.646×10^{-1}	m^3
foot^3	1 ft^3	= 2.832×10^{-2}	m^3
inch^3	1 in^3	= 1.639×10^{4}	mm^3
gallon (Brit. or Imp.)	1 gal (Brit)	= 4.546×10^{-3}	m^3
gallon (US liquid)	1 gal (US)	= 3.785×10^{-3}	m^3
litre	1 l	= 1.00×10^{-3}	m^3
Force			
dyne	1 dyn	= 1.00×10^{-5}	N
kilogram force	1 kgf	= 9.807×10^{0}	N
poundal	1 pdl	= 1.383×10^{-1}	N
pound force (avoirdupois)	1 lbf	= 4.448×10^{0}	N
ounce force (avoirdupois)	1 ozf	= 2.780×10^{-1}	N
Power			
British thermal unit/second	1 Btu/s	= 1.054×10^{3}	W
calorie/second	1 cal/s	= 4.184×10^{0}	W
foot-pound force/second	1 ft·lbf/s	= 1.356×10^{0}	W
horsepower (electric)	1 hp	= 7.46×10^{2}	W
horsepower (metric) (= ps)	1 ps	= 7.355×10^{2}	W
horsepower (550 ft·lbf/s)	1 hp	= 7.457×10^{2}	W

* Factors are given exactly or to a maximum of 4 significant figures

Multiply		*by*	*to obtain*
Density			
pound mass/inch^3	1 lbm/in^3	= 2.768 × 10^4	kg/m^3
pound mass/foot^3	1 lbm/ft^3	= 1.602 × 10^1	kg/m^3
Energy			
British thermal unit	1 Btu	= 1.054 × 10^3	J
calorie	1 cal	= 4.184 × 10^0	J
electron-volt	1 eV	≃ 1.602 × 10^{-19}	J
erg	1 erg	= 1.00 × 10^{-7}	J
foot-pound force	1 ft·lbf	= 1.356 × 10^0	J
kilowatt-hour	1 kW·h	= 3.60 × 10^6	J
Pressure			
newtons/metre^2	1 N/m^2	= 1.00	Pa
atmosphere[a]	1 atm	= 1.013 × 10^5	Pa
bar	1 bar	= 1.00 × 10^5	Pa
centimetres of mercury (0°C)	1 cmHg	= 1.333 × 10^3	Pa
dyne/centimetre^2	1 dyn/cm^2	= 1.00 × 10^{-1}	Pa
feet of water (4°C)	1 ftH$_2$O	= 2.989 × 10^3	Pa
inches of mercury (0°C)	1 inHg	= 3.386 × 10^3	Pa
inches of water (4°C)	1 inH$_2$O	= 2.491 × 10^2	Pa
kilogram force/centimetre^2	1 kgf/cm^2	= 9.807 × 10^4	Pa
pound force/foot^2	1 lbf/ft^2	= 4.788 × 10^1	Pa
pound force/inch^2 (= psi)[b]	1 lbf/in^2	= 6.895 × 10^3	Pa
torr (0°C) (= mmHg)	1 torr	= 1.333 × 10^2	Pa
Velocity, acceleration			
inch/second	1 in/s	= 2.54 × 10^1	mm/s
foot/second (= fps)	1 ft/s	= 3.048 × 10^{-1}	m/s
foot/minute	1 ft/min	= 5.08 × 10^{-3}	m/s
mile/hour (= mph)	1 mile/h	= 4.470 × 10^{-1}	m/s
		1.609 × 10^0	km/h
knot	1 knot	= 1.852 × 10^0	km/h
free fall, standard (= g)		= 9.807 × 10^0	m/s^2
foot/second^2	1 ft/s^2	= 3.048 × 10^{-1}	m/s^2
Temperature, thermal conductivity, energy/area·time			
Fahrenheit, degrees − 32	°F − 32	$\frac{5}{9}$	°C
Rankine	°R	$\frac{5}{9}$	K
1 Btu·in/ft^2·s·°F		= 5.189 × 10^2	W/m·K
1 Btu/ft·s·°F		= 6.226 × 10^1	W/m·K
1 cal/cm·s·°C		= 4.184 × 10^2	W/m·K
1 Btu/ft^2·s		= 1.135 × 10^4	W/m^2
1 cal/cm^2·min		= 6.973 × 10^2	W/m^2
Miscellaneous			
foot^3/second	1 ft^3/s	= 2.832 × 10^{-2}	m^3/s
foot^3/minute	1 ft^3/min	= 4.719 × 10^{-4}	m^3/s
rad	rad	= 1.00 × 10^{-2}	J/kg
roentgen	R	= 2.580 × 10^{-4}	C/kg
curie	Ci	= 3.70 × 10^{10}	disintegration/s

[a] atm abs: atmospheres absolute;
atm (g): atmospheres gauge.

[b] lbf/in^2 (g) (= psig): gauge pressure;
lbf/in^2 abs (= psia): absolute pressure.

HOW TO ORDER IAEA PUBLICATIONS

An exclusive sales agent for IAEA publications, to whom all orders and inquiries should be addressed, has been appointed in the following country:

UNITED STATES OF AMERICA	UNIPUB, P.O. Box 433, Murray Hill Station, New York, N.Y. 10016

In the following countries IAEA publications may be purchased from the sales agents or booksellers listed or through your major local booksellers. Payment can be made in local currency or with UNESCO coupons.

ARGENTINA	Comisión Nacional de Energía Atómica, Avenida del Libertador 8250, Buenos Aires
AUSTRALIA	Hunter Publications, 58 A Gipps Street, Collingwood, Victoria 3066
BELGIUM	Service du Courrier de l'UNESCO, 112, Rue du Trône, B-1050 Brussels
CANADA	Information Canada, 171 Slater Street, Ottawa, Ont. K1A 0S9
C.S.S.R.	S.N.T.L., Spálená 51, CS-110 00 Prague Alfa, Publishers, Hurbanovo námestie 6, CS-800 00 Bratislava
FRANCE	Office International de Documentation et Librairie, 48, rue Gay-Lussac, F-75005 Paris
HUNGARY	Kultura, Hungarian Trading Company for Books and Newspapers, P.O. Box 149, H-1011 Budapest 62
INDIA	Oxford Book and Stationery Comp., 17, Park Street, Calcutta 16; Oxford Book and Stationery Comp., Scindia House, New Delhi-110001
ISRAEL	Heiliger and Co., 3, Nathan Strauss Str., Jerusalem
ITALY	Libreria Scientifica, Dott. de Biasio Lucio "aeiou", Via Meravigli 16, I-20123 Milan
JAPAN	Maruzen Company, Ltd., P.O.Box 5050, 100-31 Tokyo International
NETHERLANDS	Marinus Nijhoff N.V., Lange Voorhout 9-11, P.O. Box 269, The Hague
PAKISTAN	Mirza Book Agency, 65, The Mall, P.O.Box 729, Lahore-3
POLAND	Ars Polona, Centrala Handlu Zagranicznego, Krakowskie Przedmiescie 7, Warsaw
ROMANIA	Cartimex, 3-5 13 Decembrie Street, P.O.Box 134-135, Bucarest
SOUTH AFRICA	Van Schaik's Bookstore, P.O.Box 724, Pretoria Universitas Books (Pty) Ltd., P.O.Box 1557, Pretoria
SPAIN	Diaz de Santos, Lagasca 95, Madrid-6 Calle Francisco Navacerrada, 8, Madrid-28
SWEDEN	C.E. Fritzes Kungl. Hovbokhandel, Fredsgatan 2, S-103 07 Stockholm
UNITED KINGDOM	Her Majesty's Stationery Office, P.O. Box 569, London SE1 9NH
U.S.S.R.	Mezhdunarodnaya Kniga, Smolenskaya-Sennaya 32-34, Moscow G-200
YUGOSLAVIA	Jugoslovenska Knjiga, Terazije 27, YU-11000 Belgrade

---

Orders from countries where sales agents have not yet been appointed and requests for information should be addressed directly to:

Division of Publications
International Atomic Energy Agency
Kärntner Ring 11, P.O.Box 590, A-1011 Vienna, Austria